UNDERSTANDING POWER QUALITY PROBLEMS

IEEE Press
445 Hoes Lane, P.O. Box 1331
Piscataway, NJ 08855-1331

IEEE Industry Applications Society, *Sponsor*
IA-S Liaison to IEEE Press, Geza Joos

IEEE Power Electronics Society, *Sponsor*
PEL-S Liaison to IEEE Press, William Hazen

IEEE Power Engineering Society, *Sponsor*
PE-S Liaison to IEEE Press, Chanan Singh

Cover design: William T. Donnelly, *WT Design*

Technical Reviewers
Mladen Kezunovic, *Texas A & M University*
Damir Novosel, *ABB Power T&D Company, Inc., Raleigh, NC*
Roger C. Dugan, *Electrotek Concepts, Inc., Knoxville, TN*
Mohamed E. El-Hawary, *Dalhousie University, Halifax, Nova Scotia, Canada*
Stephen Sebo, *Ohio State University*

UNDERSTANDING POWER QUALITY PROBLEMS

Voltage Sags and Interruptions

Math H. J. Bollen
Chalmers University of Technology
Gothenburg, Sweden

IEEE Industry Applications Society, *Sponsor*

IEEE Power Electronics Society, *Sponsor*

IEEE Power Engineering Society, *Sponsor*

IEEE PRESS SERIES ON POWER ENGINEERING

P. M. Anderson, *Series Editor*

IEEE

The Institute of Electrical and Electronics Engineers, Inc., New York

WILEY-INTERSCIENCE

A JOHN WILEY & SONS, INC., PUBLICATION

Published by John Wiley & Sons, Inc., Hoboken, New Jersey.

10 9 8 7 6 5 4

ISBN 0-7803-4713-7

Library of Congress Cataloging-in-Publication Data

Bollen, Math H. J., 1960–

Understanding power quality problems : voltage sags and interruptions

Math H. J. Bollen.

p. cm. – (IEEE Press series on power engineering)

Includes bibliographical references and index.

IBSN 0-7803-4713-7

1. Electric power system stability. 2. Electric power failures.

3. Brownouts. 4. Electric power systems—Quality control.

I. Title. II. Series.

IN PROCESS

621.319—dc21 99-23546

CIP

The master said, to learn and at due times to repeat what one has learnt, is that not after all a pleasure?

Confucius, The Analects, Book One, verse 1

BOOKS IN THE IEEE PRESS SERIES ON POWER ENGINEERING

ELECTRIC POWER APPLICATIONS OF FUZZY SYSTEMS
Edited by Mohamed E. El-Hawary, *Dalhousie University*
1998 Hardcover 384 pp IEEE Order No. PC5666 ISBN 0-7803-1197-3

RATING OF ELECTRIC POWER CABLES: Ampacity Computations for Transmission, Distribution, and Industrial Applications
George J. Anders, *Ontario Hydro Technologies*
1997 Hardcover 464 pp IEEE Order No. PC5647 ISBN 0-7803-1177-9

ANALYSIS OF FAULTED POWER SYSTEMS, Revised Printing
P. M. Anderson, *Power Math Associates, Inc.*
1995 Hardcover 536 pp IEEE Order No. PC5616 ISBN 0-7803-1145-0

ELECTRIC POWER SYSTEMS: Design and Analysis, Revised Printing
Mohamed E. El-Hawary, *Dalhousie University*
1995 Hardcover 808 pp IEEE Order No. PC5606 ISBN 0-7803-1140-X

POWER SYSTEM STABILITY, Volumes I, II, III
An IEEE Press Classic Reissue Set
Edward Wilson Kimbark, *Iowa State University*
1995 Softcover 1008 pp IEEE Order No. PP5600 ISBN 0-7803-1135-3

ANALYSIS OF ELECTRIC MACHINERY
Paul C. Krause and Oleg Wasynczuk, *Purdue University*
Scott D. Sudhoff, *University of Missouri at Rolla*
1994 Hardcover 480 pp IEEE Order No. PC3789 ISBN 0-7803-1029-2

SUBSYNCHRONOUS RESONANCE IN POWER SYSTEMS
P. M. Anderson, *Power Math Associates, Inc.*
B. L. Agrawal, *Arizona Public Service Company*
J. E. Van Ness, *Northwestern University*
1990 Softcover 282 pp IEEE Order No. PP2477 ISBN 0-7803-5350-1

POWER SYSTEM PROTECTION
P. M. Anderson, *Power Math Associates, Inc.*
1999 Hardcover 1,344 pp IEEE Order No. PC5389 ISBN 0-7803-3427-2

POWER AND COMMUNICATION CABLES: Theory and Applications
Edited by R. Bartnikas and K. D. Srivastava
2000 Hardcover 896 pp IEEE Order No. PC5665 ISBN 0-7803-1196-5

Contents

Preface

The aims of the electric power system can be summarized as "to transport electrical energy from the generator units to the terminals of electrical equipment" and "to maintain the voltage at the equipment terminals within certain limits." For decades research and education have been concentrated on the first aim. Reliability and quality of supply were rarely an issue, the argument being that the reliability was sooner too high than too low. A change in attitude came about probably sometime in the early 1980s. Starting in industrial and commercial power systems and spreading to the public supply, the power quality virus appeared. It became clear that equipment regularly experienced spurious trips due to voltage disturbances, but also that equipment was responsible for many voltage and current disturbances. A more customer-friendly definition of reliability was that the power supply turned out to be much less reliable than always thought. Although the hectic years of power quality pioneering appear to be over, the subject continues to attract lots of attention. This is certain to continue into the future, as customers' demands have become an important issue in the deregulation of the electricity industry.

This book concentrates on the power quality phenomena that primarily affect the customer: interruptions and voltage sags. During an interruption the voltage is completely zero, which is probably the worst quality of supply one can consider. During a voltage sag the voltage is not zero, but is still significantly less than during normal operation. Voltage sags and interruptions account for the vast majority of unwanted equipment trips.

The material contained in the forthcoming chapters was developed by the author during a 10-year period at four different universities: Eindhoven, Curaçao, Manchester, and Gothenburg.[1] Large parts of the material were originally used for postgraduate and industrial lectures both "at home" and in various places around the world. The material will certainly be used again for this purpose (by the author and hopefully also by others).

[1]Eindhoven University of Technology, University of the Netherlands Antilles, University of Manchester Institute of Science and Technology, and Chalmers University of Technology, respectively.

Chapter 1 of this book gives an introduction to the subject. After a systematic overview of power quality, the term "voltage magnitude event" is introduced. Both voltage sags and interruptions are examples of voltage magnitude events. The second part of Chapter 1 discusses power quality standards, with emphasis on the IEC standards on electromagnetic compatibility and the European voltage characteristics standard (EN 50160).

In Chapter 2 the most severe power quality event is discussed: the (long) interruption. Various ways are presented of showing the results of monitoring the number of interruptions. A large part of Chapter 2 is dedicated to the stochastic prediction of long interruptions—an area better known as "reliability evaluation." Many of the techniques described here can be applied equally well to the stochastic prediction of other power quality events.

Chapter 3 discusses short interruptions—interruptions terminated by an automatic restoration of the supply. Origin, monitoring, mitigation, effect on equipment, and stochastic prediction are all treated in this chapter.

Chapter 4 is the first of three chapters on voltage sags. It treats voltage sags in a descriptive way: how they can be characterized and how the characteristics may be obtained through measurements and calculations. Emphasis in this chapter is on magnitude and phase-angle jump of sags, as experienced by single-phase equipment and as experienced by three-phase equipment.

Chapter 5 discusses the effect of voltage sags on equipment. The main types of sensitive equipment are discussed in detail: single-phase rectifiers (computers, process-control equipment, consumer electronics), three-phase ac adjustable-speed drives, and dc drives. Some other types of equipment are briefly discussed. The sag characteristics introduced in Chapter 4 are used to describe equipment behavior in Chapter 5.

In Chapter 6 the theory developed in Chapters 4 and 5 is combined with statistical and stochastical methods as described in Chapter 2. Chapter 6 starts with ways of presenting the voltage-sag performance of the supply and comparing it with equipment performance. The chapter continues with two ways of obtaining information about the supply performance: power-quality monitoring and stochastic prediction. Both are discussed in detail.

Chapter 7, the last main chapter of this book, gives an overview of methods for mitigation of voltage sags and interruptions. Two methods are discussed in detail: power system design and power-electronic controllers at the equipment–system interface. The chapter concludes with a comparison of the various energy-storage techniques available.

In Chapter 8 the author summarizes the conclusions from the previous chapters and gives some of his expectations and hopes for the future. The book concludes with three appendixes: Appendix A and Appendix B give a list of EMC and power quality standards published by the IEC and the IEEE, respectively. Appendix C contains definitions for the terminology used in this book as well as definitions from various standard documents.

Math H. J. Bollen
Gothenburg, Sweden

FTP Site Information

Along with the publication of this book, an FTP site has been created containing MATLAB® files for many figures in this book. The FTP site can be reached at ftp.ieee.org/upload/press/bollen.

Acknowledgments

A book is rarely the product of only one person, and this book is absolutely no exception. Various people contributed to the final product, but first of all I would like to thank my wife, Irene Gu, for encouraging me to start writing and for filling up my tea cup every time I had another one of those "occasional but all too frequent crises."

For the knowledge described in this book I owe a lot to my teachers, my colleagues, and my students in Eindhoven, Curaçao, Manchester, and Gothenburg and to my colleagues and friends all over the world. A small number of them need to be especially mentioned: Matthijs Weenink, Wil van den Heuvel, and Wim Kersten for teaching me the profession; the two Larry's (Conrad and Morgan) for providing me with a continuous stream of information on power quality; Wang Ping, Stefan Johansson, and the anonymous reviewers for proofreading the manuscript. A final thank you goes to everybody who provided data, figures, and permission to reproduce material from other sources.

Math H. J. Bollen
Gothenburg, Sweden

Voor mijn ouders

1

Overview of Power Quality and Power Quality Standards

Everybody does not agree with the use of the term *power quality*, but they do agree that it has become a very important aspect of power delivery especially in the second half of the 1990s. There is a lot of disagreement about what *power quality* actually incorporates; it looks as if everyone has her or his own interpretation. In this chapter various ideas will be summarized to clear up some of the confusion. However, the author himself is part of the power quality world; thus part of the confusion. After reading this book the reader might want to go to the library and form his own picture. The number of books on *power quality* is still rather limited. The book "Electric Power Systems Quality" by Dugan et al. [75] gives a useful overview of the various power quality phenomena and the recent developments in this field. There are two more books with the term power quality in the title: "Electric Power Quality Control Techniques" [76] and "Electric Power Quality" [77]. But despite the general title, reference [76] mainly concentrates on transient overvoltage and [77] mainly on harmonic distortion. But both books do contain some introductory chapters on power quality. Also many recent books on electric power systems contain one or more general chapters on power quality, for example, [114], [115], and [116]. Information on power quality cannot be found only in books; a large number of papers have been written on the subject; overview papers as well as technical papers about small details of power quality. The main journals to look for technical papers are the IEEE Transactions on Industry Applications, the IEEE Transactions on Power Delivery and IEE Proceedings—Generation, Transmission, Distribution. Other technical journals in the power engineering field also contain papers of relevance. A journal specially dedicated to power quality is Power Quality Assurance. Overview articles can be found in many different journals; two early ones are [104] and [105].

Various sources use the term "power quality" with different meanings. Other sources use similar but slightly different terminology like "quality of power supply" or "voltage quality." What all these terms have in common is that they treat the interaction between the utility and the customer, or in technical terms between the power system and the load. Treatment of this interaction is in itself not new. The aim of the power system has always been to supply electrical energy to the customers.

What is new is the emphasis that is placed on this interaction, and the treatment of it as a separate area of power engineering. In Section 1.2 the various terms and interpretations will be discussed in more detail. From the discussion we will conclude that "power quality" is still the most suitable term. The various power quality phenomena will be discussed and grouped in Section 1.3. Electromagnetic compatibility and power quality standards will be treated in detail in Section 1.4. But first Section 1.1 will give some explanations for the increased interest in power quality.

1.1 INTEREST IN POWER QUALITY

The fact that power quality has become an issue recently, does not mean that it was not important in the past. Utilities all over the world have for decades worked on the improvement of what is now known as power quality. And actually, even the term has been in use for a rather long time already. The oldest mentioning of the term "power quality" known to the author was in a paper published in 1968 [95]. The paper detailed a study by the U.S. Navy after specifications for the power required by electronic equipment. That paper gives a remarkably good overview of the power quality field, including the use of monitoring equipment and even the suggested use of a static transfer switch. Several publications appeared soon after, which used the term power quality in relation to airborne power systems [96], [97], [98]. Already in 1970 "high power quality" is being mentioned as one of the aims of industrial power system design, together with "safety," "reliable service," and "low initial and operating costs" [99]. At about the same time the term "voltage quality" was used in the Scandinavian countries [100], [101] and in the Soviet Union [102], mainly with reference to slow variations in the voltage magnitude.

The recent increased interest in power quality can be explained in a number of ways. The main explanations given are summarized below. Of course it is hard to say which of these came first; some explanations for the interest in power quality given below, will by others be classified as consequences of the increased interest in power quality. To show the increased interest on power quality a comparison was made for the number of publications in the INSPEC database [118] using the terms "voltage quality" or "power quality." For the period 1969–1984 the INSPEC database contains 91 records containing the term "power quality" and 64 containing the term "voltage quality." The period 1985–1996 resulted in 2051 and 210 records, respectively. We see thus a large increase in number of publications on this subjects and also a shift away from the term "voltage quality" toward the term "power quality."

- **Equipment has become more sensitive to voltage disturbances.**

  Electronic and power electronic equipment has especially become much more sensitive than its counterparts 10 or 20 years ago. The paper often cited as having introduced the term power quality (by Thomas Key in 1978 [1]) treated this increased sensitivity to voltage disturbances. Not only has equipment become more sensitive, companies have also become more sensitive to loss of production time due to their reduced profit margins. On the domestic market, electricity is more and more considered a basic right, which should simply always be present. The consequence is that an interruption of the supply will much more than before lead to complaints, even if there are no damages or costs related to it. An important paper triggering the interest in power quality appeared in the journal Business Week in 1991 [103]. The article cited Jane

Clemmensen of EPRI as estimating that "*power-related problems cost U.S. companies $26 billion a year in lost time and revenue.*" This value has been cited over and over again even though it was most likely only a rough estimate.

- **Equipment causes voltage disturbances.**

  Tripping of equipment due to disturbances in the supply voltage is often described by customers as "bad power quality." Utilities on the other side, often view disturbances due to end-user equipment as the main power quality problem. Modern (power) electronic equipment is not only sensitive to voltage disturbances, it also causes disturbances for other customers. The increased use of converter-driven equipment (from consumer electronics and computers, up to adjustable-speed drives) has led to a large growth of voltage disturbances, although fortunately not yet to a level where equipment becomes sensitive. The main issue here is the nonsinusoidal current of rectifiers and inverters. The input current not only contains a power frequency component (50 Hz or 60 Hz) but also so-called harmonic components with frequencies equal to a multiple of the power frequency. The harmonic distortion of the current leads to harmonic components in the supply voltage. Equipment has already produced harmonic distortion for a number of decades. But only recently has the amount of load fed via power electronic converters increased enormously: not only large adjustable-speed drives but also small consumer electronics equipment. The latter cause a large part of the harmonic voltage distortion: each individual device does not generate much harmonic currents but all of them together cause a serious distortion of the supply voltage.

- **A growing need for standardization and performance criteria.**

  The consumer of electrical energy used to be viewed by most utitilies simply as a "load." Interruptions and other voltage disturbances were part of the deal, and the utility decided what was reasonable. Any customer who was not satisfied with the offered reliability and quality had to pay the utility for improving the supply.

  Today the utilities have to treat the consumers as "customers." Even if the utility does not need to reduce the number of voltage disturbances, it does have to quantify them one way or the other. Electricity is viewed as a product with certain characteristics, which have to be measured, predicted, guaranteed, improved, etc. This is further triggered by the drive towards privatization and deregulation of the electricity industry.

  Open competition can make the situation even more complicated. In the past a consumer would have a contract with the local supplier who would deliver the electrical energy with a given reliability and quality. Nowadays the customer can buy electrical energy somewhere, the transport capacity somewhere else and pay the local utility, for the actual connection to the system. It is no longer clear who is responsible for reliability and power quality. As long as the customer still has a connection agreement with the local utility, one can argue that the latter is responsible for the actual delivery and thus for reliability and quality. But what about voltage sags due to transmission system faults? In some cases the consumer only has a contract with a supplier who only generates the electricity and subcontracts transport and distribution. One could state that any responsibility should be defined by contract, so that the generation company with which the customer has a contractual agreement would be responsible for reliability and quality. The responsibility of the local distribution would only be towards the generation companies with whom they have a contract to deliver to given customers. No matter what the legal construction is, reliability and quality will need to be well defined.

- **Utilities want to deliver a good product.**
  Something that is often forgotten in the heat of the discussion is that many power quality developments are driven by the utilities. Most utilities simply want to deliver a good product, and have been committed to that for many decades. Designing a system with a high reliability of supply, for a limited cost, is a technical challenge which appealed to many in the power industry, and hopefully still does in the future.
- **The power supply has become too good.**
  Part of the interest in phenomena like voltage sags and harmonic distortion is due to the high quality of the supply voltage. Long interruptions have become rare in most industrialized countries (Europe, North America, East Asia), and the consumer has, wrongly, gotten the impression that electricity is something that is always available and always of high quality, or at least something that should always be. The fact that there are some imperfections in the supply which are very hard or even impossible to eliminate is easily forgotten. In countries where the electricity supply has a high unavailability, like 2 hours per day, power quality does not appear to be such a big issue as in countries with availabilities well over 99.9%.
- **The power quality can be measured.**
  The availability of electronic devices to measure and show waveforms has certainly contributed to the interest in power quality. Harmonic currents and voltage sags were simply hard to measure on a large scale in the past. Measurements were restricted to rms voltage, frequency, and long interruptions; phenomena which are now considered part of power quality, but were simply part of power system operation in the past.

1.2 POWER QUALITY, VOLTAGE QUALITY

There have been (and will be) a lot of arguments about which term to use for the utility–customer (system–load) interactions. Most people use the term "power quality" although this term is still prone to criticism. The main objection against the use of the term is that one cannot talk about the quality of a physical quantity like power. Despite the objections we will use the term power quality here, even though it does not give a perfect description of the phenomenon. But it has become a widely used term and it is the best term available at the moment. Within the IEEE, the term power quality has gained some official status already, e.g., through the name of SCC 22 (Standards Coordinating Committee): "Power Quality" [140]. But the international standards setting organization in electrical engineering (the IEC) does not yet use the term power quality in any of its standard documents. Instead it uses the term *electromagnetic compatibility*, which is not the same as power quality but there is a strong overlap between the two terms. Below, a number of different terms will be discussed. As each term has its limitations the author feels that power quality remains the more general term which covers all the other terms. But, before that, it is worth to give the following IEEE and IEC definitions.

The definition of power quality given in the IEEE dictionary [119] originates in IEEE Std 1100 (better known as the Emerald Book) [78]: *Power quality is the concept of powering and grounding sensitive equipment in a matter that is suitable to the operation of that equipment.* Despite this definition the term power quality is clearly used in a more general way within the IEEE: e.g., SCC 22 also covers standards on harmonic pollution caused by loads.

The following definition is given in IEC 61000-1-1: *Electromagnetic compatibility is the ability of an equipment or system to function satisfactorily in its electromagnetic environment without introducing intolerable electromagnetic disturbances to anything in that environment* [79].

Recently the IEC has also started a project group on power quality [106] which should initially result in a standard on measurement of power quality. The following definition of power quality was adopted for describing the scope of the project group: *Set of parameters defining the properties of the power supply as delivered to the user in normal operating conditions in terms of continuity of supply and characteristics of voltage (symmetry, frequency, magnitude, waveform).*

Obviously, this definition will not stop the discussion about what power quality is. The author's impression is that it will only increase the confusion, e.g., because power quality is now suddenly limited to "normal operating conditions."

From the many publications on this subject and the various terms used, the following terminology has been extracted. The reader should realize that there is no general consensus on the use of these terms.

- **Voltage quality** (the French *Qualité de la tension*) is concerned with deviations of the voltage from the ideal. The ideal voltage is a single-frequency sine wave of constant frequency and constant magnitude. The limitation of this term is that it only covers technical aspects, and that even within those technical aspects it neglects the current distortions. The term voltage quality is regularly used, especially in European publications. It can be interpreted as the quality of the product delivered by the utility to the customers.
- A complementary definition would be **current quality**. Current quality is concerned with deviations of the current from the ideal. The ideal current is again a single-frequency sine wave of constant frequency and magnitude. An additional requirement is that this sine wave is in phase with the supply voltage. Thus where voltage quality has to do with what the utility delivers to the consumer, current quality is concerned with what the consumer takes from the utility. Of course voltage and current are strongly related and if either voltage or current deviates from the ideal it is hard for the other to be ideal.
- **Power quality** is the combination of voltage quality and current quality. Thus power quality is concerned with deviations of voltage and/or current from the ideal. Note that power quality has nothing to do with deviations of the product of voltage and current (the power) from any ideal shape.
- **Quality of supply** or quality of power supply includes a technical part (voltage quality above) plus a nontechnical part sometimes referred to as "quality of service." The latter covers the interaction between the customer and the utility, e.g., the speed with which the utility reacts to complaints, or the transparency of the tariff structure. This could be a useful definition as long as one does not want to include the customer's responsibilities. The word "supply" clearly excludes active involvement of the customer.

- **Quality of consumption** would be the complementary term of quality of supply. This would contain the current quality plus, e.g., how accurate the customer is in paying the electricity bill.
- In the IEC standards the term **electromagnetic compatibility** (**EMC**) is used. Electromagnetic compatibility has to do with mutual interaction between equipment and with interaction between equipment and supply. Within electromagnetic compatibility, two important terms are used: the "emission" is the electromagnetic pollution produced by a device; the "immunity" is the device's ability to withstand electromagnetic pollution. Emission is related to the term current quality, immunity to the term voltage quality. Based on this term, a growing set of standards is being developed by the IEC. The various aspects of electromagnetic compatibility and EMC standards will be discussed in Section 1.4.2.

1.3 OVERVIEW OF POWER QUALITY PHENOMENA

We saw in the previous section that power quality is concerned with deviations of the voltage from its ideal waveform (voltage quality) and deviations of the current from its ideal waveform (current quality). Such a deviation is called a "power quality phenomenon" or a "power quality disturbance." Power quality phenomena can be divided into two types, which need to be treated in a different way.

- A characteristic of voltage or current (e.g., frequency or power factor) is never exactly equal to its nominal or desired value. The small deviations from the nominal or desired value are called "voltage variations" or "current variations." A property of any variation is that it has a value at any moment in time: e.g., the frequency is never exactly equal to 50 Hz or 60 Hz; the power factor is never exactly unity. Monitoring of a variation thus has to take place continuously.
- Occasionally the voltage or current deviates significantly from its normal or ideal waveshape. These sudden deviations are called "events." Examples are a sudden drop to zero of the voltage due to the operation of a circuit breaker (a voltage event), and a heavily distorted overcurrent due to switching of a non-loaded transformer (a current event). Monitoring of events takes place by using a triggering mechanism where recording of voltage and/or current starts the moment a threshold is exceeded.

The classification of a phenomenon in one of these two types is not always unique. It may depend on the kind of problem due to the phenomenon.

1.3.1 Voltage and Current Variations

Voltage and current variations are relatively small deviations of voltage or current characteristics around their nominal or ideal values. The two basic examples are voltage magnitude and frequency. On average, voltage magnitude and voltage frequency are equal to their nominal value, but they are never exactly equal. To describe the deviations in a statistical way, the probability density or probability distribution function should be used. Figure 1.1 shows a fictitious variation of the voltage magnitude as a function of time. This figure is the result of a so-called Monte Carlo simulation (see

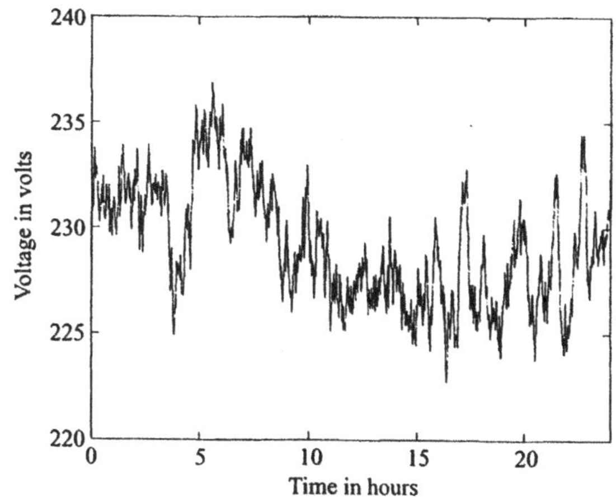

Figure 1.1 Simulated voltage magnitude as a function of time.

Section 2.5.5). The underlying distribution was a normal distribution with an expected value of 230 V and a standard deviation of 11.9 V. A set of independent samples from this distribution is filtered by a low-pass filter to prevent too large short-time changes. The probability density function of the voltage magnitude is shown in Fig. 1.2. The probability density function gives the probability that the voltage magnitude is within a certain range. Of interest is mainly the probability that the voltage magnitude is below or above a certain value. The probability distribution function (the integral of the density function) gives that information directly. The probability distribution function for this fictitious variation is shown in Fig. 1.3. Both the probability density function and the probability distribution function will be defined more accurately in Section 2.5.1.

An overview of voltage and current variations is given below. This list is certainly not complete, it merely aims at giving some example. There is an enormous range in end-user equipment, many with special requirements and special problems. In the power quality field new types of variations and events appear regularly. The following list uses neither the terms used by the IEC nor the terms recommended by the IEEE. Terms commonly used do not always fully describe a phenomenon. Also is there still

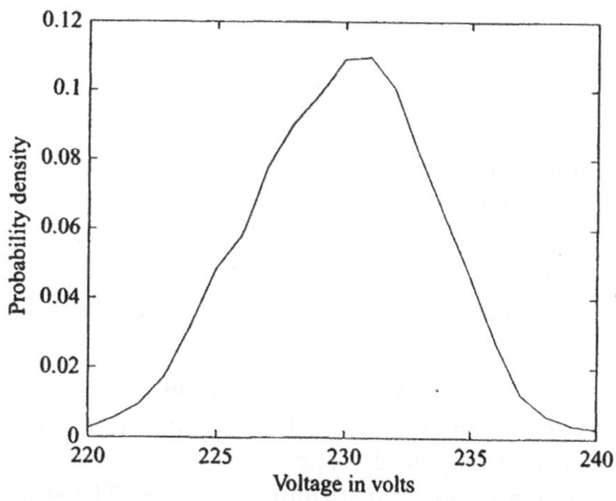

Figure 1.2 Probability density function of the voltage magnitude in Fig. 1.1.

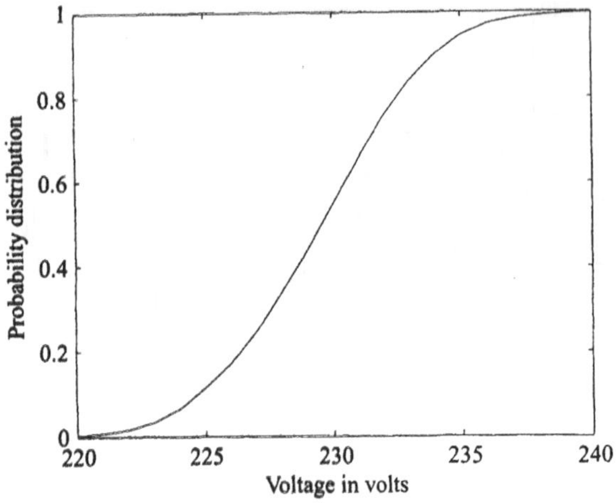

Figure 1.3 Probability distribution function of the voltage magnitude in Fig. 1.1.

some inconsistency between different documents about which terms should be used. The terms used in the list below, and in a similar list in Section 1.3.2 are not meant as an alternative for the IEC or IEEE definitions, but simply an attempt to somewhat clarify the situation. The reader is advised to continue using officially recognized terms, where feasible.

1. **Voltage magnitude variation.** Increase and decrease of the voltage magnitude, e.g., due to

- variation of the total load of a distribution system or part of it;
- actions of transformer tap-changers;
- switching of capacitor banks or reactors.

Transformer tap-changer actions and switching of capacitor banks can normally be traced back to load variations as well. Thus the voltage magnitude variations are mainly due to load variations, which follow a daily pattern. The influence of tap-changers and capacitor banks makes that the daily pattern is not always present in the voltage magnitude pattern.

The IEC uses the term "voltage variation" instead of "voltage magnitude variation." The IEEE does not appear to give a name to this phenomenon. Very fast variation of the voltage magnitude is referred to as voltage fluctuation.

2. **Voltage frequency variation.** Like the magnitude, also the frequency of the supply voltage is not constant. Voltage frequency variation is due to unbalance between load and generation. The term "frequency deviation" is also used. Short-duration frequency transients due to short circuits and failure of generator stations are often also included in voltage frequency variations, although they would better be described as events.

The IEC uses the term "power frequency variation"; the IEEE uses the term "frequency variation."

3. **Current magnitude variation.** On the load side, the current is normally also not constant in magnitude. The variation in voltage magnitude is mainly due to variation in current magnitude. The variation in current magnitude plays an important role in the design of power distribution systems. The system has to be designed for the maximum

current, where the revenue of the utility is mainly based on average current. The more constant the current, the cheaper the system per delivered energy unit.

Neither IEC nor IEEE give a name for this phenomenon.

4. **Current phase variation.** Ideally, voltage and current waveforms are in phase. In that case the power factor of the load equals unity, and the reactive power consumption is zero. That situation enables the most efficient transport of (active) power and thus the cheapest distribution system.

Neither IEC nor IEEE give a name for this power quality phenomenon, although the terms "power factor" and "reactive power" describe it equally well.

5. **Voltage and current unbalance.** Unbalance, or three-phase unbalance, is the phenomenon in a three-phase system, in which the rms values of the voltages or the phase angles between consecutive phases are not equal. The severity of the voltage unbalance in a three-phase system can be expressed in a number of ways, e.g.,

- the ratio of the negative-sequence and the positive-sequence voltage component;
- the ratio of the difference between the highest and the lowest voltage magnitude, and the average of the three voltage magnitudes; and
- the difference between the largest and the smallest phase difference between consecutive phases.

These three severity indicators can be referred to as "negative-sequence unbalance," "magnitude unbalance," and "phase unbalance," respectively.

The primary source of voltage unbalance is unbalanced load (thus current unbalance). This can be due to an uneven spread of (single-phase) low-voltage customers over the three phases, but more commonly unbalance is due to a large single-phase load. Examples of the latter can be found among railway traction supplies and arc furnaces. Three-phase voltage unbalance can also be the result of capacitor bank anomalies, such as a blown fuse in one phase of a three-phase bank.

Voltage unbalance is mainly of concern for three-phase loads. Unbalance leads to additional heat production in the winding of induction and synchronous machines; this reduces the efficiency and requires derating of the machine. A three-phase diode rectifier will experience a large current unbalance due to a small voltage unbalance. The largest current is in the phase with the highest voltage, thus the load has the tendency to mitigate the voltage unbalance.

The IEEE mainly recommends the term "voltage unbalance" although some standards (notably IEEE Std.1159) use the term "voltage imbalance."

6. **Voltage fluctuation.** If the voltage magnitude varies, the power flow to equipment will normally also vary. If the variations are large enough or in a certain critical frequency range, the performance of equipment can be affected. Cases in which voltage variation affects load behavior are rare, with the exception of lighting load. If the illumination of a lamp varies with frequencies between about 1 Hz and 10 Hz, our eyes are very sensitive to it and above a certain magnitude the resulting light flicker can become rather disturbing. It is this sensitivity of the human eye which explains the interest in this phenomenon. The fast variation in voltage magnitude is called "voltage fluctuation," the visual phenomenon as perceived by our brain is called "light flicker." The term "voltage flicker" is confusing but sometimes used as a shortening for "voltage fluctuation leading to light flicker."

To quantify voltage fluctuation and light flicker, a quantity called "flicker intensity" has been introduced [81]. Its value is an objective measure of the severity of the light flicker due to a certain voltage fluctuation. The flicker intensity can be treated as a variation, just like voltage magnitude variation. It can be plotted as a function of time, and probability density and distribution functions can be obtained. Many publications discuss voltage fluctuation and light flicker. Good overviews can be found in, among others, [141] and [142].

The terms "voltage fluctuation" and "light flicker" are used by both IEC and IEEE.

7. **Harmonic voltage distortion.** The voltage waveform is never exactly a single-frequency sine wave. This phenomenon is called "harmonic voltage distortion" or simply "voltage distortion." When we assume a waveform to be periodic, it can be described as a sum of sine waves with frequencies being multiples of the fundamental frequency. The nonfundamental components are called "harmonic distortion."

There are three contributions to the harmonic voltage distortion:

1. The voltage generated by a synchronous machine is not exactly sinusoidal due to small deviations from the ideal shape of the machine. This is a small contribution; assuming the generated voltage to be sinusoidal is a very good approximation.
2. The power system transporting the electrical energy from the generator stations to the loads is not completely linear, although the deviation is small. Some components in the system draw a nonsinusoidal current, even for a sinusoidal voltage. The classical example is the power transformer, where the nonlinearity is due to saturation of the magnetic flux in the iron core of the transformer. A more recent example of a nonlinear power system component is the HVDC link. The transformation from ac to dc and back takes place by using power-electronics components which only conduct during part of a cycle.

   The amount of harmonic distortion originating in the power system is normally small. The increasing use of power electronics for control of power flow and voltage (flexible ac transmission systems or FACTS) carries the risk of increasing the amount of harmonic distortion originating in the power system. The same technology also offers the possibility of removing a large part of the harmonic distortion originating elsewhere in the system or in the load.
3. The main contribution to harmonic voltage distortion is due to nonlinear load. A growing part of the load is fed through power-electronics converters drawing a nonsinusoidal current. The harmonic current components cause harmonic voltage components, and thus a nonsinusoidal voltage, in the system.

Two examples of distored voltage are shown in Figs. 1.4 and 1.5. The voltage shown in Fig. 1.4 contains mainly harmonic components of lower order (5, 7, 11, and 13 in this case). The voltage shown in Fig. 1.5 contains mainly higher-frequency harmonic components.

Harmonic voltages and current can cause a whole range of problems, with additional losses and heating the main problem. The harmonic voltage distortion is normally limited to a few percent (i.e., the magnitude of the harmonic voltage components

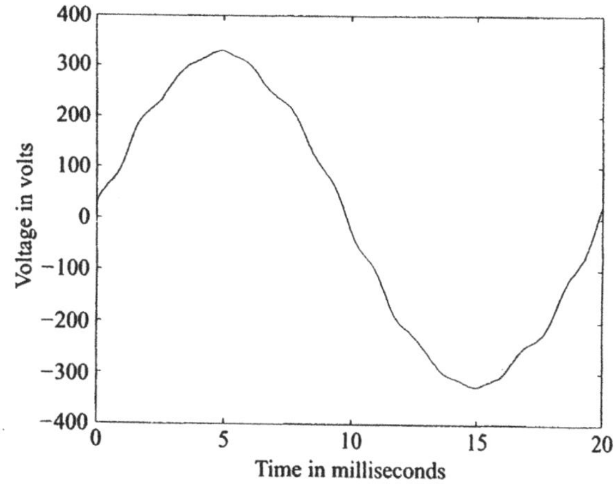

Figure 1.4 Example of distorted voltage, with mainly lower-order harmonic components [211].

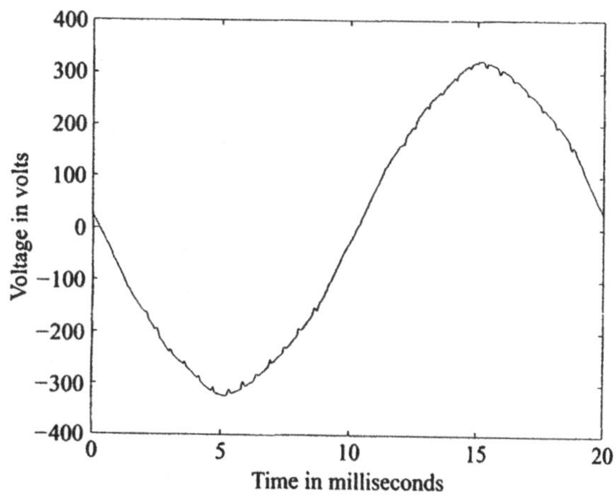

Figure 1.5 Example of distorted voltage, with higher-order harmonic components [211].

is up to a few percent of the magnitude of the fundamental voltage) in which case equipment functions as normal. Occasionally large harmonic voltage distortion occurs, which can lead to malfunction of equipment. This can especially be a big problem in industrial power systems, where there is a large concentration of distorting load as well as sensitive load. Harmonic distortion of voltage and current is the subject of hundreds of papers as well as a number of books [77], [194], [195].

The term "harmonic distortion" is very commonly used, and "distortion" is an IEC term referring to loads taking harmonic current components. Also within the IEEE the term "distortion" is used to refer to harmonic distortion; e.g., "distortion factor" and "voltage distortion."

8. **Harmonic current distortion**. The complementary phenomenon of harmonic voltage distortion is harmonic current distortion. The first is a voltage quality phenomenon, the latter a current quality phenomenon. As harmonic voltage distortion is mainly due to nonsinusoidal load currents, harmonic voltage and current distortion are strongly linked. Harmonic current distortion requires over-rating of series components like transformers and cables. As the series resistance increases with frequency, a distorted current will cause more losses than a sinusoidal current of the same rms value.

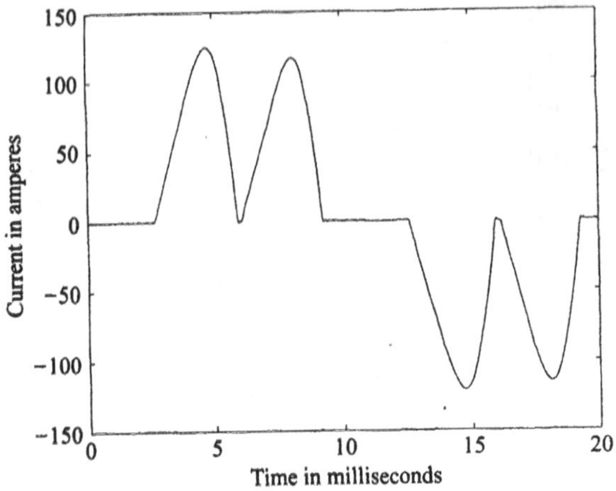

Figure 1.6 Example of distorted current, leading to the voltage distortion shown in Fig. 1.4 [211].

Two examples of harmonic current distortion are shown in Figs. 1.6 and 1.7. Both currents are drawn by an adjustable-speed drive. The current shown in Fig. 1.6 is typical for modern ac adjustable-speed drives. The harmonic spectrum of the current contains mainly 5th, 7th, 11th, and 13th harmonic components. The current in Fig. 1.7 is less common. The high-frequency ripple is due to the switching frequency of the dc/ac inverter. As shown in Fig. 1.5 this high-frequency current ripple causes a high-frequency ripple in the voltage as well.

9. **Interharmonic voltage and current components.** Some equipment produces current components with a frequency which is not an integer multiple of the fundamental frequency. Examples are cycloconverters and some types of heating controllers. These components of the current are referred to as "interharmonic components." Their magnitude is normally small enough not to cause any problem, but sometimes they can excite unexpected resonances between transformer inductances and capacitor banks. More dangerous are current and voltage components with a frequency below the fundamental frequency, referred to as "sub-harmonic distortion." Sub-harmonic currents can lead to saturation of transformers and damage to synchronous generators and turbines.

Another source of interharmonic distortion are arc furnaces. Strictly speaking arc furnaces do not produce any interharmonic voltage or current components, but a

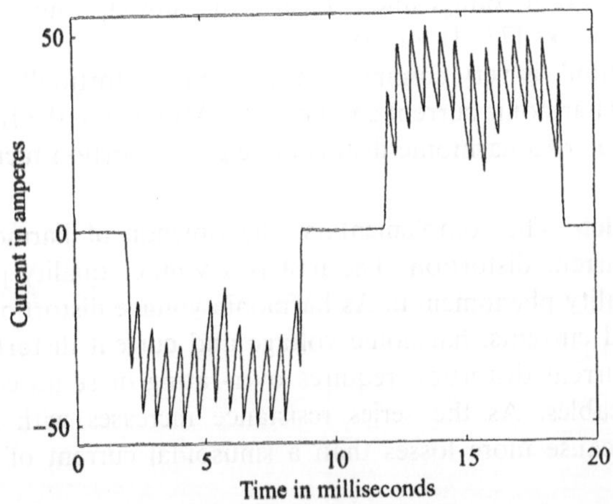

Figure 1.7 Example of distorted current, leading to the voltage distortion shown in Fig. 1.5 [211].

number of (integer) harmonics plus a continuous (voltage and current) spectrum. Due to resonances in the power system some of the frequencies in this spectrum are amplified. The amplified frequency components are normally referred to as interharmonics due to the arc furnace. These voltage interharmonics have recently become of special interest as they are responsible for serious light flicker problems.

A special case of sub-harmonic currents are those due to oscillations in the earth-magnetic field following a solar flare. These so-called geomagnetically induced currents have periods around five minutes and the resulting transformer saturation has led to large-scale blackouts [143].

10. **Periodic voltage notching.** In three-phase rectifiers the commutation from one diode or thyristor to the other creates a short-circuit with a duration less than 1 ms, which results in a reduction in the supply voltage. This phenomenon is called "voltage notching" or simply "notching." Notching mainly results in high-order harmonics, which are often not considered in power engineering. A more suitable way of characterization is through the depth and duration of the notch in combination with the point on the sine wave at which the notching commences.

An example of voltage notching is shown in Fig. 1.8. This voltage wave shape was caused by an adjustable-speed drive in which a large reactance was used to keep the dc current constant.

The IEEE uses the term "notch" or "line voltage notch" in a more general way: any reduction of the voltage lasting less than half a cycle.

11. **Mains signaling voltage.** High-frequency signals are superimposed on the supply voltage for the purpose of transmission of information in the public distribution system and to customer's premises. Three types of signal are mentioned in the European voltage characteristics standards [80]:

- *Ripple control signals*: sinusoidal signals between 110 and 3000 Hz. These signals are, from a voltage-quality point-of-view, similar to harmonic and interharmonic voltage components.
- *Power-line-carrier signals*: sinusoidal signals between 3 and 148.5 kHz. These signals can be described both as high-frequency voltage noise (see below) and as high-order (inter)harmonics.
- *Mains marking signals*: superimposed short time alterations (transients) at selected points of the voltage waveform.

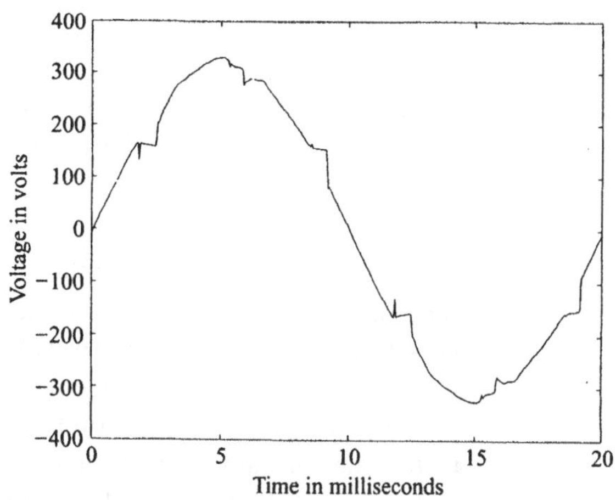

Figure 1.8 Example of voltage notching [211].

Mains signaling voltage can interfere with equipment using similar frequencies for some internal purpose. The voltages, and the associated currents, can also cause audible noise and signals on telephone lines.

The other way around, harmonic and interharmonic voltages may be interpreted by equipment as being signaling voltages, leading to wrong functioning of equipment.

12. **High-frequency voltage noise**. The supply voltage contains components which are not periodic at all. These can be called "noise," although from the consumer point of view, all above-mentioned voltage components are in effect noise. Arc furnaces are an important source of noise. But also the combination of many different nonlinear loads can lead to voltage noise [196]. Noise can be present between the phase conductors (differential mode noise) or cause an equal voltage in all conductors (common-mode noise). Distinguishing the noise from other components is not always simple, but actually not really needed. An analysis is needed only in cases where the noise leads to some problem with power system or end-user equipment. The characteristics of the problem will dictate how to measure and describe the noise.

A whole range of voltage and current variations has been introduced. The reader will have noticed that the distinction between the various phenomena is not very sharp, e.g., voltage fluctuation and voltage variation show a clear overlap. One of the tasks of future standardization work is to develop a consistent and complete classification of the various phenomena. This might look an academic task, as it does not directly solve any equipment or system problems. But when quantifying the power quality, the classification becomes less academic. A good classification also leads to a better understanding of the various phenomena.

1.3.2 Events

Events are phenomena which only happen every once in a while. An interruption of the supply voltage is the best-known example. This can in theory be viewed as an extreme voltage magnitude variation (magnitude equal to zero), and can be included in the probability distribution function of the voltage magnitude. But this would not give much useful information; it would in fact give the unavailability of the supply voltage, assuming the resolution of the curve was high enough. Instead, events can best be described through the time between events, and the characteristics of the events; both in a stochastic sense. Interruptions will be discussed in sufficient detail in Chapters 2 and 3 and voltage sags in Chapters 4, 5, and 6. Transient overvoltage will be used as an example here. A transient overvoltage recording is shown in Fig. 1.9: the (absolute value of the) voltage rises to about 180% of its normal maximum for a few milliseconds. The smooth sinusoidal curve is a continuation of the pre-event fundamental voltage.

A transient overvoltage can be characterized in many different ways; three often-used characteristics are:

1. **Magnitude**: the magnitude is either the maximum voltage or the maximum voltage deviation from the normal sine wave.
2. **Duration**: the duration is harder to define, as it often takes a long time before the voltage has completely recovered. Possible definitions are:
   - the time in which the voltage has recovered to within 10% of the magnitude of the transient overvoltage;
   - the time-constant of the average decay of the voltage;
   - the ratio of the Vt-integral defined below and the magnitude of the transient overvoltage.

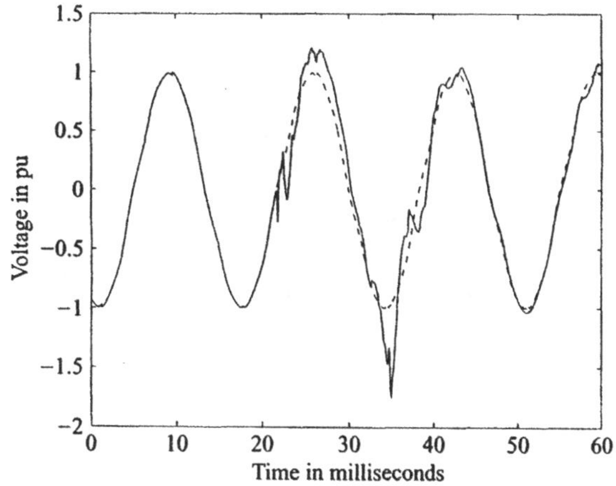

Figure 1.9 Example of transient overvoltage event: phase-to-ground voltage due to fault clearing in one of the other phases. (Data obtained from [16].)

3. **Vt-integral**: the Vt-integral is defined as

$$V_t = \int_0^T V(t)dt \tag{1.1}$$

where $t = 0$ is the start of the event, and an appropriate value is chosen for T, e.g., the time in which the voltage has recovered to within 10% of the magnitude of the transient overvoltage. Again the voltage $V(t)$ can be measured either from zero or as the deviation from the normal sine wave.

Figure 1.10 gives the number of transient overvoltage events per year, as obtained for the average low-voltage site in Norway [67]. The distribution function for the time

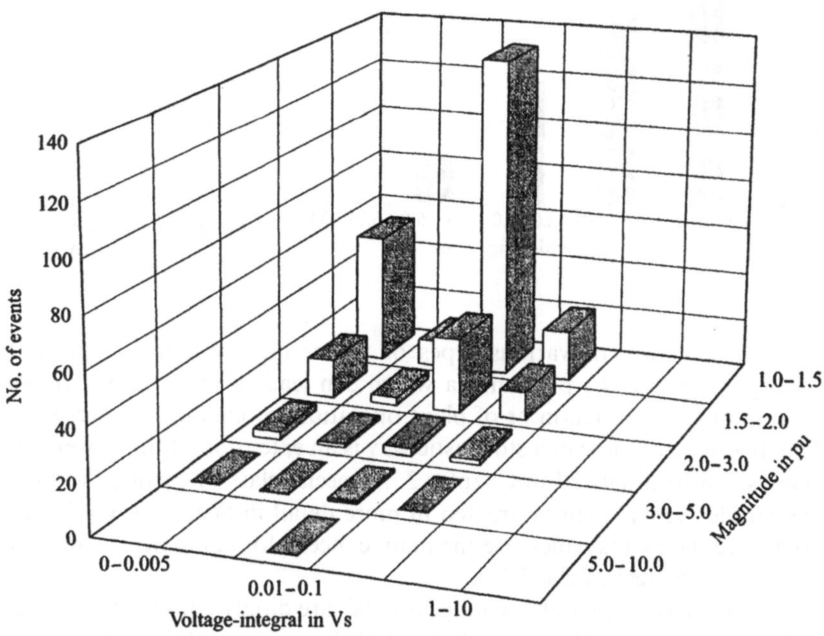

Figure 1.10 Number of transient overvoltage events per year, as a function of magnitude and voltage integral. (Data obtained from [67].)

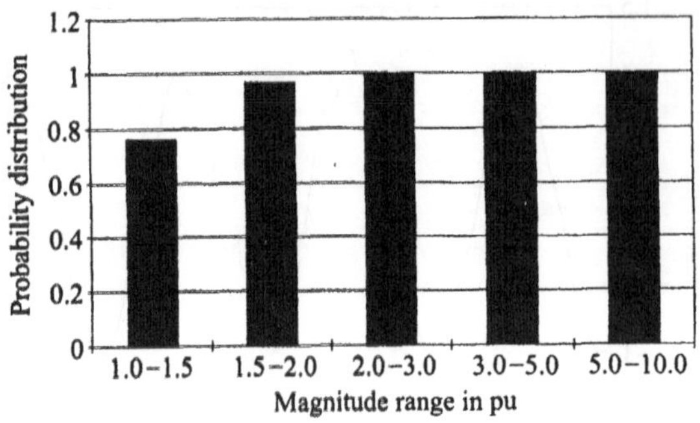

Figure 1.11 Probability distribution function of the magnitude of transient overvoltage events, according to Fig. 1.10.

between events has not been determined, but only the number of events per year with different characteristics. Note that the average time between events is the reciprocal of the number of events per year. This is the normal situation; the actual distribution function is rarely determined in power quality or reliability surveys [107].

Figures 1.11 through 1.14 give statistical information about the characteristics of the events. Figure 1.11 gives the probability distribution function of the magnitude of the event. We see that almost 80% of the events have a magnitude less than 1.5 pu. Figure 1.12 gives the corresponding density function. By using a logarithmic scale the number of events in the high-magnitude range is better visible. Figure 1.13 gives the probability distribution function of the Vt-integral; Fig. 1.14 the probability density function.

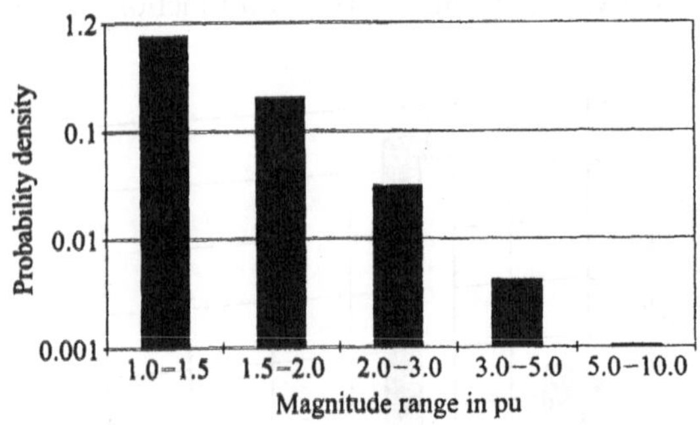

Figure 1.12 Probability density function of the magnitude of transient overvoltage events, according to Fig. 1.10.

An overview of various types of power quality events is given below. Power quality events are the phenomena which can lead to tripping of equipment, to interruption of the production or of plant operation, or endanger power system operation. The treatment of these in a stochastic way is an extension of the power system reliability field as will be discussed in Chapter 2. A special class of events, the so-called "voltage magnitude events," will be treated in more detail in Section 1.3.3. Voltage magnitude events are the events which are the main concern for equipment, and they are the main subject for the rest of this book.

Note that below only "voltage events" are discussed, as these can be of concern to end-user equipment. But similarly a list of "current events" could be added, with their possible effects on power system equipment. Most power quality monitors in use, continuously monitor the voltage and record an event when the voltage exceeds certain

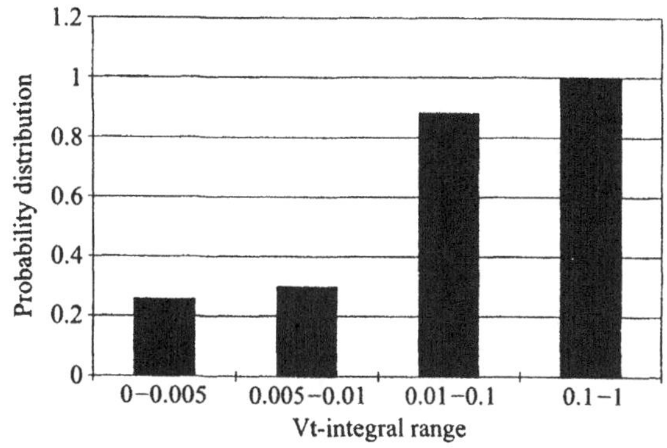

Figure 1.13 Probability distribution function of the Vt-integral of transient overvoltage events, according to Fig. 1.10.

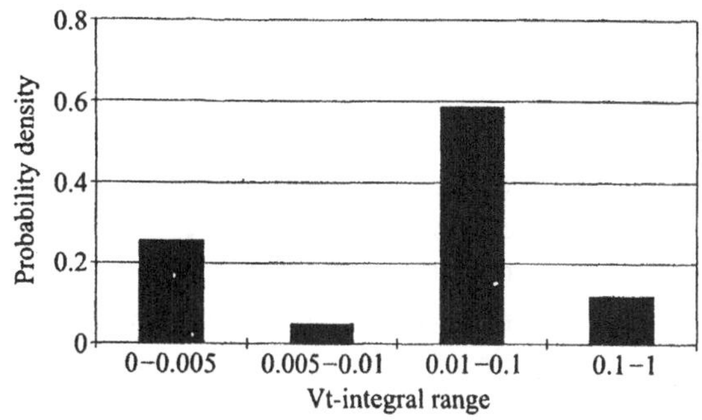

Figure 1.14 Probability density function of the Vt-integral of transient overvoltage events, according to Fig. 1.10.

thresholds, typically voltage magnitude thresholds. Although the currents are often also recorded they do not normally trigger the recording. Thus an overcurrent without an over- or undervoltage will not be recorded. Of course there are no technical limitations in using current signals to trigger the recording process. In fact most monitors have the option of triggering on current as well.

1. **Interruptions.** A "voltage interruption" [IEEE Std.1159], "supply interruption" [EN 50160], or just "interruption" [IEEE Std.1250] is a condition in which the voltage at the supply terminals is close to zero. Close to zero is by the IEC defined as "lower than 1% of the declared voltage" and by the IEEE as "lower than 10%" [IEEE Std. 1159].

Voltage interruptions are normally initiated by faults which subsequently trigger protection measures. Other causes of voltage interruption are protection operation when there is no fault present (a so-called protection maltrip), broken conductors not triggering protective measures, and operator intervention. A further distinction can be made between pre-arranged and accidental interruptions. The former allow the end user to take precautionary measures to reduce the impact. All pre-arranged interruptions are of course caused by operator action.

Interruptions can also be subdivided based on their duration, thus based on the way of restoring the supply:

- automatic switching;
- manual switching;
- repair or replacement of the faulted component.

Various terminologies are in use to distinguish between these. The IEC uses the term long interruptions for interruptions longer than 3 minutes and the term short interruptions for interruptions lasting up to 3 minutes. Within the IEEE the terms momentary, temporary, and sustained are used, but different documents give different duration values. The various definitions will be discussed in Chapter 3.

2. **Undervoltages.** Undervoltages of various duration are known under different names. Short-duration undervoltages are called "voltage sags" or "voltage dips." The latter term is preferred by the IEC. Within the IEEE and in many journal and conference papers on power quality, the term voltage sag is used. Long-duration undervoltage is normally simply referred to as "undervoltage."

A voltage sag is a reduction in the supply voltage magnitude followed by a voltage recovery after a short period of time. When a voltage magnitude reduction of finite duration can actually be called a voltage sag (or voltage dip in the IEC terminology) remains a point of debate, even though the official definitions are clear about it. According to the IEC, a supply voltage dip is a sudden reduction in the supply voltage to a value between 90% and 1% of the declared voltage, followed by a recovery between 10 ms and 1 minute later. For the IEEE a voltage drop is only a sag if the during-sag voltage is between 10% and 90% of the nominal voltage.

Voltage sags are mostly caused by short-circuit faults in the system and by starting of large motors. Voltage sags will be discussed in detail in Chapters 4, 5, and 6.

3. **Voltage magnitude steps.** Load switching, transformer tap-changers, and switching actions in the system (e.g., capacitor banks) can lead to a sudden change in the voltage magnitude. Such a voltage magnitude step is called a "rapid voltage change" [EN 50160] or "voltage change" [IEEE Std.1159]. Normally both voltage before and after the step are in the normal operating range (typically 90% to 110% of the nominal voltage).

An example of voltage magnitude steps is shown in Fig. 1.15. The figure shows a 2.5 hour recording of the voltage in a 10 kV distribution system. The steps in the voltage magnitude are due to the operation of transformer tap-changers at various voltage levels.

4. **Overvoltages.** Just like with undervoltage, overvoltage events are given different names based on their duration. Overvoltages of very short duration, and high magnitude, are called "transient overvoltages," "voltage spikes," or sometimes "voltage surges." The latter term is rather confusing as it is sometimes used to refer to overvoltages with a duration between about 1 cycle and 1 minute. The latter event is more correctly called "voltage swell" or "temporary power frequency overvoltage." Longer

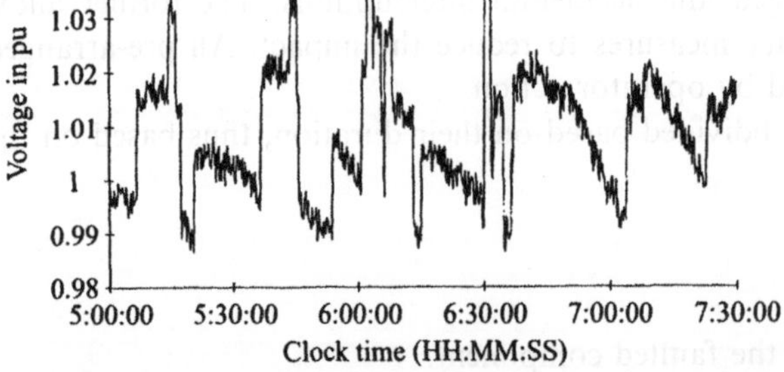

Figure 1.15 Example of voltage magnitude steps due to transformer tap-changer operation, recorded in a 10 kV distribution system in southern Sweden.

duration overvoltages are simply referred to as "overvoltages." Long and short overvoltages originate from, among others, lightning strokes, switching operations, sudden load reduction, single-phase short-circuits, and nonlinearities.

A resonance between the nonlinear magnetizing reactance of a transformer and a capacitance (either in the form of a capacitor bank or the capacitance of an underground cable) can lead to a large overvoltage of long duration. This phenomenon is called ferroresonance, and it can lead to serious damage to power system equipment [144].

5. **Fast voltage events.** Voltage events with a very short duration, typically one cycle of the power system frequency or less, are referred to as "transients," "transient (over)voltages," "voltage transients," or "wave shape faults." The term transient is not fully correct, as it should only be used for the transition between two steady states. Events due to switching actions could under that definition be called transients; events due to lightning strokes could not be called transients under that definition. But due to the similarity in time scale both are referred to as voltage transients. Even very short-duration voltage sags (e.g., due to fuse clearing) are referred to as voltage transients, or also "notches."

Fast voltage events can be divided into impulsive transients (mainly due to lightning) and oscillatory transients (mainly due to switching actions).

6. **Phase-angle jumps and three-phase unbalance.** We will see in Chapter 4 that a voltage sag is often associated with a phase-angle jump and some three-phase unbalance. An interesting thought is whether or not a jump in phase-angle without a drop in voltage magnitude should be called a voltage sag. Such an event could occur when one of two parallel feeders is taken out of operation. The same holds for a short-duration, three-phase unbalance without change in magnitude, thus where only the phase-angle of the three voltages changes.

To get a complete picture, also short-duration phase-angle jumps and short-duration unbalances should be considered as events belonging to the family of power quality phenomena.

1.3.3 Overview of Voltage Magnitude Events

As mentioned in the previous section, the majority of events currently of interest are associated with either a reduction or an increase in the voltage magnitude. We will refer to these as "voltage magnitude events."

A voltage magnitude event is a (significant) deviation from the normal voltage magnitude for a limited duration. The magnitude can be found by taking the rms of the voltage over a multiple of one half-cycle of the power-system frequency.

$$V_{rms} = \sqrt{\frac{1}{N}\sum_{k=1}^{N} V^2(k\Delta t)} \tag{1.2}$$

where $V(t)$ is the voltage as a function of time, sampled at equidistant points $t = k\Delta t$. The rms value is taken over a period $N\Delta t$, referred to as the "window length." Alternatively, the magnitude can be determined from the peak voltage or from the fundamental-frequency component of the voltage. Most power quality monitors determine the rms voltage once every cycle or once every few cycles. The moment the rms voltage deviates more than a pre-set threshold from its nominal value, the voltage as a function of time is recorded (the rms voltage, the sampled time-domain data, or both).

Most events show a rather constant rms voltage for a certain duration after which the rms voltage returns to a more or less normal value. This is understandable if one realizes that events are due to changes in the system followed by the restoration of the original system after a certain time. Before, during, and after the event, the system is more or less in a steady state. Thus the event can be characterized through one duration and one magnitude. We will see in Chapter 4 that it is not always possible to uniquely determine magnitude and duration of a voltage magnitude event. For now we will assume that this is possible, and define the magnitude of the event as the remaining rms voltage during the event: if the rms voltage during the event is 170 V in a 230 V system, the magnitude of the event is $\frac{170}{230} = 73.9\%$.

Knowing the magnitude and duration of an event, it can be represented as one point in the magnitude-duration plane. All events recorded by a monitor over a certain period can be represented as a scatter of points. Different underlying causes may lead to events in different parts of the plane. The magnitude-duration plot will come back several times in the forthcoming chapters. Various standards give different names to events in different parts of the plane. A straightforward classification is given in Fig. 1.16. The voltage magnitude is split into three regions:

- *interruption*: the voltage magnitude is zero,
- *undervoltage*: the voltage magnitude is below its nominal value, and
- *overvoltage*: the voltage magnitude is above its nominal value.

In duration, a distinction is made between:

- very short, corresponding to transient and self-restoring events;
- short, corresponding to automatic restoration of the pre-event situation;
- long, corresponding to manual restoration of the pre-event situation;
- very long, corresponding to repair or replacement of faulted components.

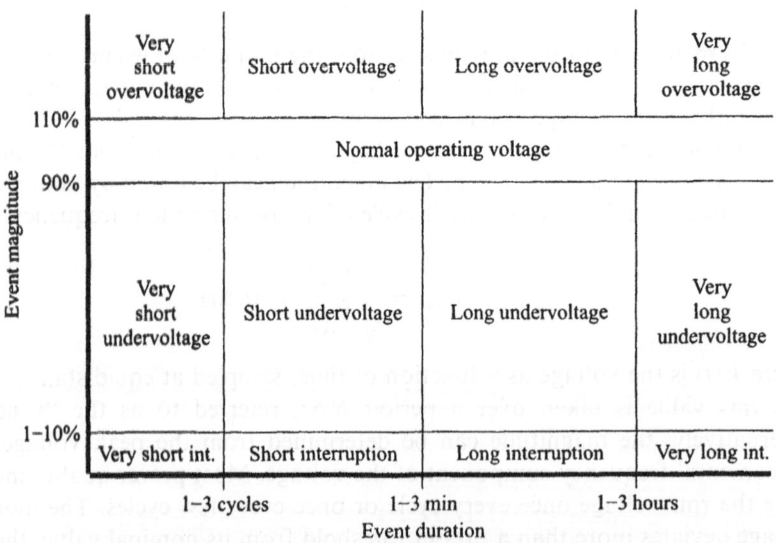

Figure 1.16 Suggested classification of voltage magnitude events.

The various borders in Fig. 1.16 are somewhat arbitrary; some of the indicated values (1–3 minutes, 1–10%, 90%, and 110%) are those used in existing IEC and IEEE standards. For monitoring purposes, strict thresholds are needed to distinguish between the different events. An example is the threshold dividing between interruptions and undervoltages. This one is placed (somewhat arbitrarily) at 1% of nominal according to the IEC and at 10% according to the IEEE (see below). Any other small value would be equally defendable.

The classification in Fig. 1.16 is only aimed at explaining the different types of events: the terms mentioned in the figures are not all used in practice. Both IEC and IEEE give different names to events in some of the regions of the magnitude-duration plane. The IEC definitions are summarized in Fig. 1.17 and the IEEE definitions in Fig. 1.18. The IEC definitions were obtained from CENELEC document EN 50160 [80], the IEEE definitions from IEEE Std.1159-1995.

The method of classifying events through one magnitude and one duration has been shown to be very useful and has resulted in a lot of information and knowledge about power quality. But the method also has its limitations, which is important to realize when using this classification. Four points should be especially kept in mind.

1. The during-event rms voltage is not always constant, leading to ambiguities in defining the magnitude of the event. It may also lead to ambiguities in defining the duration of the event.
2. Fast events (one cycle or less in duration) cannot be characterized, resulting in unrealistic values for magnitude and duration or in these disturbances simply being neglected.
3. Repetitive events can give erroneous results: they either lead to an over-estimation of the number of events (when each event in a row of events is counted as a separate event), or an under-estimation of the severity of the events (when a row of identical events is counted as one event).

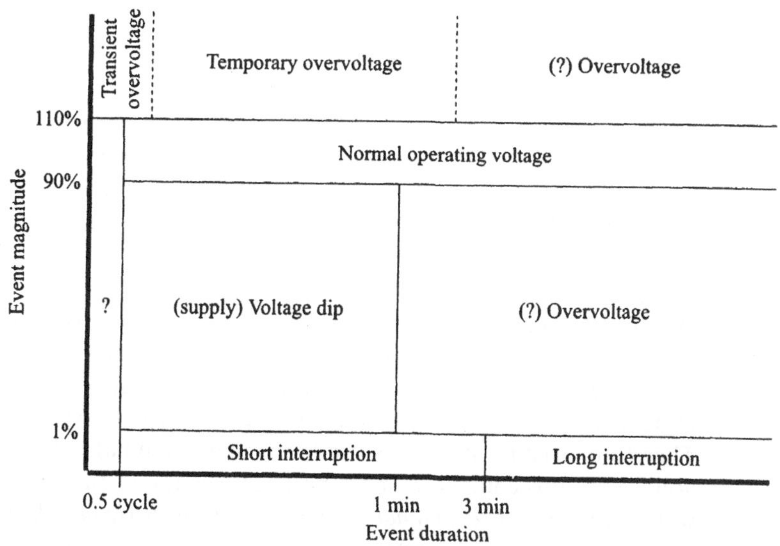

Figure 1.17 Definitions of voltage magnitude events as used in EN 50160.

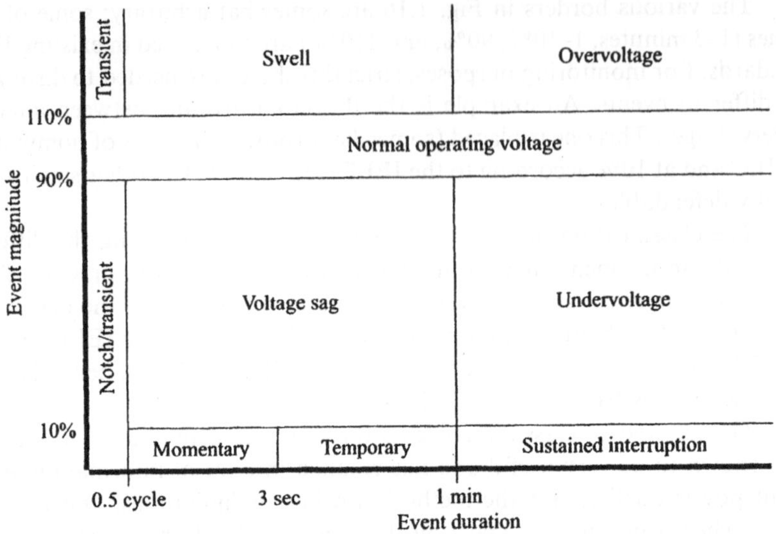

Figure 1.18 Definitions of voltage magnitude events as used in IEEE Std.1159-1995.

4. Equipment is sometimes sensitive to other characteristics than just magnitude and duration.

We will come back to these problems in more detail in Chapters 3 and 4.

Similar classifications can be proposed for voltage frequency events, for voltage phase-angle events, for three-phase voltage unbalance events, etc. But because most equipment problems are due to an increase or decrease in voltage magnitude, the emphasis is on voltage magnitude events.

1.4 POWER QUALITY AND EMC STANDARDS

1.4.1 Purpose of Standardization

Standards that define the quality of the supply have been present for decades already. Almost any country has standards defining the margins in which frequency and voltage are allowed to vary. Other standards limit harmonic current and voltage distortion, voltage fluctuations, and duration of an interruption. There are three reasons for developing power quality standards.

1. **Defining the nominal environment.** A hypothetical example of such a standard is: "*The voltage shall be sinusoidal with a frequency of 50 Hz and an rms voltage of 230 V.*" Such a standard is not very practical as it is technically impossible to keep voltage magnitude and frequency exactly constant. Therefore, existing standards use terms like "nominal voltage" or "declared voltage" in this context. A more practical version of the above standard text would read as: "*The nominal frequency shall be 50 Hz and the nominal voltage shall be 230 V,*" which comes close to the wording in European standard EN 50160 [80].

Defining nominal voltage and frequency does not say anything about the actual environment. To do this the deviations from the nominal values have to be known. Most countries have a standard giving the allowed variation in the rms voltage, a typical range being between from 90% to 110%.

2. **Defining the terminology.** Even if a standard-setting body does not want to impose any requirements on equipment or supply, it might still want to publish power quality standards. A good example is IEEE Std.1346 [22] which recommends a method for exchanging information between equipment manufacturers, utilities, and customers. The standard does not give any suggestions about what is considered acceptable.

   This group of standards aims at giving exact definitions of the various phenomena, how their characteristics should be measured, and how equipment should be tested for its immunity. The aim of this is to enable communication between the various partners in the power quality field. It ensures, e.g., that the results of two power quality monitors can be easily compared and that equipment immunity can be compared with the description of the environment. Hypothetical examples are: "*A short interruption is a situation in which the rms voltage is less than 1% of the nominal rms voltage for less than 3 minutes.*" and "*The duration of a voltage dip is the time during which the rms voltage is less than 90% of the nominal rms voltage. The duration of a voltage dip shall be expressed in seconds. The rms voltage shall be determined every half-cycle.*" Both IEEE Std.1159 and EN 50160 give these kind of definitions, hopefully merging into a future IEC standard.

3. **Limit the number of power quality problems.** Limiting the number of power quality problems is the final aim of all the work on power quality. Power quality problems can be mitigated by limiting the amount of voltage disturbances caused by equipment, by improving the performance of the supply, and by making equipment less sensitive to voltage disturbances. All mitigation methods require technical solutions which can be implemented independently of any standardization. But proper standardization will provide important incentives for the implementation of the technical solutions. Proper standardization will also solve the problem of responsibility for power quality disturbances. Hypothetical examples are:

   The current taken by a load exceeding 4 kVA shall not contain more than 1% of any even harmonic. The harmonic contents shall be measured as a 1-second average. and *Equipment shall be immune to voltage variations between 85% and 110% of the nominal voltage. This shall be tested by supplying at the equipment terminals, sinusoidal voltages with magnitudes of 85% and 110% for a duration of 1 hour. If the piece of equipment has more than one distinctive load state, it shall be tested for each load state separately, or for what are conceived the most sensitive states.*

   In this field both IEC and IEEE lack a good set of standards on power quality. The IEC has set up a whole framework on electromagnetic compatibility which already includes some power quality standards. The best example is the harmonic standard IEC-61000-2-3 which limits the amount of harmonic current produced by low-power equipment. The IEEE has a good recommended practice for the limitation of harmonic distortion: IEEE 519 [82] which gives limits both for the harmonic currents taken by the customer and for the voltages delivered by the utility.

1.4.2 The IEC Electromagnetic Compatibility Standards

Within the International Electrotechnical Committee (IEC) a comprehensive framework of standards on electromagnetic compatibility is under development. Electromagnetic compatibility (EMC) is defined as: *the ability of a device, equipment or system to function satisfactorily in its electromagnetic environment without introducing intolerable electromagnetic disturbances to anything in that environment* [79].

There are two aspects to EMC: (1) a piece of equipment should be able to operate normally in its environment, and (2) it should not pollute the environment too much. In EMC terms: immunity and emission. There are standards for both aspects. Agreement on immunity is at first a matter of agreement between the manufacturer and the customer. But the IEC sets minimum requirements in immunity standards. The third term of importance is "electromagnetic environment," which gives the level of disturbance against which the equipment should be immune. Within the EMC standards, a distinction is made between radiated disturbances and conducted disturbances. Radiated disturbances are emitted (transmitted) by one device and received by another without the need for any conduction. Conducted disturbances need a conductor to transfer from one device to another. These conducted disturbances are within the scope of power quality; radiated disturbances (although very important) are outside of the normal realm of power system engineering or power quality.

A schematic overview of the EMC terminology is given in Fig. 1.19. We see that the emission of a device may consist of conducted disturbances and radiated disturbances. Radiated disturbances can reach another device via any medium. Normally, radiated disturbances only influence another device when it is physically close to the emitting device. Conducted disturbances reach another device via an electrically conducting medium, typically the power system. The device being influenced no longer has to be physically close as the power system is a very good medium for the conduction of many types of disturbances. Of course also here the rule is that a device which is electrically closer (there is less impedance between them) is more likely to be influenced. A device connected to the power system is exposed to an electrical environment not only due to the combined emission of all other devices connected to the system but also due to all kinds of events in the power system (like switching actions, short-circuit faults, and lightning strokes). The immunity of the device should be assessed with reference to this electromagnetic environment. A special type of disturbances, not shown in the

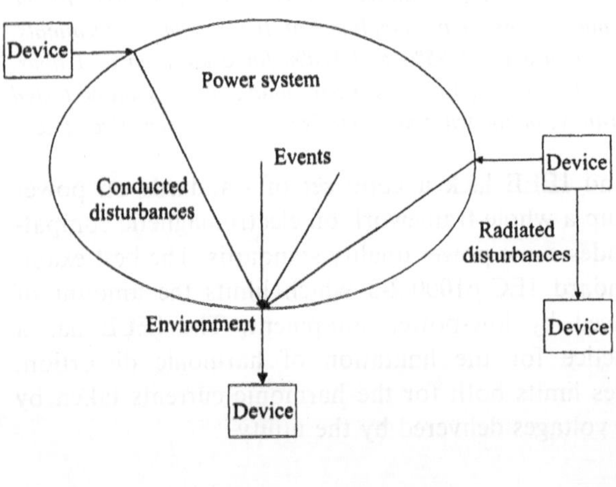

Figure 1.19 Overview of EMC terminology.

figure, are radiated disturbances which induce conducted disturbances in the power system.

Immunity Requirements. Immunity standards define the minimum level of electromagnetic disturbance that a piece of equipment shall be able to withstand. Before being able to determine the immunity of a device, a performance criterion must be defined. In other words, it should be agreed upon what kind of behavior will be called a failure. In practice it will often be clear when a device performs satisfactorily and when not, but when testing equipment the distinction may become blurred. It will all depend on the application whether or not a certain equipment behavior is acceptable.

The basic immunity standard [IEC-61000-4-1] gives four classes of equipment performance:

- Normal performance within the specification limits.
- Temporary degradation or loss of function which is self-recoverable.
- Temporary degradation or loss of function which requires operator intervention or system reset.
- Degradation or loss of function which is not recoverable due to damage of equipment, components or software, or loss of data.

These classes are general as the description should be applicable to all kinds of equipment. This classification is further defined in the various equipment standards.

Emission Standards. Emission standards define the maximum amount of electromagnetic disturbance that a piece of equipment is allowed to produce. Within the existing IEC standards, emission limits exist for harmonic currents [IEC 61000-3-2 and 61000-3-6], and for voltage fluctuations [IEC 61000-3-3, 61000-3-5, and 61000-3-7]. Most power quality phenomena are not due to equipment emission but due to operational actions or faults in the power system. As the EMC standards only apply to equipment, there are no "emission limits" for the power system. Events like voltage sags and interruptions are considered as a "fact-of-life." These events do, however, contribute to the electromagnetic environment.

The Electromagnetic Environment. To give quantitative levels for the immunity of equipment, the electromagnetic environment should be known. The electromagnetic environment for disturbances originating in or conducted through the power system, is equivalent to the voltage quality as defined before. The IEC electromagnetic compatibility standards define the voltage quality in three ways:

1. **Compatibility levels** are reference values for coordinating emission and immunity requirements of equipment. For a given disturbance, the compatibility level is in between the emission level (or the environment) and the immunity level. As both emission and immunity are stochastic quantities, electromagnetic compatibility can never be completely guaranteed. The compatibility level is chosen such that compatibility is achieved for most equipment most of the time: typically 95% of equipment for 95% of the time. It is not always possible to influence both emission and immunity: three cases can be distinguished:

- *Both emission and immunity can be affected.* The compatibility level can in principle be freely chosen. But a high level will lead to high costs of equipment immunity and a low level to high costs for limiting the emission. The compatibility level should therefore be chosen such that the sum of both costs is minimal. An example of a disturbance where both emission and immunity can be affected is harmonic distortion. A very good example of this process is described in IEEE Std.519 [82].
- *The emission level cannot be affected.* The compatibility level should be chosen such that it exceeds the environment for most equipment most of the time. An example of a disturbance where the emission level cannot be affected are voltage sags: their frequency of occurrence depends on the fault frequency and on the power system, both of which cannot be affected by the equipment manufacturer. Note that the EMC standards only apply to equipment manufacturers. We will later come back to the choice of compatibility levels for these kind of disturbances.
- *The immunity level cannot be affected.* The compatibility level should be chosen such that it is less than the immunity level for most equipment most of the time. An example of a disturbance where the immunity level cannot be affected is voltage fluctuation leading to light flicker.

2. **Voltage characteristics** are quasi-guaranteed limits for some parameters, covering any location. Again the voltage characteristics are based on a 95% value, but now only in time. They hold at any location, and are thus an important parameter for the customer. Voltage characteristics are a way of describing electricity as a product. Within Europe the EN 50160 standard defines some of the voltage characteristics. This standard will be discussed in detail in Section 1.4.3.
3. **Planning levels** are specified by the supply utility and can be considered as internal quality objectives of the utility.

These ideas were originally developed for disturbances generated by equipment, for which other equipment could be sensitive: mainly radio frequency interference. These ideas have been extended towards variations like harmonic distortion or voltage fluctuations. The concept has not yet been applied successfully towards events like voltage sags or interruptions.

EMC and Variations. Variations can be stochastically described through a probability distribution function, as shown in Fig. 1.20. The curve gives the probability that the disturbance level will not exceed the given value. The compatibility level can, according to the recommendations in the IEC standards, be chosen at the 95% percentile, as indicated in Fig. 1.20. The curve can hold for one site or for a large number of sites. When the curve represents a large number of sites it is important that it gives the disturbance level not exceeded for most of the sites (typically 95% of the sites). Consider as an example that the compatibility level of total harmonic distortion (THD) is 0.08. Suppose the THD is measured at 100 sites during 1000 10-minute intervals. A compatibility level of 0.08 implies that at 95 sites (out of 100) at least 950 THD samples (out of 1000) have a value of 0.08 or less.

In case a higher reliability is required for the successful operation of a device, a higher level than 95% should be chosen, e.g., 99.9%.

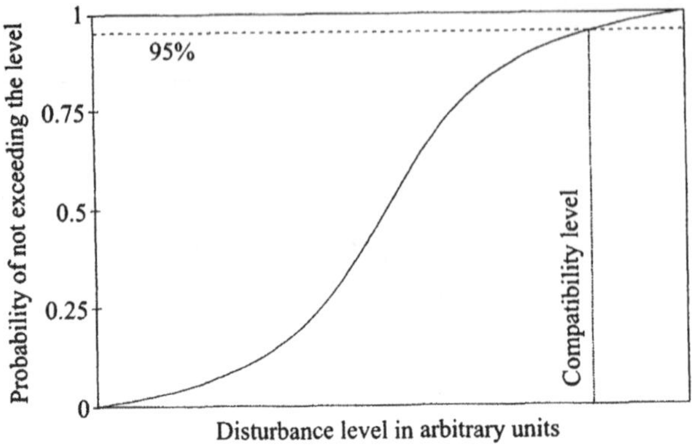

Figure 1.20 Probability distribution function for a variation, with the compatibility level indicated.

EMC and Events. The EMC framework has not been developed for events and its application to them has not been defined yet. For important power quality phenomena like voltage sags and interruptions, the EMC standards can thus not be used. This explains for a large part why the EMC standards are not (yet) well known in the power quality field. Still an attempt should be made at applying the concepts of electromagnetic compatibility to events.

Events only happen occasionally and are not present all of the time; applying a 95% criterion is therefore no longer possible. An immunity to 95% of voltage sags would depend on the way of counting the sags. Counting all sags below 200 V (in a 230 V supply) would give a much higher number than counting all sags below 150 V. The immunity requirement in the latter case would be much stricter than in the former.

In some power quality monitoring surveys a 95% criterion in space is applied. The electromagnetic environment is defined as the level of disturbance (number of events) not exceeded for 95% of the sites. But the knowledge of the environment in itself does not say anything about equipment immunity requirements. The immunity requirement should be based on the minimum time between events exceeding the immunity level. Figure 1.21 shows the time between events exceeding a certain disturbance level as a function of the disturbance level (the severity of the event). The more severe the event the more the time between events (the lower the event frequency). A piece of equipment or an industrial process to which the equipment belongs will have a certain reliability requirement, i.e., a certain minimum time between events leading to tripping of the equipment or interruption of the process. By using the curve in Fig. 1.21 this can be translated into an immunity requirement. As we will see later, the actual situation is more complicated: the severity of an event is a multidimensional quantity as at least magnitude and duration play a role.

A possible compatibility level would be the level not exceeded more than ten times a year by 95% of the customers. This can be done for any dimension of the event, leading to a multidimensional compatibility level. This concept has been applied to the results of the Norwegian power quality survey [67]. The frequency of transient overvoltage events, for the 95% site, is shown in Fig. 1.22. The 95% site is chosen such that 95% of the sites have less transient overvoltage events per year than this site. From Fig. 1.22 we can see that reasonable compatibility levels are:

- 2.5 pu for the magnitude of the transients.
- 0.3 Vs for the Vt-integral.

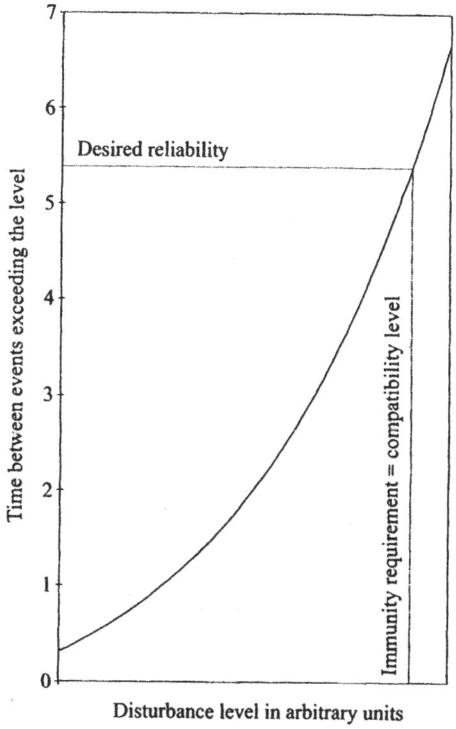

Figure 1.21 Time between events as a function of the disturbance level.

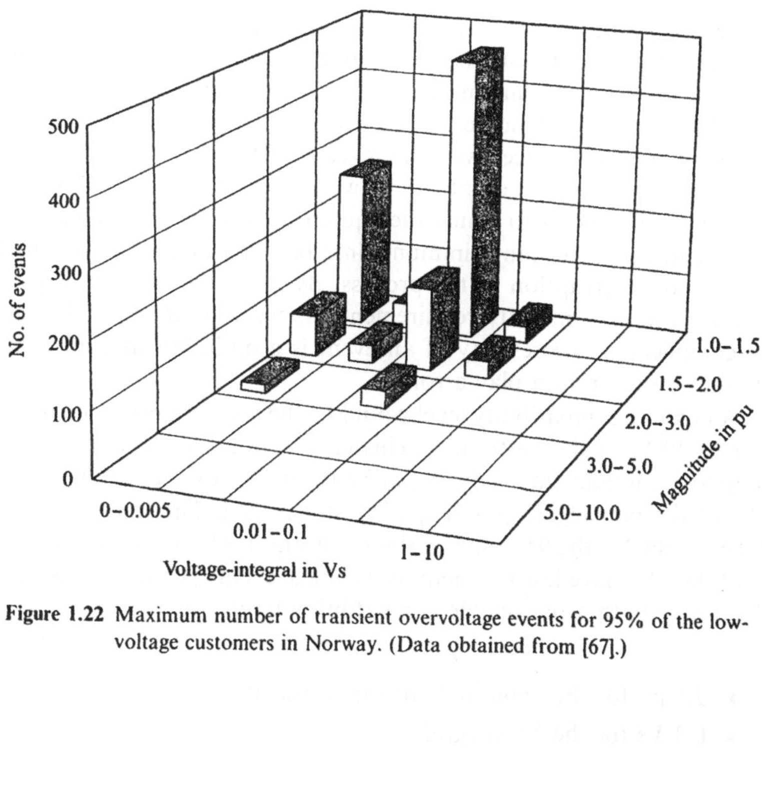

Figure 1.22 Maximum number of transient overvoltage events for 95% of the low-voltage customers in Norway. (Data obtained from [67].)

As a next step, these levels could be used as a basis for equipment immunity requirements. This concept could be worked out further by giving compatibility levels for 10 events and 1 event per year. Compatibility levels for 1 event per year cannot be obtained from Fig. 1.22 because of the short monitoring period (about one year).

1.4.3 The European Voltage Characteristics Standard

European standard 50160 [80] describes electricity as a product, including its shortcomings. It gives the main characteristics of the voltage at the customer's supply terminals in public low-voltage and medium-voltage networks under normal operating conditions.

Some disturbances are just mentioned, for others a wide range of typical values are given, and for some disturbances actual voltage characteristics are given.

Voltage Variations. Standard EN 50160 gives limits for some variations. For each of these variations the value is given which shall not be exceeded for 95% of the time. The measurement should be performed with a certain averaging window. The length of this window is 10 minutes for most variations; thus very short time scales are not considered in the standard. The following limits for the low-voltage supply are given in the document:

- **Voltage magnitude**: 95% of the 10-minute averages during one week shall be within ±10% of the nominal voltage of 230 V.
- **Harmonic distortion**: For harmonic voltage components up to order 25, values are given which shall not be exceeded during 95% of the 10-minute averages obtained in one week. The total harmonic distortion shall not exceed 8% during 95% of the week. The limits have been reproduced in Table 1.1. These levels appear to originate from a study after harmonic distortion performed by a CIGRE working group [83], although the standard document does not refer to that study. In reference [83] two values are given for the harmonic voltage distortion:

  — **low value**: the value likely to be found in the vicinity of large disturbing loads and associated with a low probability of causing disturbing effects;

  — **high value**: value rarely found in the network and with a higher probability of causing disturbing effects.

TABLE 1.1 Harmonic Voltage Limits According to EN 50160

Order	Relative Voltage	Order	Relative Voltage
3	5%	15	0.5%
5	6%	17	2%
7	5%	19	1.5%
9	1.5%	21	0.5%
11	3.5%	23	1.5%
13	3%	25	1.5%

TABLE 1.2 Harmonic Voltage Levels in Europe [83]

Order	Low	High	Order	Low	High
3	1.5%	2.5%	15	$\leq$0.3%	
5	4%	6%	17	1%	2%
7	4%	5%	19	0.8%	1.5%
9	0.8%	1.5%	21	$\leq$0.3%	
11	2.5%	3.5%	23	0.8%	1.5%
13	2%	3%	25	0.8%	1.5%

The values found by the CIGRE working group have been summarized in Table 1.2. The values used in EN 50160 are obviously the values rarely exceeded anywhere in Europe. This is exactly what is implemented by the term "voltage characteristics."

- **Voltage fluctuation**: 95% of the 2-hour long-term flicker severity values obtained during one week shall not exceed 1. The flicker severity is an objective measure of the severity of light flicker due to voltage fluctuations [81].
- **Voltage unbalance**: the ratio of negative- and positive-sequence voltage shall be obtained as 10 minute averages, 95% of those shall not exceed 2% during one week.
- **Frequency**: 95% of the 10 second averages shall not be outside the range 49.5 .. 50.5 Hz.
- **Signaling voltages**: 99% of the 3- second averages during one day shall not exceed 9% for frequencies up to 500 Hz, 5% for frequencies between 1 and 10 kHz, and a threshold decaying to 1% for higher frequencies.

Events. Standard EN 50160 does not give any voltage characteristics for events. Most event-type phenomena are only mentioned, but for some an indicative value of the event frequency is given. For completeness a list of events mentioned in EN 50160 is reproduced below:

- **Voltage magnitude steps**: these normally do not exceed ±5% of the nominal voltage, but changes up to ±10% can occur a number of times per day.
- **Voltage sags**: frequency of occurrence is between a few tens and one thousand events per year. Duration is mostly less than 1 second, and voltage drops rarely below 40%. At some places sags due to load switching occur very frequently.
- **Short interruptions** occur between a few tens and several hundreds times per year. The duration is in about 70% of the cases less than 1 second.
- **Long interruptions of the supply voltage**: their frequency may be less than 10 or up to 50 per year.
- **Voltage swells** (short overvoltages in Fig. 1.16) occur under certain circumstances. Overvoltages due to short-circuit faults elsewhere in the system will generally not exceed 1.5 kV rms in a 230 V system.
- **Transient overvoltage** will generally not exceed 6 kV peak in a 230 V system.

The 95% Limits. One of the recurring criticisms on the EN 50160 standard is that it only gives limits for 95% of the time. Nothing is said about the remaining 5% of the time. Looking at the voltage magnitude as an example: 95% of the time the voltage is between 207 V and 253 V (10% variation around the nominal voltage of 230 V), but during the remaining 5% of the time the voltage could be zero, or 10 000 V, and the voltage would still conform with the voltage characteristics.

The voltage magnitude (rms value) is obtained every 10 minutes—that gives a total of $7 \times 24 \times 6 = 1008$ samples per week; all but 50 of those samples should be in the given range. If we only consider normal operation (as is stated in the document) it would be very unlikely that these are far away from the ±10% band. Understanding this requires some knowledge of stochastic theory. In normal operation, the voltage at the customer is determined by a series of voltage drops in the system. All of those are of a stochastic character. According to stochastic theory, a variable which is the sum of a sufficient number of stochastic variables, can be described by a normal distribution. The normal distribution is one of the basic distributions in stochastic theory: its probability density function is

$$f(v) = \frac{1}{\sqrt{2\pi}\sigma} e^{-\frac{(v-\mu)^2}{2\sigma^2}} \tag{1.3}$$

where v is the value of the stochastic variable, μ its expected value, and σ its standard deviation. The well-known bell-shape of this function is shown in Fig. 1.23 for $\mu = 230$ V and $\sigma = 11.7$ V.

There is no analytical expression for the probability distribution function, but it can be expressed in the so-called error function Φ:

$$F(v) = \int_0^v f(\phi) d\phi = \Phi\left[\frac{v-\mu}{\sigma}\right] \tag{1.4}$$

The voltage characteristics standard gives the expected value (230 V) and the 95% interval (207 .. 253 V). Assuming that the voltage is normally distributed we can calculate the standard deviation which results in the given 95% confidence interval. As 95% of the voltage samples are between 207 and 253 V, 97.5% is below 253 V, thus:

$$\Phi\left[\frac{253\text{ V} - 230\text{ V}}{\sigma}\right] = 0.975 \tag{1.5}$$

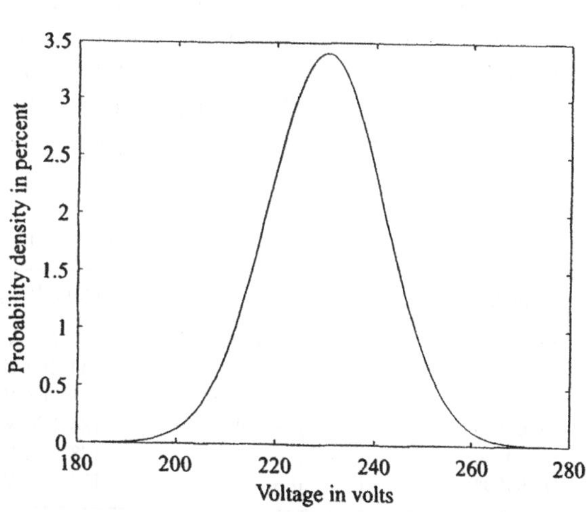

Figure 1.23 Probability density function of the normal distribution.

From a table of the error function, which can be found in almost any book on statistics or stochastic theory, we find that $\Phi(1.96) = 0.975$ which gives $\sigma = 11.7$ V. Knowing expected value and standard deviation of the normal distribution, the whole distribution is known. It is thus no longer difficult to calculate the probability that the voltage deviates more than 10% from its nominal value. The results of this calculation are given in Table 1.3. The first column gives the probability that the voltage is within the voltage range in the second, third, and fourth columns. The voltage range is given in standard deviations, in volts and as a percentage of the nominal voltage. The voltage is thus between 200 and 260 V for 99% of the time. The last column indicates how often the voltage is outside of the range, assuming all samples to be stochastically independent. In reality there is strong correlation between the samples which makes that large deviations become even more unlikely. Further, there are voltage regulation mechanisms (capacitor banks, transformer tap-changers) which become active when the voltage deviates too much from its nominal value. Finally, one should realize that the 95% value given in the standard does not hold for the average customer but for the worst-served customer. All this leads to the conclusion that voltage magnitude variations of much more than 10% are extremely unlikely.

From this reasoning one should absolutely not draw the conclusion that the voltage magnitude will never be lower than a value like 80%. The main assumption used is that the voltage variations are due to the sum of a number of small voltage drops. During, e.g., a voltage sag, this no longer holds. This brings us back to the principal difference between "events" and "variations": for variations the normal distribution can be used; for events it is the time between events which is of main importance. The probabilities in Table 1.3 thus only hold for voltage magnitude variations; absolutely nothing is said yet about voltage magnitude events.

Scope and Limitations. Standard EN 50160 contains some well-defined limits and measurement protocols, but it falls short of putting responsibility with any party. This is of course understandable when one realizes that the document describes the "voltage characteristics" which is the electromagnetic environment as it is now, not as it should be, and not even as it will be in future. Of course the underlying thought is that the situation will not become worse and that it is up to the utilities to ensure this.

When interpreting this standard it is also very important to realize that it only applies under "normal operating conditions." The document specifies a list of situations to which the limits do not apply. This list includes "operation after a fault," but also "industrial actions" and such vague terms as "force majeure" and "power shortages due to external events." This list removes a lot of the potential value from the document. A description of the electromagnetic environment should include all events and

TABLE 1.3 Probability of Voltage Exceeding Certain Levels

Probability	Voltage Range			Frequency
95%	$\mu \pm 1.96\sigma$	207–253 V	±10%	50 per week
99%	$\mu \pm 2.58\sigma$	200–260 V	±13%	10 per week
99.9%	$\mu \pm 3.29\sigma$	193–268 V	±17%	1 per week
99.99%	$\mu \pm 3.90\sigma$	184–276 V	±20%	5 per year
99.999%	$\mu \pm 4.42\sigma$	178–282 V	±23%	1 per 2 years
99.9999%	$\mu \pm 4.89\sigma$	173–287 V	±25%	1 per 20 years

variations to which a customer is exposed, not just those which occur during "normal operating conditions." A voltage sag during a severe lightning storm (exceptional weather) is equally damaging as a sag during a sunny afternoon in May.

Looking at the document in a more positive light, one can say that it only gives limits for what we have called "variations"; voltage quality "events" are not covered by the document.

What Next? Despite all its shortcomings, EN 50160 is a very good document. It is probably the best that could be achieved under the circumstances. One should realize that it is the first time that the electromagnetic environment has been described in such detail in an official document. Although limits are only given for some of the phenomena, and although the standard only applies during normal operation, and although absolutely no guarantees are given, at least a first step is set. Based on this standard one can see a number of developments:

- Utilities all over Europe have started to characterize their voltage quality by using the measurements as defined in EN 50160; thus 10-minute averages are taken of the rms voltage, 10-minute averages of the harmonic voltages, etc. The values not exceeded during 95% of the time are then used to characterize the local voltage quality. A problem is that some utilities then compare the results with the EN 5160 limits and state that their voltage quality confirms with the European standards. Understanding the concept of voltage characteristics, it is

TABLE 1.4 Voltage Characteristics as Published by Göteborg Energi

Phenomenon	Basic Level
	Voltage Variations
Magnitude variations	Voltage shall be between 207 and 244 V
Harmonic voltages	Up to 4% for odd harmonic distortion
	Up to 1% for even harmonic distortion
	Up to 6% THD
	Up to 0.3% for interharmonic voltages
Voltage fluctuations	Not exceeding the flicker curve
Voltage unbalance	Up to 2%
Frequency	In between 49.5 and 50.5 Hz
	Voltage Events
Magnitude steps	Frequent events shall be less than 3% in magnitude
Voltage sags	No limits
Short interruptions	No limits
Long interruptions	
accidental	On average less than one in three years
	On average shorter than 20 minutes
	Individual interruptions shorter than 8 hours
planned	On average less than one in 18 years
	On average shorter than 90 minutes
	Individual interruptions shorter than 8 hours
Transients	The utility tries to minimize size and frequency of transients which influence customers

no surprise that the local voltage quality is better than the limits given in the standard. This result should thus absolutely not be used by a utility to show that their supply is good enough. The statement "our supply confirms with EN 50160" is nonsense, as the standard does not give requirements for the supply, but only existing characteristics of the worst supply in Europe.

- Some utilities have come up with their own voltage characteristics document, which is of course better than the one described in the standard. The local utility in Gothenburg, Sweden has distributed a flyer with the limits given in Table 1.4. The term "voltage characteristics" is actually not used in the flyer; instead the term "basic level" is used [108].
- Measurements are being performed all over Europe to obtain information about other power quality phenomena. For voltage sags, interruptions, and transient voltages no limits are given in the existing document. A voltage characteristic for voltage sags, and for other events, is hard to give as already mentioned before. An alternative is to give the maximum number of events below a certain severity, for 95% of the customers. Figure 1.22 gives this voltage characteristic for transient overvoltage, as obtained through the Norwegian Power Quality survey [67]. Such a choice of voltage characteristic would be in agreement with the use of this same 95% level for the definition of the compatibility level.

2

Long Interruptions and Reliability Evaluation

2.1 INTRODUCTION

2.1.1 Interruptions

A long interruption is a power quality event during which the voltage at a customer connection or at the equipment terminals drops to zero and does not come back automatically. Long interruptions are one of the oldest and most severe power quality concerns. The official IEC definition mentions three minutes as the minimum duration of a long interruption. An interruption with a duration of less than three minutes should be called a "short interruption." Within the IEEE standards the term "sustained interruption" is used for interruptions lasting longer than 3 seconds [IEEE Std.1159] or longer than 2 minutes [IEEE Std.1250]. In this chapter the term "long interruption" will be used as an interruption which is terminated through manual action, thus not automatic. An interruption terminated through automatic reclosure or switching, is called a "short interruption" and will be treated in detail in Chapter 3.

2.1.2 Reliability Evaluation of Power Systems

An area of research called "power system reliability" has developed, in which number and duration of long interruptions are stochastically predicted. This area has long been confined to universities and to industrial power systems, but the recent interest in power quality in all its aspects has caused increased activities in reliability both at universities and in utilities. An additional reason for the increased interest in reliability is the availability of cheap fast computers. In the past reliability evaluation studies of realistic power systems required large computers, gross simplifications, and long calculation times. Many ideas proposed in the past can only now be implemented. Some of the basics behind reliability evaluation of power systems will be discussed in Sections 2.4 and 2.5; some examples will be presented in Section 2.8.

2.1.3 Terminology

In this chapter three terms will appear regularly: failure, outage, and interruption. In daily life their meanings are interchangeable, but in the reliability evaluation of power systems, there are clear and important differences.

- **Failure.** The term failure is used in the general meaning of the term: a device or system which does not operate as intended. Thus we can talk about a failure of the protection to clear a fault, but also of the failure of a transformer, and even about the failure of the public supply.
- **Outage.** An outage is the removal of a primary component from the system, e.g., a transformer outage or the outage of a generator station. A failure does not necessarily have to lead to an outage, e.g., the failure of the forced cooling of a transformer. And the other way around, an outage is not always due to a failure. A distinction is therefore made between "forced outages" and "scheduled outages." The former are directly due to failures, the latter are due to operator intervention. Scheduled outages are typically to allow for preventive maintenance, but also the aforementioned failure of the forced cooling of a transformer could initiate the scheduling of a transformer outage.
- **Interruption.** The term interruption has already been used before. It is the situation in which a customer is no longer supplied with electricity due to one or more outages in the supply. In reliability evaluation the term interruption is used as the consequence of an outage (or a number of overlapping outages), which is in most cases the same as the definition used in the power quality field (a zero-voltage situation).

2.1.4 Causes of Long Interruptions

Long interruptions are always due to component outages. Component outages are due to three different causes:

1. A fault occurs in the power system which leads to an intervention by the power system protection. If the fault occurs in a part of the system which is not redundant or of which the redundant part is out of operation the intervention by the protection leads to an interruption for a number of customers or pieces of equipment. The fault is typically a short-circuit fault, but situations like overloading of transformers or underfrequency may also lead to long interruptions. Although the results can be very disturbing to the affected customers, this is a correct intervention of the protection. Would the protection not intervene, the fault would most likely lead to an interruption for a much larger group of customers, as well as to serious damage to the electrical equipment.

   As distribution systems are often operated radially (i.e., without redundancy) and transmission systems meshed (with redundancy), faults in transmission systems do not have much influence on the reliability of the supply, but faults in distribution systems do.
2. A protection relay intervenes incorrectly, thus causing a component outage, which might again lead to a long interruption. If the incorrect tripping (or maltrip) occurs in a part of the system without redundancy, it will always lead

to an interruption. If it occurs in a part of the system with redundancy the situation is different. For a completely random maltrip, the chance that the redundant component is out of operation is rather small. Random maltrips are thus not a serious reliability concern in redundant systems. However maltrips are often not fully random, but more likely when the system is faulted. In that case there will be two trips by the protection: a correct intervention and an incorrect one. The maltrip trips the redundant component just at the moment that redundance is needed. Fault-related maltrips are a serious concern in redundant systems.

3. Operator actions cause a component outage which can also lead to a long interruption. Some actions should be treated as a backup to the power system protection, either correct or incorrect. But an operator can also decide to switch off certain parts of the system for preventive maintenance. This is a very normal action and normally not of any concern to customers. There is in most cases at least some level of redundancy available so that the maintenance does not lead to an interruption for any of the customers. In some low-voltage networks there is no redundancy present at all, which implies that preventive maintenance and repair or changes in the system can only be performed when the supply to a part of the customers is interrupted. These interruptions are called "scheduled interruptions" or "planned interruptions." The customer can take some precautions that make the consequences of the interruption less than for a nonscheduled interruption. This of course assumes that the utility informs the customer well in advance, which is unfortunately not always the case.

2.2 OBSERVATION OF SYSTEM PERFORMANCE

Long interruptions have long been considered as something worth preventing: the number and duration of long interruptions was viewed as the measure of how good the supply was. Today we would call it a power quality indicator or, in IEC terms, a voltage characteristic.

Many utilities have records of number and duration of interruptions, but mostly for internal use. The amount of published material is relatively low. That not only makes it hard to get information about supply performance for education and research purposes, but even for customers it is often hard to find out what the reliability of the supply is. The former is just an inconvenience, the latter is a serious concern. A positive exception to this is the privatized electricity industry in the United Kingdom. The data presented in the remainder of this section has mainly been obtained from the reports published by the British Office of Electricity Regulation (OFFER) [109]. Some additional information has been obtained for The Netherlands [110], [111].

2.2.1 Basic Indices

As already mentioned in Section 1.3.2 the main stochastic characteristic of any voltage magnitude event is the time between events, or (which is in effect the same) the number of events per year. The latter is indeed one of the main characteristics collected for long interruptions. Figure 2.1 shows the average number of interruptions per customer for six consecutive one-year periods. When the U.K. electricity industry was privatized in December 1990 there was a serious concern that the reliability of the

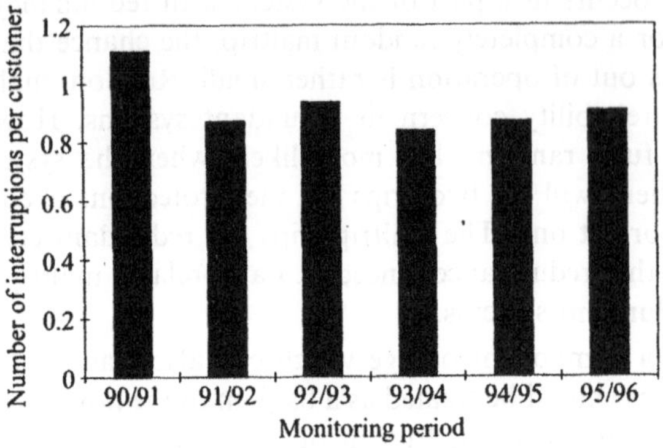

Figure 2.1 Number of interruptions per customer, average for Great Britain. (Data obtained from [109].)

supply would deteriorate. Figure 2.1 clearly shows that this has not been the case; the number of supply interruptions has stayed remarkably constant.

Individual interruptions are characterized through their duration, i.e., the time it takes until the supply is restored. Often the average duration of an interruption is not published but instead the total duration of all interruptions during one year is provided. This value is referred to as the "minutes lost per connected customer" or more correctly as the unavailability of the supply. The data for Great Britain (Wales, Scotland, and England) is shown in Fig. 2.2. We again see that the reliability of the supply remained constant, with the exception of the year 1990/91, during which severe blizzards made it impossible to restore the supply within a few hours. The number of interruptions due to this severe weather was relatively small, as can be concluded from Fig. 2.1, but its duration had a serious impact on the unavailability of the supply.

The collection of this data is less trivial than it may look. One should realize that most utilities do not automatically become aware that the supply to one or more customers is interrupted. It is typically the customers that report an interruption to the utility. The starting moment of an interruption, and thus the duration, is therefore not always easy to determine. The total number of long interruptions in the service territory of a utility can be obtained simply by counting them, as each interruption requires an operator action for the supply to be restored. The number of customers affected by an interruption requires a study of customer records which is often time consuming. Some utilities just assume a fixed amount of customers connected to each feeder, while other utilities link the interruption records with their customer database.

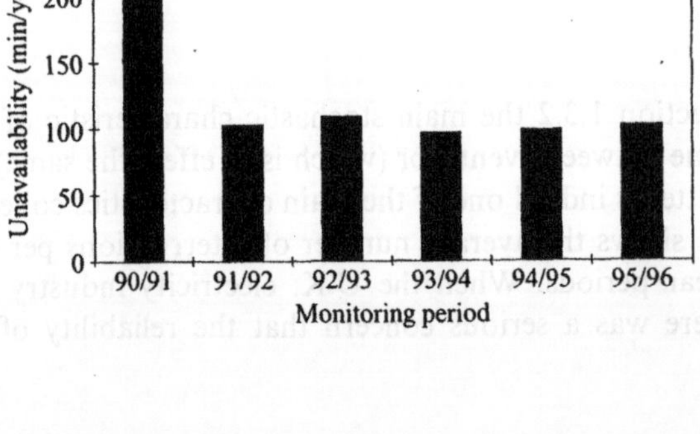

Figure 2.2 Unavailability of the supply, average for Great Britain. (Data obtained from [109].)

The calculation of the indices from the collected data could proceed as follows. Consider a utility serving N_{tot} customers. During the reporting period (typically one year) a total of K outages in the system lead to an interruption for one or more customers. Interruption i affects N_i customers and has a duration of D_i minutes. The average number of interruptions per customer per year $\overline{\lambda}$ is given by

$$\overline{\lambda} = \frac{\sum_{i=1}^{K} N_i}{N_{tot}} \tag{2.1}$$

The underlying assumption often used in the interpretation of this data is that the system average over 1 year, equals the customer average over many years. Thus $\overline{\lambda}$ would also be the expected number of interruptions per year for each customer. But variations in customer density, system design and operation, and weather patterns, make that not all customers are equal from a reliability point of view.

The average unavailability per customer $\overline{q}$, in minutes per year, may be calculated as

$$\overline{q} = \frac{\sum_{i=1}^{K} N_i D_i}{N_{tot}} \tag{2.2}$$

The average duration of an interruption $\overline{D}$ is

$$\overline{D} = \frac{\sum_{i=1}^{K} N_i D_i}{\sum_{i=1}^{K} N_i} \tag{2.3}$$

This value is redundant, as it may be calculated from (2.1) and (2.2) by using the following relation:

$$\overline{D} = \frac{\overline{q}}{\overline{\lambda}} \tag{2.4}$$

Utilities often publish two of these three values, $\overline{\lambda}$, $\overline{q}$, $\overline{D}$.

Note that (2.3) gives the average duration of an interruption from a customer perspective. From a utility perspective another value is also of interest: the average duration per interruption, D_{int}, calculated as

$$D_{int} = \frac{\sum_{i=1}^{K} D_i}{K} \tag{2.5}$$

This value gives information about how fast a utility is able to restore an interruption. The outcome of (2.4) and (2.5) is certainly not the same. Interruptions serving more customers, originating at higher voltage levels, tend to have a shorter duration. Thus the average duration per customer is likely to be shorter than the average duration per interruption. Which value should be used is open for discussion.

2.2.2 Distribution of the Duration of an Interruption

We will later see that the costs of an interruption increase nonlinearly with the duration of the interruption. The average duration of an interruption will thus not give the average cost. To calculate the latter, information about the distribution of the duration should be available. The U.K. utilities publish information about the percentage of interruptions restored within 3 hours and the percentage of interruptions restored within 24 hours. This is part of the so-called "overall standards of service" which we will discuss in Section 2.3. The assumption made in almost all reliability evaluation studies is that the component outage duration as well as the supply interruption duration are exponentially distributed. The exponential distribution, also called "negative-exponential distribution," is the basic distribution of most reliability evaluation techniques, as we will see in Section 2.5. The probability distribution function of the exponential distribution can be expressed as

$$F(t) = 1 - e^{-\frac{t}{T}} \tag{2.6}$$

where T is the expected value of the stochastic variable, which will be estimated by the average duration. Knowing the average duration, e.g., from Table 2.2 and Table 2.3, the percentage of interruptions restored within a time t_1 may be determined as

$$F(t_1) = 1 - e^{-\frac{t_1}{T}} \tag{2.7}$$

Table 2.1 gives the percentage of interruptions restored within 3 hours for a number of British distribution companies. The values in the columns labeled "practice" have been obtained from [109], the values in the columns labeled "theory" have been obtained from (2.7) by using the average duration of supply interruptions for the same year. Using the average duration and assuming an exponential distribution will overestimate the impact of interruptions: the number of interruptions longer than 3 hours is significantly less than would be expected from the measured average. This is clearly a case for more detailed reporting of the distribution of the duration of both component outages and supply interruptions. It also calls for including nonexponential distributions in the reliability evaluation.

Figure 2.3 shows the probability density function of the duration of all interruptions obtained for The Netherlands between 1991 and 1994 [112]. We see that the majority of interruptions has a duration between 30 minutes and 2 hours, with a

TABLE 2.1 Distribution of Interruption Duration, 1996/97 Values for Various British Utilities: Theory and Practice

Company	Average Duration in Hours	Supply Not Restored Within 3 Hours	
		Theory	Practice
A	2.38	28.4%	19.3%
B	1.38	11.4%	9.8%
C	1.42	12.1%	7.3%
D	1.45	12.6%	7.0%
E	1.63	15.9%	11.5%
F	1.62	15.7%	8.6%
G	2.27	26.7%	13.4%
H	1.38	11.4%	7.1%

Source: Data obtained from [109].

TABLE 2.2 Number of Interruptions per Customer per Year $\bar{\lambda}$ for Some British Utilities

Distribution Company	Reporting Year					
	90/91	91/92	92/93	93/94	94/95	95/96
A	0.41	0.47	0.38	0.37	0.40	0.33
B	0.58	0.62	0.57	0.56	0.70	0.61
C	1.70	1.11	1.29	1.25	1.21	1.39
D	0.76	0.68	0.96	0.59	0.65	0.85
E	2.85	2.29	1.95	2.14	2.20	2.23
F	1.46	1.29	1.18	1.19	1.24	1.16
G	0.82	0.74	0.86	0.89	0.70	0.62
H	1.69	0.82	0.75	0.92	0.96	0.97

Source: Data obtained from [109].

TABLE 2.3 Supply Unavailability $\bar{q}$ for Some British Utilities

Distribution Company	Reporting Year					
	90/91	91/92	92/93	93/94	94/95	95/96
A	51	67	53	52	58	54
B	88	75	77	69	70	67
C	398	118	122	144	128	151
D	76	65	91	63	94	85
E	325	212	212	200	212	233
F	185	176	184	167	133	111
G	185	108	129	121	102	88
H	1004	87	87	97	105	95

Source: Data obtained from [109].

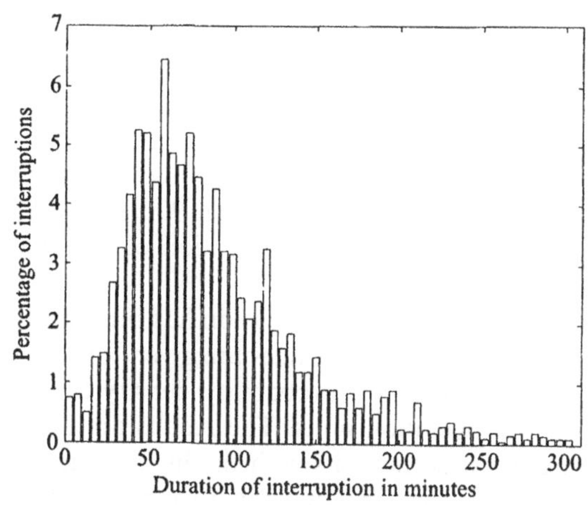

Figure 2.3 Distribution of duration of interruption, The Netherlands, 1991–1994. (Reproduced from Hendrik Boers and Frenken [112].)

long tail up to 5 hours. What is a more important conclusion is that the distribution is absolutely not exponential. (The density function of the exponential distribution has its maximum for zero duration and continues to decay after that.) To estimate the expected costs of interruption it is important to take this distribution into account. However, most studies still assume an exponential distribution.

2.2.3 Regional Variations

Both Fig. 2.1 and Fig. 2.2 give the average supply reliability for the whole of Great Britain. An old question is, how useful is this data for an individual customer. No information about individual customers is available, but separate data are available for each of the 12 distribution companies [109]. Some of this data is shown in Table 2.2 and Table 2.3. In Great Britain the distribution companies operate the voltage levels of 132 kV and lower. As will be shown in Table 2.4 their systems are responsible for 97% of the number of interruptions, as well as for 97% of the unavailability. The comparison between the different utilities can give information about how differences in system design and operation influence the supply performance. Apart from the adverse-weather year 90/91 the number of interruptions and the supply unavailability have remained remarkably constant. An accurate stochastic prediction method should thus be capable of reproducing these numbers, an interesting challenge. We will come back to the comparison between observation and prediction in Section 2.7.

TABLE 2.4 Contributions to the Supply Performance in Great Britain, 1995/96

	Number of Interruptions per Customer	Unavailability per Customer per Year		Average Duration of an Interruption
Total	1.03	158 min		150 min
Low voltage (240/415 V)	0.06	22 min	14%	370 min
6.6 and 11 kV	0.63	81 min	52%	130 min
33 kV	0.13	12 min	8%	90 min
132 kV	0.06	7 min	4%	120 min
Other	0.03	4 min	3%	130 min
Scheduled	0.12	32 min	20%	270 min

Source: Data obtained from [109].

From Table 2.2 and Table 2.3 we can also see that companies C, E, and H suffered most from the severe weather in 90/91. It is possible to calculate the average duration of an interruption for each of the distribution companies, by using (2.4). For company H we obtain for the year 90/91: $\overline{D} = \frac{1004}{1.69} = 594$ minutes, almost 10 hours. For the year 91/92 the average duration of an interruption was only 106 minutes for the same company.

An even further subdivision has been made in [109]: for each so-called "operation unit" within the utility values are given for number of interruptions and unavailability. Based on this data a probability density function has been obtained for the unavailability of operation units. The results are shown in Fig. 2.4 and Fig. 2.5. The latter figure includes the units with the highest unavailability. We see that 50% of the units have an unavailability between 50 and 100 minutes per year. The 95% percentile of the distribution is at 350 minutes. It is obvious from this graph that the average unavailability does not give any information about the unavailability which can be expected by a specific customer. One should note that this is not the distribution for the customers, as not all operation units have the same number of customers and not all customers within one operation unit have the same unavailability. Getting such a graph for all customers would require a much more intensive data collection effort than currently being done.

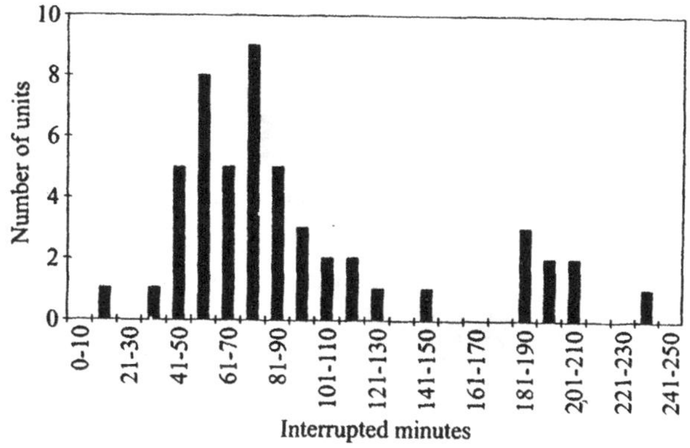

Figure 2.4 Probability density function for the average unavailability in Great Britain. (Data obtained from [109].)

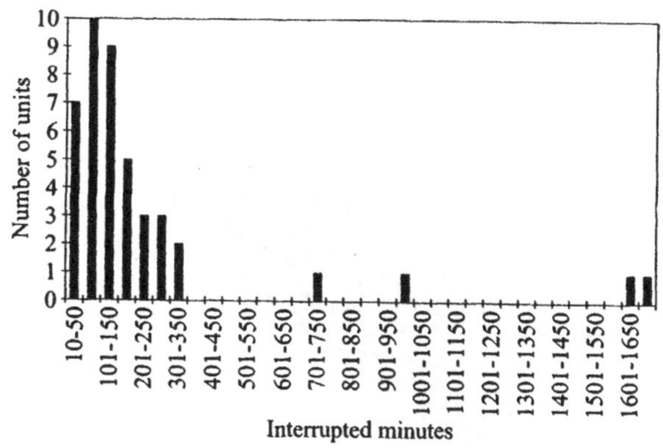

Figure 2.5 Extension of Fig. 2.4 toward higher values.

2.2.4 Origin of Interruptions

The data on number and duration of interruptions might be very interesting by itself, especially for customers, but they do not directly lead to any understanding of the causes of interruptions. For that purpose, additional data collection is required. A first step is to obtain data on the voltage level at which the outage occurred which led to the interruption. Table 2.4 gives this data for Great Britain over the year 1995/96. The values for other years are very similar. We see that the major contribution to the number of interruptions, as well as to the unavailability, comes from the medium voltage network (6.6 and 11 kV). An explanation for this is not too difficult to give. These networks have no redundancy so that a component outage immediately leads to a supply interruption. The 33 kV network is partly operated as a loop, hence its lower contribution. The low voltage network is also operated radially, thus without any redundancy, still its contribution is rather small. This is because a low voltage customer is exposed to much more (kilo)meters of medium voltage feeder than of low voltage feeder. Thus, there will thus be much more outages affecting the customer at medium voltage than at low voltage. An additional factor is that a larger part of the low voltage network is underground, which accounts for a lower failure rate. The data in Table 2.4 are shown graphically in Fig. 2.6 and Fig. 2.7. These figures again confirm that an increased reliability of the supply can only be achieved through investment at distribution level. An important conclusion from Table 2.4, Fig. 2.6, and Fig. 2.7 is that the longest interruptions are due to scheduled outages and outages at low voltage level. But

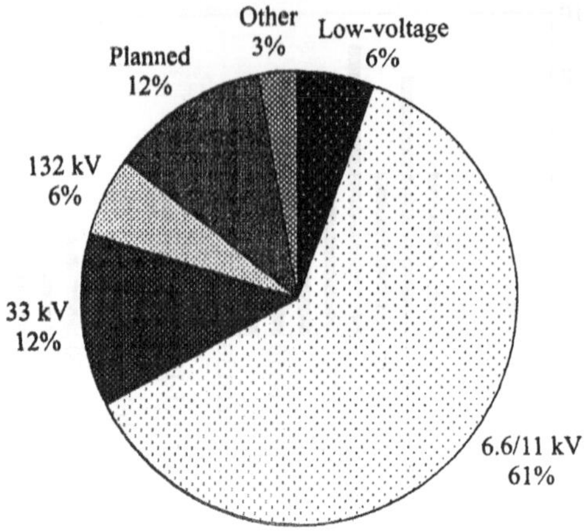

Figure 2.6 Contributions to the number of supply interruptions in Great Britain. (Data obtained from [109].)

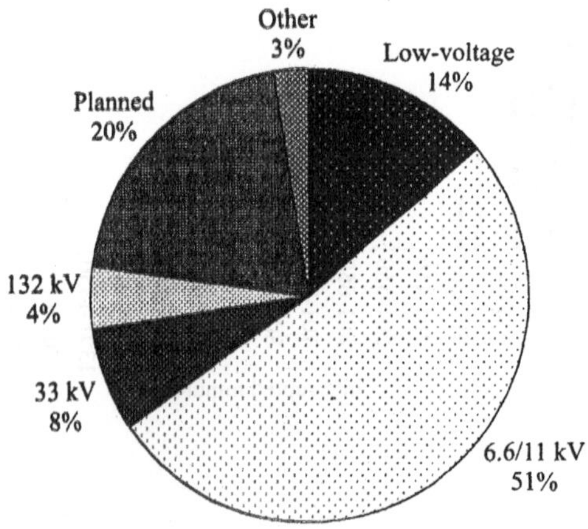

Figure 2.7 Contributions to the unavailability of the supply in Great Britain. (Data obtained from [109].)

as these occur less often than interruptions due to outages at medium voltage level, the latter make the largest contribution to the unavailability of the supply.

Surveys in other countries confirm that the majority of interruptions is due to outages at medium voltage level. Table 2.5 gives interruption data obtained in The Netherlands over the period 1991 through 1995 [110]. ("High voltage" is typically 150 kV and 380 kV, "medium voltage" 10 kV, and "low voltage" 400 V.) Here we see the somewhat remarkable phenomenon that about one third of the interruptions for urban customers are due to outages in high voltage networks. This is due to the large consumer density in the cities, and due to the fact that all low voltage and medium voltage distribution is underground. The number of outages in medium voltage networks is therefore simply very low. The high voltage networks are mainly overhead, which makes them comparable to the U.K. situation. We see 6 interruptions per 100 customers in The Netherlands and 9 per 100 customers in the U.K. ("132 kV" and "others"), indeed a similar number. Like in the U.K., the unavailability of the power supply in The Netherlands is mainly due to the medium voltage distribution network.

Figure 2.8 shows the contributions of the three voltage levels to the interruption frequency, between 1976 and 1995, for the average low voltage customer in The Netherlands. The contribution of the low voltage and medium voltage systems to the interruption frequency is rather constant. The contribution of the high voltage network

TABLE 2.5 Supply Performance in The Netherlands, 1991–1995

	Urban Customers						
	High Voltage		Medium Voltage		Low Voltage		Total
Number of interruptions	0.06/year	29%	0.12/year	58%	0.01/year	5%	0.21/year
Unavailability	2 minutes	15%	9.5 minutes	73%	1.5 minutes	12%	13 minutes
Interruption duration	26 minutes		75 minutes		198 minutes		62 minutes
	All Customers						
	High Voltage		Medium Voltage		Low Voltage		Total
Number of interruptions	0.06/year	22%	0.20/year	74%	0.01/year	4%	0.27/year
Unavailability	2 minutes	11%	15 minutes	79%	2 minutes	11%	19 minutes
Interruption duration	26 minutes		75 minutes		199 minutes		70 minutes

Source: Data obtained from [110].

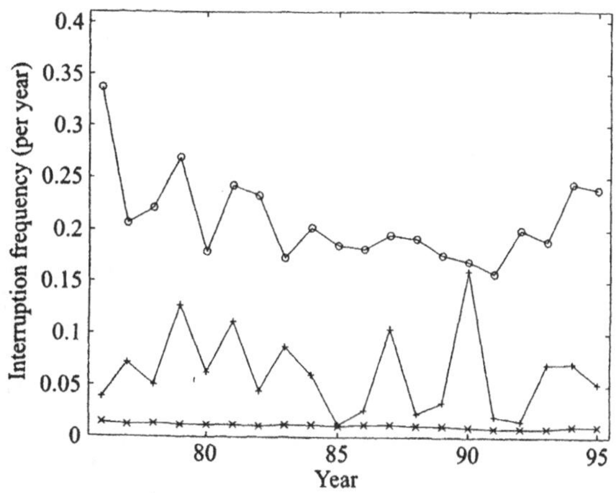

Figure 2.8 Number of interruptions per year for the average low voltage customer in The Netherlands, 1976–1995, with contributions from low voltage (x), medium voltage (o), and high voltage (+) systems. (Reproduced from van Kruining et al. [110].)

varies much more. In some years (1985, 1991) its contribution is negligible, while in other years (1990) they make up half of the number of interruptions. This large variation is partly of a stochastic nature (the number of outages of high voltage components leading to an interruption is very small) but also due to weather variations having more influence on the (mainly overhead) high voltage network than on the (mainly underground) medium voltage and low voltage networks.

Figure 2.9 shows the probability density function for the duration of interruptions originating at different voltage levels in The Netherlands [111]. For interruptions due to high voltage component outages, the majority of durations is short: about 75% is shorter than 30 minutes. Outages in the medium voltage and low voltage networks (typically 10 kV and 400 V, respectively, in The Netherlands) lead to longer interruptions. For medium voltage only about 15% of the interruptions is shorter than 30 minutes, for low voltage this value is even lower: about 5%. This has to do with the methods used for restoration of the supply. Outages in the high voltage networks are normally restored via operator intervention from a central control room. In medium voltage and low voltage networks there is no such control room and both fault localization and restoration of the supply has to take place locally. From the density functions in Fig. 2.9 it is clear that 30 minutes is about the minimum time needed

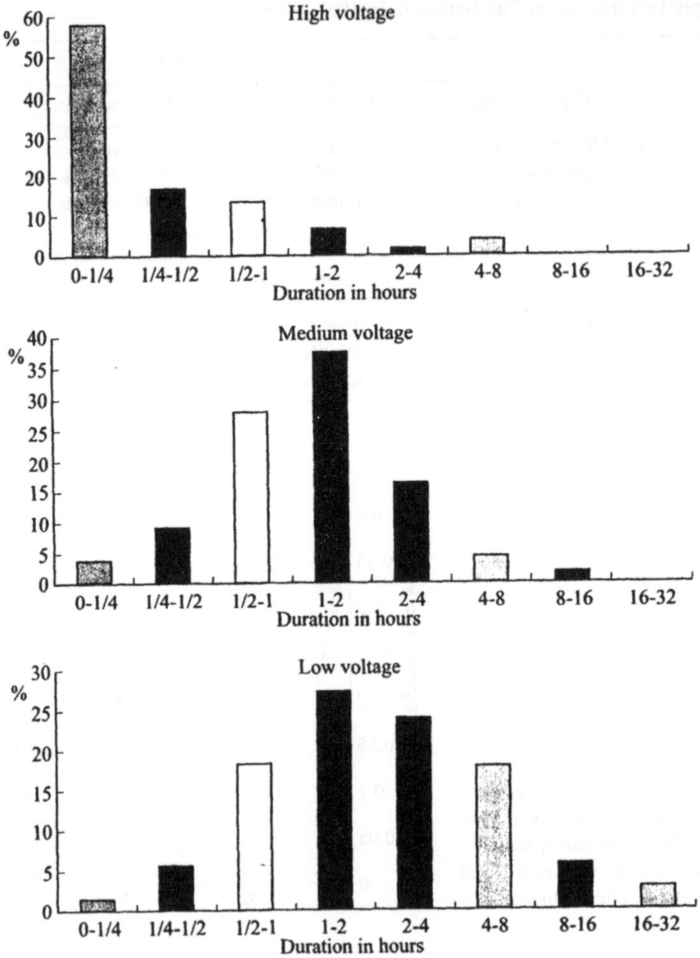

Figure 2.9 Probability density function for duration of interruptions, originating at three voltage levels in The Netherlands power systems. (Reproduced from Waumans [111].)

for this. Almost 100% of medium and low voltage networks in The Netherlands are underground. Restoration of the supply takes place normally via switching in radially operated loops.

2.2.5 More Information

From recording interruption events, much more information can be obtained than just average duration and frequency. We already saw origin of the interruption and the probability distribution of the duration as examples of additional information. The amount of information that can be obtained depends on how detailed the record of the interruption is. There are two applications for the recorded information, each with their own requirements:

1. **Increase the quality of supply.** This mainly requires information about the origin of interruptions and the way in which the supply is restored. For

example, the knowledge that most interruptions originate at medium voltage level teaches us that most gain can be obtained by improvements there. But suppose that for a certain customer interruption costs are small for interruption durations up to 2 hours, e.g., because essential equipment is supplied through a battery backup (an uninterruptable power supply or UPS). By using Fig. 2.9 it is shown that improvements in the low voltage network become more appropriate. To make such a decision it is obvious that more data is needed than just interruption frequency and unavailability.

2. **Serve as input data for reliability evaluation studies.** This requires a lot more data, not just about interruptions but also about outages not leading to interruptions. Most utilities and industries do keep information about outage frequencies and durations of components, but not much of it is openly available. Some large surveys have been performed to obtain outage frequencies: e.g., by the IEEE Industry Applications Society for industrial power systems [21], and by CIGRE for components of high voltage networks [197]. What is clearly still missing are data on failure of the power system protection, and probability distributions for time between outages and time to restore the component. Especially the latter could become very important in future reliability evaluation studies, as the interruption costs, and thus the interruption duration, becomes the desired output. A detailed literature survey performed by the author in the early 90s resulted in suggestions for expected component lifetimes [107]. The results of that study are reproduced in Table 2.6.

TABLE 2.6 Suggested Values for Number of Component Outages and Failures

Component Type	Number of Outages per 1000 Components per Year	Number of Outages per Component per Year	Failure Probability
MV/LV transformers	1–2		
MV/MV transformers	10–12		
HV/MV transformers	14–25		
MV and LV circuit breakers	0.2–1		
Disconnect switches	1–4		
Electromagnetic relays	1–4		
Electronic relays (single function)	5–10		
Electronic relay systems	30–100		
Fuses	0.2–1		
Voltage and current transformers	0.3–0.5		
Standby generators		20–75	
failure to start			0.5–2%
Continuous generators		0.3–1	
UPS inverter		0.5–2	
UPS rectifier	30–100		
Underground cable (1000 meters)	13–25		
Cable terminations	0.3–1		
Cable joints	0.5–2		
Busses (one section)	0.5–2		
Large motors	30–70		

Source: [107].

2.3 STANDARDS AND REGULATIONS

2.3.1 Limits for the Interruption Frequency

Long interruptions are by far the most severe power quality event; thus any document defining or guaranteeing the quality of supply should contain limits on frequency and duration of interruptions. The international standards on power quality do not yet give any limitations for interruption frequency or duration. The European voltage quality standard EN 50160 (see Section 1.4.3) comes closest by stating that "*under normal operating conditions the annual frequency of voltage interruptions longer than three minutes may be less than 10 or up to 50 depending on the area.*" The document also states that "*it is not possible to indicate typical values for the annual frequency and durations of long interruptions.*"

Many customers want more accurate limits for the interruption frequency. Therefore, some utilities offer their customers special guarantees, sometimes called "power quality contracts." The utility guarantees the customer that there will be no more than a certain number of interruptions per year. If this maximum number of interruptions is exceeded in a given year, the utility will pay a certain amount of money per interruption to the customer. This can be a fixed amount per interruption, defined in the contract, or the actual costs and losses of the customer due to the interruption. Some utilities offer various levels of quality, with different costs. The number of options is almost unlimited: customer willingness to pay extra for higher reliability and utility creativity are the main influencing factors at the moment. Technical considerations do not appear to play any role in setting levels for the maximum number of interruptions or the costs of the various options. For a customer to make a decision about the best option, data should be available, not only about the average interruption frequency but also on the probability distribution of the number of interruptions per year.

Contractual agreements about the voltage quality are mainly aimed at industrial customers. But also for domestic customers, utilities offer compensation. Utilities in the U.K. have to offer a fixed amount to each customer interrupted for longer than 24 hours. In The Netherlands a court has ruled that utilities have to compensate the customers for all interruption costs, unless the utility can prove that they are not to blame for the interruption. Also in Sweden some utilities offer customers compensation for an interruption.

2.3.2 Limits for the Interruption Duration

The inconvenience of an interruption increases very fast when its duration exceeds a few hours. This holds especially for domestic customers. Therefore it makes sense to not reduce the number of interruptions (which might be very expensive) but their duration. Limiting the duration of interruptions is a basic philosophy in power system design and operation in almost any country. In the U.K., as an example, the duration of interruptions is limited in three ways:

1. The Office of Electricity Regulation (OFFER) sets targets for the percentage of interruptions lasting longer than 3 hours and for the percentage of interruptions lasting longer than 24 hours. These are so-called "overall standards of service" [109].

2. The distribution company pays all customers whose supply is interrupted for longer than 24 hours. This is a so-called "guaranteed standard of service" [109].
3. The design of the systems is such that a supply interruption is likely to be restored within a certain time.

The OFFER regulations contain, for each distribution company, a target for the percentage of interruptions that is restored within 3 hours, and a target for the percentage restored within 24 hours. At the end of each year the distribution companies report back to OFFER, which publishes the targets together with the actual achievement. Table 2.7 shows targets and achievement over 1996/97 for some of the utilities. We see that most utilities meet their targets. All targets for 24 hours are at least 99%, and the 3-hour targets are no lower than 80%.

The maximum duration of interruption is also an important part of the design of systems. As we will see in Chapter 7 the concept of "redundancy" plays a very important role in that. To achieve a certain reliability of supply, the power system should contain a certain amount of redundancy. A common rule in the design of public systems is that the larger the number of customers that would be affected by the outage of a component, the more redundancy there should be present and the faster this redundancy should be available. Table 2.8 summarizes the way this is implemented in the U.K. [119]. These rules used to be part of a so-called engineering recommendation, and it has been in use in the U.K. for many years. When the utilities were privatized this recommendation became part of the license agreement. Depending on the load size, maximum durations of interruption are given. The larger the amount of

TABLE 2.7 Performance of U.K. Utilities over 1996/97

	3 hours		24 hours	
	Target	Achieved	Target	Achieved
A	80%	80.7%	100%	100%
B	85%	90.2%	99%	100%
C	95%	92.7%	100%	99.9%
D	93%	93.0%	100%	100%
E	80%	88.5%	99%	100%
F	80%	91.4%	99%	100%
G	85%	86.6%	99%	99.3%
H	85%	92.9%	99%	100%

Source: Data obtained from [109].

TABLE 2.8 Design Recommendations for the U.K. Supply System

	Amount of Load Restored			
Load Size	Immediately	Within 15 Min	Within 3 Hours	In Repair Time
0–1 MW	—	—	—	Total load
1–12 MW	—	—	Load – 1 MW	Total load
12–60 MW	—	Load – 12 MW or 2/3 load	Total load	
60–300 MW	Load – 20 MW	Total load		

Source: U.K. Engineering Recommendation P2/5 [119].

load affected, the faster the restoration of the supply. In terms of power system operation and design, this requires parallel supply for loads above 60 MW, automatic or remote manual transfer for loads above 12 MW, and local manual transfer for loads above 1 MW. The relation between reliability and power system design is discussed in detail in Chapter 7.

2.4 OVERVIEW OF RELIABILITY EVALUATION

A number of books and hundreds of papers have been written on power system reliability. The most well-known books are those by Billinton and Allan [84], [85], [86], but also the book by Endreyni [87] and the IEEE Gold Book [21] treat this subject in considerable detail. The latter publication does not give detailed theoretical considerations, but a useful set of basic calculations. It also gives a set of component outage rates, which is somewhat missing in the other books. Interesting books on power system reliability have also been written in the German language: [88], [89], and probably in other languages as well. An overview of publications on power system reliability in the international refereed literature, is published about once every five years in the IEEE Transactions on Power Systems [90], [91], [92]. Other sources of information are reports on power system reliability issued by national and international organizations [93], [94]. Also more and more books on power system analysis, design, or operation contain chapters on power system reliability. In the remainder of this section, and in Section 2.5, some general thoughts will be presented about reliability evaluation of power systems. For more details, the reader is referred to the literature.

The power system is often divided into three functional parts, each with its own specific design and operation problems and solutions:

- generation
- transport (transmission)
- distribution

In the reliability analysis a similar distinction is made between three so-called hierarchical levels of reliability:

- level I: generation
- level II: generation and transport
- level III: generation, transport, and distribution

Virtually all books and papers on reliability use this classification, either implicitly or explicitly, but nor everybody actually uses the term "hierarchical levels." This being a useful educational concept, it is used in this section to discuss the various techniques. The concept of hierarchical levels remains an approximation, as most classifications. The reliability of a generation station depends in part on the auxiliary supply, which must be treated as a distribution system, thus level III. Also, a substantial part of the generation has become embedded in the distribution system, in some countries well over 10% [120]. The amount of embedded generation is likely to grow further, with more industrial combined heat and power (CHP), a growth in the use of small-scale renewable energy and possibly so-called micro-CHPs appearing with domestic customers.

Another disadvantage of this concept of hierarchical levels is that it is developed for the large public supply system in industrialized countries. For developing countries, for small insular systems, and for industrial power systems, different thoughts might be needed. At the end of this section an equivalent of hierarchical levels for large industrial power systems will be proposed.

Despite the shortcomings of the classification in hierarchical levels, it still gives a good insight into the subject. New developments are most likely to appear at those places where the classification no longer holds, but to understand those the classification should be understood first.

2.4.1 Generation Reliability

As we saw from the observation results presented in Section 2.2, outages of generators have no influence whatsoever on the interruption frequency nor on the supply availability experienced by a customer. Thus, for a customer, level I reliability studies do not appear very important. This conclusion is correct for an existing, well-planned, and well-operated power system. But in the planning stage, level I studies are extremely important. In modern power systems, generation of power takes place at the highest voltage level; thus a lack of generation becomes immediately a national or even international problem. Such a situation should be avoided as much as possible. Because a suitable reserve in generation capacity has been planned and is available during operation, the customer does not have to worry about lack of generation anymore.

Annual Peak Load. The rule that the total generation capacity in a power system should exceed the annual peak load is likely to be the most important planning criterion in power systems. Planning and building of large power stations take between 5 and 10 years, thus decisions about these have to be made several years in advance. The most basic level I reliability study is to calculate the probability that the available generation capacity is less than the annual peak load in a certain year (e.g., 7 years ahead of the decision date). The input data for such a study consists of the expected annual peak load, the capacity of each generator unit, and its forced unavailability. The forced unavailability is the fraction of time during which a unit is not available due to forced outages, i.e., during which it is in repair. The assumption to be made is that the probability that the unit is not available during the annual peak is equal to the forced unavailability. This gives us sufficient information to calculate the probability that the available capacity is less than the annual peak load. This probability is called the "loss of load expectation" (LOLE) of the annual peak. Note that scheduled outages are not considered in peak load studies. It is assumed that preventive maintenance will not be scheduled during the period of the year in which the peak load can be expected.

Preventive Maintenance. Preventive maintenance of generators contributes significantly to their unavailability. The unavailability consists of two terms: the above-mentioned "forced unavailability" and the "scheduled unavailability." The latter is the fraction of time during which a unit is not available due to scheduled outages (i.e., maintenance). The scheduled unavailability of a unit may exceed its forced unavailability. The scheduled unavailability should not be treated as a probability, like the forced unavailability. Generator maintenance can be planned several months or even more than a year ahead. The maintenance planning will be such that the supply of the daily peak load will not be endangered. Typically, maintenance is scheduled away from the annual peak: if the annual peak occurs in winter, generator maintenance is done in summer and the other way around. In tropical areas, where the temperature and thus the load do not vary much during the year, this kind of scheduling of maintenance is not possible. The consequence is that a higher LOLE needs to be accepted part of the time, or that additional units are needed. The problem can be especially stringent in small systems (insular or isolated systems) where the unit size is a large fraction of the total load.

A way of including preventive maintenance in the level I evaluation is to split the year into periods of, e.g., 1 week. For each period a LOLE is calculated for the peak load over that period. The generation capacity for each period excludes the units that are in maintenance. Such a study is typically performed as an aid in maintenance scheduling.

The maintenance frequency (i.e., how often maintenance is performed) is normally assumed given in level I studies. When varying the maintenance frequency it is very important to realize that this will influence the component failure rate. An accurate model requires knowledge about the aging of the component and the influence which preventive maintenance has on this. This is an aspect of reliability evaluation which is seldom considered in power systems. We will come back to component aging in Section 2.5.6.

Load-Duration Curve. The loss-of-load expectation (LOLE) quantifies the risk that the generator capacity is not sufficient to supply the (annual) peak load. It does not quantify the unavailability of the supply due to insufficient generation capacity. To obtain the level I contribution to the unavailability, a more detailed study is required. Not only the unavailability of each generator unit needs to be known, but also its outage frequency and the repair time distribution. The load variation with time and scheduled maintenance have to be taken into account here as well. A simple method is to use the load-duration curve, approximate this through a number of steps, and calculate a LOLE for each load level. The application of such calculations is rather limited as they are too complicated to be of use in planning studies, but the influence on the customer is too small to be of any importance there. Exceptions are power systems in underdeveloped or very fast developing countries, where lack of generation can seriously contribute to the supply unavailability.

Derated States. The simplest LOLE calculations assume two states for a generator unit: available and outage (unavailable). In reality this is a gross oversimplification, especially for the larger units. It is very common that due to an auxiliary failure the unit will reach a so-called "derated state" in which it is only able to generate part of its maximum capacity. An example is the failure of one of the burners—this limits the combustion capacity and thus the power capacity. Considering such a failure as a complete outage of the unit underestimates the level I reliability. In the planning process this leads to an overestimation of the number of units that have to be built. As costs reduction became important several years ago, the interest in derated state models increased. An additional factor explaining the use of more detailed models is again the availability of faster computers enabling the implementation of these more detailed models.

Operating Reserve. Reliability studies are typically performed for planning purposes, where questions like "how many generating capacity should be available ten years from now" are addressed. In that case it is assumed that all generating plants and lines that are not in repair or in maintenance are available for generation and transport. For operational reserve studies the situation is different: one needs to take into account only those plants that are actually running or which can be brought online at short notice and assess the risk that the total load cannot be supplied within the next few hours.

2.4.2 Transmission Reliability

Level II reliability concerns the availability of power at so-called bulk supply points: typically transmission substations where power is transformed down to distribution voltage. The power not only has to be generated but also transported to the customers. The availability of sufficient lines or cables has to be taken into account. Level II reliability studies are much more difficult than level I studies, and are still under considerable development. Some of the difficulties and suggested solutions are discussed in the remaining parts of this section.

Overloading of Lines. Due to the outage of a transmission line the flow of active and reactive power through the transmission system changes. This can lead to overloading of other lines. The standard example is the overloading of a parallel line. Normally parallel lines will be rated such that the outage of one of them will not lead to overloading of the other. Thus two lines feeding a 200 MVA load should each be able to transport 200 MVA. This so-called $(n - 1)$ criterion has been an important part in the design of transmission systems: a system consisting of n components should be able to operate with any combination of $(n - 1)$ components, thus for any single-component outage. In important parts of the system, more strict criteria are used: $(n - 2)$, $(n - 3)$, etc.

Large transmission networks have become so complex that it is hard to realize the actual loction of the parallel paths. In systems that are meshed across several voltage levels, overloading due to an outage is a serious risk as some recent interruptions and "almost-interruptions" have taught us. The risk has been increased by the growing transport of power over large distances.

For level II studies in large systems, a load flow calculation has to be performed for each outage. These calculations make level II studies very time consuming. The processing of overload events depends on the policy used by the utility to rectify the overload. Typically two models for this are used in reliability studies.

1. The overload leads to an outage of the overloaded component, either immediately or after a certain delay which could depend on the amount of overload. As this second outage can lead to further overloads a cascade effect may occur.
2. The overload is assumed to be alleviated through the shedding of load.

Reliability of the Protection. The power system protection's aim is to remove faulted components from the system so as to limit the damage as much as possible. Failure of the protection to remove the faulted component can lead to significantly more damage, including an interruption for customers which would normally not be interrupted. It will be clear that the reliability of the protection is an important part of the reliability of the supply. Protection failure is already mentioned as one of the underlying causes of component outages. The power system protection can fail in several ways.

1. The protection fails to operate when required. In that case the backup protection will operate and clear the fault. This backup protection often clears more than only the faulted component making the impact on the system much bigger. As the transmission system often has only single redundancy, such a protection failure can potentially eliminate the redundancy and lead to an unnecessary interruption.
2. The protection operates when not supposed to. If this happens independently of another event it will simply lead to an outage of the protected component. The redundancy in transmission systems makes that these maltrips do not have a big influence on the reliability of the supply.
3. The power system protection shows a maltrip when another relay is supposed to operate. This leads to the loss of two components at the same time. The large currents flowing through the system during a short circuit make this an event which has to be considered in the calculations. Accurate models for it have not been developed yet. The main problem is that each fault can in theory lead to a maltrip of any of the other relays in the power system.
4. The power system protection shows a maltrip due to another event in the system, e.g., a switching action. Although the event itself does not lead to any required protection intervention, it can still potentially eliminate the redundancy. The reason is that several relays will experience a similar disturbance and thus all might show a maltrip at the same moment.

The reliability of power system protection is often split into two aspects, "dependability" and "security." The dependability is the degree of certainty that the protection will operate correctly (point 1 above); the security is the degree of certainty that the relay will not operate incorrectly. As shown above this neglects the different aspects within the "security."

Dynamic System Behavior. Most component outages are due to short-circuit faults. Occurrence and clearing of a fault lead to dynamic oscillations in the system. These can lead to overloading or tripping of components. This so-called security aspect of level II reliability is often not taken into account. To include it, detailed dynamic models of the system are needed. In the reliability literature a distinction is made between adequacy (static evaluation) and security (dynamic evaluation). The adequacy part is taken care of by most evaluation techniques, but security is often forgotten. In a well-designed transmission system a short circuit should not lead to loss of any generator, or overloading of any component. But one can think of several situations in which the dynamic system behavior can have a significant influence on the level II reliability.

- The system might be secure for each short circuit in an otherwise undisturbed system, but not for short circuits in a system in which already one or more components are out of operation. Both the states before and after the fault (i.e., after removal of the faulted component) might be healthy, but the transition between the two might not be healthy due to large dynamic oscillations. The system could appear to have double or triple redundancy where in reality it only has single redundancy.

- Failure of the protection can lead to fault clearing by the backup protection; this leads to a longer fault-clearing time and thus to more adverse dynamic effects. The system might be stable when the fault is cleared by the primary protection but not when the fault is cleared by the secondary protection.
- In small power systems with two centers of generation, a fault close to a generator might lead to some generators accelerating, while others slow down. The difference between their rotor increases very fast, leading to large instabilities. This phenomenon is especially severe for systems with a transmission grid at voltages of 10 to 30 kV with mainly underground cables [113]. A reliability evaluation study for such a system should not just consider cable outages but also the underlying short-circuit faults.
- In industrial power systems the maximum motor load connected to a bus is limited to a certain fraction of the short-circuit level of the bus. The actual motor load is often rather close to this limit. If in the course of time the amount of motor load grows, some faults can lead to loss-of-synchronism of synchronous motors or to stalling of induction motors.

Common-Mode Outages. The components in a level II study are often considered independent, i.e., the outage of one component does not depend on the state of the others. But sometimes two or more component outages occur at the same time. Classical examples are the collapse of a tower carrying two circuits and excavation leading to damage of two parallel cables. Several of the other aspects of level II reliability studies (failure of the protection, overloading of a parallel line) are sometimes also considered common-mode failures. For example, a maltrip of a relay during a fault on the parallel line will lead to an outage of both lines. By modeling this as a common-mode failure, no detailed protection model is needed.

Weather-Related Outages. The outage rate is in most studies considered constant, but in reality this is not the case. Many outages are weather related (lightning, storm, snow) and thus strongly time dependent. For nonredundant systems this does not matter at all, but for parallel systems it will significantly increase the interruption rate, even if the average component outage rate stays the same. Some numerical examples of this effect are given in Section 2.8.

The IEEE standard for collecting outage data [198] recommends to distinguish between three levels of outage rate:

- normal weather
- adverse weather
- major storm disaster

The contribution of adverse weather on the outage of transmission and distribution system components, for a U.K. utility, is shown in Table 2.9 [199]. Different utilities will have different contributing phenomena, especially when they are in different climates (snow storms are more likely in Scotland than in Texas), but the general impression is that adverse weather related outages are the biggest contribution to the outage rate.

TABLE 2.9 Various Contributions to the Outage Rate of Transmission and Distribution Components

Cause of Outage	Transmission System	Distribution System
Lightning strikes	9%	12%
Snow/ice on lines	52%	11%
High winds	32%	7%
Plant failures	5%	39%
Line interference	2%	21%
Animal/bird strikes	—	8%
Adjacent loads	—	2%

Source: Data obtained from [199].

2.4.3 Distribution Reliability

Most published work on power system reliability concerns the generation and transmission systems, what has been called level I and level II before. Level III (distribution) reliability studies are rather rare, although this is changing in the last few years. The lack of interest in distribution reliability is clearly not due to the high reliability of the distribution system. In fact, both interruption frequency and unavailability are mainly determined by events at distribution level, both medium voltage and low voltage. A number of reasons can be given for the lack of interest in distribution system reliability:

- The interest in distribution system research is in general (much) lower than that in transmission and generation.
- Reliability of power transmission and generation is of national interest, and thus requires more effort. An interruption originating at the transmission level will affect a large part of the system, and is thus more likely to lead to newspaper headlines.
- Investments in transmission systems are easier than in distribution systems because there are much more of the latter. This means that a reliability analysis of various distribution alternatives is not attractive.
- Reliability studies in distribution systems are relatively simple, which make them less attractive to the academic world.
- A reliability analysis would only be of interest to the customer if it would give an absolute value of the interruption frequency or availability. A widely held belief used to be that the results of reliability studies can only be used in a relative sense (i.e., to compare alternatives); such a study would therefore be of no use to the customer.

But, as already said, the interest in distribution system reliability is growing, probably due to the increasing attention for the customers' interests. Distribution system reliability has its own problems and solutions, some of which we will discuss below.

Radial Systems. Distribution systems are most often radially operated. The consequence of this is that each component outage will lead to a supply interruption. To obtain the interruption frequency one only needs to sum the outage rates of all components between the bulk supply point and the customer. Occasionally, parts of the system are operated in parallel or meshed. As this concerns small parts of the system, the mathematical difficulties for calculating the interruption frequency remain limited.

Duration of an Interruption. The main problem in distribution system reliability concerns the duration of the interruption. As we will see later, the costs of interruption increases nonlinearly with its duration. The probability distribution function of the interruption duration is of great influence on the expected costs. It is further important to realize that the restoration time depends on the position in the network. The average interruption duration, and thus the interruption costs, can therefore vary significantly throughout the network. The duration of an interruption consists of a number of terms, each of which has a stochastic character. A list of contributing terms is given, e.g., in [121] and [122]; the most relevant ones are

- receive alarm, contact or travel to affected substation;
- find fault location or faulted section;
- perform required switching actions;
- restore supply.

A well-known law in stochastic theory is that the sum of a sufficient number of stochastic terms has a normal distribution. Thus the distribution of the interruption duration because of its stochastic nature is likely to be normal and not exponential as assumed in most calculation methods. This could give unrealistic values for the interruption costs.

The Availability of the Alternative Supply. The list of terms given above, contributing to the duration of an interruption, assumes that the alternative supply is available. Thus, the moment the fault is located (or the faulted section is identified) the supply can be restored. But this is not always the case, as the alternative supply can also be interrupted, or the alternative supply is only able to take over part of the load. In that case the supply can only be completely restored after repair or replacement of the faulted component. When the supply can be restored by switching, the customer experiences a "long interruption." When the supply can only be restored through repair/replacement, the customer experiences a "very long interruption" as defined in Section 1.3.3. The frequency of very long interruptions will be rather small in most distribution systems (with the exception of remote rural networks), but the interruption costs may become very large, which makes it important that they become an essential part of the reliability evaluation results. Another reason for putting special emphasis on very long interruptions may be that the utility has to publish the number of interruptions not restored within a certain time, or has to pay damages for these "very long interruptions."

To get exact details of the distribution of the duration of interruptions, complicated stochastic models of the system are needed. But a two-step approach can be used if one is only interested in the frequency of very long interruptions. For very long interruptions, the time-scale of interest is longer than the time needed for the alternative supply to be made available. For the assessment of the number of very long interruptions the switches used to restore the supply can be considered in a closed position already. To evaluate the reliability of the resulting system, techniques developed for transmission systems may be used. The models required for this are much more complicated than for predicting the total interruption frequency.

Some of the before-mentioned aspects of transmission system reliability (common-mode failures, adverse weather, overloading) have to be incorporated in a level III study if the number of very long interruptions and/or the interruption duration distribution are of interest.

Adverse Weather. Adverse weather not only influences the number of very long interruptions (by increasing the probability that both a feeder and its backup are not available) but it also makes repair much more difficult. Blizzards and heavy thunderstorms cause a substantial fraction of outages. During the storm, repair is very difficult, if not impossible, and after the storm the large number of outages can make this process more difficult given that repair crews have to handle the outages one after the other. Such aspects of the reliability of the supply are extremely difficult to take into account in a stochastic model. As already mentioned before, one of the problems is the lack of data, but certainly not the only one. But despite the mathematical difficulties, more data collection must be encouraged. Also, the collected data should be made available for a wider public.

Embedded Generation. The presence of embedded generation somewhat complicates the reliability calculations. But the amount of embedded generation is seldom large enough to have a significant influence on the reliability of the supply. Industrial power systems are an exception because in such cases embedded generation can be used to obtain a very high level of reliability.

Embedded generation in public distribution systems consist mainly of wind turbines and CHP units. In all cases the design of the distribution system is such that the outage of one generator unit will not lead to an overload, and thus not to an interruption of the supply for any of the customers. Therefore the presence of the embedded generation does not influence the interruption frequency. An exception are those cases where outage of a generator leads to an interruption indirectly, e.g., when the heat production of a CHP unit is essential for an industrial process, or when a contract with the utilities requires load shedding upon a generator outage.

The presence of embedded generation can have some influence on the availability of the alternative supply, and thus on the frequency of very long interruptions. The interruption will normally lead to the loss of all embedded generation connected to the affected feeder. Thus the alternative supply also has to supply this additional load. Further, embedded generation connected to the alternative feeder can have tripped on the voltage sag due to the fault which led to the interruption. The speed with which this generation becomes available again will influence the probability that the alternative supply is able to take over all load from the affected feeder.

2.4.4 Industrial Power Systems

Large industrial and commercial users own and operate their own medium voltage distribution system. The largest users even own and operate a high voltage network. The point-of-connection to the public supply is somewhere at distribution or transmission level: the customer is responsible for the further distribution to the various points of utilization. In these so-called industrial power systems the general structure is often somewhat different than in public systems. Also there is no need for separate studies at separate hierarchical levels; all that matters is the continuity (or whatever word one likes to use) of the supply to the equipment essential for the production process. A possible list of questions that need to be addressed for a reliability study in an industrial power system is given below. We will only discuss interruption frequency below. Restoration of the supply will often take place faster compared to the time it takes to restart the production process. Of course this is not always the case, and for some industrial systems, the questions need to be modified. The list below should not be blindly followed, but be used as a basis for a specific study.

Each of the questions gives feedback on the design of the system. The starting point may be the existing system, or detailed design based on past experience. The whole "design process" is shown in Fig. 2.10. The term "layer" has been used here to distinguish from the "hierarchical levels" used for the reliability analysis of the public supply, but in fact both terms denote exactly the same.

1. **How often will the available generation not be enough to supply the load?**

   This layer corresponds to hierarchical level I in the public supply, for which a large number of tools are available. Some aspects of the calculations are already mentioned in Section 2.4.1. A few points of special interest to industrial systems need to be mentioned.

   - Maintenance on generator units can play a very important role in industrial systems. The load does not show much variation through the year, thus maintenance cannot be scheduled during a period of low load. This means that the generation capacity will influence the scheduling of

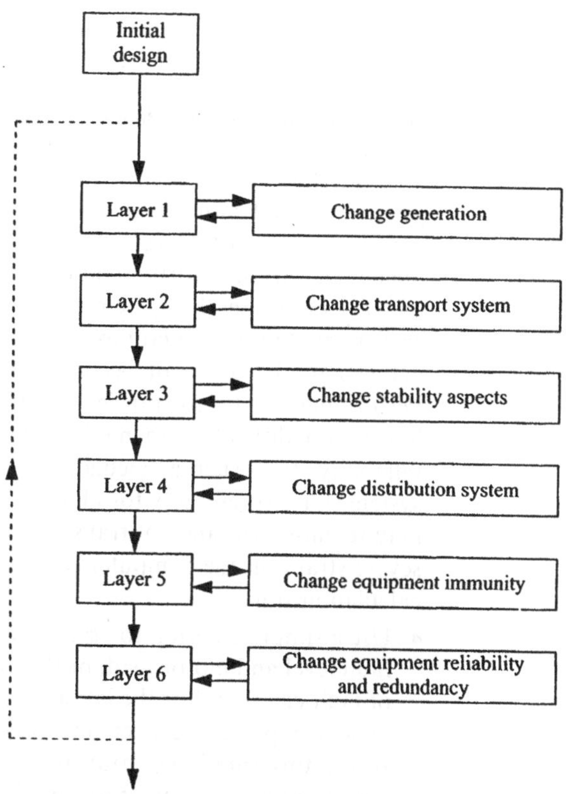

Figure 2.10 Reliability layers in industrial power systems and their role in system design.

maintenance. The lower the reserve (difference between load and capacity) the less likely that maintenance can be performed.

- The influence of maintenance on aging can only be assessed rather qualitatively as accurate models are still lacking. Therefore a constant failure rate will often be used. In that case one should realize that the calculation results cannot be used to optimize the maintenance frequency.
- Power generation units may be linked, e.g., through the use of a common steam circuit. This needs to be taken into account in the reliability studies as it might increase the probability that two or more units have an outage at the same time.
- During capacity shortages or when the capacity margin is low a load shedding policy is often in place. This needs to be incorporated in the reliability calculations.
- When the plant is connected to the public supply (which is mostly the case), its reliability needs to be considered. When the plant is fed via multiple infeeds, common-mode failures need to be considered.

2. **How often will a situation occur that the generation is available but that it cannot be transported to the load?**

   This layer corresponds to hierarchical level II in the public supply. The various considerations are very similar, but with some difference in emphasis.

   - Component loading is higher in industrial systems, and more constant. Therefore assessment of overloads due to outages becomes more important, but load variation often does not need to be considered.
   - Distances between substations are much smaller, which makes substation failures to play a larger role (relatively speaking) than in the public supply.

3. **How often will transient instability lead to a plant trip?**

   This is a rather new subject, corresponding somewhat to the security part of hierarchical level II. In industrial systems, with large motor load, on-site generation, and short distances between them, transient stability aspects can play a very important role. What is needed first is a prediction of the frequency of various short-circuit events, and next an assessment of the effects of each event on the system stability. The event frequencies follow from earlier reliability calculations. Assessing the effects of the event requires a detailed model and can become a severe strain on the computer power. Performing a detailed transient stability calculation for a large system is no longer too difficult with modern computer speed and memory, but for a reliability study such a calculation needs to be performed for many possible system states (preferably for all possible states). Even a medium-sized system may require thousands of transient stability calculations, which still places a severe strain on the computer power. Two options are available to limit the calculation time.

   - Use a simple criterion to assess the system stability, e.g., the ratio between fault level and motor load, or the differences in rotor angles at the moment of fault clearing. For the latter, simple models can be used, e.g., the change in active power between the pre-event and during-event steady states. Apply this simple criterion to all (or at least many) system states. The criteria might appear gross simplifications, which would never be acceptable for a conventional transient-stability calculation. But all we need to know here is the sum of the frequencies of all events leading to an unstable situation.

- Use a detailed system model, but limit the number of events to be studied. A first pruning is to look only at those events which involve a short circuit and for which both the initial steady state and the resulting steady state are stable. A second pruning is to stop looking for states with more components out, when a state has unstable events associated with it. As an example, if a fault leads to transient instability when two of six generators are out of operation, there is no need to study a fault when three generators are out.

One should note that it is not the actual instability limit which matters, but whether generators or motors will be tripped by their protection (undervoltage, overcurrent, reverse-power, under- or overfrequency). This can happen in systems which are in principle still stable. Thus a detailed model would also require sufficient details of the protection present in the system.

4. **How often will the distribution system fail to transport power to the plant?**

Layer 4 of industrial power system reliability corresponds to level III in the public supply. We can thus apply similar techniques, with the difference that the duration of an interruption is often not so important in industrial systems. As it is the assessment of the interruption duration which makes reliability analysis in distribution systems complicated, the calculations in industrial distribution systems will be simpler than in public systems.

The distribution system starts at the transport system studied in layer 2 and layer 3, and ends at the equipment terminals. The various distribution systems are normally considered independent of each other. An industrial distribution system can be extremely complex: many pieces of equipment with many levels of redundancy and importance. Some kind of pruning needs to be made before a study can be started with a reasonable chance of success. A first pruning is to only consider the supply to equipment which is essential for the operation of the plant.

A decision to be made beforehand is where the transmission system stops and the distribution system begins. The answer to this will again depend on the details of the study. For smaller systems it might be appropriate to not make any distinction between transmission and distribution, while for large systems each plant is considered as a separate distribution system.

5. **How often will the plant operation be interrupted due to insufficient voltage or current quality?**

In this layer all other power quality phenomena (i.e., apart from interruptions which were discussed in layers 1 through 4) have to be assessed. Examples of voltage quality events to be studied are:

- Transient overvoltages.
- Voltage sags and swells.
- Notching and bursts of harmonic distortion.
- High-frequency conducted disturbances.

To study all these in as much depth as for the long interruptions would lead to extremely long studies without much hope of useful results. The level of detail again depends on the system. An appropriate choice is to only look at first or second order events (first order events are short circuits in the normal system, second order events are short circuits when already one other component is out of operation).

These kind of studies are extremely rare, and where they are done do not contain much quantitative details. Still, even the decision to not study a certain type of event in detail because it is not likely to be of influence is already much better than simply forgetting about it.

To actually determine the number of equipment trips is not possible without a detailed knowledge of equipment immunity. In the design phase of the system, this information is simply not available. It will then be easier to determine the electromagnetic environment which the equipment will experience and to propose immunity requirements for the equipment to be used. Here it becomes important to distinguish between (voltage) variations and (voltage) events, as described in Section 1.3.

Current quality events will not directly lead to tripping of the plant, but utility requirements might force a plant shutdown, e.g., when the harmonic current distortion exceeds a certain level. If such a shutdown will have severe consequences, it needs to be considered in the reliability study.

6. **How often will the plant operation be interrupted due to the failure of essential equipment?**

   Equipment failure is normally not considered as part of supply reliability, but in an industrial system it is equally important. There is no need to build a very reliable power system if the plant will stop twice a week due to equipment problems. Industrial customers often use the term interruption in a more general meaning than the utility. The descriptive terms "voltage interruption" and "interruption of plant operation" indicate the difference in interpretation rather well.

   Detailed knowledge of the plant process is needed to perform a study like this. Like in several of the steps before, some serious pruning will be needed to make the study feasible. It might even be that only a qualitative assessment is feasible.

   Note that there is some overlap with layer 4 (distribution systems) and layer 5 (equipment trips due to voltage quality events).

   Additional aspects to be considered are:

   - redundancy of equipment, e.g., the function of a motor being taken over by another one;
   - linkage between plants on the production side, e.g., the steam production by one plant which is needed to operate another plant.

2.5 BASIC RELIABILITY EVALUATION TECHNIQUES

2.5.1 Basic Concepts of Reliability Evaluation Techniques

Stochastic Components. For a reliability evaluation study, the power supply system is split into stochastic components. The choice of components is rather arbitrary: the whole transmission system might be one component, but a single relay could be several components. Each component can be in at least two states: healthy and nonhealthy, the latter often referred to as the outage state. For a two-state component, two events can occur: the transition from the healthy to the nonhealthy state, an outage or failure event; and the reverse transition (i.e., from the nonhealthy to the healthy state), the repair or restore event.

The system state is a combination of all event states; if the state of one of the components changes, the system state changes. The system state for a system with N components can be thought of as a vector of rank N. The value of each element is the state of the corresponding component. An event is a transition between two system states, due to the change in state of one or more components.

EXAMPLE Consider, as an example, the system in Fig. 2.11: a generator with generator transformer, feeding into a large system via two parallel transmission lines and a transformer. We are interested in the reliability of the supply into the large system, thus, at point C in the figure.

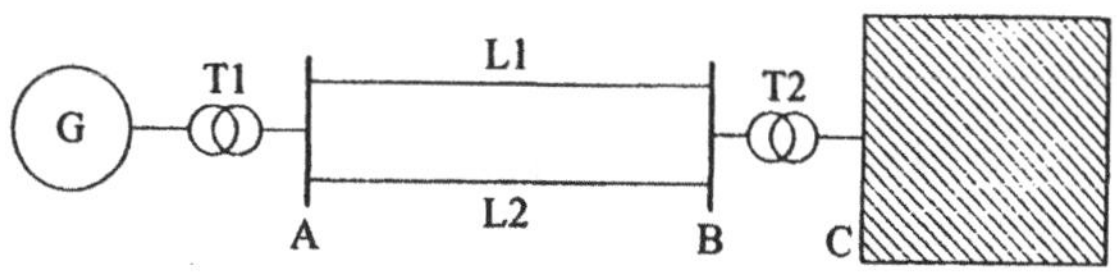

Figure 2.11 Power system example, for choice of stochastic components.

A possible subdivision into stochastic components is as follows:

1. generator plus generator transformer T1
2. substation A
3. line L1
4. line L2
5. substation B
6. transformer T2

In case a detailed study is needed of the generator plus the generator transformer, component 1 may be subdivided into stochastic components as follows:

1. the mechanical side of the generator, including the fuel availability
2. the electrical side of the generator, including its protection
3. the generator circuit breaker
4. the auxiliary supply
5. the generator transformer
6. the protection of the generator transformer

The Interruption Criterion. For each system state or for each event, an "interruption criterion" is used to determine if this state or event should be counted as an interruption or not. In most studies the interruption criterion is rather trivial, but for more detailed studies, especially for Monte Carlo simulation, the definition of the interruption criterion becomes an important part of the modeling effort. It is recommended to spend at least some time on defining the interruption criterion for a reliability evaluation study. Some simple examples of interruption criteria are given below. Note that these are just examples, and certainly not the only possibilities.

- In a level I study a state is an interruption state if the generator capacity is less than the load demand. Note that there is only one interruption criterion for the whole system. Each customer is equal at this level.
- In a level II study a state is an interruption state for a given transmission substation if the maximum power that can be transported to this substation is less than the demand. For level II studies, each substation has its own interruption criterion, thus its own reliability.
- In a level II security study an event is an interruption event if the transient phenomenon due to the event leads to tripping of generators and/or load.
- In a level III study a state is an interruption state for a given customer if the voltage at the customer terminals is zero.
- In a level III power quality study an event is an interruption event for a given device if it leads to the voltage at the device terminals to exceed certain magnitudes and durations.

The General Component Model. Two quantities are normally used to describe the behavior of a stochastic component: the failure rate and the (expected or average) repair time. The meaning of the term "expected repair time" is obvious: the expected value of the time the component resides in the nonhealthy state. The failure rate λ gives the average probability that the component will fail in the next small period of time:

$$\lambda = \lim_{\Delta t \to 0} \frac{Pr(\text{failure in period } \Delta t)}{\Delta t} \tag{2.8}$$

For components representing primary parts of the power system, which are the majority of the components in most studies, the term outage rate might be used. Here we shall use the general term failure rate.

The definition of failure rate in (2.8) is rather mathematical. It will become of use below. A more practical way of defining the failure rate is through the number of failures in a population. Consider a population of N similar components (e.g., distribution transformers). During a period n, this population shows K component failures. The failure rate may be determined as

$$\lambda = \frac{K}{nN} \tag{2.9}$$

The two definitions of failure rate are equivalent under a number of assumptions. The most important of which is that the component is repaired (within a short time) after every failure. The definition according to (2.9) is used to obtain failure rates from observed failures.

Some other quantities which are in use will be described below.

- The expected time to failure T is the reciprocal of the failure rate:

$$T = \frac{1}{\lambda} \tag{2.10}$$

- The repair rate μ is the reciprocal of the expected time to repair R:

$$\mu = \frac{1}{R} \tag{2.11}$$

Note that expected time to failure can be defined in a similar way as the expected repair time, and the repair rate similarly as the failure rate according to (2.8).

- The availability of the component is the probability to find the component in the healthy state:

$$P = \frac{T}{R+T} \tag{2.12}$$

- The unavailability is the probability that the component is in the nonhealthy state:

$$Q = \frac{R}{R+T} \tag{2.13}$$

- The expected time between failures (ETBF) is the sum of the expected time to failure (ETTF) and the expected repair time. As the repair time is normally much smaller than the time to failure, ETBF and ETTF are about equal and as a consequence often mixed up. From a mathematical point of view, this is a serious mistake, but in engineering these kind of errors are common and not considered very seriously.

EXAMPLE A distribution company operates 7500 distribution transformers. Over a period of 10 years, 140 of these transformers fail for various reasons. A small fraction of them can be repaired, but most failures require replacement with a spare transformer. Records have been kept of the repair or replacement time needed. Adding all these for the 140 failures gives a total of 7360 hours. From these observation data, the values of the above parameters are obtained:

$$\lambda = \frac{140}{10 \times 7500} = 0.0019\,\text{yr}^{-1} \tag{2.14}$$

$$T = \frac{1}{0.0019} = 530\,\text{yr} \tag{2.15}$$

$$R = \frac{7360}{140} = 52.6\,\text{h} = 0.006\,\text{yr} \tag{2.16}$$

$$\mu = \frac{1}{R} = 167\,\text{yr}^{-1} \tag{2.17}$$

$$P = \frac{530}{0.006 + 530} = 0.999989 \tag{2.18}$$

$$Q = \frac{0.006}{0.006 + 530} = 0.000011 = 6\,\text{min/yr} \tag{2.19}$$

This can be interpreted in normal words, as follows:

- Each transformer has a probability of 0.0019 to fail in the coming year. In the whole population, 14 transformers are expected to fail.
- After such a failure, the repair or replacement of the transformer is expected to last 52.6 hours.
- Each transformer will be out of operation, on average, 6 minutes per year.

Note that we have used past-performance data to predict future behavior. This is the basis for all reliability analysis: the assumption that the average performance in the past gives the expected behavior for the future.

The Detailed Component Model. Describing a stochastic component by means of two quantities (e.g., failure rate and repair time) is a gross simplification of the actual situation. Still this model is used in more than 95% of all reliability evaluation studies. To understand the reasons for this, we first need to introduce the general component model. The component is again assumed to be in one of two states. The theory can be extended to multi-state models, but describing those would not lead to better understanding. For each of the two states a probability distribution is defined for the time the component stays in that state. There is thus one probability distribution function for the repair time (the time in the nonhealthy state) and one for the lifetime (the time in the healthy state). Let T be the lifetime (expected time to failure) of the component. The probability distribution function of the lifetime $F(t)$ is the probability that the component fails before it reaches an age t:

$$F(t) = Pr(T \leq t) \tag{2.20}$$

The probability density function is the derivative of the probability distribution function:

$$f(t) = \frac{dF}{dt} = \lim_{\Delta t \to 0} \frac{Pr(t < T \leq t + \Delta t)}{\Delta t} \tag{2.21}$$

The probability density function $f(t)$ is a measure for the probability that the component will fail around an age t:

$$f(t)\Delta t \approx Pr(t < T \leq t + \Delta t) \tag{2.22}$$

The failure rate $\lambda(t)$ is defined as the probability that the component fails soon after the age t assuming that it has not failed before age t:

$$\lambda(t) = \lim_{\Delta t \to 0} \frac{Pr(T \leq t + \Delta t | T > t)}{\Delta t} \tag{2.23}$$

The failure rate can be calculated from the probability density function $f(t)$ and the probability distribution function $F(t)$:

$$\lambda(t) = \frac{f(t)}{1 - F(t)} \tag{2.24}$$

We will discuss the failure rate and its relation to component aging in more detail in Section 2.5.6.

Similar definitions can be given for the repair time, resulting e.g., in the repair rate $\mu(t)$, a probability density function $g(t)$ and a probability distribution function $G(t)$.

The Weibull Distribution. A distribution often used in stochastic theory is the so-called Weibull distribution. The probability distribution function for a Weibull distributed variable T is

$$F(t) = 1 - \exp\left\{-\left(\frac{t}{\theta}\right)^m\right\} \tag{2.25}$$

For $m = 1$ we obtain the exponential distribution discussed before. We refer to m as the shape factor and to θ as the characteristic time of the Weibull distribution. The probability density function $f(t)$ is

$$f(t) = m\frac{t^{m-1}}{\theta^m}\exp\left\{-\left(\frac{t}{\theta}\right)^m\right\} \tag{2.26}$$

The failure rate $\lambda(t)$ for a Weibull distribution is obtained from (2.24):

$$\lambda(t) = m\frac{t^{m-1}}{\theta^m} \tag{2.27}$$

We see that the failure rate increases for $m > 1$ and decreases for $m < 1$. From a relatively simple expression it is possible to generate a whole range of lifetime distributions.

The Exponential Distribution—Lifetime. As already stated before, over 95% of reliability evaluation studies use the simple model with a single failure rate and a single repair rate. The underlying assumption is that both repair time and lifetime are exponentially distributed. The exponential distribution (also called "negative exponential distribution") is defined through the following probability distribution function:

$$F(t) = 1 - e^{-\lambda t} \tag{2.28}$$

From the above equations it follows easily that λ in (2.28) is the failure rate according to (2.24). Thus, the negative exponential distribution has a constant failure (repair) rate and the general component model can be used. There are a number of reasons why this distribution is almost the only one used:

- Using nonexponential distributions makes that most reliability evaluation techniques currently available can no longer be used. For many years the choice was between using the exponential distribution or not doing any reliability evaluation at all.
- Even the small number of studies which are able to use nonexponential distributions (the so-called Monte Carlo simulations which we will discuss below) often still use exponential distributions, because of the lack of data. Collection schemes of component failure data normally only provide failure rates and average repair times.
- The lack of experience with nonexponential distributions makes that the results of such a study are rather hard to interpret.
- In an actual power system there is a mixture of components with different ages for three reasons: preventive maintenance is performed on components at different times; components are replaced after failure; and the system is not built at once but has grown over time. The mixture of ages makes that the system behavior, being a kind of average of the component behavior, can be described by assuming all components to have a constant failure rate.
- Most components in use are in their so-called "useful operating time": they have passed the wear-in time, and have not yet reached the time of serious wear-out. This is based on the assumption that the failure rate of a component versus time can be described through a "bathtub curve." During most of the

operating time of a component, it resides in the flat part of the bathtub curve where the failure rate is constant.

The Exponential Distribution—Repair Time. For repair time distributions, the above reasonings do not hold. We already saw in Table 2.1 that the duration of an interruption is nonexponentially distributed. If we assume the interruption duration to be Weibull distributed, the shape factor in (2.25) can be calculated from the available data:

$$m = \frac{\ln(-\ln(Fr_3))}{\ln\left(\frac{3}{\theta}\right)} \tag{2.29}$$

with Fr_3 the fraction of interruptions not restored within three hours and θ the characteristic repair time. If we take the average repair time as the characteristic repair time, we only make a small error as long as $m > 1$. Including the effect of the shape factor on the average repair time would make the calculation too complicated. The resulting shape factors for the interruption durations are given in Table 2.10. We find shape factors somewhat in excess of unity.

The IEEE Gold Book [21] gives, among others, repair times for large electrical motors in an industrial environment. As both the average and the median value are given, it is again possible to estimate the shape factor assuming a Weibull distribution. In most cases the median value is much larger than the average, which indicates a shape factor less than one. An alternative explanation is the combination of two Weibull distributions, both with shape factor greater than one, but with significantly different characteristic or average repair times.

More theoretical modeling and observation work is needed to validate the use of the exponential distribution in power system reliability evaluation. Based on the evidence presented, the following preliminary conclusions can be drawn:

- The exponential model appears an acceptable approximation for lifetime distributions, with the exception of studies in which the effect of preventive maintenance is evaluated.
- The exponential model is not correct for the repair time.

A short discussion on component aging will be given in Section 2.5.6.

TABLE 2.10 Shape Factor for Weibull Distribution of Interruption Duration

θ	Fr_3	Shape Factor
2.38	0.193	2.15
1.38	0.098	1.09
1.42	0.073	1.29
1.45	0.070	1.35
1.63	0.115	1.27
1.62	0.086	1.46
2.27	0.134	2.50
1.38	0.071	1.25

2.5.2 Network Approach

When using the so-called network approach, the system is modeled as a "stochastic network." The stochastic behavior of the system is represented graphically by means of a number of network blocks, connected in parallel or in series. Each block refers to a stochastic component in the system. The model is such that the system is healthy (i.e., the supply is available) as long as there is a path through the network. This graphical character of the method makes it very suitable to get an overview of the reliability of the system. An additional advantage of the network approach is the similarity with the electrical network. Electrically parallel components are often modeled as a parallel connection in the stochastic network. An electrical series connection in most cases results in a stochastic series connection.

When the reliability is quantified by using a stochastic network, a number of mathematical approximations are needed. The calculations assume that the repair time and the lifetime are exponentially distributed for all components.

Each block (network element) is characterized through an outage rate λ and an expected repair time r. For each element we further define the "availability" P and the "unavailability" Q.

$$P = 1 - \lambda r \tag{2.30}$$

$$Q = \lambda r \tag{2.31}$$

Sometimes a different form of these expressions is used: the outage rate is given in failures per year, and the repair time in hours, leading to the following (mathematically not fully correct, but very handy) expressions for availability and unavailability:

$$P = 1 - \frac{\lambda r}{8760} \tag{2.32}$$

$$Q = \frac{\lambda r}{8760} \tag{2.33}$$

EXAMPLE Consider the supply system in Fig. 2.12. A possible stochastic network for this system is shown in Fig. 2.13 where the numbers refer to the following types of failure:

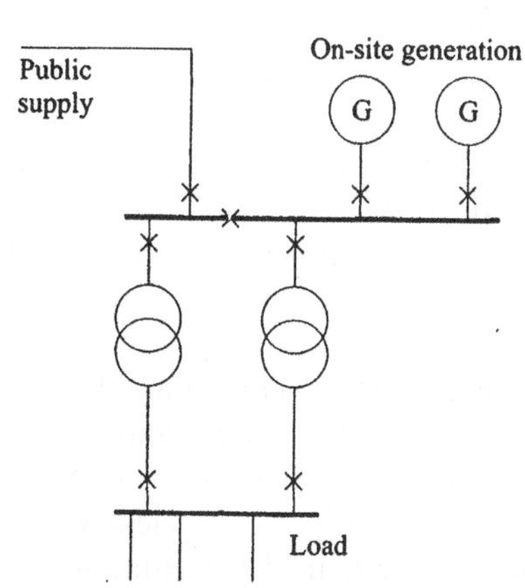

Figure 2.12 Single-line diagram of a supply system.

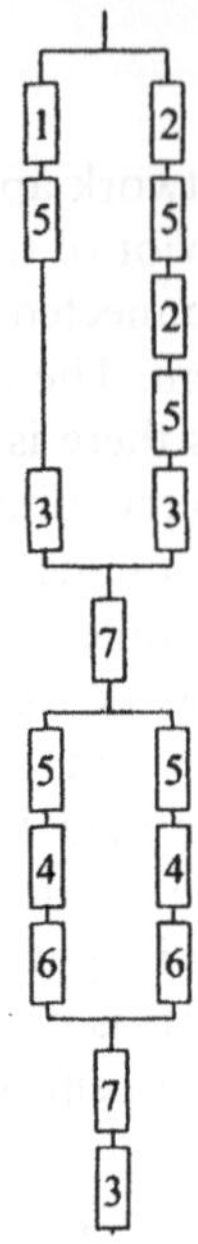

Figure 2.13 Stochastic network representation of the system shown in Fig. 2.12.

1. outage of the public supply
2. outage of a generator
3. bus outage
4. transformer outage
5. circuit breaker failure (maltrip or short circuit)
6. circuit breaker failure (maltrip)
7. circuit breaker failure (short circuit)

All components in the network in Fig. 2.13 are stochastically independent, so that simple mathematics can be applied. Note that the capacity of one generator (5 MW) is not enough to supply the load (7 MW). To supply the load the public supply needs to be present, or both on-site generators need to be in operation. In the network diagram this is shown as the "public supply" in parallel with both "on-site generators" in series. Also note the difference between a circuit breaker maltrip and a short circuit in the breaker. In the latter case the protection on both sides of the breaker will trip leading to the loss of two primary components at the same time.

Various methods are available to calculate interruption rate and expected interruption duration from component failure rate and repair time; all these methods replace the whole network by one equivalent component.

An obvious method for network reduction is to find series and parallel components. A parallel connection represents redundant components, where the supply is not interrupted until all of them are in the outage state. A series connection represents the situation where each component outage leads to an interruption of the supply. The correspondence with electrical series and parallel connections is clear but not one-to-one. Consider as an example two transformers in parallel. If one of them fails the other one can take over the supply. This is clearly a stochastic parallel connection. But if the total load is much more than the maximum loading of one transformer, the other one will also soon fail or be tripped by its overload protection. In that case a stochastic series connection is a better representation.

Stochastic Series Connections. Consider the series connection of two stochastic components with outage rates λ_1 and λ_2 and repair time r_1 and r_2, as shown in Fig. 2.14. We want to derive expressions for outage rate λ_s and repair time r_s of the series connection, so that the series connection can be replaced by one equivalent component.

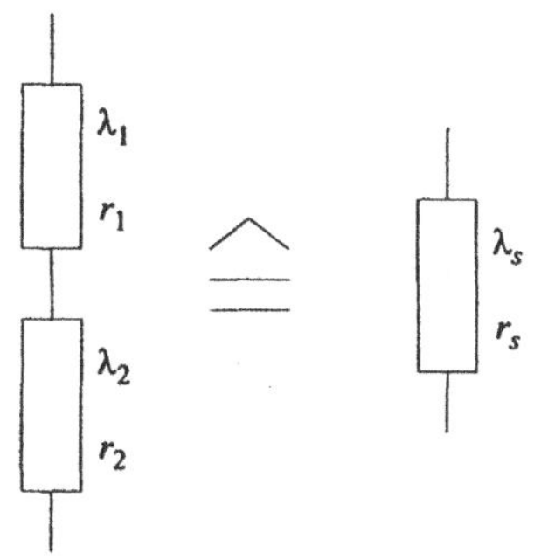

Figure 2.14 Stochastic series connection.

A series connection fails when either of the components fails. The outage rate for the series connection is thus the sum of the outage rates of the components:

$$\lambda_s = \lambda_1 + \lambda_2 \tag{2.34}$$

The series connection is not available when one of the components is not available, giving for the unavailability of the series connection:

$$Q_s = Q_1 + Q_2 \tag{2.35}$$

Using the definition of unavailability (2.31) gives an expression for the equivalent repair time of the series connection:

$$r_s = \frac{\lambda_1 r_1 + \lambda_2 r_2}{\lambda_1 + \lambda_2} \tag{2.36}$$

For n components in series, the following expressions can be derived:

$$\lambda_s = \sum_{i=1}^{n} \lambda_i \tag{2.37}$$

$$r_s = \frac{\sum_{i=1}^{n} \lambda_i r_i}{\sum_{j=1}^{n} \lambda_j} \tag{2.38}$$

In deriving the expressions for equivalent outage rate and repair time a number of assumptions have been made, all coming back to the system being available most of the time, thus $\lambda r \ll 1$. Exact expressions will be derived in Section 2.5.3.

Stochastic Parallel Connections. A parallel connection of two stochastic components is shown in Fig. 2.15.

A parallel connection fails when one of the components is not available and the other one fails: thus when 1 is unavailable and 2 fails or when 2 is unavailable and 1 fails. The outage rate of the parallel connection is

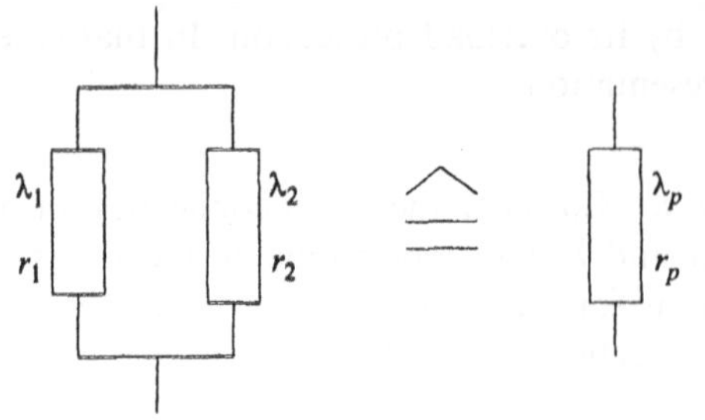

Figure 2.15 Stochastic parallel connection.

$$\begin{aligned}\lambda_p &= Q_1\lambda_2 + Q_2\lambda_1 \\ &= \lambda_1\lambda_2(r_1 + r_2)\end{aligned} \tag{2.39}$$

The parallel connection is not available when both components are not available. The unavailability of the parallel connection is

$$Q_p = Q_1 \times Q_2 \tag{2.40}$$

The repair time of the parallel connection is obtained from (2.39) and (2.40):

$$r_p = \frac{r_1 r_2}{r_1 + r_2} \tag{2.41}$$

The equations can be extended to a system with three components in parallel by considering it as the parallel connection of one component and the equivalent of the parallel connection of the two other components. This results in the following expressions for outage rate and repair time:

$$\begin{aligned}\lambda_p &= \lambda_1 r_1 \lambda_2 r_2 \lambda_3 r_3 (r_1 r_2 + r_1 r_3 + r_2 r_3) \\ &= \lambda_1\lambda_2\lambda_3\left(\frac{1}{r_1} + \frac{1}{r_2} + \frac{1}{r_3}\right)\end{aligned} \tag{2.42}$$

$$\frac{1}{r_p} = \frac{1}{r_1} + \frac{1}{r_2} + \frac{1}{r_3} \tag{2.43}$$

The same process can be repeated several times, resulting in the following general expressions for a system consisting of n components in parallel:

$$\lambda_p = \prod_{i=1}^{n} \lambda_i r_i \sum_{j=1}^{n} \frac{1}{r_j} \tag{2.44}$$

$$\frac{1}{r_p} = \sum_{j=1}^{n} \frac{1}{r_j} \tag{2.45}$$

Minimum Cut-Sets. A second method for analysis of stochastic methods is the so-called "minimum-cut-set method." A "cut-set" is a combination of components whose combined outage would lead to an interruption. In the stochastic network in Fig. 2.16 the combinations {1, 2, 3} and {4, 5} are examples of cut-sets. A cut-set is a "minimum cut-set" if the removal of any one of the components from the cut-set would make it no longer a cut-set. In other words, if the repair of any one component would restore the supply. In Fig. 2.16 the cut-set {1, 2, 3} is not a minimum

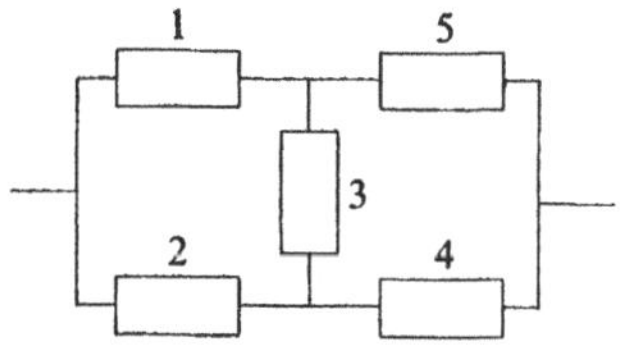

Figure 2.16 Example of stochastic network, for explaining the minimum cut-set method.

cut-set because repair of component 3 does not restore the supply, even though repair of component 1 or component 2 does. The cut-set {4, 5} is a minimum cut-set because both repair of component 4 and repair of component 5 restore the supply. For each network there are a limited number of minimum cut-sets. Finding all minimum cut-sets is the first step of the minimum-cut-set method.

The network in Fig. 2.16 has the following minimum cut-sets:

$$\{1, 2\}$$
$$\{4, 5\}$$
$$\{1, 3, 4\}$$
$$\{2, 3, 5\}$$

The supply is interrupted when any combination of these components is not available. The system behavior can thus also be described as a series connection of four parallel connections, representing the four minimum cut-sets. This is shown for this example in Fig. 2.17. After having found the minimum cuts-sets, the calculation proceeds straightforward: outage rates and repair times are determined, first for the parallel connections, next for the resulting series connection. The latter gives the interruption rate and expected interruption duration for the supply.

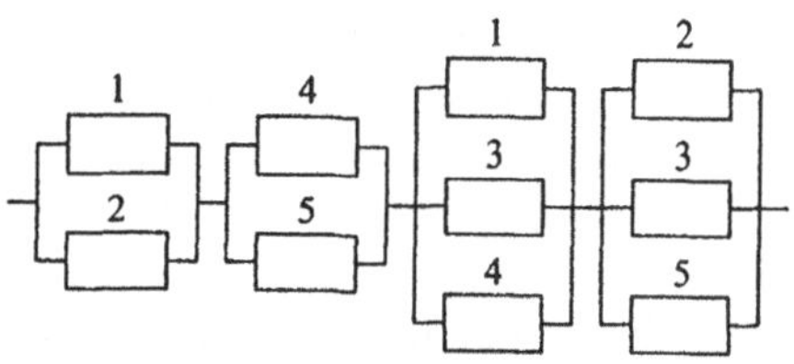

Figure 2.17 Alternative drawing of the network in Fig. 2.16: series connection of parallel connections.

EXAMPLE Consider the following outage rates and repair times for the network elements in Fig. 2.16:

$$\begin{array}{lll} \lambda_1 = 1 & r_1 = 0.2 & \lambda_1 r_1 = 0.2 \\ \lambda_2 = 2 & r_2 = 0.1 & \lambda_2 r_2 = 0.2 \\ \lambda_3 = 0.5 & r_3 = 0.1 & \lambda_3 r_3 = 0.05 \\ \lambda_4 = 0.8 & r_4 = 0.15 & \lambda_4 r_4 = 0.12 \\ \lambda_5 = 1.5 & r_5 = 0.2 & \lambda_5 r_5 = 0.3 \end{array}$$

Equations (2.44) and (2.45) give equivalent failure rate and repair time for the parallel connections representing the four cut-sets.

$$\begin{aligned}
\lambda_{c1} &= \lambda_1\lambda_2(r_1 + r_2) = 0.6 \\
\lambda_{c2} &= \lambda_4\lambda_5(r_4 + r_5) = 0.42 \\
\lambda_{c3} &= \lambda_1 r_1 \times \lambda_3 r_3 \times \lambda_4 r_4 \times \left(\frac{1}{r_1} + \frac{1}{r_3} + \frac{1}{r_4}\right) = 0.026 \\
\lambda_{c4} &= \lambda_2 r_2 \times \lambda_3 r_3 \times \lambda_5 r_5 \times \left(\frac{1}{r_2} + \frac{1}{r_3} + \frac{1}{r_5}\right) = 0.075
\end{aligned} \tag{2.46}$$

$$\begin{aligned}
r_{c1} &= \left(\frac{1}{r_1} + \frac{1}{r_2}\right)^{-1} = 0.067 \\
r_{c2} &= \left(\frac{1}{r_4} + \frac{1}{r_5}\right)^{-1} = 0.086 \\
r_{c3} &= \left(\frac{1}{r_1} + \frac{1}{r_3} + \frac{1}{r_4}\right)^{-1} = 0.046 \\
r_{c4} &= \left(\frac{1}{r_2} + \frac{1}{r_3} + \frac{1}{r_5}\right)^{-1} = 0.04
\end{aligned} \tag{2.47}$$

The failure rate λ and repair time r of the whole system can be calculated by considering it as a series connection of the four cut-sets:

$$\lambda = \lambda_{c1} + \lambda_{c2} + \lambda_{c3} + \lambda_{c4} = 1.121 \tag{2.48}$$

$$r = \frac{\lambda_{c1}r_{c1} + \lambda_{c2}r_{c2} + \lambda_{c3}r_{c3} + \lambda_{c4}r_{c4}}{\lambda_{c1} + \lambda_{c2} + \lambda_{c3} + \lambda_{c4}} = 0.072 \tag{2.49}$$

A second example of the use of the network approach is shown in Fig. 2.18 and Fig. 2.19. The first figure shows part of a subtransmission system. The transmission grid is assumed to be fully reliable. Also substations A, B, and C are assumed not to fail. The load of interest is connected to substation D. The network representation for the system in Fig. 2.18 is shown in Fig. 2.19. Component 8 represents outages in the local substation (D) which lead to an interruption for the load of interest. This network can no longer be reduced through series and parallel connections, but the minimum cut-set method can still be used.

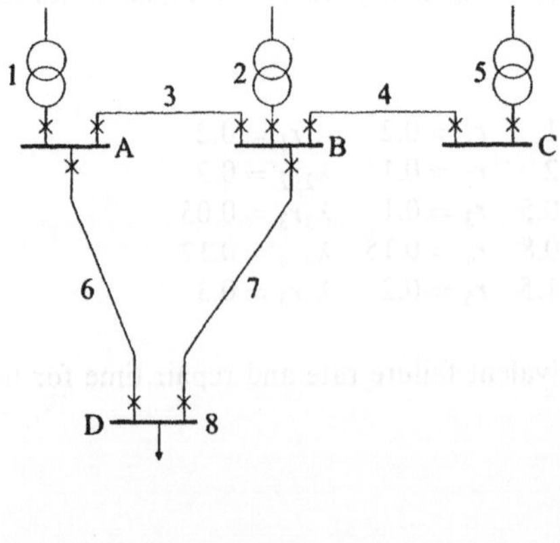

Figure 2.18 Example of public supply, with single redundancy.

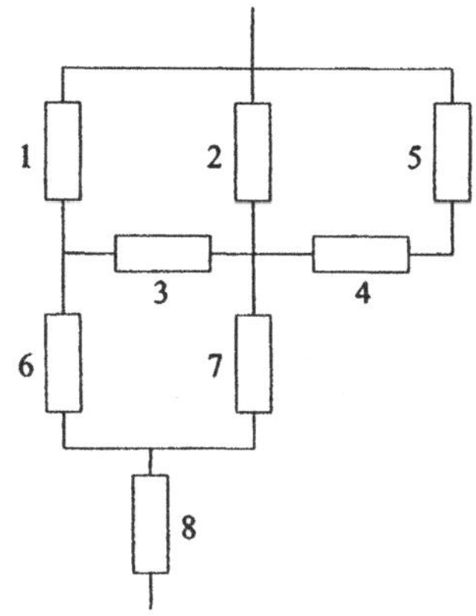

Figure 2.19 Network representation of the supply in Fig. 2.18.

The following minimum cut-sets can be found for this network:

$$\{8\}$$
$$\{6, 7\}$$
$$\{1, 2, 4\}$$
$$\{1, 2, 5\}$$
$$\{1, 3, 7\}$$
$$\{2, 3, 4, 6\}$$
$$\{2, 3, 5, 6\}$$

These minimum cut-sets are shown in Fig. 2.20 from where the term cut-set becomes clear. A cut-set cuts all paths between the source and the load. A minimum cut-set can be described as a "shortest cut."

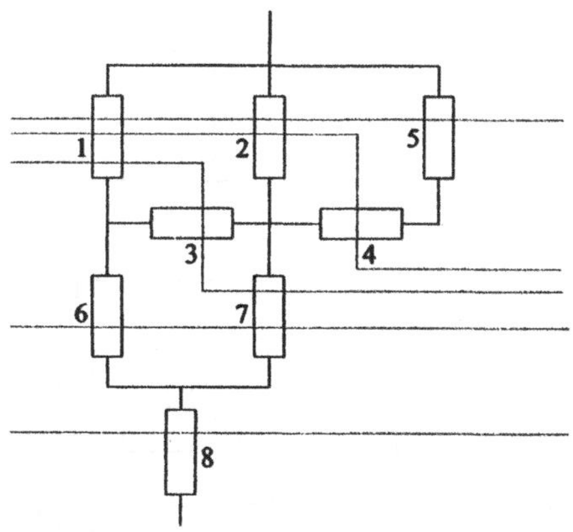

Figure 2.20 Network representation of the supply in Fig. 2.18, with minimum cut-sets indicated as dotted lines.

A third example is shown in Fig. 2.21. This supply system contains a substation with a third bus (4), a configuration used in industrial systems to prevent a circuit breaker failure from leading to loss of the whole substation. The various components have been numbered in the figure. Translating this to a network diagram is not obvious, as component 3 is in series with 1, 4, and 6, but 1 and 4 are in parallel. A possible solution is shown in Fig. 2.22. Components 3 and 5, representing bus outages, are now placed in a triangle with themselves. The network might seem somewhat artificial, the list of minimum cut-sets can be obtained in a normal way, resulting in

$\{8\}$
$\{1, 2\}$
$\{1, 5\}$
$\{2, 3\}$
$\{3, 5\}$
$\{3, 7\}$
$\{5, 6\}$
$\{6, 7\}$
$\{1, 4, 7\}$
$\{2, 4, 6\}$

The advantage of the network approach is that it gives a fast understanding of the reliability of the system. It also enables reliability calculations in large systems and provides, through minimum cut-set techniques, an insight into the weak points of the supply system. Drawing the stochastic network is a useful exercise in itself, often more useful than the actual results. The disadvantage is that approximation errors are made in each step of the calculation process. This could lead to serious errors in the results,

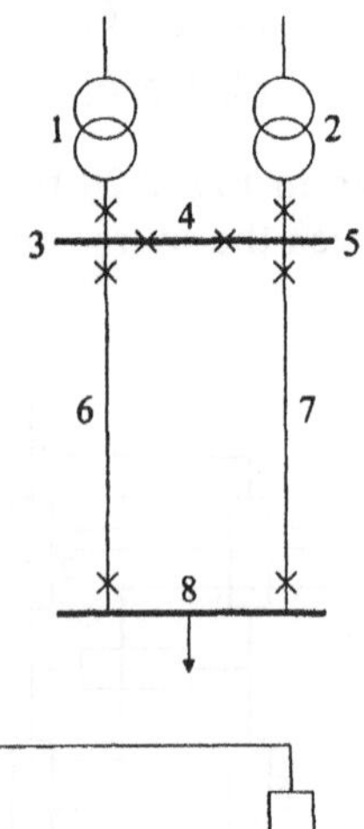

Figure 2.21 Industrial system with three-bus substation.

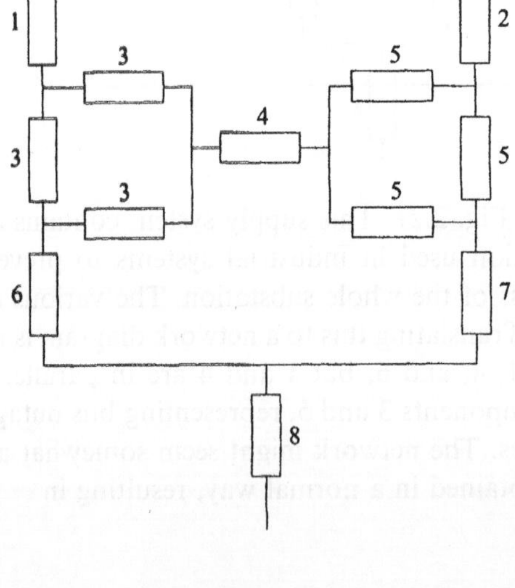

Figure 2.22 Network representation of the system in Fig. 2.21.

especially for large systems. The errors are due to the assumptions made when replacing series and parallel connections by one element. The assumptions made are that the unavailability of the element is small and that the elements are stochastically independent. The latter assumption is no longer fully correct when the series connection of minimum cut-sets is replaced by one element. As the same network component can appear in more than one minimum cut-set, the minimum cut-sets will become stochastically dependent.

2.5.3 State-Based and Event-Based Approaches

In the state-based approach the system behavior is described via states and transitions between states. A state is either healthy or nonhealthy. A healthy state is a state in which the supply is available, a nonhealthy state one in which the system is not available. The probability of all the nonhealthy states is calculated and added. This sum is the probability that the supply is not available. In addition to probability it is also possible to calculate other parameters, like the expected number of interruptions per year, or the average duration of an interruption.

In the event-based approach the system behavior is described by means of events. For each event the consequences for the supply are determined. In case analytical techniques are used the system is often still modeled as a collection of states and transitions. But now the transitions are either healthy or nonhealthy. A transition between two healthy states is NOT necessarily healthy.

A Four-State Component Model. The basic component model for a state-based approach consists of two states: { in operation }; and { not in operation }, often shortened to { in } and { out }. A more detailed model is shown in Fig. 2.23. This model consists of four states: { healthy }, { faulted }, { out of operation for repair }, and { out of operation for maintenance }. We can see from the figure that the component cannot fail while in maintenance, but that maintenance can start while the component is in repair. We also see that a faulted component will first be repaired before it becomes "healthy" again. The faulted state represents a short-circuit fault, the duration of which is much smaller than of the other states. Therefore this state is often combined with the repair state. But in studies of power system protection, the faulted state plays an essential role.

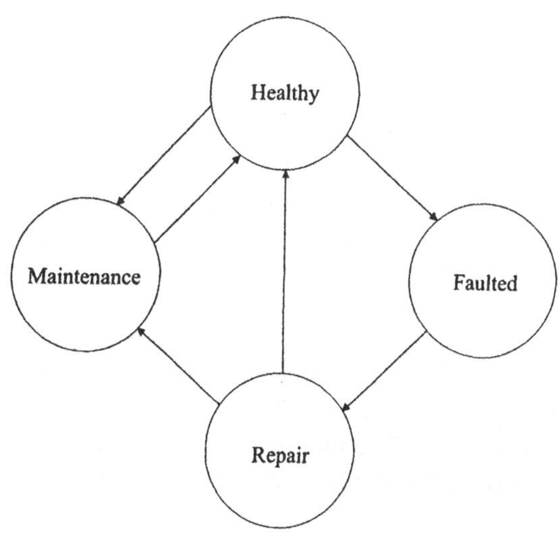

Figure 2.23 Four-state component model.

A Protective Relay. An example of a state model for a protective relay is shown in Fig. 2.24. We see the same healthy, repair, and maintenance states as in Fig. 2.23, but now the component can fail in three different ways. A maltrip leads directly to an outage of the component to be protected, after which the relay needs to be repaired. A hidden failure (also called "dormant fail to trip") means that the relay will no longer trip when it needs to. This failure will only reveal itself when the relay needs to trip, thus when there is a short circuit in the component to be protected. A potential maltrip is a situation where the relay will send an incorrect trip signal under certain system conditions. Maintenance can bring the relay from the "hidden failure" or "potential maltrip" states back to the "healthy" state.

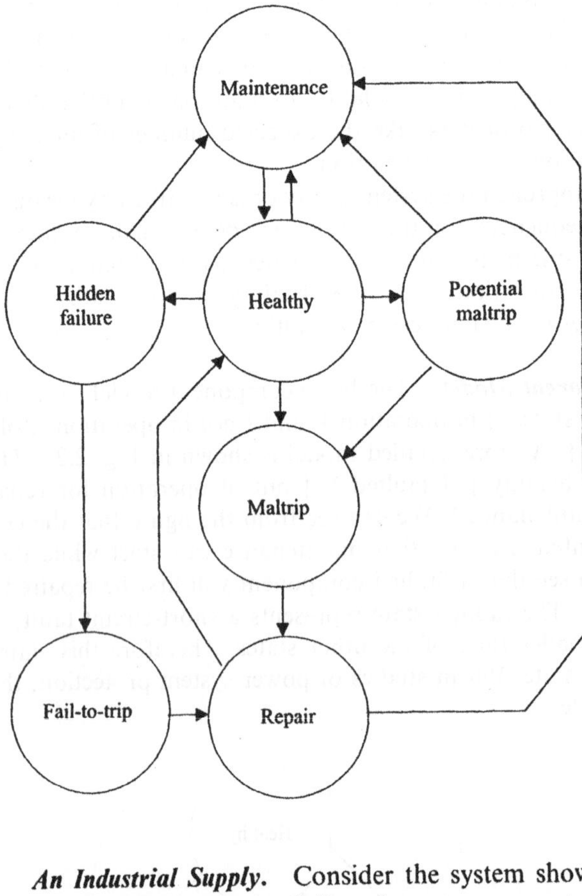

Figure 2.24 Model for protective relay, consisting of one healthy and six nonhealthy states.

An Industrial Supply. Consider the system shown in Fig. 2.25. The industrial load is fed via three overhead lines from two generator units plus the public supply. The rating of the components is such that one line is sufficient to supply the whole load; also one generator or the public supply are sufficient. We further assume that a failure of a line and a failure of the public supply are associated with a short circuit, but that a generator failure only involves the tripping of the unit.

It is assumed that each component can be in one of two states. Only failures of the public supply, the on-site generators and the overhead lines, are considered. This results in the system states as shown in Fig. 2.26. The system consists of 6 components, each with two states. The number of system states is therefore equal to $2^6 = 64$, but only 23 states are shown in Fig. 2.26. By assuming that the three lines are identical, and the two

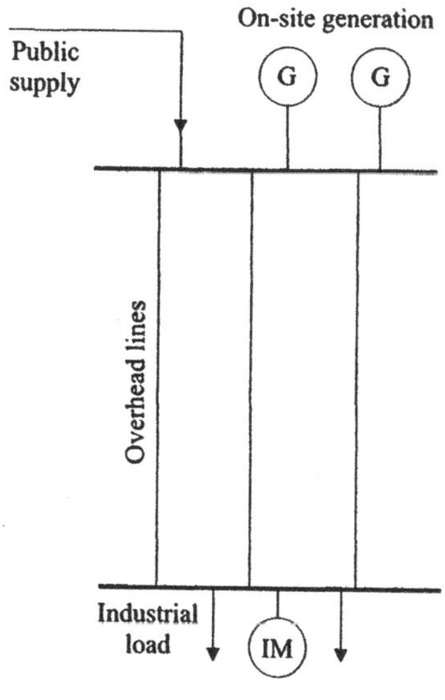

Figure 2.25 Example of industrial supply with double redundancy.

on-site generators also, states can be aggregated. For example, state 2 {1 line out} represents three basic states {line 1 out, line 2 out, line 3 out}; state 5 {2 lines out} also represents three basic states: {line 1 and line 2 out}, {line 1 and line 3 out}, {line 2 and line 3 out}. The state shown on top is the one with all components in operation. From this state the system can reach three other states:

- One line out of operation.
- One generator out of operation.
- The public supply out of operation.

An interruption of the supply can be due to the system being in an unhealthy state (e.g., three lines out), but also due to an unhealthy transition between two healthy states. A state-based study would only consider the states, not the transitions between states. To include interruptions due to unhealthy transitions, an event-based approach is more suitable.

In this system it can be assumed that only short-circuit faults lead to unhealthy transitions, thus only line failures and failures of the public supply. These potentially unhealthy transitions are indicated by an arrow in Fig. 2.26. From the state {2 lines out}, again, three transitions are possible:

- The failure of the last remaining line will anyway result in an interruption as the final state is an unhealthy one. This transition does not need to be further studied.
- The failure of a generator leads to the state {2 lines and one generator out} which is a healthy state. The transition is not associated with a short circuit and does not require further study.
- The failure of the public supply is associated with a short circuit and it leads to a healthy state. This transition requires further study.

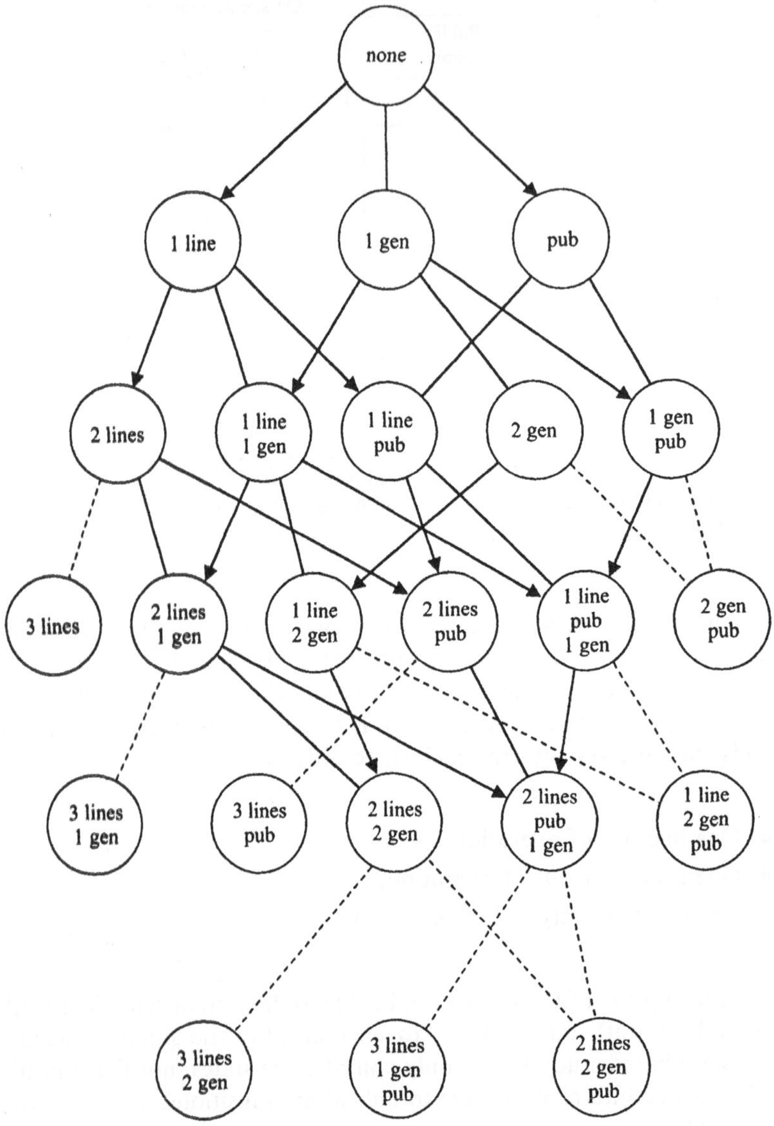

Figure 2.26 States and transitions for the system shown in Fig. 2.25. The solid lines indicate transitions between healthy states, the dotted lines indicate transitions between a healthy state and a nonhealthy state, the arrows indicate transitions associated with a short-circuit event.

2.5.4 Markov Models

Markov models are a mathematical way of calculating state probabilities and event frequencies in stochastic models. In Markov-model calculations all lifetimes and repair times are assumed exponentially distributed. A Markov model consists of a number of states, with transitions in between them; several examples will be given below.

One-Component Two-State Model. The simplest Markov model is shown in Fig. 2.27: a two-state model of one component. In state 1 the component is healthy,

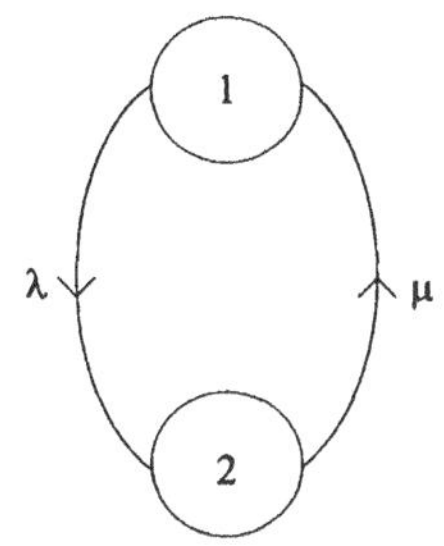

Figure 2.27 Two-state Markov model.

in state 2 it is nonhealthy. The transition rates are λ and μ, as indicated. This model will be used to introduce some of the basic concepts and calculation techniques.

To derive the expressions for the state probabilities, one should consider an infinite number of stochastically identical systems. At a time t a fraction p_1 of the systems is in state 1 and a fraction p_2 in state 2, with $p_1 + p_2 = 1$. In mathematical terms: the probability of finding the system in state 1 is equal to p_1. The transition rate from state 1 to state 2 is λ. Thus in a very short period Δt a fraction $\lambda \Delta t$ of the systems in state 1 transits to state 2. In the mean time a fraction $\mu \Delta t$ of the systems in state 2 transits to state 1. The fraction of systems in state 1 at time $t + \Delta t$ becomes

$$p_1(t + \Delta t) = p_1 - p_1 \lambda \Delta t + p_2 \mu \Delta t \tag{2.50}$$

A similar expression is obtained for the probability to find the system in state 2. Making the transition for $\Delta t \to 0$ gives the following differential equations for the state probabilities:

$$\frac{dp_1}{dt} = -\lambda p_1 + \mu p_2 \tag{2.51}$$

$$\frac{dp_2}{dt} = \mu p_2 - \lambda p_1 \tag{2.52}$$

We see that $\frac{dp_1}{dt} + \frac{dp_2}{dt} = 0$, which is understandable if one realizes that

$$p_1 + p_2 = 1 \tag{2.53}$$

i.e., the sum of state probabilities equals certainty. To calculate the state probabilities only one of the expressions (2.51) and (2.52) is needed, together with (2.53).

From (2.51) and (2.53) we can solve the probability that the system is in state 1, thus that the component is healthy. It is assumed that the component is healthy for $t = 0$ which corresponds to $p_1(0) = 1$.

$$p_1(t) = \frac{\mu}{\lambda + \mu} + \frac{\lambda}{\lambda + \mu} e^{-t(\lambda+\mu)} \tag{2.54}$$

We see that the probability reaches a constant value after an exponentially decaying transient with a time constant $\frac{1}{\lambda+\mu}$. For almost any engineering system we may assume that repair is much faster than failure, thus $\lambda \ll \mu$. When we also realize that $\frac{1}{\mu}$ is the average repair time, we see that the probability reaches a constant value within a time scale equal to the repair time. The time period of interest is normally much larger than the repair time (years versus hours) so that we can consider the system states and transition frequencies constant. This holds not only for a two-component model but for every Markov model in which repair takes place much faster than failure.

Steady-State Calculation. As the transition between the initial condition and the steady-state probabilities can be neglected, we can directly calculate the steady-state probabilities. In steady state, the state probabilities are constant as a function of time; thus,

$$\frac{dp_i}{dt} = 0 \tag{2.55}$$

The equations which describe the state probabilities become algebraic equations, which can be easily solved. For the two-state model we obtain

$$\begin{aligned} 0 &= -\lambda p_1 + \mu p_2 \\ 0 &= \lambda p_1 - \mu p_2 \\ p_1 + p_2 &= 1 \end{aligned} \tag{2.56}$$

One of the equations in this set is redundant, so that only one of the first two equations is needed. From this one and the third equation, the steady-state probability becomes

$$p_1 = \frac{\mu}{\lambda + \mu} \tag{2.57}$$

$$p_2 = \frac{\lambda}{\lambda + \mu} \tag{2.58}$$

Operating Reserve. We mentioned before that we can neglect the transition to the steady state and thus calculate steady-state probabilities directly. Two exceptions to this rule must be mentioned: one for very short time scales, and one with a very long repair time. When looking at a very short time scale the exponentially decaying component of the state probability can no longer be neglected. Very short time scales are of interest in operating reserve studies, where one knows that a component is in operation, and wants to know the probability that it fails within a time Δt. For a two-state model we derived before:

$$p_2(\Delta t) = 1 - p_1(\Delta t) = \frac{\lambda}{\lambda + \mu} - \frac{\lambda}{\lambda + \mu} e^{-\Delta t(\lambda + \mu)} \tag{2.59}$$

Assuming that $\Delta t \ll \frac{1}{\mu} \ll \frac{1}{\lambda}$ we obtain the following approximated expression:

$$p_2(\Delta t) \approx \frac{\lambda}{\mu} \mu \Delta t = \lambda \Delta t \tag{2.60}$$

Note that the same result is obtained if we assume that the component may fail but that it is not repaired within the period Δt.

Hidden Failures in a Protective Relay. A second example in which the exponentially decaying term cannot be neglected is a protective relay with hidden failures. Hidden failures of protective relays have already been discussed in Section 2.4.2. If we neglect all other failures of the relay, and assume that repair takes place instantaneously when the hidden failure is detected, we obtain the state model shown in Fig. 2.28. In state 1 the relay is healthy and a fault in the primary component to be protected is cleared as intended. If the relay is in state 2, the fault will not be cleared by this relay, but instead some backup protection needs to take over. The third state shown in Fig. 2.28 is the repair state. The failure rate λ_2 is the fault frequency in the primary component. We will initially assume that no preventive maintenance is performed on the relay.

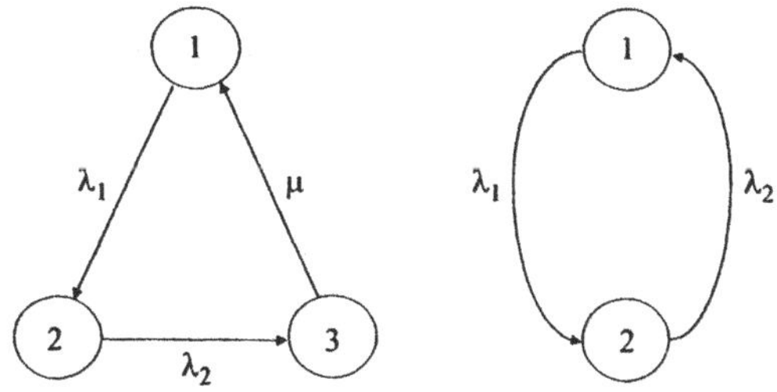

Figure 2.28 Model for relay with hidden failure (left); the relay is healthy in state 1 and contains a hidden failure in state 2. The figure on the right gives the two-state model which is obtained by neglecting the repair time μ.

From the three-state model in Fig. 2.28 we obtain the following set of equations for the state probabilities:

$$\begin{aligned}\frac{dp_1}{dt} &= -\lambda_1 p_1 + \mu p_3 \\ \frac{dp_2}{dt} &= \lambda_1 p_1 - \lambda_2 p_2 \\ \frac{dp_3}{dt} &= \lambda_2 p_2 - \mu p_3 \\ p_1 + p_2 + p_3 &= 1\end{aligned} \tag{2.61}$$

From this it is possible to obtain expressions for the state probabilities p_1, p_2, and p_3 and for the transition frequencies $\lambda_1 p_1$, $\lambda_2 p_2$, and μp_3.

Neglecting the transient to steady state gives the following equations for the state probabilities in steady state

$$\begin{aligned}\lambda_1 p_1 &= \mu p_3 \\ \lambda_2 p_2 &= \lambda_1 p_1 \\ \mu p_3 &= \lambda_2 p_2 \qquad (2.62) \\ p_1 + p_2 + p_3 &= 1 \qquad (2.63)\end{aligned}$$

Eliminating p_1 and p_3 from the first three expressions and substituting this in the fourth one results in

$$\left(\frac{\lambda_2}{\lambda_1} + 1 + \frac{\lambda_2}{\mu}\right) p_2 = 1 \tag{2.64}$$

The frequency of fail-to-trip events in steady state is

$$p_2 \lambda_2 = \frac{\lambda_1 \lambda_2 \mu}{(\lambda_1 + \lambda_2)\mu + \lambda_1 \lambda_2} \tag{2.65}$$

If we assume that repair (the transition from state 3 to state 1) takes place much faster than detection of the hidden fault (from state 2 to state 3), we can neglect state 3 and obtain the two-state system shown on the right of Fig. 2.28. This model results in the following equations:

$$\frac{dp_1}{dt} = -\lambda_1 p_1 + \lambda_2 p_2 \tag{2.66}$$

$$p_1 + p_2 = 1 \tag{2.67}$$

which corresponds to the equations for the two-state single-component model in Fig. 2.27 and (2.51) through (2.53). The resulting probability of being in the hidden-failure state is

$$p_2(t) = \frac{\lambda_1}{\lambda_1 + \lambda_2}\left[1 - e^{-t(\lambda_1+\lambda_2)}\right] \tag{2.68}$$

The fail-to-trip frequency is equal to $\lambda_2 p_2$ and reaches its steady-state value with a time constant $\frac{1}{\lambda_1+\lambda_2}$. This holds if we assume that hidden failures only reveal themselves during a fault in the primary component. In case maintenance is performed with a frequency λ_3 the transition rate from state 2 to state 1 is $\lambda_2 + \lambda_3$. The probability that the relay is in state 2 becomes

$$p_2(t) = \frac{\lambda_1}{\lambda_1 + \lambda_2 + \lambda_3}\left[1 - e^{-t(\lambda_1+\lambda_2+\lambda_3)}\right] \tag{2.69}$$

Maintenance reduces the time constant with which the steady-state probability is reached, and (more importantly) it reduces the steady-state probability. The number of fail-to-trip events per year n_{mt} remain equal to $\lambda_2 p_2$, thus given by the following expression:

$$n_{mt}(t) = \frac{\lambda_1 \lambda_2}{\lambda_1 + \lambda_2 + \lambda_3}\left[1 - e^{-t(\lambda_1+\lambda_2+\lambda_3)}\right] \tag{2.70}$$

We see that for maintenance to be effective, the maintenance frequency needs to be higher than the sum of the fault frequency in the primary component and the hidden-failure rate of the relay

$$\lambda_3 > \lambda_1 + \lambda_2 \tag{2.71}$$

Two-Component Model. Consider a system that consists of two components: component 1 and component 2, with failure rates λ_1 and λ_2, and repair rates μ_1 and μ_2. If we model each component through two states, this system has four states:

- State 1 with both components in operation.
- State 2 with only component 2 in operation.
- State 3 with only component 1 in operation.
- State 4 with none of the components in operation.

The resulting state model is shown in Fig. 2.29. The equations for the state probabilities are

$$\frac{dp_1}{dt} = -(\lambda_1 + \lambda_2)p_1 + \mu_1 p_2 + \mu_2 p_3 \tag{2.72}$$

$$\frac{dp_2}{dt} = \lambda_1 p_1 - (\mu_1 + \lambda_2)p_2 + \mu_2 p_4 \tag{2.73}$$

$$\frac{dp_3}{dt} = \lambda_2 p_1 - (\mu_2 + \lambda_1)p_3 + \mu_1 p_4 \tag{2.74}$$

$$\frac{dp_4}{dt} = \lambda_2 p_2 + \lambda_1 p_3 - (\mu_1 + \mu_2)p_4 \tag{2.75}$$

$$p_1 + p_2 + p_3 + p_4 = 1 \tag{2.76}$$

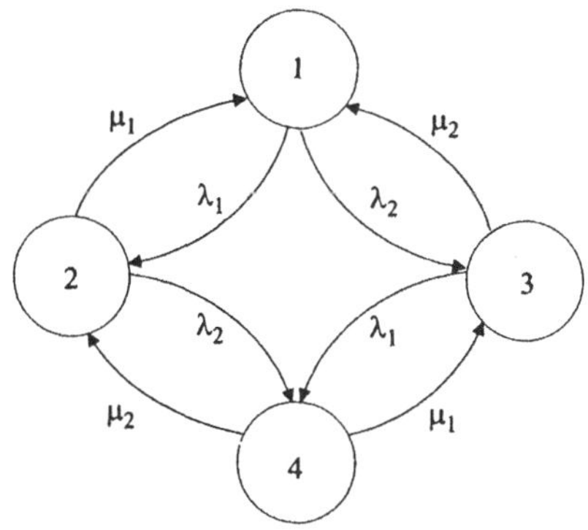

Figure 2.29 Two-component, two-state Markov model.

These can be solved again like for the previous examples, but there is an alternative solution method. We have assumed that the two components are stochastically independent. This assumption has not been made explicitly but by making the failure and repair rates of the components independent of the state of the other component. If the components are stochastically dependent, the transition rate from state 1 to state 2 is not the same as the one from state 3 to state 4 (both represent failure of component 1), etc. For stochastically independent components we can multiply the component state probabilities to get the system state probabilities. Thus, with p_{idown} and p_{iup} the probabilities that component i is in the "up" and in the "down" state, respectively, we obtain for the state probabilities

$$p_1 = p_{1up} \times p_{2up} \tag{2.77}$$

$$p_2 = p_{1down} \times p_{2up} \tag{2.78}$$

$$p_3 = p_{1up} \times p_{2down} \tag{2.79}$$

$$p_4 = p_{1down} \times p_{2down} \tag{2.80}$$

These equations hold for each moment in time, thus for the transient to steady state, as well as for the steady state. Using the expressions for the state probabilities in the one-component model the steady-state probabilities in the two-component model become

$$p_1 = \frac{\mu_1\mu_2}{(\lambda_1 + \mu_1)(\lambda_2 + \mu_2)} \tag{2.81}$$

$$p_2 = \frac{\lambda_1\mu_2}{(\lambda_1 + \mu_1)(\lambda_2 + \mu_2)} \tag{2.82}$$

$$p_3 = \frac{\mu_1\lambda_2}{(\lambda_1 + \mu_1)(\lambda_2 + \mu_2)} \tag{2.83}$$

$$p_4 = \frac{\lambda_1\lambda_2}{(\lambda_1 + \mu_1)(\lambda_2 + \mu_2)} \tag{2.84}$$

Series and Parallel Connections. We can use these results to obtain exact expressions for the failure rate and repair time of series and parallel connections, approximations for which were given in Section 2.5.2. For a series connection of components 1 and 2, state 1 is the healthy state. System failure is a transition from state 1 to state 2, or from state 1 to state 3. The system failure rate λ_s is the sum of these two transition rates:

$$\lambda_s = p_1\lambda_1 + p_1\lambda_2 = \frac{\mu_1\mu_2(\lambda_1 + \lambda_2)}{(\lambda_1 + \mu_1)(\lambda_2 + \mu_2)} \tag{2.85}$$

The system is unavailable when it is not in state 1. The system repair time r_s is found from the unavailability Q_s:

$$\lambda_s r_s = Q_s = 1 - p_1 \tag{2.86}$$

The average repair time for the series connection is

$$r_s = \frac{\lambda_1\mu_2 + \lambda_2\mu_1 + \lambda_1\lambda_2}{\mu_1\mu_2(\lambda_1 + \lambda_2)} \tag{2.87}$$

In a similar way expressions can be derived for the parallel connection. For a parallel connection, states 1, 2, and 3 are healthy, and system failure is a transition from state 2 to state 4 or from state 3 to state 4. The resulting expressions for failure rate λ_p and repair time r_p are

$$\lambda_p = p_2\lambda_2 + p_3\lambda_1 = \frac{\lambda_1\lambda_2(\mu_1 + \mu_2)}{(\lambda_1 + \mu_1)(\lambda_2 + \mu_2)} \tag{2.88}$$

$$r_p = \frac{1}{\mu_1 + \mu_2} \tag{2.89}$$

Exact Solution of Large Markov Models. For a system with a large number of states, the underlying equations can be derived in the same way as shown in the above example. The set of differential equations can be written in the following matrix form:

$$\frac{d\overline{P}}{dt} = A\overline{P}(t) \tag{2.90}$$

with A the matrix of state transitions and $\overline{P}$ the vector of state probabilities. For the Markov model in Fig. 2.29 we get

$$\overline{P}(t) = \begin{bmatrix} p_1(t) \\ p_2(t) \\ p_3(t) \\ p_4(t) \end{bmatrix} \tag{2.91}$$

and

$$A = \begin{bmatrix} -\lambda_1 - \lambda_2 & \mu_1 & \mu_2 & 0 \\ \lambda_1 & -\mu_1 - \lambda_2 & 0 & \mu_2 \\ \lambda_2 & 0 & -\mu_2 - \lambda_1 & \mu_1 \\ 0 & \lambda_2 & \lambda_1 & -\mu_1 - \mu_2 \end{bmatrix} \tag{2.92}$$

The off-diagonal element A_{ij} is the transition rate from state j to state i. The diagonal element A_{ii} is minus the sum of all transition rates away from state i:

$$A_{ij} = \lambda_{ji} \tag{2.93}$$

$$A_{ii} = -\sum_j \lambda_{ij} \tag{2.94}$$

Together with an initial condition for the state probability vector

$$\overline{P}(0) = \overline{P}_0 \tag{2.95}$$

we obtain the following solution for this initial value problem:

$$\overline{P}(t) = S\exp[-\Lambda t]S^{-1}\overline{P}_0 \tag{2.96}$$

with S the matrix of eigenvectors of A and Λ the diagonal matrix of eigenvalues of A. Because A is a singular matrix (the sum of all transitions is zero) one of the eigenvalues is zero. That leads to a constant term in the solution

$$\overline{P}(t) = \overline{P}_{ss} + \sum_{i>1} \overline{P}_i e^{-\frac{t}{\tau_i}} \tag{2.97}$$

In most cases we can neglect the transients and are only interested in the steady-state solution $\overline{P}_{ss}$. Note that the steady-state solution is independent of the initial values. The steady-state solution can be obtained directly from the transition rates by setting the time derivatives to zero:

$$A\overline{P} = 0 \tag{2.98}$$

$$\sum_i p_i = 1 \tag{2.99}$$

Approximate Solution of Large Markov Models. The main problem with the exact solution of large systems is that all state probabilities have to be calculated at the same time, even those with a very low probability. For an N-state model, an $N \times N$ matrix has to be inverted to find the steady-state probabilities. Assuming that all components have two states (up and down) an n-component system requires 2^n states. Thus, a 10-component system already requires 1000 states, and a 150-component model requires the inversion of a matrix of size 10^{45}. In other words, this method has serious limitations. We might be able to somewhat reduce the number of states, but exact solutions for systems with more than 10 components are in practice not possible to obtain. To overcome these limitations, one can use an approximated method, which gives recursive expressions for the state probabilities [145]. The assumptions made are as follows:

- The state with all components in operation has a probability equal to one.
- The repair rate of a component is much larger than its failure rate.
- The probability of a state with k components out of operation is much lower than the probability of a corresponding state with $(k-1)$ components out of operation.

All these assumptions can be brought back to one basic assumption: the components are repaired much faster than they fail. This is a reasonable assumption for most engineering systems. An exception are the so-called "hidden failures" discussed before. For hidden failures the model requires some adjustments.

Consider again the state model for an industrial supply, as shown in Fig. 2.26. Part of this figure has been reproduced in Fig. 2.30. Here λ and μ are failure and repair rates, respectively. The index 1 refers to lines, the index 2 to generators, and the index 3 to the public supply.

The exact expressions for the state probabilities of states 1 through 4 are

$$(3\lambda_1 + 2\lambda_2 + \lambda_3)p_1 = \mu_1 p_2 + \mu_2 p_3 + \mu_3 p_4 \tag{2.100}$$

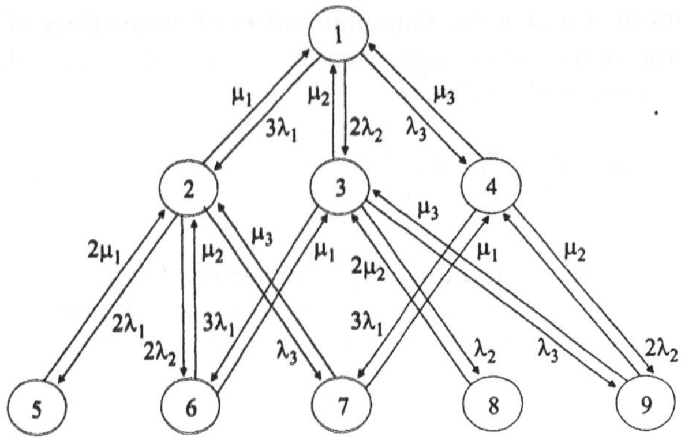

Figure 2.30 Part of a multistate Markov model. (Reproduced from Fig. 2.26.)

$$(2\lambda_1 + 2\lambda_2 + \lambda_3 + \mu_1)p_2 = 3\lambda_1 p_1 + 2\mu_1 p_5 + \mu_2 p_6 + \mu_3 p_7 \tag{2.101}$$
$$(3\lambda_1 + \lambda_2 + \lambda_3 + \mu_2)p_3 = 2\lambda_2 p_1 + \mu_1 p_6 + 2\mu_2 p_8 + \mu_3 p_9 \tag{2.102}$$
$$(3\lambda_1 + 2\lambda_2 + \mu_3)p_4 = \lambda_3 p_1 + \mu_1 p_7 + \mu_2 p_9 \tag{2.103}$$

The approximated method starts with assuming that the system is almost certainly healthy, thus

$$p_1 = 1 \tag{2.104}$$

According to the third assumption, we neglect the terms with p_5, p_6, p_7, p_8, and p_9 on the right-hand side of (2.101) through (2.103). That gives the following equations for the states 2 through 4:

$$(2\lambda_1 + 2\lambda_2 + \lambda_3 + \mu_1)p_2 = 3\lambda_1 p_1 \tag{2.105}$$
$$(3\lambda_1 + \lambda_2 + \lambda_3 + \mu_2)p_3 = 2\lambda_2 p_1 \tag{2.106}$$
$$(3\lambda_1 + 2\lambda_2 + \mu_3)p_4 = \lambda_3 p_1 \tag{2.107}$$

As p_1 is known we obtain the state probabilities of these three states without having to know the other state probabilities:

$$p_2 = \frac{3\lambda_1}{2\lambda_1 + 2\lambda_2 + \lambda_3 + \mu_1} \tag{2.108}$$
$$p_3 = \frac{2\lambda_2}{3\lambda_1 + \lambda_2 + \lambda_3 + \mu_2} \tag{2.109}$$
$$p_4 = \frac{\lambda_3}{3\lambda_1 + 2\lambda_2 + \mu_3} \tag{2.110}$$

A correction can be made by recalculating the probability p_1 from

$$p_1 = 1 - \sum_{i>1} p_i \tag{2.111}$$

The same method can be applied to states 5 through 15, each time resulting in an equation in which only one state probability is unknown. Instead of having to solve all state probabilities at the same time, this procedure allows solving state probabilities sequentially. For very large systems, not all states are of equal interest, which can further reduce the computational requirements. The recursive procedure can, e.g., be stopped when the state probability drops below a certain value.

2.5.5 Monte Carlo Simulation

Basic Principles. In all preceding examples, the unknown quantities were actually calculated. We saw several times that approximations and assumptions were needed to obtain a solution. In a Monte Carlo simulation, or simply simulation, these assumptions and approximations are no longer needed. The Monte Carlo simulation method does not solve the equations describing the model; instead the stochastic behavior of the model is simulated and observed.

The behavior of the system (stochastic process is actually a better term) is observed many times or for a long period of time. The average observation is used as an estimate for the expected behavior of the system.

The basis of each Monte Carlo simulation involves using a so-called random-number generator. The random-number generator is needed to bring the stochastic element in the calculations. One could use a physical random-number generator like a dice or a coin, but a numerical random-number generator is more suitable for computer-based calculations.

A coin can be used to model a state with a probability of 50%. Consider as an example a three-component system with 50% availability for each component. The coin is used to generate component states, with the second column in Table 2.11 the resulting sequence. This represents the state of one of the components over 24 consecutive 1-hour periods. The same is done for component 2 and component 3, resulting in columns 3

TABLE 2.11 Monte Carlo Simulation with 50% Probabilities

Hour	Component 1	Component 2	Component 3	System 1	System 2	System 3
1	up	up	down	up	up	up
2	up	down	up	up	up	up
3	down	down	up	down	up	down
4	down	up	down	down	down	down
5	up	down	up	up	down	up
6	down	down	down	down	down	down
7	up	up	up	up	up	up
8	up	down	up	up	up	down
9	down	up	down	down	up	down
10	down	up	down	down	down	down
11	up	up	down	up	down	dow
12	up	down	up	up	down	down
13	up	up	down	up	up	down
14	down	up	down	down	up	down
15	down	down	up	down	down	down
16	up	up	down	up	down	down
17	up	down	up	up	down	down
18	down	up	up	up	up	down
19	up	down	down	down	up	down
20	up	up	down	up	up	up
21	down	up	down	down	up	down
22	up	up	up	up	up	up
23	down	down	up	down	up	down
24	down	up	down	down	down	down

and 4, respectively. The column labeled "system 1" gives the state of a system which is available if at least two components are available.

One can make this Monte Carlo simulation as complicated as one wants. In the column labeled "system 2" the system is down if less than two components are available for two consecutive 1-hour periods, and if the system is down it remains down for at least 3 hours. For "system 3" the system needs three components to be available in the hourly periods 8 through 18, but only two for the other periods. As a second example consider three components whose lifetime is uniformly distributed between 0 and 6 years. To generate the lifetime of these components we can use a dice. By using this we simulate the behavior of this three-component system during 10 years. In Fig. 2.31 three possible outcomes of this "experiment" are shown. Each possible outcome is called a "sequence." During sequence 1, the first component fails after 3 years and again after 6 years; the second component fails after 2, 6, 7, 9, and 10 years, etc.

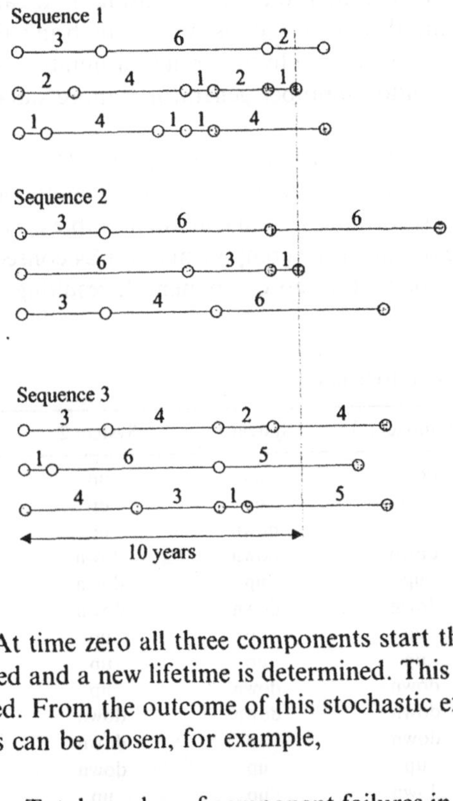

Figure 2.31 Three sequences of a Monte Carlo simulation. The circles indicate failures followed by repair; the numbers in between indicate times-to-failure.

At time zero all three components start their first lifetime. Upon failure they are repaired and a new lifetime is determined. This process is repeated until $t = 10$ years is reached. From the outcome of this stochastic experiment, many different output parameters can be chosen, for example,

- Total number of component failures in a 10-year period. In this case the values 11, 7, and 8 are found.
- Total number of events with two or more component failures in the same year, with values 3, 2, and 1 being found.
- Probability distribution function of the component lifetime.

Numerical Random-Number Generators. In practice one never uses physical random-number generators like dice or coins. The reason is that it is difficult to actually use them in a computer program and hand-calculations of Monte Carlo simulations are very complicated, as will be clear after the preceding examples.

A numerical random-number generator creates a row of integers of pseudo-random nature. The row is not really random as a numerical algorithm is used to calculate it—therefore, the term "pseudo-random number generator." Most computer simulations use random-number generators of the following form:

$$U_{i+1} = (aU_i) mod N \tag{2.112}$$

where a and N have to be chosen. The output of this is a row of integers with values between 1 and $(N-1)$.

EXAMPLE Consider the values $N = 11$ and $a = 7$. That gives the following row of integers:

1, 7, 5, 2, 3, 10, 4, 6, 9, 8, 1, 7, 5, 2, 3, 10, 4, 6, 9, 8, 1, etc.

The row repeats itself after 10 elements, which is understandable if one realizes that there are only 10 possible outcomes of (2.112). A cycle length 10 (in general $(N-1)$) is the longest possible value. To show that shorter cycle lengths are also possible, consider the random-number generator with $N = 11$ and $a = 5$ which has two possible rows, each of cycle length 5:

1, 5, 3, 4, 9, 1

2, 10, 6, 8, 7, 2

The random-number generators in use in Monte Carlo simulations have much longer cycles lengths, and therefore much higher values of N. A popular value is $N = 2^{31} - 1 = 2\,147\,483\,647$. Most values of a give a cycle length less than $N - 1$. A value of a which gives the maximum cycle length is $a = 950\,706\,376$. Starting from $U = 1$ we get the following row of integers:

1, 950 706 376, 129 027 171, 1 782 259 899, 365 181 143, 1 966 843 080, etc.

The resulting integer is often divided by N to get a random number between 0 and 1, which leads to a slightly different version of (2.115):

$$U_{i+1} = \frac{aNU_i modN}{N} \tag{2.113}$$

The result of (2.113) is a random draw from the uniform distribution on the interval (0,1). Neither zero nor one can be obtained through this method, which is often an advantage as it prevents dividing by zero in further processing of the result. This standard uniform distribution is the basis for all Monte Carlo simulations.

EXAMPLE For $N = 11$ and $a = 7$, (2.113) results in the following row of samples: 0.09, 0.63, 0.45, 0.18, 0.27, 0.91, 0.36, 0.55, 0.82, 0.73, 0.09, etc.

Simulating a Probability—Random Monte Carlo Simulation. Two types of Monte Carlo simulation can be distinguished: random simulation and sequential simulation. An example of random simulation is the simulation shown in Table 2.11. In a random Monte Carlo simulation each component has a probability of being in a certain state. The simulation generates combinations of component states. For each resulting combination the system state (healthy or nonhealthy) is evaluated. This whole process is repeated until a certain accuracy is obtained.

The basis of a random Monte Carlo simulation is the probability: an event takes place with a certain probability, a quantity has a certain value with a certain probability, or a component is in a certain state with a certain probability. A probability is simulated by drawing a value from the standard uniform distribution introduced before. Let p be the probability that the component is in state S_1; otherwise, the component will be in state S_2, then the Monte Carlo simulation proceeds as follows:

- Draw a value U from the standard uniform distribution.
- If $U \leq p$ the component is in state S_1.
- If $U > p$ the component is in state S_2.

Note that for $U = p$ the component state is actually not defined. In this example this situation is attributed to state S_1 but it could equally have been attributed to state S_2. This ambiguity has to do with our discretization of the uniform distribution. For a continuous distribution the probability that $U = p$ is zero. For a random-number generator with a cycle length of $2^{31} - 1$ this probability (5×10^{-10}) is small enough to neglect in all practical cases.

Simulating a Time Distribution. The basis of a sequential simulation is the time distribution. Therefore we need a method of obtaining other distributions than just the standard uniform distribution, i.e., the uniform distribution on the interval (0,1).

The uniform distribution on an interval (T_1, T_2) is obtained from a sample of the standard uniform distribution U as follows:

$$X = T_1 + (T_2 - T_1)U \tag{2.114}$$

where X is a sample from the uniform distribution on the interval (T_1, T_2). More general: a stochastic variable S with a distribution function $F(s)$ is obtained from

$$S = F^{-1}(U) \tag{2.115}$$

where U is a stochastic variable with a standard uniform distribution. To prove this, we look at the probability distribution function of the stochastic variable S according to (2.115), thus at the probability that S is less than a certain value s.

$$Pr\{S < s\} = Pr\{F^{-1}(U) < s\} \tag{2.116}$$

As F is a non-decreasing function, we can write this as

$$Pr\{S < s\} = Pr\{U < F(s)\} \tag{2.117}$$

The stochastic variable U has a standard uniform distribution; thus,

$$Pr\{U < x\} = x, \text{ for } 0 < x < 1 \tag{2.118}$$

As $0 < F(s) < 1$ we get the intended expression, which proves that S is distributed according to $F(s)$.

$$Pr\{S < s\} = F(s) \tag{2.119}$$

Consider as an example, the Weibull distribution introduced by (2.25). From (2.115) it follows that a sample W from the Weibull distribution with characteristic time θ and shape factor m is obtained from a sample U from the standard uniform distribution by

$$W = \theta \sqrt[m]{-\ln(1 - U)} \tag{2.120}$$

For $m = 1$ we obtain the exponential distribution as a special case of the Weibull distribution. A sample E from the exponential distribution with expected time θ is obtained by

$$E = -\theta \ln(1 - U) \tag{2.121}$$

Sequential Monte Carlo Simulation. The examples in Fig. 2.31 show a sequential simulation. In a sequential Monte Carlo simulation, the whole time behavior of a system is simulated, with failure and repair of components the main subject in a reliability study. But also other events, like load switching and weather changes, can be part of the simulation. This kind of simulation offers the most opportunities of obtaining output, but it also requires the most programming and computing efforts.

The details of a sequential Monte Carlo simulation vary widely and depend on the particular application, the kind of programming language available, and on personal taste of the program developers. Below, a possible structure is given which was used successfully by the author for evaluating the reliability of industrial power systems [61], [62], [63]. Only one sequence of a given length is described here. This sequence should be repeated a large number of times to get statistically relevant results.

1. **Set up an initial event list**. At the start of each sequence, times for the first event are drawn for each component. The first event is typically a failure or start of maintenance. These events are sorted on time of occurrence and placed in a so-called "event list." Part of an event list would typically look as follows:

0.15 years	component 2	failure
1.74 years	component 5	maintenance
3.26 years	component 1	hidden failure
4.91 years	component 5	failure
5.67 years	component 2	maintenance
6.21 years	component 1	maltrip

This event list should be interpreted as follows: at $t = 0.15$ years, component 2 will fail; at $t = 1.74$ years, maintenance on component 5 is planned, etc. Not all events in the list will actually occur. We will see below that events may be removed from the event list and that events may be inserted. Further on in the simulation of this sequence, it will always be the event on top of the list which will be processed, after which the event list will be updated. When the event list is empty the simulation of this sequence is over.

2. **Process the event on top of the event list**. Processing of the event on top of the event list (thus the next event to happen in the system) is the main part of the simulation, which will take up most time in programming and deciding about. This is where the stochastic model of the power system and its components is implemented. The processing of an event typically consists of making changes in the event list and making changes in the electrical model of the power system. Changes in the power system can be the removal

of a component (e.g., due to the intervention of the protection) or the insertion of a branch (e.g., repair of a component or due to a short-circuit fault). To assess the effect of the event on the load, either the new steady state or the electrical transient due to the event need to be evaluated. The interruption criterion needs to be applied to decide if this event leads to an interruption or not. The changes in the event list will be discussed below for different events.

(a) **Short-circuit event.** The next event after a short-circuit event will be an intervention of the protection. Some rules are needed to decide which relays will intervene: the relay or relays which need to clear this fault; those which incorrectly intervene; and those which take over the protection in case one or more of the primary relays fails to trip. For each relay a time until tripping needs to be determined. Tripping of the fault normally takes place very soon after the short-circuit event. Therefore one can decide to treat fault initiation (short-circuit event) and fault clearing (protection intervention event) as one event. Here they are considered as two events.

(b) **Protection intervention event.** During the processing of this event one needs to distinguish between the last relay to trip and all the other protection intervention events. After the last relay has tripped the repair of the faulted component can start and also the switching needed to restore the nonfaulted components tripped by the protection. For the Monte Carlo simulation this means that times to repair and times to switching need to be determined. Alternatively one can determine all these times when processing the short-circuit event.

(c) **Repair event.** When a component is repaired, it can fail again. Therefore a time to failure needs to be determined for all its failure modes: short circuit, maltrip, hidden failure, etc. Different failure modes will typically have different lifetime distributions.

(d) **Maltrip event.** A maltrip event is associated with the power system protection, either with a circuit breaker or with a protective relay. The next events to be determined are repair of failed component and restoration of the primary component tripped.

(e) **Hidden failure event.** A hidden failure event will not reveal itself immediately. Therefore it will only change the way the relay will in future react to a short-circuit event. Only when a hidden failure reveals itself, either due to a short circuit or due to maintenance, will the repair start.

(f) **Start of maintenance event.** Start of maintenance will require the scheduling of an end of maintenance event. For an accurate maintenance model, one needs to introduce an additional event called "maintenance attempt." Maintenance attempts are scheduled and either immediately lead to a start of maintenance event or to a new maintenance attempt event. Some rules are needed to decide if the system state is suitable for maintenance to be performed. The rules will depend among others on the company rules for performing maintenance. Some examples are

- Maintenance cannot be performed at more than one component at the same time, e.g., because there is only one maintenance crew available.
- Maintenance will not be performed if it leads to an interruption of the supply for any of the loads.

- Maintenance will not be performed when a parallel or redundant component is out of operation.

When processing the start of maintenance event, the time for an end of maintenance event needs to be determined.

(g) **End of maintenance event.** When the maintenance is finished a new maintenance attempt or start of maintenance event needs to be determined. Also some future fail events will be influenced by the maintenance. Typically the component is assumed to be "as-good-as-new" after maintenance. In that case all future fail events are removed from the event list and new ones are drawn from appropriate distribution functions.

Some additional rules might be needed to control the processing of events. One might, for instance, decide that a component cannot fail while it is out of operation (for any reason). One can make a check during a failure event to see if a component is in operation and simply draw a new failure event without any additional processing if the component is not in operation. One can also decide to shift all failure events belonging to a component further into the future with a time equal to the time during which it is out of operation.

3. **Update the event list**. All new events which occur before the end of the sequence are placed in the event list; the event just processed is removed; the event list is sorted again; after which the event that appears on top of the event list is processed.

Errors in the Monte Carlo Simulation. An example of the result of a Monte Carlo simulation is shown in Fig. 2.32. The figure has been obtained by taking samples from the uniform distribution on the interval (0,1), followed by calculating the average over all the proceeding samples. For an increasing number of samples, the average value approaches 0.5. As we can see from the figure, the error is still rather large after 100 samples.

Figure 2.33 gives the behavior for a much larger number of samples. After 10 000 samples, the error has become less than 1%, but is still not zero. An important property of the Monte Carlo simulation is that the error approaches zero, but never becomes zero. Figure 2.33 also shows another property of the Monte Carlo simulation: the fact

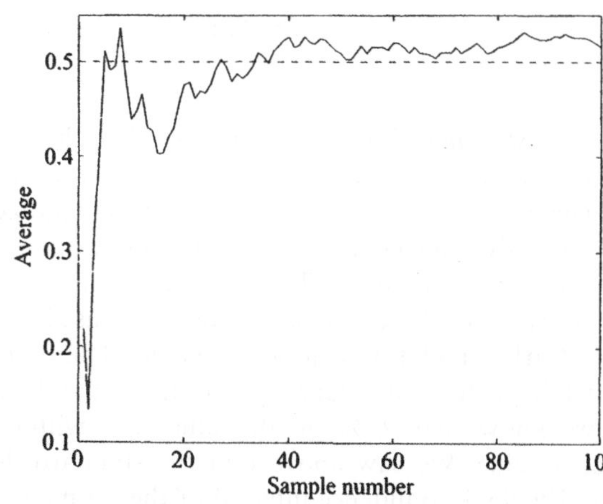

Figure 2.32 Outcome of a Monte Carlo simulation.

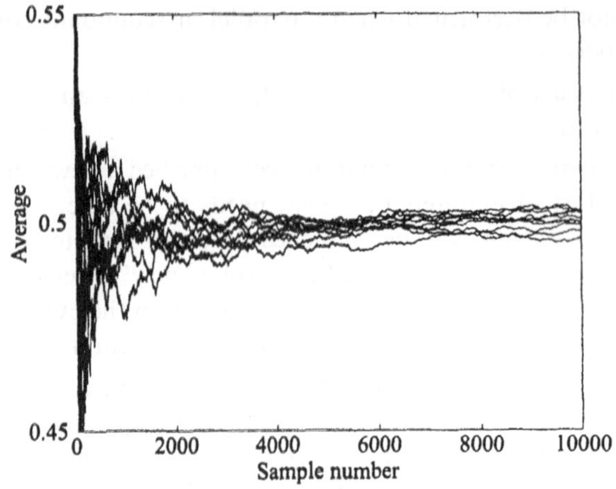

Figure 2.33 Outcome of 10 identical Monte Carlo simulations.

that each simulation may give a different result. The figure gives the result of 10 simulations, each using a different starting value of the random-number generator. Note that exactly the same results are obtained if the same starting value is used for the random-number generator.

The error in the result of a Monte Carlo simulation can be estimated by using the so-called central-limit theorem. This theorem states that the sum of a large number of stochastic variables has a normal distribution. Suppose that each sequence of a simulation gives a value X_i for a certain stochastic variable X. This value can be the total number of interruptions during 20 years, but also the fraction of interruptions with durations between 1 and 3 hours. What we are interested in is the expected value of such a variable. To estimate the expected value we use the average value, which is a standard procedure in statistics. Let $\overline{X}$ be the average of N samples of X_i:

$$\overline{X} = \frac{1}{N}\sum_{i=1}^{N} X_i \tag{2.122}$$

For sufficiently large N, $\overline{X}$ is normally distributed with expected value μ_X and standard deviation $\frac{\sigma_X}{\sqrt{N}}$, where μ_X and σ_X are expected value and standard deviation of X_i. Thus, $\overline{X}$ is an estimate for μ_X (the expected value of X). The error in the estimate is proportional to the standard deviation. Note that obtaining the value of μ_X is the aim of the simulation.

The Stopping Criterion. The fact that the error in a Monte Carlo simulation will never become zero means that we have to accept a certain uncertainty in the result. This is sometimes mentioned as a disadvantage of the Monte Carlo simulation, but also analytical calculations are uncertain, due to the assumptions and approximations made in the model. Where the error in an analytical calculation is often impossible to estimate (unless a better model is used), the uncertainty in the result of a Monte Carlo simulation can be estimated. The outcome of any Monte Carlo simulation will be a stochastic quantity with a normal distribution. For the normal distribution we know that 95% of all values are within two standard deviations of the expected value. We saw above that the standard deviation after N samples is equal to $\frac{\sigma_X}{\sqrt{N}}$. The 95% confidence interval of the estimate is thus,

$$\overline{X} - 2\frac{\sigma_X}{\sqrt{N}} < \mu_X < \overline{X} + 2\frac{\sigma_X}{\sqrt{N}} \tag{2.123}$$

The standard deviation of the stochastic quantity X, σ_X, can be estimated through the following expression:

$$\sigma_X \approx \sqrt{\frac{1}{N-1}\sum_{i=1}^{N} X_i^2 - \left[\frac{1}{N}\sum_{i=1}^{N} X_i\right]^2} \tag{2.124}$$

At regular moments during the simulation, e.g., after every 100 sequences, the error in the estimates may be calculated and compared with the required accuracy. When the required accuracy is reached the simulation can be stopped. Note that to determine the error, one needs not only keep a record of the sum of the X_i values but also of the sum of their squares.

Convergence Tests. Because of the slow convergence process of a Monte Carlo simulation it is hard to recognize a case in which the average no longer converges to the expected value. Such a situation arises, e.g., when the random-number generator has a short cycle length.

Consider again (2.123), which shows that the error $(\overline{X} - \mu_X)$ decreases as $\frac{1}{\sqrt{N}}$. One can conclude from this that the function

$$C = (\overline{X} - \mu_X) \times \sqrt{N} \tag{2.125}$$

neither converges nor diverges. The convergence parameter C has been plotted in Fig. 2.34 for 10 simulations of 10 000 samples each. The underlying simulation is the same as in Figs. 2.32 and 2.33. We see that the plotted quantity remains within a band around zero, thus that the average $\overline{X}$ indeed converges to the expected value μ_X.

In Fig. 2.35 the same convergence parameter is plotted for a simulation which does not converge. The divergence is clearly visible. (From sample 2000 onward, the random-number generator was given a cycle length of 1000 samples.)

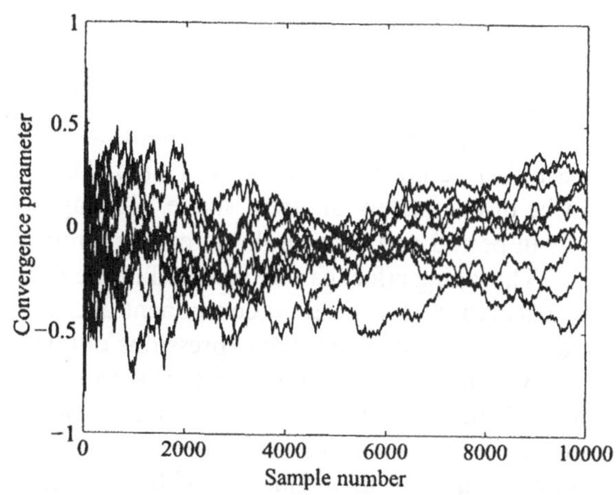

Figure 2.34 Convergence parameter for 10 identical Monte Carlo simulations.

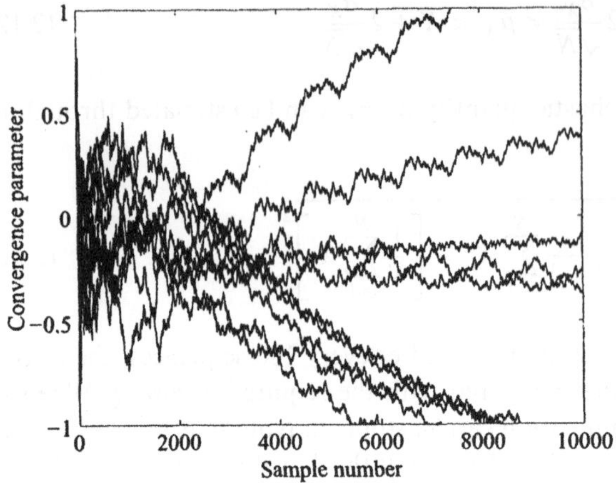

Figure 2.35 Convergence parameter for a non-convergence case.

2.5.6 Aging of Components

In most studies it is assumed that both failure rate and repair rate are constant. The underlying reasons are a lack of data and a lack of evaluation techniques. At the moment, only the Monte Carlo simulation is capable of incorporating nonexponential distributions for nontrivial systems. But despite the lack of application of nonexponential lifetime distributions, it is still worthwhile to have a closer look at the various aging phenomena. Nonexponential repair time distributions are easier to understand, although equally difficult to incorporate in the reliability evaluation.

Two Types of Aging. Aging is used in daily life as the phenomenon that the failure rate of a component increases with its age. Here it will be used in a slightly more general sense: aging is the phenomenon that the failure rate of a component is dependent on:

- the actual age of the component.
- the time since the last repair or maintenance.

To quantify the dependence of the failure rate on the age of the component, the so-called "bathtub curve" is often used. A common way of drawing the bathtub curve is shown in Fig. 2.36. The period between 0 and T_1 is called the wear-in period, after T_2 the wear-out period, and between T_1 and T_2 the useful life or the period of random failures. One should realize that the bathtub curve is only a stylized version of what can be a rather complicated function of time. The actual failure rate as a function of time can be of completely different shape, although it is likely to contain at least an initial wear-in period and an overall increasing failure rate for older components [146].

This aging effect can be included in the reliability evaluation models, by repeating the calculations for different component age. For each age one assumes that all failure rates are constant. From the expressions obtained by using Markov models in Section 2.5.4, we know that the time constant with which the system reacts to changes is of the order of the repair times. For such short time scales we can safely assume the failure rate to be constant. That way one can assess the aging of the system, e.g., the interruption frequency as a function of time. When performing such a study one should

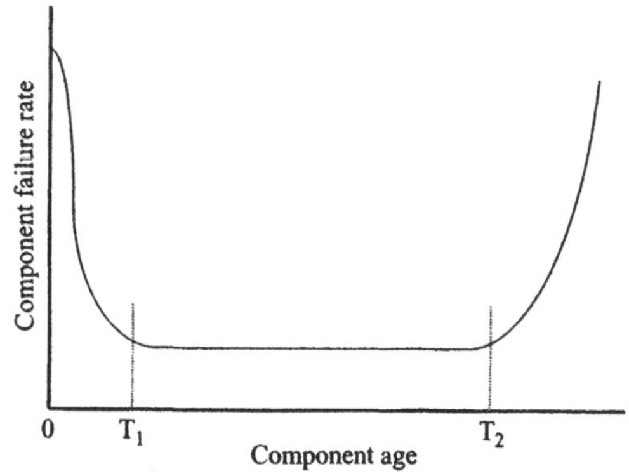

Figure 2.36 Bathtub curve: component failure rate versus age.

realize that also the repair time and the duration of maintenance are likely to increase when the component grows older. The second type of aging, the fact that the failure rate depends on the time elapsed since the last repair or maintenance, is more difficult to consider in a reliability evaluation study. Here it is essential that nonexponential distributions are used for the component lifetimes. Techniques like Markov modeling and network representations can no longer be used. For smaller systems one might use highly mathematical techniques like renewal theory [123], [215]; for larger systems only Monte Carlo simulation remains as a practical tool.

As an example of the second type of aging, assume that the failure rate only depends on the time until maintenance and that maintenance is performed at regular intervals. The failure rate as a function of time is as shown in Fig. 2.37: the failure rate increases until maintenance is performed on the component, at which instant the failure rate drops to its initial value again. The dotted line in Fig. 2.37 represents a kind of average failure rate.

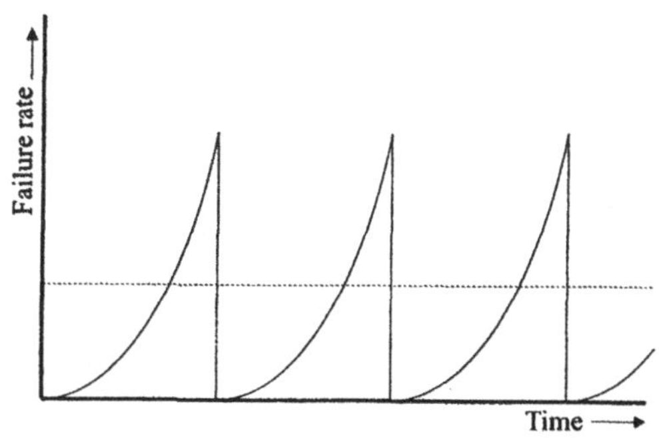

Figure 2.37 Failure rate versus time for regular maintenance intervals.

In Fig. 2.38 the failure rates of two components are plotted (the dashed and the dotted line), plus the average of the two failure rates (the solid line). It is assumed here that maintenance on the second component takes place in between two maintenance instants for the first component. We see that the average of the two failure rates varies less than each of the failure rates. It is easy to imagine that the failure rate of a large number of components becomes constant when maintenance on them is performed at different times.

In reality the failure rate not only depends on the time elapsed since the last maintenance but also on the time elapsed since the last maintenance or repair.

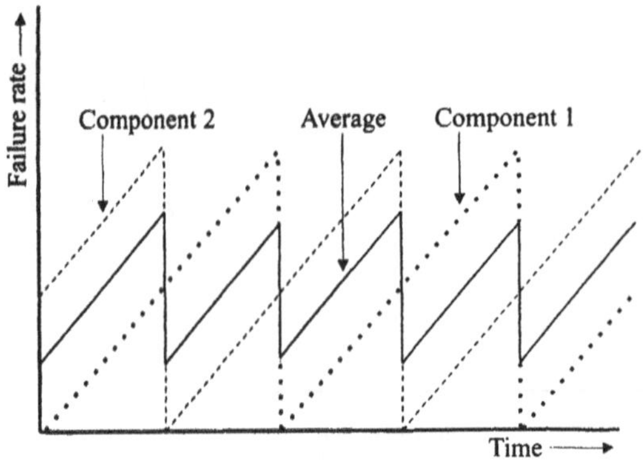

Figure 2.38 Failure rate versus time for two components.

Similar reasonings as given for maintenance can be used for failure, with the difference that the failure instants are less regularly positioned than maintenance instants.

As-Good-As-New or As-Bad-As-Old. In Fig. 2.37 and Fig. 2.38 it was assumed that the failure rate dropped to its original value after maintenance. This model is called maintenance (or repair) "as-good-as-new." The opposite model is called maintenance (or repair) "as-bad-as-old." In the latter case the maintenance or repair has no influence on the failure rate; thus the failure rate just after maintenance is the same as just before. The two models are shown in Fig. 2.39. For repair as-bad-as-old the failure rate depends on the age of the component, for repair as-good-as-new it depends on the time since last repair.

The actual failure rate is normally a combination between as-good-as-new and as-bad-as-old. This can be modeled as the sum of two failure rate, thus two components in series: one being repaired as-good-as-new and the other being repaired as-bad-as-old. The latter one will lead to an average increase in failure rate which leads to the wear-out phase in the bathtub curve.

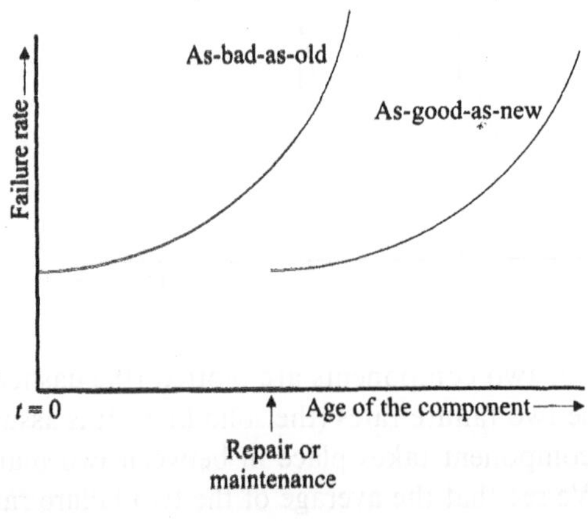

Figure 2.39 Repair as-good-as-new and as-bad-as-old.

Failure Rate Increase due to Maintenance. Something that should also be considered in reliability evaluation is that maintenance and repair can lead to an actual increase in failure rate. The standard example is the screwdriver left inside the switchgear. But also more subtle effects are possible. In a maintenance optimization study one has to take this into account one way or the other. Also during maintenance the chance of an outage of another component is increased: its loading is higher and there is activity in the neighborhood with the associated risk of errors.

Many aspects of aging are extremely difficult to quantify, but should at least be considered in a qualitative way in reliability evaluation studies. A serious difficulty in including component aging is the lack of available data: not just component failure data is needed, but also repair and maintenance records of all the components.

Aging Data. Information on aging of power system components is hard to find. A few examples of good data are given below. There are more publications addressing this problem [107], but the total amount of data is not enough to include aging with sufficient confidence into the reliability evaluation.

- A number of Dutch utilities published "expert opinions" on the age of a component at which the failure rate significantly starts to increase [124]. A group of experts was asked to give their estimation of this age for components operated under "good circumstances," "average circumstances," and "bad circumstances."
- Bathtub curves for transformers are presented in reference [125]. One of the conclusions is that newer generations of transformers have not only a lower overall failure rate but also a longer useful life. The useful life is the period during which the failure rate is more or less constant. Newer production techniques have however not been able to significantly reduce the number of wear-in failures.
- Another interesting study is published in reference [126]. By using purchasing records an assessment is made of the age at which transformers fatally fail, i.e., a failure severe enough for them to be scrapped. It turned out that the failure rate stayed constant, at about 0.01 per year, for the first 12 years. After that, the failure rate increased until 1 per year at an age of 29 years.
- A bathtub curve for circuit breakers is presented in [127], based on the observation of a large number of breakers. The failure rate decreases from 0.2 for age zero through 0.05 for 8 years after which it rises to 0.15 for 10-year-old breakers.
- In reference [128] the fail behavior of circuit breakers is studied by dividing the cause of failures into three categories:

  — initial failures.

  — random failures.

  — wear-out failures.

  By plotting the failure rate as a function of age for each category, it is shown that the failure rate of initial failures decreases, that the failure rate of random failures stays constant, and that the failure rate of wear-out failures increases with time.

2.6 COSTS OF INTERRUPTIONS

To consider interruptions of the supply in the design and operation of power systems, the inconvenience due to interruptions needs to be quantified one way or the other. The term inconvenience is rather vague and broad. Any serious quantification requires a

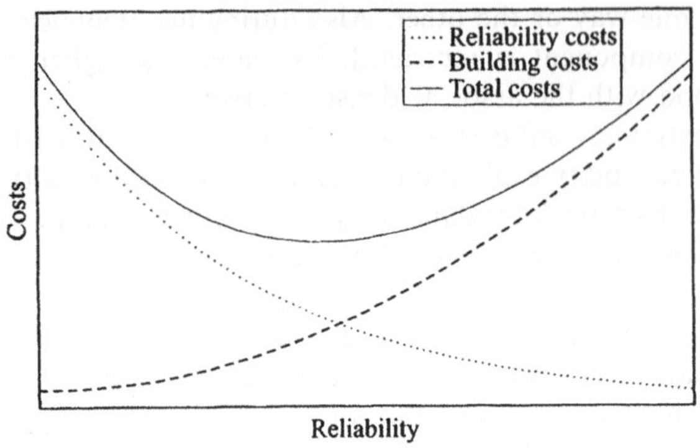

Figure 2.40 Costs versus reliability: costs of building and operation (dashed curve), costs of supply interruptions (dotted curve), and total costs (solid curve).

translation of all inconvenience into amount of money. In the remainder of this section we will consider costs of interruptions in dollars, but any other currency can be used of course.

Many publications on costs of interruption show a graph with costs against reliability. Such a curve is reproduced in Fig. 2.40. The idea behind this curve is that a more reliable system is more expensive to build and operate, but the costs of interruption (either over the lifetime of the system, or per year) are less. The total costs will show a minimum, which corresponds to the optimal reliability. Even if we assume that both cost functions can be determined exactly, the curve still has some serious limitations. Figure 2.40 should only be used as a qualitative demonstration of the trade-off between costs and reliability.

- Additional investment does not always give a more reliable system: an increase in the number of components could even decrease the reliabiity.
- Reliability is not a single-dimensional quantity. Both interruption frequency and duration of interruption influence the interruption costs.
- There is no sliding scale of reliability and costs. The system designer can choose between a limited number of design options; sometimes there are just two options available. The choice becomes simply a comparison of advantages and disadvantages of the two options.
- The two cost terms cannot simply be added. One term (building and operational costs) has a small uncertainty, the other term (interruption costs) has a large uncertainty due to the uncertainty in the actual number and duration of interruptions. A more detailed risk analysis is needed than just adding the expected costs.

The cost of an interruption consists of a number of terms. Each term has its own difficulty in being assessed. Again simply adding the terms to obtain the total costs of an interruption is not the right way, but due to lack of alternatives it is often the only feasible option.

1. **Direct costs.** These are the costs which are directly attributable to the interruption. The standard example for domestic customers is the loss of food in the refrigerator. For industrial customers the direct costs consist, among others, of lost raw material, lost production, and salary costs during the non-productive period. For commercial customers the direct costs are the

loss of profit and the salary costs during the non-productive period. When assessing the direct costs one has to be watchful of double-counting. One should at first subtract the savings made during the interruption. The obvious savings are in the electricity costs, but for industrial processes there is also a saving in use of raw material. An example of double-counting is adding the lost sales and the salary costs (as the price of the product already includes the salary costs). Also to be subtracted from the costs of interruption is the lost production which can be recovered later. Some plants only run part of the time. Extra salary during overtime needed to recover lost production should be added to the direct costs.

2. **Indirect costs.** The indirect costs are much harder to evaluate, and in many cases not simply to express in amount of money. A company can lose future orders when an interruption leads to delay in delivering a product. A domestic customer can decide to take an insurance against loss of freezer contents. A commercial customer might install a battery backup. A large industrial customer could even decide to move a plant to an area with less supply interruptions. The main problem with this cost term is that it cannot be attributed to a single interruption, but to the (real or perceived) quality of supply as a whole.
3. **Non-material inconvenience.** Some inconvenience cannot be expressed in money. Not being able to listen to the radio for 2 hours can be a serious inconvenience, but the actual costs are zero. In industrial and commercial environments, the non-material inconvenience can also be big without contributing to the direct or indirect costs. A way of quantifying these costs is to look at the amount of money a customer is willing to pay for not having this interruption.

To evaluate the costs of supply interruptions, different methods have been proposed. For large industrial and commercial customers an inventory of all the direct and indirect costs can be made, and this can then be used in the system design and operation. Even for small customers such a study could be made, e.g., to decide about the purchase of equipment to mitigate interruptions. However, for small and domestic customers it is often the non-material inconvenience which has a larger influence on the decision than the direct or indirect costs. For a group of customers, such an individual assessment is no longer possible. The only generally accepted method is the large survey among customers. Customers get asked a number of questions. Based on the answers the average costs of interruption are estimated. These results are typically the ones used by utilities in decision making. When comparing the results of different surveys, it is important to realize that they not all ask the same questions. Some surveys ask a very specific question: "What are the costs of an interruption of 2 hours on a Monday afternoon in January?" Other surveys use more indirect questioning: "What is a reasonable compensation for an interruption" or "What would you be willing to pay to reduce the interruption frequency from 4 per year to 3 per year?" Different questions obviously lead to different estimates for the costs of interruption.

To quantify the costs of an interruption, again different methods are in use. Some values can be easily calculated into each other, with some values a certain amount of care is needed. Worse is that it is not always clear from the context which method is actually used.

- **Costs per interruption.** For an individual customer the costs of an interruption of duration d can be expressed in dollars. There is no confusion possible about this. For simplicity, we neglect the fact that the costs not only depend on the duration but on many other factors as well. The costs per interruption can be determined through an inventory of all direct and indirect costs.
- **Costs per interrupted kW.** Let $C_i(d)$ be the costs of an interruption of duration d for customer i, and L_i the load of this customer when there would not have been an interruption. The costs per interrupted kW are defined as

$$\frac{C_i(d)}{L_i} \tag{2.126}$$

and are expressed in \$/kW. For a group of customers experiencing the same interruption, the costs per interrupted kW are defined as the ratio of the total costs of the interruption and the total load in case there would not have been an interruption:

$$\frac{\sum_i C_i(d)}{\sum_i L_i} \tag{2.127}$$

- **Costs per kWh not delivered.** In many studies the assumption is made that the cost of an interruption is proportional to the duration of the interruption. The cost per kWh not delivered is defined as

$$\frac{C_i(d)}{dL_i} \tag{2.128}$$

and is constant under the assumption. The cost per kWh is expressed in \$/kWh. For a group of customers the cost per kWh not delivered is defined as

$$\frac{\sum_i C_i(d)}{d\sum_i L_i} \tag{2.129}$$

Some utilities obtain an average cost per kWh not delivered for all their customers. This value is assumed constant and used as a reference value in system operation and design. The term "value of lost load" is sometimes used for the cost per kWh not delivered averaged over all customers.
- **Costs of interruption rated to the peak load.** A problem in surveys is that the actual load of individual customers in case there would not have been an interruption is often not known. One should realize that surveys consider hypothetical interruptions, rarely actual ones. For industrial and commercial customers the peak load is much easier to obtain, as it is typically part of the supply contract. The cost of an interruption can be divided by the peak load, to get a value in \$/kW. Some care is needed when interpreting this value, as it is not the same as the cost per kW interrupted (also in \$/kW). For planning purposes the cost of interruption rated to the peak load can still be a useful value. The design of a system is based for a large part on peak load, so that rating the cost to the peak load gives a direct link with the design.
- **Costs per interruption rated to the annual consumption.** For domestic customers it is easier to obtain the annual consumption than the peak load. Rating the

cost of an interruption to the annual consumption gives a value in $/kWh. Note that this has no relation to the costs per kWh not delivered.

Some of the results of a Swedish survey after costs of interruptions [200] are given in Figs. 2.41 and 2.42. The survey was conducted among 4000 customers in 1993 and resulted in interruption costs per kW of peak load for interruption duration of 2 minutes, 1 hour, 4 hours, and 8 hours. Figure 2.41 gives the costs for a forced inter-

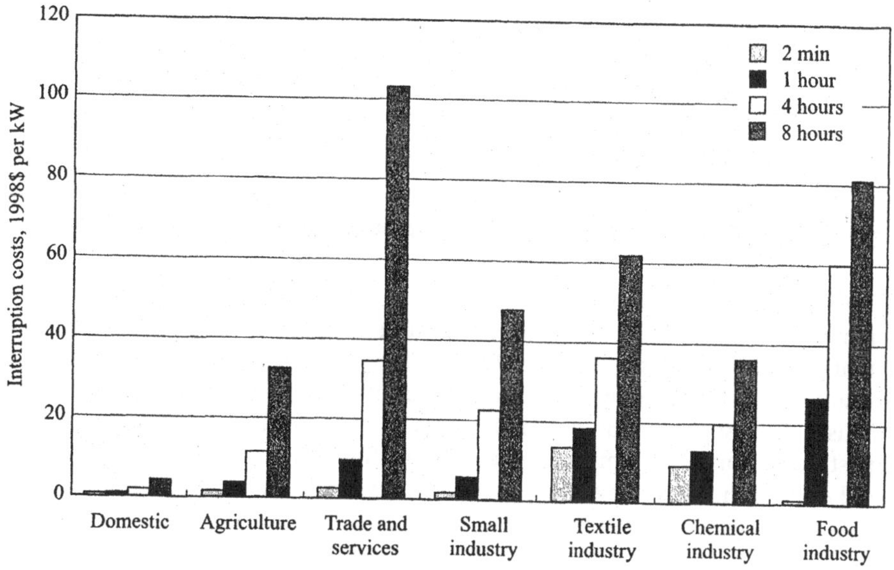

Figure 2.41 Interruption costs in $/kW for different customers, for forced interruptions. Results from a Swedish study in 1993 [200].

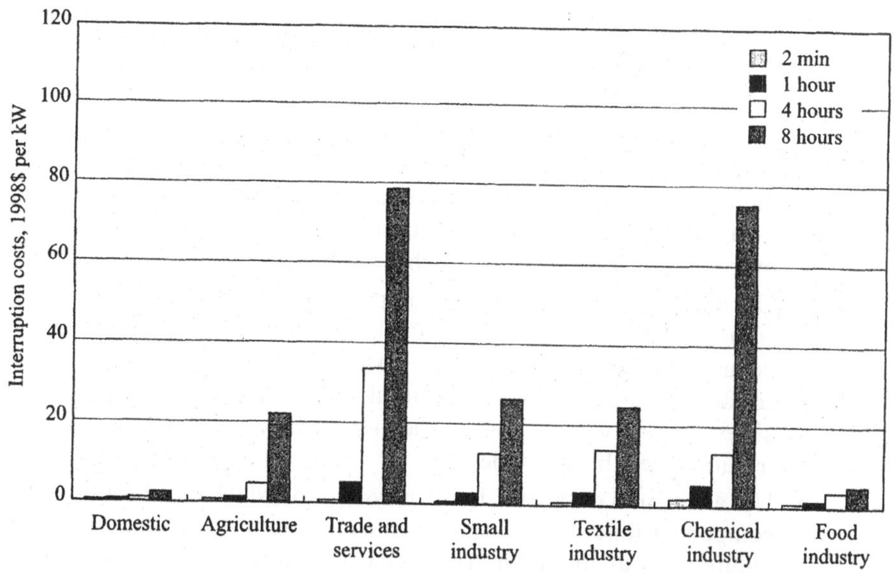

Figure 2.42 Interruption costs in $/kW for different customers, for scheduled interruptions. Results from a Swedish study in 1993 [200].

ruption, i.e., in case the customer receives no pre-warning of the interruption. Figure 2.42 relates to scheduled interruptions where the customer receives sufficient pre-warning. An exchange rate of 7.32 Swedish crowns per U.S. dollar has been used and an inflation rate of 2.5% per year, to obtain the costs in 1998 dollars.

The values indicated are averages over a number of customers. Surveys have shown that the range between different customers is very large, even within one type of industry. Ranges of interruption cost within one type of industry are given by Skof [147]. For a 1-minute interruption the cost for automobile factories varies between 0.001 \$/kW and 6 \$/kW. For a 1-hour interruption the range is from 0.3 to 40 \$/kW. Thus, an industry average should be treated with care when assessing the cost of interruption for a specific industrial customer. Where possible, it is recommended to use customer-specific data instead of industry national averages. Several other publications give survey results and results of other ways to estimate the interruption costs; an admittedly incomplete list is [21], [129], [130], [131], [132], [216].

2.7 COMPARISON OF OBSERVATION AND RELIABILITY EVALUATION

Despite all the reliability analysis tools available, simple past-performance records remain the main source of information on system performance. This does not imply that reliability analysis has no value. To the contrary, analysis techniques can obtain results much faster and with a higher degree of accuracy than past performance records. This holds especially for individual sites. For the evaluation of operational reserve, past-performance is simply not available. Stochastic prediction techniques are the only option here. However, comparison between stochastic prediction techniques and past performance measures is a highly undervalued area. Very little work has been done on this often with the justification that it is not possible.

Some kind of verification of stochastic prediction techniques remains needed, especially as many engineers remain, rightly or wrongly, skeptical about the outcome of reliability evaluations. The emphasis on past-performance records is, in the author's view, also determined by the skepticism toward stochastic prediction techniques. A number of ways of comparing observations and the results of reliability evaluation are given in the following list:

- Apply stochastic prediction techniques to a system that has not changed too much over a longer period, and for which data are available on the number and durations of supply interruptions over this period. As the transmission networks in most industrialized countries have remained more or less the same over the last 10 years or so, such a verification technique could be used here.
- Use a large number of observation points, e.g., all urban distribution networks within one utility. Some further selection might be needed to get a homogeneous group of systems. Apply stochastic prediction techniques to a typical configuration and compare the results with the average observation results of all existing networks. This verification technique is suitable for level III (distribution) reliability studies.
- Use a common data set. Choose a system for which interruption data as well as component failure data are known over a number of years. Use the observed failure rates as input for the stochastic prediction, thus eliminating the data uncertainty. Any differences between observed and predicted number of interruptions can be contributed to model limitations.

- Perform detailed analysis of the underlying events of interruptions. Assess whether these events or combinations of events are part of the stochastic prediction model. This technique might be somewhat trivial for distribution systems, but it appears especially useful for transmission and generation systems where only multiple events lead to interruptions.

2.8 EXAMPLE CALCULATIONS

2.8.1 A Primary Selective Supply

Consider an industrial customer with a so-called primary selective supply, as shown in Fig. 2.43. Primary selective supplies and other ways of improving the reliability are discussed in detail in Chapter 7.

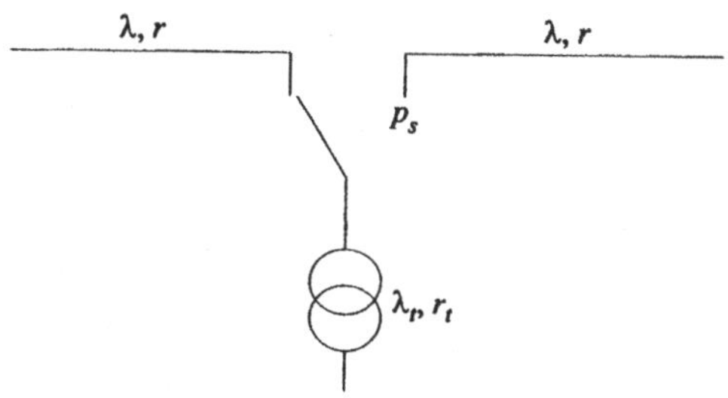

Figure 2.43 Example of reliability calculation: primary selective supply.

For this example we use the following component data:

- $\lambda = 5\,\text{year}^{-1}$, failure rate of each of the two public supplies.
- $r = 0.00025$ years = 2 hours and 11 minutes, average repair time of the public supply.
- $\lambda_t = 0.02\ \text{year}^{-1}$, transformer failure rate.
- $r_t = 0.0114$ year = 100 hours, transformer repair time.
- $p_s = 3\%$, transfer switch failure probability.

The frequency of interruptions due to overlapping outages is obtained from the equation for the failure rate of two parallel components (2.39):

$$\lambda_p = 2r\lambda^2 = 2 \times 0.00025 \times 5^2 = 0.0125 \text{ interruptions per year} \tag{2.130}$$

The average duration of an interruption is the equivalent repair time of the parallel connection as obtained from (2.41):

$$r_p = \frac{r}{2} = 0.000125 \text{ years} = 1.1 \text{ hours} \tag{2.131}$$

In other words, the second of two overlapping outages starts on average in the middle of the first outage. From the interruption rate and the interruption duration, we can obtain the unavailability due to overlapping outages:

$$Q_p = \lambda_p r_p = 1.56 \times 10^{-6} = 0.014 \text{ hours per year} \tag{2.132}$$

In a primary selective supply, a transformer outage can also lead to an interruption. The transformer outage rate (0.02 year^{-1}) is of the same order of magnitude as the outage rate due to overlapping outages in the supply. The duration of transformer outages is much longer. The unavailability due to transformer outages is

$$Q_t = \lambda_t r_t = 2.28 \times 10^{-4} = 2 \text{ hours per year} \tag{2.133}$$

When very long interruptions are a concern, a second transformer should be placed in parallel with the existing one and the switching should be performed on secondary side. This leads to the so-called secondary selective supply. The interruption frequency due to overlapping transformer outages is very small:

$$\lambda_{tp} = 2r_t\lambda_t^2 = 9.1 \times 10^{-6} \text{ interruptions per year} \tag{2.134}$$

Apart from overlapping supply outages and transformer failures, interruptions can be due to a failure of both supplies at the same time and due to a failure of the transfer switch. Failure of both supplies at the same time is mainly due to outages at a higher voltage level, either medium voltage distribution or transmission, depending on the supply configuration. Interruption rates associated with this vary significantly, with a typical range between 0.05 and 0.5 interruptions per year. A separate study is needed for each supply configuration, or alternatively information needs to be obtained from the utility. The probability that the transfer switch fails was given as $p_s = 3\%$, which means that the switch will not transfer the load correctly in 3% of the cases for which it is supposed to do so. The frequency of cases in which the transfer switch is supposed to transfer the load is equal to the outage rate of one of the supplies. The interruption frequency due to transfer switch failure is thus,

$$p_s \times \lambda_s = 0.15 \text{ per year} \tag{2.135}$$

We see that the transfer switch is apparently a weak part in the supply. To obtain a reliable supply it is thus essential to choose a reliable transfer switch. Also maintenance on the transfer switch plays an important role.

2.8.2 Adverse Weather

Consider again the primary selective supply in Fig. 2.43. We consider the fact that the failure rate is not constant during the year. Most overhead line outages are due to adverse weather like snow, storm, or lightning. Overhead line outages are much more likely during adverse weather than during normal weather. The failure rate as a function of time will look like in Fig. 2.44: the failure rate is low most of the year, but high during a number of short periods of adverse weather.

The adverse weather periods are not fixed but stochastic in time as well. A Monte Carlo simulation would be an appropriate tool, if sufficient data and model details were available. To enable a simplified analysis, we consider a two-state model, as shown in Fig. 2.45. The failure rate during adverse weather is λ_1 and during normal weather λ_2. The adverse weather is present during a fraction T_1 of the time and normal weather during a fraction T_2. The average failure rate λ is obtained from

$$\lambda = \lambda_1 T_1 + \lambda_2 T_2 \tag{2.136}$$

For both states an interruption frequency can be determined, after which the annual interruption frequency is the average of these two. Suppose as an example that 75% of supply outages are due to adverse weather which takes place during 100 hours per year. The failure rates during adverse and normal weather are, respectively: $\lambda_1 = 329$ per year

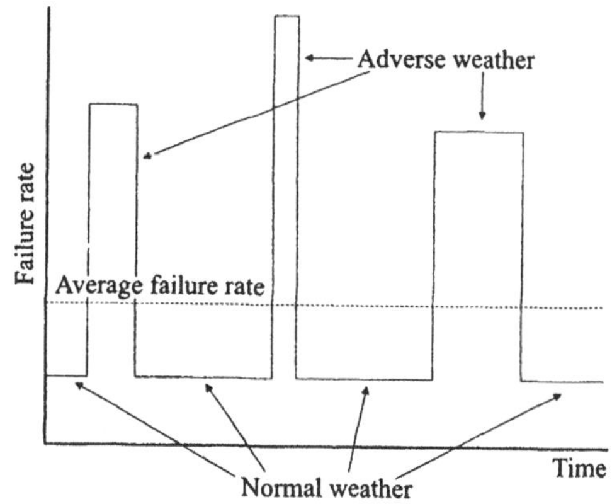

Figure 2.44 Failure rate as a function of time—normal and adverse weather.

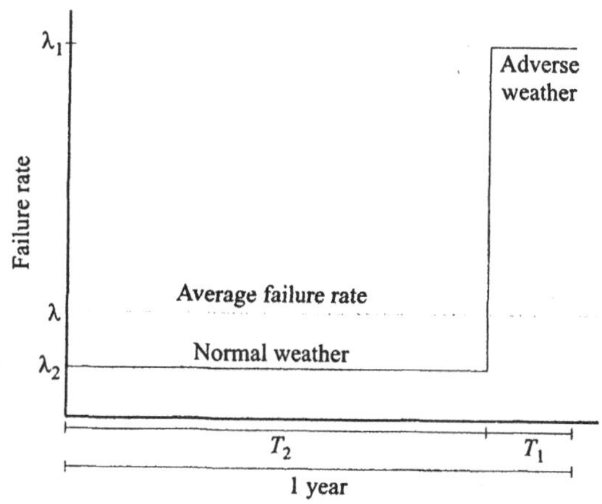

Figure 2.45 Two-state model with normal and adverse weather.

and $\lambda_2 = 1.25$ per year. The average failure rate is the same as in the previous example: $\lambda = 5$ per year. The repair time is also likely to be affected by the adverse weather. We use the following repair times: $r_1 = 2.59$ hour (during adverse weather) and $r_2 = 1$ hour (during normal weather) leading to the same average repair time as before ($\bar{r}=2$ hours 11 min).

$$\bar{r} = \frac{\lambda_1 T_1 r_1 + \lambda_2 T_2 r_2}{\lambda_1 T_1 + \lambda_2 T_2} \tag{2.137}$$

The normal weather interruption rate is found by using the same expression for the parallel connection as before, with the exception that failure rate and repair time during normal weather are used instead of the average values.

$$\lambda_{p2} = 2r_2\lambda_2^2 = 0.0003566 \text{ per year} \tag{2.138}$$

Normal weather is present during a fraction $T_2 = \frac{8660}{8760}$ of the year, which gives for the expected number of interruptions per year due to normal weather:

$$T_2\lambda_{p2} = 0.0003525 \text{ interruptions per year} \tag{2.139}$$

The adverse weather interruption rate is

$$\lambda_{p1} = 2r_1\lambda_1^2 = 64 \text{ per year} \tag{2.140}$$

This is a very high value, but normal weather is only present during a fraction $T_1 = \frac{100}{8760} = 0.0114$ of the year. The contribution of adverse weather to the annual interruption frequency is

$$T_1 \lambda_{p1} = 0.73 \text{ interruptions per year} \tag{2.141}$$

The annual interruption frequency is therefore very much affected by adverse weather. Note the large difference with the interruption frequency found before by assuming a constant failure rate (0.0125 per year). It is clear that the influence of adverse weather cannot be neglected in reliability evaluation studies with parallel connections. For series connections the interruption rate is the sum of the component failure rates and the average interruption rate is the sum of the average component failure rates. Only for parallel connections do we need to explicitly consider adverse weather.

2.8.3 Parallel Components

Consider a system consisting of n identical components in parallel. Each component has an outage rate λ and an average repair time r. The interruption rate of the system can be calculated from expression (2.44), resulting in

$$\lambda_1 = n\frac{(\lambda r)^n}{r} \tag{2.142}$$

Apart from interruptions due to these overlapping outages, the system can be interrupted when a failure in one component leads to the outage of all components. One can think of failure of the protection, tripping of equipment on the voltage sag or on another transient, or transient instability. Suppose that there is a probability α that the underlying failure of a component outage leads to a system interruption. For an n-component system with a component outage rate λ, this gives an additional interruption frequency of

$$\lambda_2 = \alpha n \lambda \tag{2.143}$$

The total number of interruptions is given by the following expression:

$$\lambda_{tot} = \lambda_1 + \lambda_2 = \alpha n \lambda + n\frac{(\lambda r)^n}{r} \tag{2.144}$$

For most components $\lambda r \ll 1$ so that the second term reduces very fast for increasing n, while the first term increases linearly with the number of parallel components. The first term will rather quickly start to dominate after which an increase of the number of parallel components only decreases the reliability. Assume the following component data: $\lambda = 1$ per year, $r = 0.001$ year, $\alpha = 1\%$. The resulting interruption rates are given in Table 2.12. We note the somewhat surprising result that three components in parallel is less reliable than two components in parallel.

TABLE 2.12 Influence of Number of Parallel Components on Interruption Rate

n	Individual	Overlapping Outages	Total Interruption Frequency
1	0.01 per year	1 per year	1.01 per year
2	0.02 per year	2×10^{-3} per year	0.022 per year
3	0.03 per year	3×10^{-6} per year	0.030 per year

To justify a three-component model, the interruption frequency for $n = 3$ needs to be less than for $n = 2$, thus

$$3\alpha\lambda + 3\frac{(\lambda r)^3}{r} < 2\alpha\lambda + 2\frac{(\lambda r)^2}{r} \tag{2.145}$$

resulting in the following upper bound for the probability that a component outage leads directly to a system outage

$$\alpha < 2\lambda r - 3(\lambda r)^2 \tag{2.146}$$

For the previous example this gives $\alpha < 0.002$. Thus, a three-component system is only justified if the protection of the component is very reliable, the risk of transient instability is low, etc.

2.8.4 Two-Component Model with Aging and Maintenance

To assess the effect of aging and maintenance on a parallel connection, we consider two components with a time-dependent outage rate:

$$\lambda_1 = \lambda_2 = \lambda = 0.01t^3 \tag{2.147}$$

with t the time since last maintenance. For maintenance performed every 4 years, the average outage rate is

$$\overline{\lambda} = \frac{1}{4}\int_0^4 \lambda(t)dt = 0.16 \text{ outages per year} \tag{2.148}$$

We will calculate the interruption frequency of the parallel connection of these two components. We assume that both repair time r and maintenance duration m are on average 100 hours. For each of the models to be discussed we will calculate both the interruption rate due to overlapping outages (λ_{oo}), and the interruption rate due to outage during maintenance (λ_{om}).

Average Failure Rate—Overlapping Outages. Using the average failure rate for the two components, we can calculate the interruption rate of the parallel connection due to overlapping outages:

$$\lambda_{oo} = \overline{\lambda}^2 2r = 5.84 \times 10^{-4} \text{ interruptions per year} \tag{2.149}$$

The expected number of interruptions due to overlapping outages during a 4-year period is equal to 2.34×10^{-3}.

Average Failure Rate—Outage During Maintenance. When one component is being maintained, an outage of the other component will lead to an interruption. One of the two components is in scheduled outage during a period $2m$ every 4 years. An outage during this period leads to an interruption. The expected number of outages during maintenance during a 4-year period is thus,

$$4\lambda_{om} = 2m\overline{\lambda} = 3.65 \times 10^{-3} \text{ outages per 4 years} \tag{2.150}$$

Maintenance Every Four Years—Overlapping Outages. When the failure rate of the components is time dependent, it is still possible to determine the interruption rate due to overlapping outages. The only difference with the previous case is that the outage rates are time dependent and therefore the interruption rate as well:

$$\lambda_{oo}(t) = \lambda(t)^2 2r = 2.28 \times 10^{-6} t^6 \text{ interruptions per year} \tag{2.151}$$

The average interruption frequency is 1.334×10^{-3} interruptions per year, and the maximum interruption frequency (just before maintenance) is 9.34×10^{-3} interruptions per year. The expected number of interruptions due to overlapping outages, during a 4-year period, is equal to 5.34×10^{-3}.

Maintenance Every Four Years—Outage During Maintenance. Normally maintenance will not be performed on both components at the same time because that would lead to an interruption. Maintenance is performed first on one component and then on the other. During maintenance on the second component the first one is as-good-as-new, has a failure rate close to zero, and the risk of an outage can be neglected. The situation is completely different for maintenance on the first component, because the other component has its highest failure rate. The probability that the second component will fail while the first one is being maintained is

$$4\lambda_{om} = m\lambda(4) = \frac{100}{8760} \times 0.64 = 7.31 \times 10^{-3} \text{ interruptions per maintenance interval} \tag{2.152}$$

Maintenance Every Two Years—Overlapping Outages. Above it was assumed that maintenance on the two components is done immediately after each other. An alternative is to spread the maintenance over time; that is, by performing maintenance every 2 years and each time only on one component. Suppose that maintenance has been performed on component 1 at $t = 0$ and on component 2 at $t = -2$. The component failure rates become

$$\lambda_1(t) = 0.01t^3 \tag{2.153}$$

$$\lambda_2(t) = 0.01(t+2)^3 \tag{2.154}$$

The interruption rate due to overlapping outages is

$$\lambda_{oo}(t) = \lambda_1(t)\lambda_2(t)2r = 2.28 \times 10^{-6} t^3 (t+2)^3 \text{ interruptions per year} \tag{2.155}$$

Note that this expression is valid between $t = 0$ and $t = 2$ after which component 1 and component 2 switch roles. The average interruption rate over this 2-year period is

$$\overline{\lambda}_p = \frac{1}{2}\int_0^2 \lambda_p(t)dt = 2.18 \times 10^{-4} \text{ interruptions per year} \tag{2.156}$$

The expected number of interruptions due to overlapping outages, during a 4-year period, is equal to 0.87×10^{-3}.

Maintenance Every Two Years—Outage During Maintenance. Failure during maintenance can happen for each of the two components. When maintenance is performed on one component, the other component has an "age" of 2 years; thus, its failure rate is 0.08 per year. The expected number of outages of the parallel component during maintenance on the other component is equal to

$$m\lambda(2) = \frac{100}{8760} \times 0.08 = 0.913 \times 10^{-3} \text{ interruptions per maintenance} \tag{2.157}$$

Such a situation occurs twice during a 4-year period, so that the expected number of interruptions due to outage during maintenance, over a 4-year period, is 1.83×10^{-3}.

Overview. The results of the various models are summarized in Table 2.13. We see that the aging/maintenance model influences the interruption frequency over almost a factor of 10. Also note that the number of interruptions due to outages during maintenance is, for each of the models, higher than the number of interruptions due to overlapping outages. Further optimization studies would be needed to assess if the total interruption rate can be brought down. An obvious choice is to reduce the duration of maintenance, as the number of interruptions due to outages during maintenance is directly proportional to the duration of maintenance. One should take a certain care with that, as the quality of the maintenance might also become less. In the above calculations it has been assumed that the outage rate is brought back to zero after maintenance, and that the outage rate of the parallel component is not increased during the maintenance.

Without any optimization study, it is obvious, however, that maintenance should be scheduled as much as possible during periods with low interruption costs.

TABLE 2.13 Influence of Aging and Maintenance Model on Interruption Rate

	Constant Failure Rate	Maintenance Every 4 Years	Maintenance Every 2 Years
Interruptions due to overlapping outages	2.34×10^{-3} per 4 years	5.34×10^{-3} per 4 years	0.85×10^{-3} per 4 years
Interruptions due to failure during maintenance	3.65×10^{-3} per 4 years	7.31×10^{-3} per 4 years	1.83×10^{-3} per 4 years

3

Short Interruptions

3.1 INTRODUCTION

A short interruption has the same causes as a long interruption: fault clearing by the protection, incorrect protection intervention, etc. When the supply is restored automatically, the resulting event is called a short interruption. Long interruptions and very long interruptions result when the supply is restored manually. Automatic restoration can take place by reclosing the circuit breaker which cleared the fault or by switching to a healthy supply. The former takes place mainly in overhead distribution networks, the latter is a typical solution in industrial systems.

Short interruptions in the public supply are due to attempts by the utility to limit the duration of interruptions. We saw already in Section 2.3 that the duration of an interruption is an important aspect of distribution and transmission system design. By using automatic reclosing the duration of an interruption can be brought back from typically about 1 hour, to typically less than 1 minute. For many years interruptions shorter than several minutes were not considered as a cause of concern to most customers. Recently this has changed: more and more equipment is sensitive to very short duration events, and more and more customers (domestic as well as industrial) view short interruptions as a serious imperfection of the supply. This is part of the trends mentioned in Section 1.1 for the increased interest in power quality in general. Short interruptions also occur in industrial power systems due to the operation of automatic transfer switches. We discuss this in Chapter 7.

3.2 TERMINOLOGY

There is some serious confusion about terminology on interruptions of different duration. Terms like short interruptions, momentary interruptions, temporary interruptions, instantaneous interruptions, and transient outages are all used with more or less the same meaning. The definition of short interruptions used for this chapter is not based on duration but on the method of restoring the supply. This chapter (short interruptions) discusses automatic restoration, where Chapter 2 (long interruptions) discusses manual restoration.

Below, an overview is given of the various terms and definitions used in the European standard EN 50160 and in three IEEE standards. The definitions used in EN 50160 are identical to the IEC definitions.

- **EN 50160**
  - **Long interruption**: longer than three minutes.
  - **Short interruption**: up to three minutes.
- **IEEE Std.1159-1995**

  This standard is considered by many as providing the basic power quality definitions. It distinguishes between momentary, sustained, and temporary interruptions. Note the overlap between sustained and temporary interruptions.
  - **Momentary interruption**: between 0.5 cycles and 3 seconds.
  - **Sustained interruption**: longer than 3 seconds.
  - **Temporary interruption**: between 3 seconds and 1 minute.
- **IEEE Std.1250-1995**

  This standard was published at about the same time as IEEE Std.1159-1995, but it uses somewhat different definitions. The difference is especially striking for interruptions.
  - **Instantaneous interruption**: between 0.5 and 30 cycles (half a second).
  - **Momentary interruption**: between 30 cycles and 2 seconds.
  - **Temporary interruption**: between 2 seconds to 2 minutes.
  - **Sustained interruption**: longer than 2 minutes.
- **IEEE Std.859-1987**

  This somewhat older standard document gives definitions for terms related to power system reliability. A distinction is made between different types of outages based on the duration of the outage. This standard does not give specific time ranges but uses the restoration method to distinguish the different types. Although outages and interruptions are different phenomena (see Section 2.1.3) they are related closely enough to compare the terminology.
  - **Transient outages** are restored automatically.
  - **Temporary outages** are restored by manual switching.
  - **Permanent outages** are restored through repair or replacement.

3.3 ORIGIN OF SHORT INTERRUPTIONS

3.3.1 Basic Principle

Figure 3.1 shows an example of an overhead distribution network. Each feeder consists of a main feeder and a number of lateral conductors. Most faults on overhead lines are transient: they require operation of the protection, but do not cause permanent damage to the system. A typical cause of a transient fault is a lightning stroke to an overhead line. The lightning stroke injects a very high current into the line causing a very fast rising voltage. The lightning current varies between 2 and 200 kA in peak value. The typical lightning current has a peak value of $I_{peak} = 20\,\text{kA}$ which is reached within 1 μs after its initiation. If the wave impedance Z_{wave} of the line is 200 Ω, the voltage can theoretically reach a value of

$$V_{peak} = \frac{Z_{wave}}{2} I_{peak} = 100\,\Omega \times 20\,\text{kA} = 2\,\text{MV} \tag{3.1}$$

The voltage will never reach such a value in reality (with the possible exception of transmission systems with operating voltages of 400 kV or higher), because a flashover to ground or between two phases will result long before the voltage reaches such a high value. The result is an arcing fault between one phase and ground or between two or more phases with or without ground. Soon after the protection removes the faulted line from the system, the arc disappears. Automatic reclosing will restore the supply without any permanent damage to the system.

Also, smaller objects causing a temporary path to ground will only cause a transient short circuit. The object (e.g., a small branch fallen from a tree) will either drop to the ground or evaporate due to the high current during the fault, leaving only an arc which disappears again soon after the protection intervenes. The duration of an interruption due to a transient fault can thus be enormously reduced by automatically restoring the supply after an interruption. In case of a fault somewhere on the feeder, the circuit breaker opens instantaneously and closes again after a "reclosing interval" or "dead time" ranging from less than one second up to several minutes. There is of course a risk that the fault was not a transient one but permanent. In that case the protection will again notice a large overcurrent after reclosure leading to a second trip signal. Often the recloser gives the fault a second chance at extinguishing, by means of a longer tripping time and/or a longer reclosing interval.

3.3.2 Fuse Saving

A practice associated with reclosing and short interruptions is "fuse saving." In Fig. 3.1 the laterals away from the main feeder are protected by means of expulsion fuses. These are slow fuses which will not trigger when a transient fault is cleared by the main breaker/recloser. Thus, a transient fault will be cleared by the recloser and the supply will be automatically restored.

A permanent fault can also be cleared by the main breaker, but that would lead to a long interruption for all customers fed from this feeder. Instead, a permanent fault is

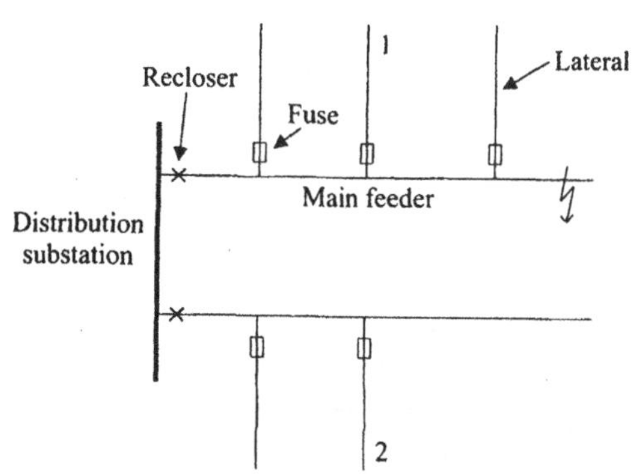

Figure 3.1 Overhead distribution network with fuses and reclosers.

cleared by an expulsion fuse. To achieve this, the recloser has two settings: an instantaneous trip and a delayed trip. The protection coordination should be such that the instantaneous trip is faster than the expulsion fuse and the delayed trip slower, for all possible fault currents.

From the above description we can conclude that the following trade-off has been made: a short interruption for all customers (fed from this feeder) instead of a long interruption for some customers. The alternative would be more long interruptions; however, not every short interruption would become a long interruption.

3.3.3 Voltage Magnitude Events due to Reclosing

The combination of reclosing and fuse saving, as decribed above, leads to different voltage magnitude events for different customers. Figure 3.2 shows the events due to one reclosing action as experienced by a customer on the faulted feeder (indicated by "1" in Fig. 3.1) and by a customer on another feeder fed from the same substation bus (indicated by "2"). In Fig. 3.2, A is the fault-clearing time and B the reclosing interval. The customer on the faulted feeder (solid line) will experience a decrease in voltage during the fault, similar in cause and magnitude to a voltage sag. The difference between the two customers is in the effect of the fault clearing. For the customer on the nonfaulted feeder, the voltage recovers to its pre-event value. The customer will only experience a voltage sag. For the customer on the faulted feeder, the voltage drops to zero.

The customer on a neighboring feeder (dashed line) will see a voltage sag with a duration equal to the fault-clearing time. The moment the recloser opens, the voltage recovers. If the fault is still present at the first reclosure, the customer on the nonfaulted feeder will experience a second voltage sag. Customers on the faulted feeder will experience a second short interruption or a long interruption.

Figure 3.3 [11] shows an actual recording of a short interruption. The top figure corresponds to the dashed line in Fig. 3.2 (customer on a nonfaulted feeder). The bottom figure is for a customer on the faulted feeder (solid line in Fig. 3.2). The fault-clearing time is about two cycles, the dead time about two seconds. The first reclosure is not successful, the second one is. The top figure shows a voltage sag to about 75% of two-cycle duration, the bottom figure a voltage reduction to 50% for two cycles followed by zero voltage for about two seconds.

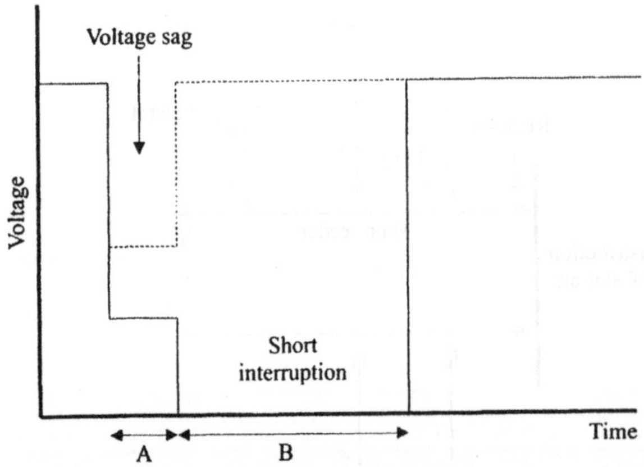

Figure 3.2 RMS voltage during a reclosure sequence on the faulted feeder (solid line) and on the nonfaulted feeder (dashed line). A = fault-clearing time; B = reclosing interval.

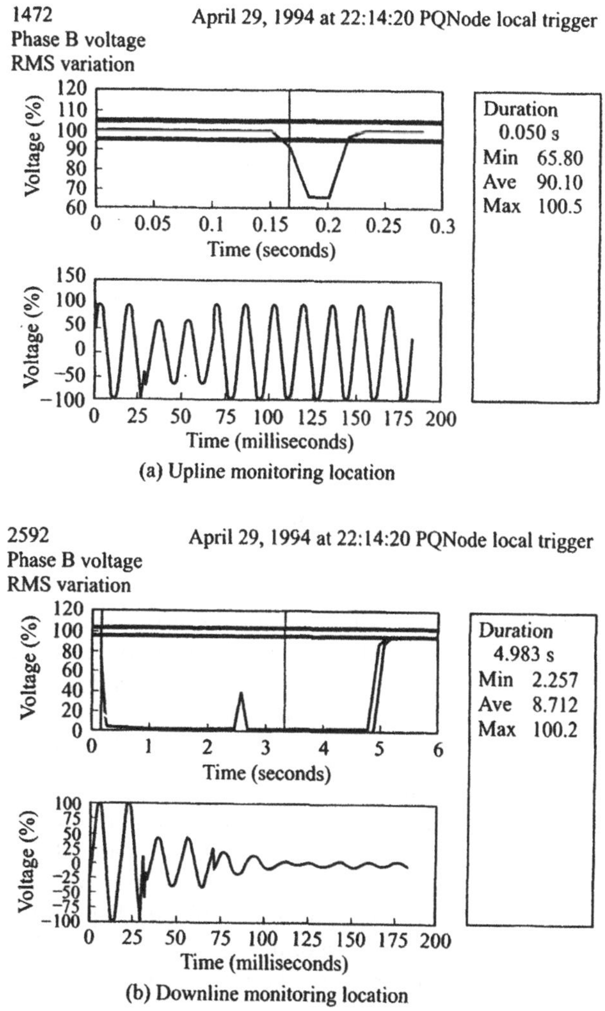

Figure 3.3 Recorded rms voltage during a short interruption. (Reproduced from Dugan et al. [11].)

When comparing Fig. 3.2 and Fig. 3.3, note that the horizontal axis of Fig. 3.2 is not to scale, B is much larger than A. This is the typical situation. The fault-clearing time (A) is only a few cycles, whereas the reclosing time (B) can be up to several minutes.

Another example of the initiation of a short interruption is shown in Fig. 3.4 [3]. We see that the voltage magnitude initially drops to about 25% of nominal and to almost zero after three cycles. The spikes in the voltage are due to the arc becoming instable around the current zero-crossing. Apparently the arc gets more stable after two cycles.

3.3.4 Voltage During the Interruption

The moment the circuit breaker in Fig. 3.1 opens, the feeder and the load fed from it are no longer supplied. The effect of this is normally that the voltage drops to zero very fast. There are, however, situations in which the voltage drops to zero relatively slow, or even remains at a nonzero value. The latter would strictly speaking not be an

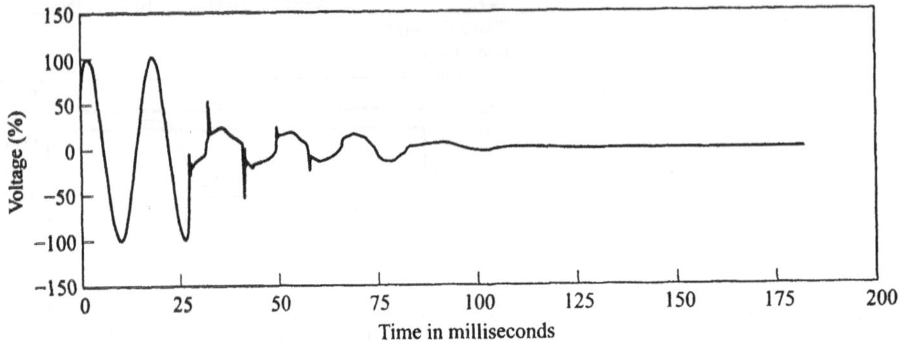

Figure 3.4 Recorded voltage during the initiation of a short interruption. (Reproduced from IEEE Std.1159 [3].)

interruption, but the origin is similar to that of an interruption so that a short description of the phenomenon is appropriate here.

- Induction motor load is able to maintain some voltage in the system for a short time. This contribution is typically rather small because the motors have already been feeding into a short circuit for a few cycles; thus, part of the rotor field of the induction motors will be gone already. Most induction motors will thus only give a small voltage contribution and only for a few cycles.
- Synchronous motors maintain their field even when the supply voltage disappears. They will be able to maintain some system voltage until their load has come to a standstill, which can take several seconds. If there is a significant amount of synchronous motor load present, its fault contribution could make fault extinguishing difficult. Typically synchronous motors will be tripped by their undervoltage protection after about 1 second, after which they no longer contribute to the feeder voltage.
- Synchronous and induction generators connected to the feeder (e.g., wind turbines or combined-heat-and-power installations) are capable of maintaining the feeder voltage at a nonzero value even during a long interruption. This could be a potential problem when large amounts of generation are connected to the feeder. This so-called embedded generation is often not equipped with any voltage or frequency control (relying on the grid to maintain voltage and current within limits) so that an islanding situation can occur in which voltage and frequency deviate significantly from their nominal values. Especially overvoltage and overfrequency can lead to serious damage. To prevent such a situation, most embedded generation is equipped with a loss-of-grid protection that disconnects the generator when an unusual voltage or frequency is detected.

All this assumes that the short-circuit fault is no longer present on the feeder. As long as the fault is present, all above-mentioned machines feed into the fault so that the feeder voltage remains low. The fault-current contribution makes that the arc is less likely to extinguish, but after extinguishing of the arc there will be a chance of a remaining voltage on the feeder.

For interruptions due to incorrect protection intervention there is no short-circuit fault present on the feeder and the machines connected to the feeder may cause a temporary or permanent nonzero voltage. Also the contribution of induction motors will be larger.

3.4 MONITORING OF SHORT INTERRUPTIONS

As short interruptions are due to automatic switching actions, their recording requires automatic monitoring equipment. Unlike long interruptions, a short interruption can occur without anybody noticing it. That is one of the reasons why utilities do not yet collect and publish data on short interruptions on a routine basis. One of the problems in collecting this data on a routine basis is that some kind of monitoring equipment needs to be installed on all feeders. A number of surveys have been performed to obtain statistical information about voltage magnitude variations and events. With those surveys, monitors were installed at a number of nodes spread through the system. The surveys will be discussed in more detail in Chapter 6. As with long interruptions, interruption frequency and duration of interruption are normally presented as the outcome of the survey. Again like with long interruptions much more data analysis is possible, e.g., interruption frequency versus time of day or time of year, distributions for the time between events, variation among customers.

3.4.1 Example of Survey Results

Figures 3.5, 3.6, and 3.7 show some results of analysis of the data obtained by a large North American survey [68]. Figure 3.5 gives the interruption frequency as a function of the interruption duration. Each vertical bar gives the average number of interruptions per year, with a duration in the given interval. The average number of interruptions has been obtained as follows:

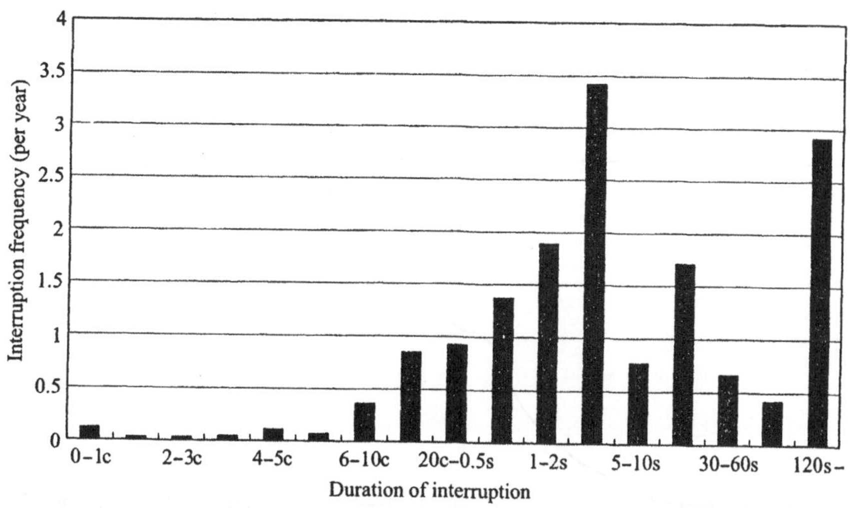

Figure 3.5 Interruption frequency (number of interruptions per year) as a function of interruption duration. (After data obtained from Dorr [68].)

$$\overline{N^{(r)}} = \frac{\sum_{i=1}^{k} N_i^{(r)}}{\sum_{i=1}^{k} T_i} \tag{3.2}$$

where $N_i^{(r)}$ is the number of events in range r observed by monitor i during a monitoring interval T_i, and $\overline{N^{(r)}}$ the resulting average as plotted in Fig. 3.5. We see from Fig. 3.5 that the typical event has a duration between 1 and 30 seconds. Events shorter than six cycles (100 ms) are very unlikely. These "very short interruptions" are most likely due to short-circuit faults close to the monitor position. One should realize that in this survey an event is recorded as an interruption if the rms voltage somewhere during the event drops below 10% of nominal. Note also that the horizontal scale is non-homogeneous. From the data shown in Fig. 3.5 one can calculate the probability density function of the interruption duration by dividing each value by the sum of all values:

$$f(r) = \frac{\overline{N^{(r)}}}{\sum_{(k)} \overline{N^{(k)}}} \tag{3.3}$$

The probability distribution function of the interruption duration can be obtained by adding the values of the density function up to a certain duration.

$$F(t) = \sum_{(r)<t} f(r) \tag{3.4}$$

The resulting probability distribution function is presented in Fig. 3.6. This curve gives the fraction of interruptions with a duration not exceeding the indicated value. We see that 10% of interruptions lasts less than 20 cycles, and 80% of interruptions less than 2 minutes (thus 20% more than 2 minutes). From an equipment point of view the reverse data are of more interest, the fraction of interruptions (or the absolute number) lasting longer than a given duration. This will give information about the number of times a device will trip or (for a given maximum trip frequency) about the immunity requirements of the device. Figure 3.7 plots the number of interruptions per year lasting longer than the indicated value. Apart from a small shift (due to the discretization of the data)

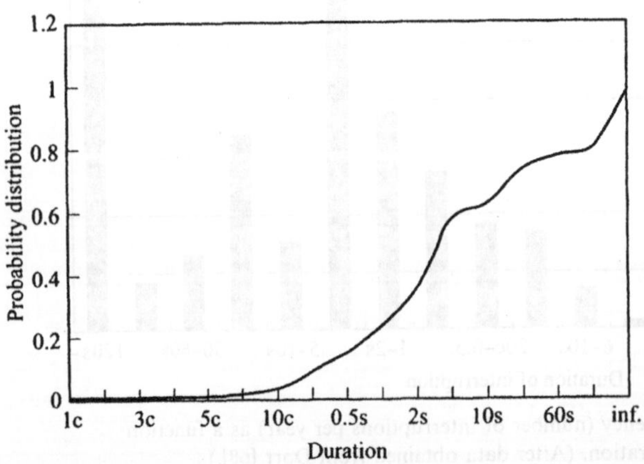

Figure 3.6 Probability distribution function of interruption duration. (From the data in Fig. 3.5.)

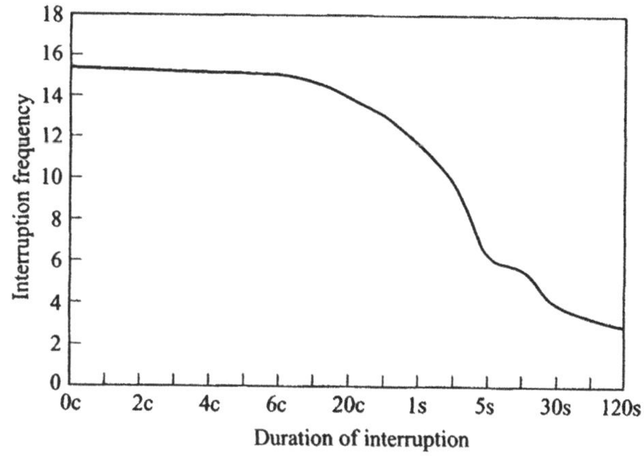

Figure 3.7 Number of interruptions lasting longer than the indicated value. (From the data in Fig. 3.5.)

and a multiplication factor equal to the total number of interruptions, the curve is the complement of the curve in Fig. 3.6. We can conclude from the figure that equipment which trips for an interruption of 20 cycles will trip on average 14 times per year. To limit the equipment trip frequency to four per year, the equipment should be able to tolerate interruptions up to 30 seconds in duration.

3.4.2 Difference between Medium- and Low-Voltage Systems

The number of short interruptions has been obtained by various power quality surveys. Comparison of the numbers obtained by each survey gives information about the average voltage quality in the various areas. A comparison between the number of short interruptions counted at various places in the system can teach us how the interruptions "propagate" in the system. Such a comparison is made in Table 3.1 for two large North American surveys: the EPRI survey and the NPL survey [54]. The EPRI survey monitored both distribution substations and distribution feeders.

From Table 3.1 we see that the overall trend is for the number of short interruptions to increase when moving from the source to the load. This is understandable as there are more possible tripping points the further one moves towards the load. Especially interruptions lasting several seconds and longer mainly originate in the low-voltage system. For interruptions less than one second in duration, the frequency remains about the same, which makes us conclude that they probably originate in the distribution substation or even higher up in the system. The large number of very short

TABLE 3.1 Interruption Frequency (number of events per year) for Three Points in the U.S. Distribution System

	Duration							
Survey	1–6c	6–10c	10–20c	20–30c	0.5–1 sec	1–2 sec	2–10 sec	> 10 sec
EPRI substation	0.2	0.1	0.4	0.8	0.5	0.9	1.1	1.3
EPRI feeder	1.6	0.1	0.2	0.6	0.5	1.1	2.3	1.7
NPL low voltage	0.2	0.3	0.7	0.8	1.2	1.5	3.3	4.2

Source: After data obtained from [54].

TABLE 3.2 Interruption Frequency (per year) for Primary and Secondary Systems in Canada

Survey	Duration							
	1–6c	6–10c	10–20c	20–30c	0.5–1 sec	1–2 sec	2–10 sec	> 10 sec
CEA primary side	1.9	0.0	0.1	0.0	0.4	0.0	0.0	0.7
CEA secondary side	3.7	0.0	0.0	0.0	0.2	0.5	0.5	2.1

Source: After data obtained from [69].

TABLE 3.3 Interruption Frequency (per year) for Distribution and Low-voltage Systems in Norway

Survey	Duration					
	0.01–0.1 sec	0.1–0.5 sec	0.5–1.0 sec	1–3 sec	3–20 sec	> 20 sec
EFI distribution	1.5	0.0	0.0	0.0	0.5	5.2
EFI low-voltage	1.1	0.7	0.0	0.7	0.9	5.9

Source: After data obtained from [67].

interruptions (less than six cycles) on distribution feeders is hard to explain, especially as they do not show up in the low-voltage data.

Similar conclusions can be drawn from the CEA survey [69] and from the EFI survey [67], some results of which are shown in Tables 3.2 and 3.3. We again see a larger number of interruptions, mainly of 1 second and longer, for low-voltage than for medium-voltage systems. Both the Canadian (CEA) and the Norwegian (EFI) data show a considerable number of very short interruptions, for which no explanation has been found yet.

3.4.3 Multiple Events

A direct consequence of reclosing actions is that a customer may experience two or more events within a short interval. When the short-circuit fault is still present upon the first reclosure, the customers fed from the faulted feeder will experience a second event. This is another short interruption if a second attempt at reclosing is made. Otherwise the second event will be a long interruption. A customer fed from a non-faulted feeder experiences two voltage sags in a short period of time.

For a few years a discussion has been going on about whether to count this as one event or as multiple events [20]. The most recent publications of North American surveys consider a 1-minute or 5-minute window. If two or more events take place within such a window, they are counted as one event. The severity of the multiple event (i.e., magnitude and duration) is the severity of the most severe single event within the window. Some examples of the working of a "five-minute filter" are shown in Fig. 3.8.

Using such a "filter" is suitable for assessment of the number of equipment trips, as the equipment will trip on the most severe event or not at all. The cumulative effect of the events is neglected, but the general impression is that this effect is small. This has however not been confirmed by measurements yet. In some cases it could still be needed to know the total event frequency, thus counting all events even if they come very close. Two possible applications are: (1) components which show accelerated aging due to short undervoltage events; and (2) equipment which only trips during a certain fraction

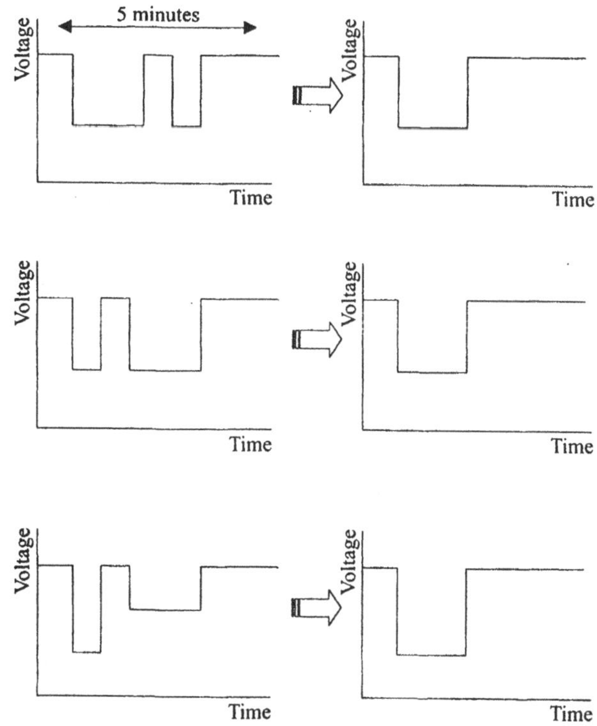

Figure 3.8 Effect of a "five-minute filter" on the voltage magnitude events. The figures on the left show the recorded rms voltages; the figures on the right show the equivalent event after the filter.

TABLE 3.4 Number of Single and Multiple Interruptions per Year, NPL Low-Voltage Survey

Survey	Duration							
	1–6c	6–10c	10–20c	20–30c	0.5–1 sec	1–2 sec	2–10 sec	> 10 sec
No filter	0.3	0.3	0.8	0.9	1.4	1.9	4.2	5.7
5-min filter	0.2	0.3	0.7	0.8	1.2	1.5	3.3	4.2
Percent reduction	33%	—	12%	11%	14%	21%	21%	26%

Source: After data obtained from [54].

of its load cycle. In the latter case the equipment has a probability to trip during each of the three events, and the total probability is of course larger than the probability to trip during the most severe event only.

The NPL low-voltage data for short interruptions have been presented with and without the above-mentioned filter in Table 3.4 [54]. The three rows give, from top to bottom: the number of short interruptions when each event is counted as one event no matter how close it is to another event; the number of events when multiple events within a 5-minute interval are counted as one event; the reduction in number of events due to the application of this filter.

3.5 INFLUENCE ON EQUIPMENT

During a short interruption the voltage is zero; thus, there is no supply of power at all to the equipment. The temporary consequences are that there is no light, that motors slow down, that screens turn blank, etc. All this only lasts for a few seconds, but the consequences can last much longer: disruption of production processes, loss of contents of computer memory, evacuation of buildings due to fire alarms going off, and sometimes damage when the voltage comes back (uncontrolled starting).

For most sensitive equipment, there is no strict border between a voltage sag and an interruption: an interruption can be seen as a severe sag, i.e. one with zero remaining voltage. The effect of voltage sags on equipment is discussed in detail in Chapter 5. Many of the conclusions in that chapter also hold for short interruptions. In this section only some general aspects of the load behavior are pointed out.

3.5.1 Induction Motors

The effect of a zero voltage on an induction motor is simple: the motor slows down. The mechanical time constant of an induction motor plus its load is in the range of 1 to 10 seconds. With dead times of several seconds, the motor has not yet come to a standstill but is likely to have slowed down significantly. This reduction in speed of the motors might disrupt the industrial process so much that the process control trips it.

The motor can re-accelerate when the voltage comes back, if the system is strong enough. For public distribution systems re-acceleration is seldom a problem.

Also the setting of the undervoltage protection should be such that it does not trip before the voltage comes back. This calls for a coordination between the undervoltage setting of the motor protection and the reclosure interval setting on the utility feeder.

Induction motors fed via contactors are disconnected automatically as the contactor drops out. Without countermeasures this would always lead to loss of the load. In some industrial processes the induction motors are automatically reconnected when the voltage comes back: either instantaneously or staged (the most important motors first, the rest later).

3.5.2 Synchronous Motors

Synchronous motors can normally not restart on full load. They are therefore equipped with undervoltage protection to prevent stalling when the voltage comes back. For synchronous motors the delay time of the undervoltage protection should be less than the reclosing interval. Especially for very fast reclosure this can be a problem. We see here a situation where an interruption causes a more serious threat to the synchronous motors the faster the voltage comes back. With most other load the situation is the other way around: the shorter the interruption, the less severe it is to the load.

3.5.3 Adjustable-Speed Drives

Adjustable-speed drives are very sensitive to short interruptions, and to voltage sags as we will see in Chapter 5. They normally trip well within 1 second, sometimes even within one cycle; thus even the shortest interruption will cause a loss of the load. Some of the more modern drives are able to automatically reconnect the moment the voltage comes back. But being disconnected from the supply for several seconds will often have disrupted the process behind the drive so much that reconnection does not make much sense anymore.

3.5.4 Electronic Equipment

Without countermeasures electronics devices will trip well within the reclosing interval. This leads to the infamous "blinking-clock syndrome": clocks of video recorders, microwave ovens, and electronic alarms start blinking when the supply is interrupted; and they keep on blinking until manually reset. An easy solution is to install a small rechargeable battery inside of the equipment, to power the internal memory during the interruption.

Computers and process control equipment have basically the same problem. But they require more than a simple battery. An uninterruptible power supply (UPS) is a much-used solution.

3.6 SINGLE-PHASE TRIPPING

Single-phase tripping is used in transmission systems to maintain synchronicity between both sides of a line. Single-phase tripping is rarely used in distribution or low-voltage systems. Not only will it require more expensive equipment, but it will also reduce the chance of a successful reclosure. The fault current continues to flow via the nonfaulted phases. This reduces the chance that the fault will extinguish and thus increases the number of reclosure attempts and the number of long interruptions. But if the reclosure is successful, single-phase tripping has clear advantages over three-phase tripping and therefore justifies being discussed here. We will have a look at the voltages experienced by the customer during single-phase tripping. A distinction is made between two distinctly different situations, both assuming a single-phase-to-ground fault followed by tripping of the faulted phase.

- The low-impedance path between the faulted phase and ground (the fault) is still present so that the voltage in the faulted phase remains zero or close to zero. We will call this the "during-fault period."
- The fault has extinguished, the short circuit has now become an open circuit because the breaker in that phase is still open. This we will call the "post-fault period."

3.6.1 Voltage—During-Fault Period

The phase-to-neutral voltages in the during-fault period are, with a the faulted phase:

$$
\begin{aligned}
V_a &= 0 \\
V_b &= \left(-\frac{1}{2} - \frac{1}{2}j\sqrt{3}\right)E \\
V_c &= \left(-\frac{1}{2} + \frac{1}{2}j\sqrt{3}\right)E
\end{aligned}
\tag{3.5}
$$

with E the magnitude of the pre-event voltage. It has been assumed here that the pre-event voltages form a balanced three-phase set, and that the voltage in the faulted phase is exactly equal to zero. We will in most of the remainder of this book use per unit voltages, with the pre-event voltage magnitude as base. In that case we get $E = 1$ and (3.5) becomes

$$V_a = 0$$
$$V_b = -\frac{1}{2} - \frac{1}{2}j\sqrt{3} \tag{3.6}$$
$$V_c = -\frac{1}{2} + \frac{1}{2}j\sqrt{3}$$

Figure 3.9 shows the phase-to-neutral voltages as a phasor diagram. In this and subsequent phasor diagrams the during-event voltage is indicated via solid lines, the pre-event voltage (i.e., the balanced three-phase voltage) via dotted lines, if different from the during-event voltage. If single-phase tripping would take place in a low-voltage network, the voltages in Fig. 3.9 would be the voltages experienced by the customers. Only one out of three customers would experience an interruption. The others would not notice anything. Single-phase tripping would thus reduce the number of interruption events by a factor of three.

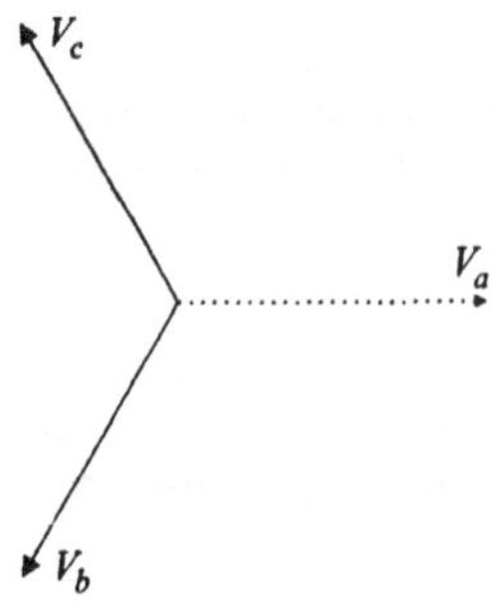

Figure 3.9 Phase-to-neutral voltages for single-phase tripping.

For tripping taking place on medium-voltage feeders, the phase-to-phase voltages are of more importance. Large equipment fed at medium-voltage level is in most cases connected in delta; small single-phase equipment tends to be connected between a phase and neutral but at a lower voltage level fed via a delta-star connected transformer. In both cases the equipment experiences the pu value of the phase-to-phase voltage at the medium-voltage level.

The phase-to-phase voltages in pu are obtained from the phase-to-neutral voltages as follows:

$$V'_a = j\frac{V_b - V_c}{\sqrt{3}}$$
$$V'_b = j\frac{V_c - V_a}{\sqrt{3}} \tag{3.7}$$
$$V'_c = j\frac{V_a - V_b}{\sqrt{3}}$$

The factor $\sqrt{3}$ is needed because 1 pu of the line (phase-to-phase) voltage is $\sqrt{3}$ times as big as 1 pu of the phase (phase-to-neutral) voltage. The multiplication with j results in a rotation over 90° such that the axis of symmetry of the disturbance remains along phase a and along the real axis. The transformation in (3.7) will be the basis of a detailed analysis of unbalanced voltage sags in the forthcoming chapters. When we leave away the prime $'$, we obtain the following expressions for the voltages due to single-phase tripping at the terminals of delta-connected equipment:

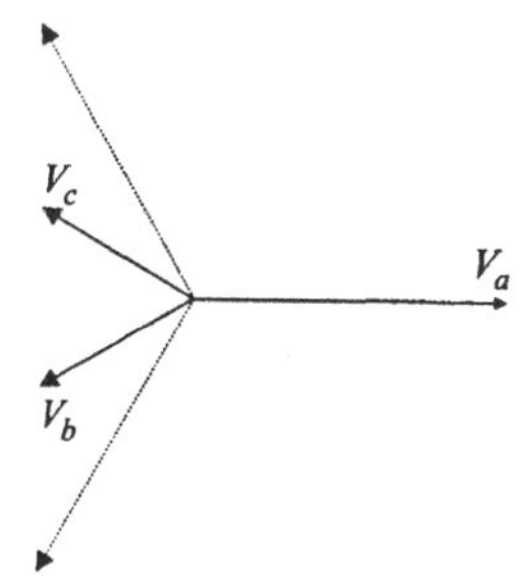

Figure 3.10 Phase-to-phase voltages for single-phase tripping.

$$
\begin{aligned}
V_a &= 1 \\
V_b &= -\frac{1}{2} - \frac{1}{6}j\sqrt{3} \\
V_c &= -\frac{1}{2} + \frac{1}{6}j\sqrt{3}
\end{aligned}
\tag{3.8}
$$

Figure 3.10 again shows the voltages at the equipment terminals in phasor-diagram form. Using the definitions given in the various standards this should not be called a short interruption but a voltage sag. It would again bring up the discussion between consequence-based terminology and cause-based terminology. In the first case this event would have to be called a voltage sag, in the latter case it would be a short interruption. But no matter which name is given to the event, it is clearly less severe than the effect of three-phase tripping, when all three phase voltages go down to zero. An exception to this might have to be made for induction motors. The voltages during single-phase tripping contain a large negative sequence voltage component (0.33 pu) which may lead to overheating of induction motors. With a negative sequence impedance 5 through 10 times as small as the positive sequence impedance, the negative sequence current would become 170 through 330% of the rated (positive sequence) current. It is unlikely that induction motor load is able to withstand such an unbalance for longer than several seconds.

Low-voltage customers also experience the voltages in Fig. 3.10. None of the customers experiences a zero voltage, but two-thirds of the customers experience an event with a during-event voltage of 58% magnitude with a change in voltage phase-angle of 30°.

3.6.2 Voltage—Post-Fault Period

When the fault extinguishes, the situation in the faulted phase changes from a short circuit to an open circuit. In many cases a change in voltage occurs, thus the resulting voltage is no longer equal to zero. The voltage in the faulted phase depends on the type of load connected. To calculate this voltage we need to consider the coupling between the phases or use the theory of symmetrical components. The latter, which is normally used for the analysis of nonsymmetrical faults, is described in detail in many reference books. A good and detailed description of the use of symmetrical components for the analysis of nonsymmetrical faults is, e.g., given in reference [24], and is not repeated here.

To analyze an open circuit, the system has to be modeled as seen from the open-circuit point. This results in three equivalent circuits: for the positive sequence, for the

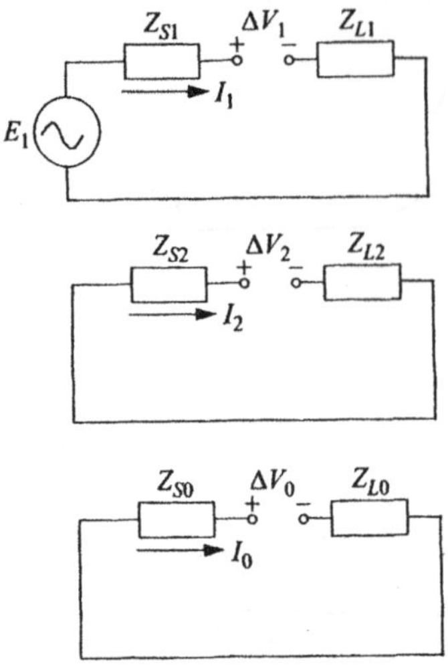

Figure 3.11 Sequence networks for the analysis of single-phase open-circuit faults: positive sequence (top), negative sequence (center), and zero sequence (bottom).

negative sequence, and for the zero sequence. These three networks are shown in Fig. 3.11: Z_{S1}, Z_{S2}, and Z_{S0} are positive, negative, and zero-sequence impedance of the source; Z_{L1}, Z_{L2}, and Z_{L0} are positive, negative, and zero-sequence impedance of the load; ΔV_1, ΔV_2, and ΔV_0 are positive, negative, and zero-sequence voltage drop at the open-circuit point; and E_1 is the positive-sequence source voltage. Negative and zero-sequence source voltages are assumed zero, and the load is assumed not to contain any sources. Below we again assume $E_1 = 1$.

Sequence voltages and currents at the open-circuit point can be calculated for different types of open-circuit faults, by connecting the three sequence networks in different ways. For a single-phase open circuit, the voltage difference in the two non-faulted phases is zero and the current in the faulted phase is zero:

$$\begin{aligned} \Delta V_b &= 0 \\ \Delta V_c &= 0 \\ I_a &= 0 \end{aligned} \tag{3.9}$$

where a is the faulted (open-circuited) phase. Transforming these equations to symmetrical components gives the following set of equations:

$$\begin{aligned} I_1 + I_2 + I_0 &= 0 \\ \Delta V_1 &= \Delta V_2 \\ \Delta V_1 &= \Delta V_0 \end{aligned} \tag{3.10}$$

These expressions correspond to a connection of the sequence networks, as shown in Fig. 3.12. From Fig. 3.12 the positive-sequence voltage drop at the open-circuit point can be written as

$$\Delta V_1 = \Delta V_2 = \Delta V_0 = \frac{1}{1 + \dfrac{Z_{L1} + Z_{S1}}{Z_{L0} + Z_{S0}} + \dfrac{Z_{L1} + Z_{S1}}{Z_{L2} + Z_{S2}}} \tag{3.11}$$

and the voltage drop in the faulted phase is

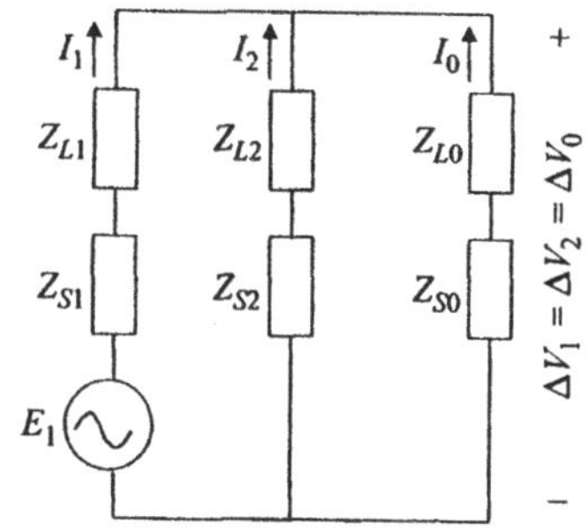

Figure 3.12 Connection of the sequence networks in Fig. 3.11 for a single-phase open circuit.

$$\Delta V_a = \Delta V_1 + \Delta V_2 + \Delta V_0 = \frac{3}{1 + \frac{Z_{L1} + Z_{S1}}{Z_{L0} + Z_{S0}} + \frac{Z_{L1} + Z_{S1}}{Z_{L2} + Z_{S2}}} \tag{3.12}$$

Normally the load impedance dominates over the source impedance ($Z_{Li} \gg Z_{Si}$, $i = 0, 1, 2$) so that we can write with good approximation:

$$\Delta V_a = \frac{3}{1 + \frac{Z_{L1}}{Z_{L0}} + \frac{Z_{L1}}{Z_{L2}}} \tag{3.13}$$

The voltage at the load side of the open phase is

$$V_a = 1 - \frac{3}{1 + \frac{Z_{L1}}{Z_{L0}} + \frac{Z_{L1}}{Z_{L2}}} \tag{3.14}$$

which can be written as an expression using admittances by introducing $Y_{L1} = \frac{1}{Z_{L1}}$, $Y_{L2} = \frac{1}{Z_{L2}}$, and $Y_{L0} = \frac{1}{Z_{L0}}$, resulting in

$$V_a = 1 - \frac{Y_{L1}}{\frac{1}{3}(Y_{L1} + Y_{L2} + Y_{L0})} \tag{3.15}$$

From (3.15) the voltage experienced by the load during the interruption can be found for different types of load. As can be seen it is the ratio between the sequence impedances of the load which determines the voltage. The source impedance does have a small influence as the load current will give a voltage drop between the load and the open-circuit point. This influence was neglected when going from (3.12) to (3.13).

3.6.2.1 Star-connected Static Load. For star-connected static load, the three sequence impedances are equal: $Y_{L1} = Y_{L2} = Y_{L0}$, (3.15) gives

$$V_a = 0 \tag{3.16}$$

In other words, this type of load does not affect the voltage in the open phase. Single-phase, low-voltage load can normally be represented in this way.

3.6.2.2 Delta-connected Static Load. Delta-connected static load is found in medium-voltage public distribution networks. The delta-star connected transformer feeding the low-voltage customers can be considered a delta-connected static load, as long as mainly single-phase load is present. For this kind of load, positive and negative sequence impedances are equal and the zero-sequence impedance is infinite

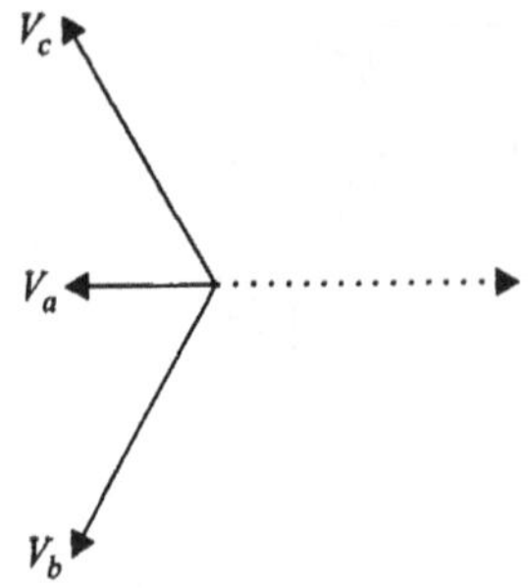

Figure 3.13 Phase-to-ground voltages during single-phase reclosure with delta-connected load.

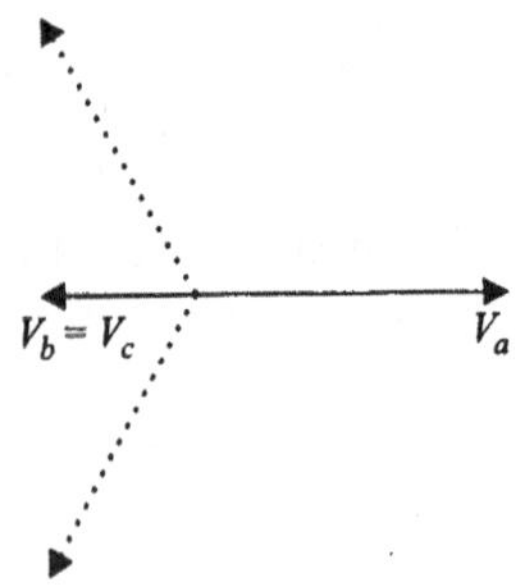

Figure 3.14 Phase-to-phase voltages during single-phase reclosure with delta-connected load.

because of the lack of any return path; in admittance terms, $Y_{L1} = Y_{L2}$ and $Y_{L0} = 0$, resulting in

$$V_a = -\frac{1}{2} \tag{3.17}$$

In high-impedance grounded or isolated-neutral systems, the zero-sequence source impedance is very large or even infinite. From the above equations it is easy to prove that the resulting voltage in the open phase is again equal to $-\frac{1}{2}$. The phase voltages and the line voltages for delta-connected static load are shown in Fig. 3.13 and Fig. 3.14, respectively.

3.6.2.3 Motor Load. For motor load, a typical load in industrial systems and in some public systems, the zero-sequence impedance is again infinite, and the negative sequence impedance is smaller than the positive-sequence impedance: $Y_{L2} > Y_{L1}$ and $Y_{L0} = 0$. The resulting expression for the open-phase voltage is, with $Y_{L2} = \gamma Y_{L1}$

$$V_a = \frac{\gamma - 2}{\gamma + 1} \tag{3.18}$$

For $\gamma = 1$, which corresponds to static delta-connected load, we again obtain $V_a = -\frac{1}{2}$, for $\gamma = 2$ we obtain $V_a = 0$. A typical range of the ratio between positive and negative sequence impedance is: $\gamma = 3 \cdots 10$ resulting in $V_a = 0.25 \cdots 0.73$. When the induction motors slow down, the negative sequence impedance stays about the same while the positive sequence impedance becomes smaller, until they are equal when the motor has come to a standstill. From equation (3.18) we can conclude that the open-phase voltage decays when γ gets smaller, thus when the motors slow down. The open-phase voltage for a system with motor load is initially between 50% and 70% of the pre-fault voltage, decaying to −50% of pre-fault voltage (i.e., 50% of magnitude, but with opposite phase).

From the above examples, we can conclude that the voltage in the open phase varies between −0.50 and +0.75 times the pre-fault voltage. When we use the symbol V to indicate this voltage, we get the following phasor expression for the voltages in the three phases:

$$\begin{aligned} V_a &= V \\ V_b &= -\frac{1}{2} - \frac{1}{2} j\sqrt{3} \\ V_c &= -\frac{1}{2} + \frac{1}{2} j\sqrt{3} \end{aligned} \tag{3.19}$$

Using the transformation as defined by (3.7), we get for the line voltages (i.e., the voltages experienced by a delta-connected load)

$$\begin{aligned} V_a &= 1 \\ V_b &= -\frac{1}{2} - \frac{1}{2} j\sqrt{3}\left(\frac{1}{3} + \frac{2}{3} V\right) \\ V_c &= -\frac{1}{2} + \frac{1}{2} j\sqrt{3}\left(\frac{1}{3} + \frac{2}{3} V\right) \end{aligned} \tag{3.20}$$

We see that a delta-connected load experiences a voltage drop in two phases, but this voltage drop is smaller than the voltage drop in the open phase as experienced by a star-connected load. Also the load is less influenced by single-phase tripping than by three-phase tripping.

3.6.2.4 Transfer to Lower Voltage Levels. Transfer to lower voltage levels often takes place through delta-star connected transformers. The first transformer simply changes line into phase voltages, resulting in expression (3.20) but for the phase voltages instead of for the line voltages.

To obtain the line voltages after a delta-star connected transformer, or the phase voltages after two such transformers, the transformation (3.7) has to be applied a second time, to (3.20), resulting in

$$\begin{aligned} V_a &= \frac{1}{3} + \frac{2}{3} V \\ V_b &= -\frac{1}{2}\left(\frac{1}{3} + \frac{2}{3} V\right) - \frac{1}{2} j\sqrt{3} \\ V_c &= -\frac{1}{2}\left(\frac{1}{3} + \frac{2}{3} V\right) + \frac{1}{2} j\sqrt{3} \end{aligned} \tag{3.21}$$

The resulting voltages for different types of load are summarized in Table 3.5. The transfer of this kind of voltage events to lower voltage levels is discussed in much more detail in Section 4.4. There we will denote the voltage events in (3.19), (3.20), and (3.21) as sags of type B with magnitude V, of type C with magnitude $\frac{1}{3} + \frac{2}{3} V$, and of type D with magnitude $\frac{1}{3} + \frac{2}{3} V$, respectively.

TABLE 3.5 Voltages Due to Single-Phase Tripping, for Various Types of Load

	Star-connected Load	Delta-connected Load	Induction Motor Load: Initial	Induction Motor Load: Motor Slowed Down
Voltage in the Open Phase				
	$V_a = 0$	$V_a = -0.5$	$V_a = 0.75$	$V_a = 0.25$
Voltages After the First Dy-transformer				
Phasors	$V_a = 1$	$V_a = 1$	$V_a = 1$	$V_a = 1$
	$V_b = -\frac{1}{2} - \frac{1}{6}j\sqrt{3}$	$V_b = -\frac{1}{2}$	$V_b = -\frac{1}{2} - \frac{5}{12}j\sqrt{3}$	$V_b = -\frac{1}{2} - \frac{1}{4}j\sqrt{3}$
	$V_c = -\frac{1}{2} + \frac{1}{6}j\sqrt{3}$	$V_c = -\frac{1}{2}$	$V_c = -\frac{1}{2} + \frac{5}{12}j\sqrt{3}$	$V_c = -\frac{1}{2} + \frac{1}{4}j\sqrt{3}$
Magnitudes	100%, 57.7%, 57.7%	100%, 50%, 50%	100%, 87.8%, 87.8%	100%, 66.1%, 66.1%
Voltages After the Second Dy-transformer				
Phasors	$V_a = \frac{1}{3}$	$V_a = 0$	$V_a = \frac{5}{6}$	$V_a = \frac{1}{2}$
	$V_b = -\frac{1}{6} - \frac{1}{2}j\sqrt{3}$	$V_b = -\frac{1}{2}j\sqrt{3}$	$V_b = -\frac{5}{12} - \frac{1}{2}j\sqrt{3}$	$V_b = -\frac{1}{4} - \frac{1}{2}j\sqrt{3}$
	$V_c = -\frac{1}{6} + \frac{1}{2}j\sqrt{3}$	$V_c = -\frac{1}{2}j\sqrt{3}$	$V_c = -\frac{5}{12} + \frac{1}{2}j\sqrt{3}$	$V_c = -\frac{1}{4} + \frac{1}{2}j\sqrt{3}$
Magnitudes	33.3%, 88.2%, 88.2%	0, 86.6%, 86.6%	83.3%, 96.1%, 96.1%	50%, 90.1%, 90.1%

3.6.3 Current—During-Fault Period

As we have seen in the previous section, the voltage in the faulted phase during the post-fault period is not necessarily zero. A nonzero voltage after fault extinguishing implies a nonzero current while the fault is present. This makes fault extinguishing more difficult.

To calculate the fault current after single-phase tripping but before the fault extinguishes, we consider the circuit in Fig. 3.15. Source and load impedances are indicated by the same symbols as before. Voltages and currents at the system side of the open point are indicated as V_a, V_b, etc., and at the load side as V'_a, V'_b, etc.

The electrical behavior of this system can be described through 12 equations, three equations describing the source (with again $E_1 = 1$):

$$\begin{aligned} 1 - Z_{S1}I_1 &= V_1 \\ -Z_{S2}I_2 &= V_2 \\ -Z_{S0}I_0 &= V_0 \end{aligned} \tag{3.22}$$

three equations describing the load:

$$\begin{aligned} V'_1 &= Z_{L1}I'_1 \\ V'_2 &= Z_{L2}I'_2 \\ V'_0 &= Z_{L0}I'_0 \end{aligned} \tag{3.23}$$

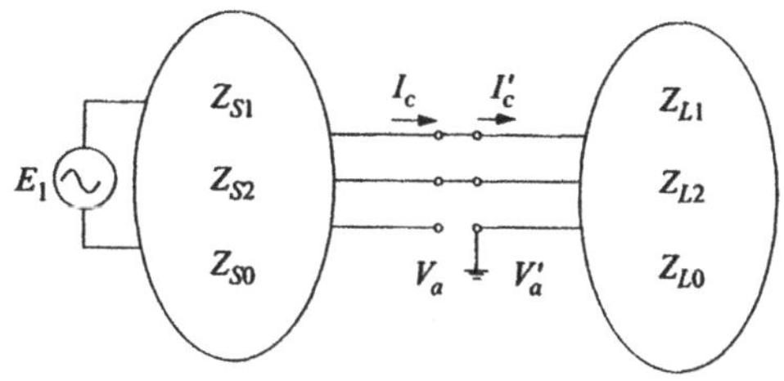

Figure 3.15 Single-phase tripping with the short circuit still present.

three voltage equations at the open point:

$$\begin{aligned} V'_a &= 0 \\ V'_b &= V_b \\ V'_c &= V_c \end{aligned} \tag{3.24}$$

and three current equations at the open point:

$$\begin{aligned} I_a &= 0 \\ I_b &= I'_b \\ I_c &= I'_c \end{aligned} \tag{3.25}$$

If we neglect the source impedances, the voltages at the system side of the open point are equal to the source voltages:

$$\begin{aligned} V_1 &= 1 \\ V_2 &= 0 \\ V_0 &= 0 \end{aligned} \tag{3.26}$$

From (3.24) relations can be obtained between the component voltages on both sides of the open point:

$$\begin{aligned} V'_1 &= \frac{2}{3}V_1 - \frac{1}{3}V_2 - \frac{1}{3}V_0 \\ V'_2 &= -\frac{1}{3}V_1 + \frac{2}{3}V_2 - \frac{1}{3}V_0 \\ V'_0 &= -\frac{1}{3}V_1 - \frac{1}{3}V_2 + \frac{2}{3}V_0 \end{aligned} \tag{3.27}$$

With (3.26), the component voltages at the load side of the open point can be found. Together with (3.23) and $I'_a = I'_0 + I'_1 + I'_2$ we obtain an expression for the fault current after single-phase tripping:

$$I'_a = \frac{2}{3Z_{L1}} - \frac{1}{3Z_{L2}} - \frac{1}{3Z_{L0}} \tag{3.28}$$

We see that the current depends on the load impedances in positive, negative, and zero sequence. As these impedances are significantly larger than the source impedances (typically a factor of 10 to 20) the current becomes much smaller than the original fault current. This certainly helps the extinguishing of the fault, but still the fault is most likely to extinguish when the current is close to zero, thus when: $2Y_{L1} \approx Y_{L2} + Y_{L0}$ with $Y_{L1} = \frac{1}{Z_{L0}}$, etc. Not surprisingly this is also the condition for which the voltage after fault extinguishing is zero, according to (3.15).

3.7 STOCHASTIC PREDICTION OF SHORT INTERRUPTIONS

To stochastically predict the number of short interruptions experienced by a customer fed from a certain feeder, the following input data is required:

- Failure rate per km of feeder, different values might be used for the main and for the lateral conductors.
- Length of the main feeder and of the lateral conductors.
- Success rate of reclosure, if multiple reclosure attempts are used: success rate of the first reclosure, of the second reclosure, etc.
- Position of reclosing breakers and fuses.

We will explain the various steps in a stochastic prediction by using the system shown in Fig. 3.16. Note that this is a hypothetical system. Stochastic prediction studies in larger, albeit still hypothetical, systems have been performed by Warren [139]. The following data is assumed for the system in Fig. 3.16:

- The failure rate of the main feeder is: 0.1 faults per year per km of feeder.
- The failure rate of the lateral conductors is: 0.25 faults per year per km of feeder.
- The success rate of the first reclosure is 75%; thus, in 25% of the cases a second trip and reclosure are needed.
- The success rate of the second attempt is 10% of the number of faults. Thus, for 15% of the faults the second attempt does not clear the fault. Those faults are "permanent faults" leading to a long interruption.

The reclosing procedure used is as follows:

1. The circuit breaker opens instantaneously on the overcurrent due to the fault.
2. The circuit breaker remains open for a short time (1 sec); 75% of the faults clears in this period.
3. The circuit breaker closes. If the fault is still present the breaker again opens instantaneously on overcurrent. This is required in 25% of the cases.
4. The circuit breaker now leaves a longer dead time (5 sec). Another 10% of the faults clear in this period.

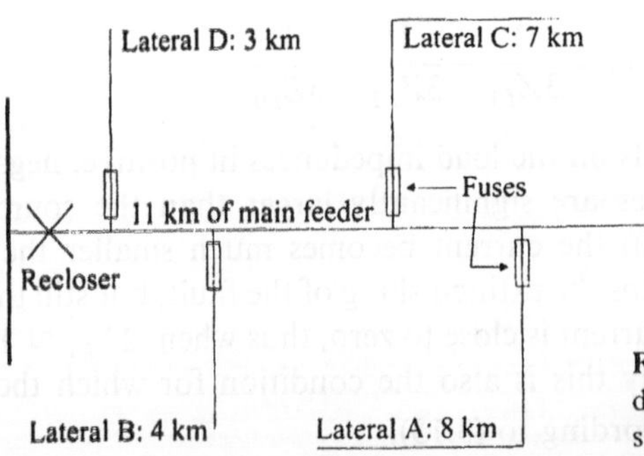

Figure 3.16 Example of overhead distribution feeder, for stochastic prediction study.

5. The circuit breaker closes for a second time. If the fault is still present the breaker remains closed until the fuse protecting the lateral conductor has had time to blow.
6. If the fault is still present (i.e., if the current magnitude still exceeds its threshold) after the time needed for the fuse to clear the fault, the breaker opens for a third time and now remains open. Further reclosure has to take place manually and the whole feeder will experience a long interruption.

The total number of faults on the feeder is

$$11\,\text{km} \times 0.1\,\text{faults/km year} + 22\,\text{km} \times 0.25\,\text{faults/km year} = 6.6\,\text{faults/year} \quad (3.29)$$

Each fault will lead to a voltage magnitude event. There are four different events possible:

- a short interruption of 1 second duration.
- two short interruptions; one of 1 second duration and one of 5 seconds duration.
- two short interruptions followed by a voltage sag.
- two short interruptions followed by a long interruption.

Due to short-circuit faults on this feeder, 6.6 events per year occur, of which

- 75% = 5.0 per year need one trip, leading to one short interruption for all customers.
- 10% = 0.7 per year need two trips, leading to two short interruptions for all customers.
- 15% = 1.0 per year are permanent, leading to two short interruptions followed by a voltage sag or followed by a long interruption.

The number of short interruptions is equal for every customer connected to this feeder:

5.0/year of 1 second duration.
0.7/year of 1 + 5 seconds duration.

The number of long interruptions depends on the position at the feeder. A permanent fault on the main feeder leads to a long interruption for all customers. A permanent fault on one of the laterals leads to a long interruption only for customers fed from this lateral. The number of permanent faults is, for the different parts of the feeder:

- lateral A: 8 km × 0.25 faults/km year × 0.15 = 0.3 faults per year
- lateral B: 4 km × 0.25 faults/km year × 0.15 = 0.15 faults per year
- lateral C: 7 km × 0.25 faults/km year × 0.15 = 0.26 faults per year
- lateral D: 3 km × 0.25 faults/km year × 0.15 = 0.11 faults per year
- main: 11 km × 0.1 faults/km year × 0.15 = 0.17 faults per year

The number of long interruptions experienced by customers connected to different parts of the feeder, is

- main: 0.17/year
- lateral A: 0.17 + 0.3 = 0.47/year
- lateral B: 0.17 + 0.15 = 0.32/year
- lateral C: 0.17 + 0.26 = 0.43/year
- lateral D: 0.17 + 0.11 = 0.28/year

Getting rid of the reclosure scheme and letting a fuse clear all faults on the lateral conductors would lead to long interruptions only.

- main: 1.1/year
- lateral A: 3.1/year
- lateral B: 2.1/year
- lateral C: 2.9/year
- lateral D: 1.9/year

Table 3.6 compares the number of long and short interruptions for systems with and without a reclosure scheme. For equipment or production processes sensitive to long interruptions only, the system with a reclosure scheme is clearly preferable. It leads to a reduction of the number of long interruptions by 85%. But when equipment/production process is sensitive to short and to long interruptions, it is better to abolish the reclosure scheme and trip permanently on every fault. That would reduce the number of equipment trips by a factor between 2 and 5, depending on the position of the load on the feeder. In reality this decision is not that easy to make, as some customers prefer more short interruptions above a few long ones, while for others only the number of interruptions matters. The first group is mainly the domestic customers, the second one the industrial customers. A financial assessment will almost always be in the favor of the industrials. An assessment on numbers of customers or on kWh will be in favor of the domestic customers.

TABLE 3.6 Number of Short and Long Interruptions per Year on an Overhead Distribution Feeder, With and Without Automatic Reclosure

	Long Interruptions Only		All Interruptions	
	With Reclosure	Without Reclosure	With Reclosure	Without Reclosure
Main feeder	0.2	1.1	6.6	1.1
Lateral A	0.5	3.1	6.6	3.1
Lateral B	0.3	2.1	6.6	2.1
Lateral C	0.4	2.9	6.6	2.9
Lateral D	0.3	1.9	6.6	1.9

4

Voltage Sags—Characterization

4.1 INTRODUCTION

Voltage sags are short duration reductions in rms voltage, caused by short circuits, overloads, and starting of large motors. The interest in voltage sags is mainly due to the problems they cause on several types of equipment: adjustable-speed drives, process-control equipment, and computers are notorious for their sensitivity. Some pieces of equipment trip when the rms voltage drops below 90% for longer than one or two cycles. In this and the two following chapters, it will become clear that such a piece of equipment will trip tens of times a year. If this is the process-control equipment of a paper mill, one can imagine that the damage due to voltage sags can be enormous. Of course a voltage sag is not as damaging to industry as a (long or short) interruption. But as there are far more voltage sags than interruptions the total damage due to sags is still larger. Short interruptions and most long interruptions originate in the local distribution network. However, voltage sags at equipment terminals can be due to short-circuit faults hundreds of kilometers away in the transmission system. A voltage sag is thus much more of a "global" problem than an interruption. Reducing the number of interruptions typically requires improvements on one feeder. Reducing the number of voltage sags requires improvements on several feeders, and often even at transmission lines far away.

An example of a voltage sag due to a short-circuit fault is shown in Fig. 4.1. We see that the voltage amplitude drops to a value of about 20% of the pre-event voltage for about two cycles. After these two cycles the voltage comes back to about the pre-sag voltage. This magnitude and duration are the main characteristics of a voltage sag. Both will be discussed in more detail in the forthcoming sections. We can also conclude from Fig. 4.1 that magnitude and duration do not completely characterize the sag. The during-sag voltage contains a rather large amount of higher frequency components. Also the voltage shows a small overshoot immediately after the sag.

Most of the current interest in voltage sags is directed to voltage sags due to short-circuit faults. These voltage sags are the ones which cause the majority of equipment trips. But also the starting of induction motors leads to voltage sags. Figure 4.2 gives an

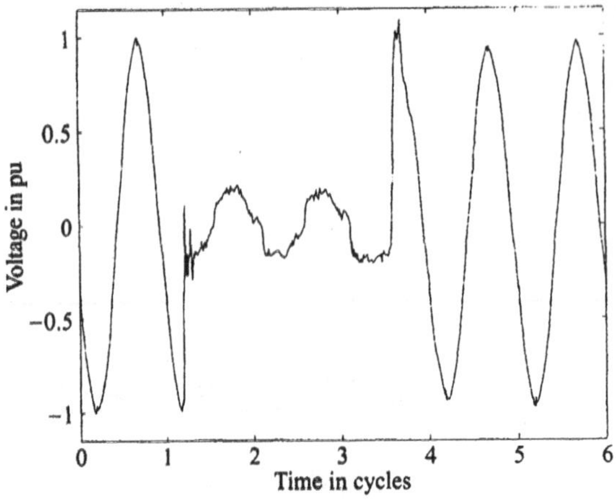

Figure 4.1 A voltage sag due to a short-circuit fault—voltage in one phase in time domain. (Data obtained from [16].)

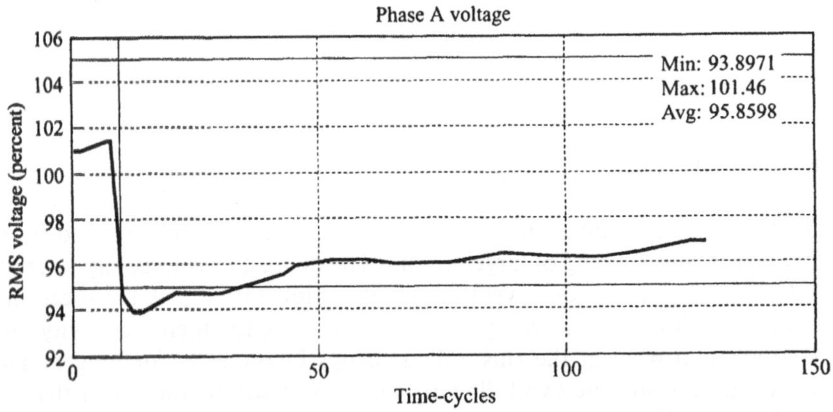

Figure 4.2 A voltage sag due to induction motor starting. (Data obtained from Electrotek Concepts [19].)

example of such a voltage sag [19]. Comparing this figure with Fig. 4.1 shows that no longer the actual voltage as a function of time is given but the rms voltage versus time. The rms voltage is typically calculated every cycle or half-cycle of the power system frequency. Voltage sags due to induction motor starting last longer than those due to short circuits. Typical durations are seconds to tens of seconds. The remainder of this chapter will concentrate on voltage sags due to short circuits. Voltage sags due to motor starting will be discussed in short in Section 4.9.

4.2 VOLTAGE SAG MAGNITUDE

4.2.1 Monitoring

The magnitude of a voltage sag can be determined in a number of ways. Most existing monitors obtain the sag magnitude from the rms voltages. But this situation might well change in the future. There are several alternative ways of quantifying the voltage level. Two obvious examples are the magnitude of the fundamental (power frequency) component of the voltage and the peak voltage over each cycle or half-cycle. As long as the voltage is sinusoidal, it does not matter whether rms voltage, fundamental voltage, or peak voltage is used to obtain the sag magnitude. But especially during a voltage sag this is often not the case.

4.2.1.1 Rms Voltage. As voltage sags are initially recorded as sampled points in time, the rms voltage will have to be calculated from the sampled time-domain voltages. This is done by using the following equation:

$$V_{rms} = \sqrt{\frac{1}{N}\sum_{i=1}^{N} v_i^2} \tag{4.1}$$

where N is the number of samples per cycle and v_i are the sampled voltages in time domain.

The algorithm described by (4.1) has been applied to the sag shown in Fig. 4.1. The results are shown in Fig. 4.3 and in Fig. 4.4. In Fig. 4.3 the rms voltage has been calculated over a window of one cycle, which was 256 samples for the recording used. Each point in Fig. 4.3 is the rms voltage over the preceeding 256 points (the first 255 rms values have been made equal to the value for sample 256):

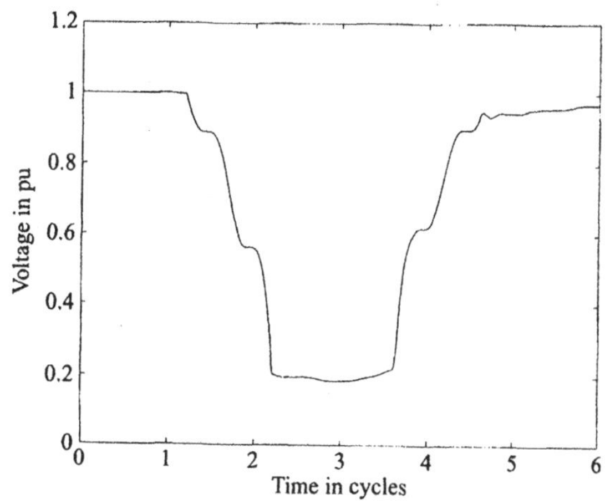

Figure 4.3 One-cycle rms voltage for the voltage sag shown in Fig. 4.1.

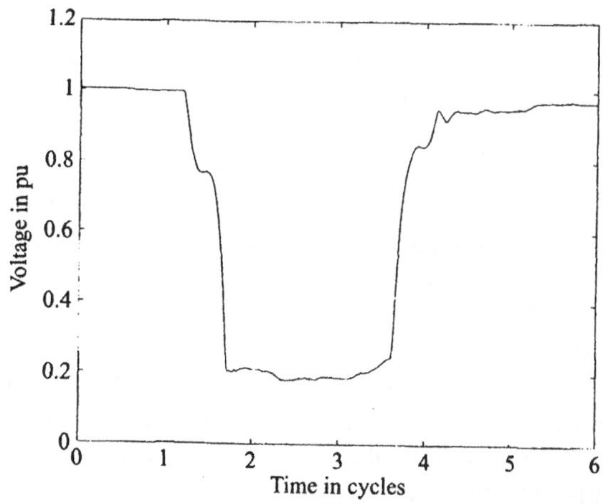

Figure 4.4 Half-cycle rms voltage for the voltage sag shown in Fig. 4.1.

$$V_{rms}(k) = \sqrt{\frac{1}{N}\sum_{i=k-N+1}^{i=k} v_i^2} \tag{4.2}$$

with $N = 256$. We see that the rms voltage does not immediately drop to a lower value but takes one cycle for the transition. We also see that the rms value during the sag is not completely constant and that the voltage does not immediately recover after the fault. A surprising observation is that the rms voltage immediately after the fault is only about 90% of the pre-sag voltage. We will come back to this phenomenon in Section 4.9. From Fig. 4.1 one can see that the voltage in time domain shows a small over-voltage instead. In Fig. 4.4 the rms voltage has been calculated over the preceeding 128 points, $N = 128$ in (4.2). The transition now takes place in one half-cycle. A shorter window than one half-cycle is not useful. The window length has to be an integer multiple of one half-cycle. Any other window length will produce an oscillation in the result with a frequency equal to twice the fundamental frequency. For both figures the rms voltage has been calculated after each sample. In power quality monitors, this calculation is typically made once a cycle:

$$V_{rms}(kN) = \sqrt{\frac{1}{N}\sum_{i=(k-1)N+1}^{i=kN} v_i^2} \tag{4.3}$$

It is thus very likely that the monitor will give one value with an intermediate magnitude before its rms voltage value settles down. We will come back to this when discussing sag duration.

4.2.1.2 Fundamental Voltage Component. Using the fundamental component of the voltage has the advantage that the phase-angle jump can be determined in the same way. The phase-angle jump will be discussed in detail in Section 4.5. The fundamental voltage component as a function of time may be calculated as

$$V_{fund}(t) = \frac{2}{T}\int_{t-T}^{t} v(\tau)e^{j\omega_0\tau}d\tau \tag{4.4}$$

where $\omega_0 = \frac{2\pi}{T}$ and T one cycle of the fundamental frequency. Note that this results in a complex voltage as a function of time. The absolute value of this complex voltage is the voltage magnitude as a function of time; its argument can be used to obtain the phase-angle jump. In a similar way we can obtain magnitude and phase angle of a harmonic voltage component as a function of time. This so-called "time-frequency analysis" is a well-developed area within digital signal processing with a large application potential in power engineering.

The fundamental component has been obtained for the voltage sag shown in Fig. 4.1. The absolute value of the fundamental component is shown in Fig. 4.5. Each point represents the magnitude of the (complex) fundamental component of the previous cycle (256 points). The fundamental component of the voltage has been obtained through a fast-Fourier transform (fft) algorithm [148]. A comparison with Fig. 4.3 shows that the behavior of the fundamental component is very similar to the behavior of the rms voltage.

The rms voltage has the advantage that it can be applied easily to a half-cycle window. Obtaining the fundamental voltage from a half-cycle window is more complicated. A possible solution is to take a half-cycle window and to calculate the second half-cycle by using

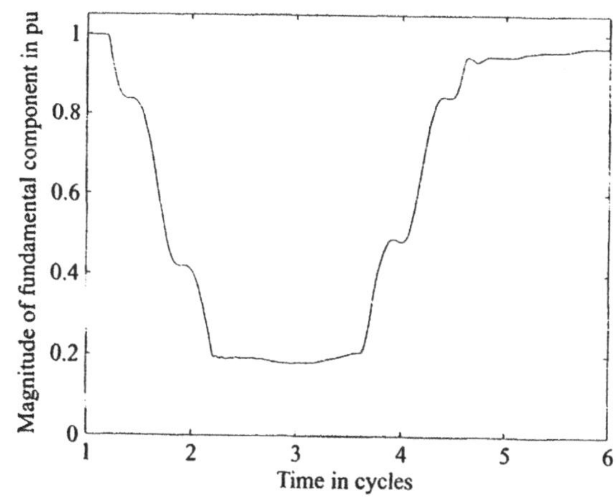

Figure 4.5 Magnitude of the fundamental component of the voltage sag in Fig. 4.1.

$$\cos(\omega t + \phi + \pi) = -\cos(\omega t + \phi) \tag{4.5}$$

Let $v_i, i = 1 \cdots \frac{N}{2}$ be the samples voltages over a half-cycle window. The fundamental voltage is obtained by taking the Fourier transform of the following series:

$$v_1 \cdots v_{\frac{N}{2}}, -v_1 \cdots - v_{\frac{N}{2}} \tag{4.6}$$

This algorithm has been applied to the voltage sag shown in Fig. 4.1, resulting in Fig. 4.6. The transition from pre-fault to during-voltage is clearly faster than in Fig. 4.5. Note that this method assumes that there is no dc voltage component present. The presence of a dc voltage component will lead to an error in the fundamental voltage.

An alternative method of obtaining the fundamental voltage component is discussed in Section 4.5.

4.2.1.3 Peak Voltage. The peak voltage as a function of time can be obtained by using the following expression:

$$V_{peak} = \max_{0 < \tau < T} |v(t - \tau)| \tag{4.7}$$

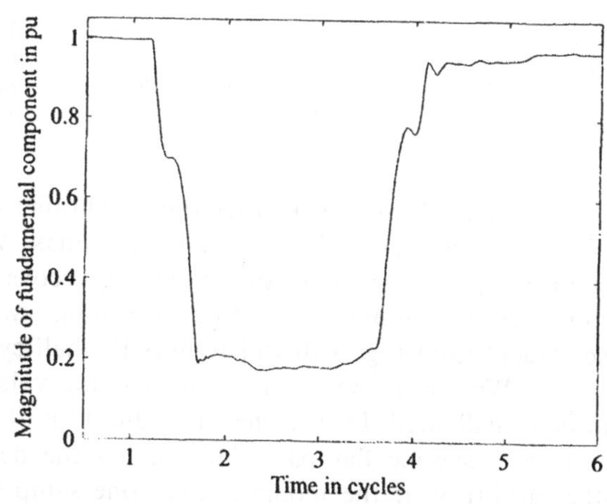

Figure 4.6 Magnitude of the fundamental component of the voltage sag in Fig. 4.1, obtained by using a half-cycle window.

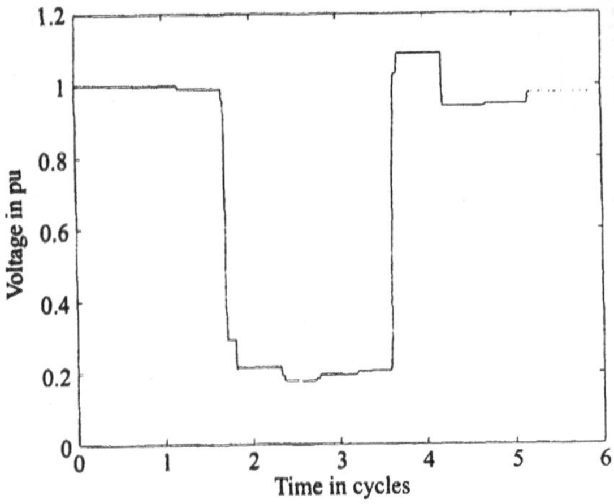

Figure 4.7 Half-cycle peak voltage for the voltage sag shown in Fig. 4.1.

with $v(t)$ the sampled voltage waveform and T an integer multiple of one half-cycle. In Fig. 4.7, for each sample the maximum of the absolute value of the voltage over the preceding half-cycle has been calculated. We see that this peak voltage shows a sharp drop and a sharp rise, although we will see later that they do not coincide with commencement and clearing of the sag. Contrary to the rms voltage, the peak voltage shows an overshoot immediately after the sag, which corresponds to the overvoltage in time domain. The two methods are compared in Fig. 4.8. We see that the peak voltage tends to be higher most of the time with the exception of the end of the deep part of the sag.

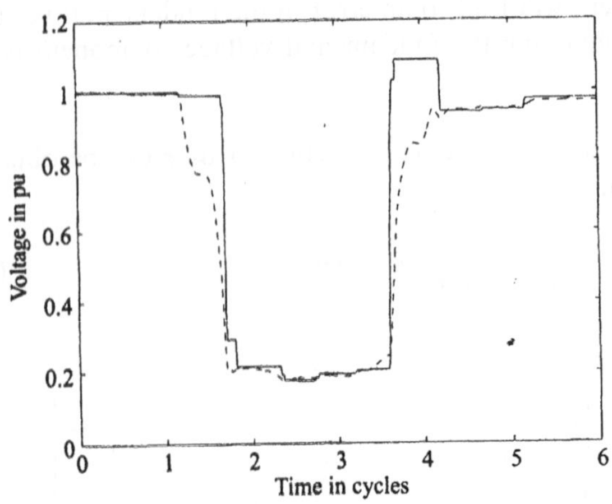

Figure 4.8 Comparison between half-cycle peak (solid line) and half-cycle rms voltage (dashed line) for the voltage sag shown in Fig. 4.1.

4.2.1.4 A One-Cycle Voltage Sag. Another example of a voltage sag is shown in Fig. 4.9; contrary to Fig. 4.1, all three phase voltages are shown. The voltage is low in one phase for about one cycle and recovers rather fast after that. The other two phases show some transient phenomenon, but no clear sag or swell. The latter is also evident from Fig. 4.10 which gives the half-cycle rms value for the sag shown in Fig. 4.9. We see in the latter figure that the voltage in the two non-faulted phases shows a small swell. Due to the short duration of the sag the rms voltage curve does not have a specific flat part. This makes the determination of the sag magnitude rather arbitrary. If the monitor takes one sample every half-cycle the resulting sag

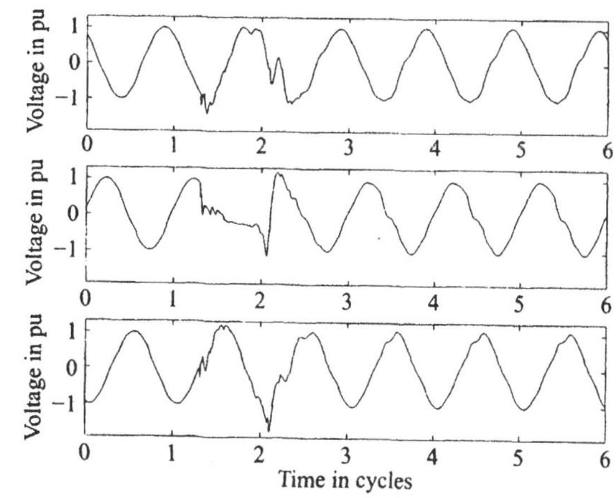

Figure 4.9 Time-domain plot of a one-cycle sag, plots of the three phase voltages. (Data obtained from [16].)

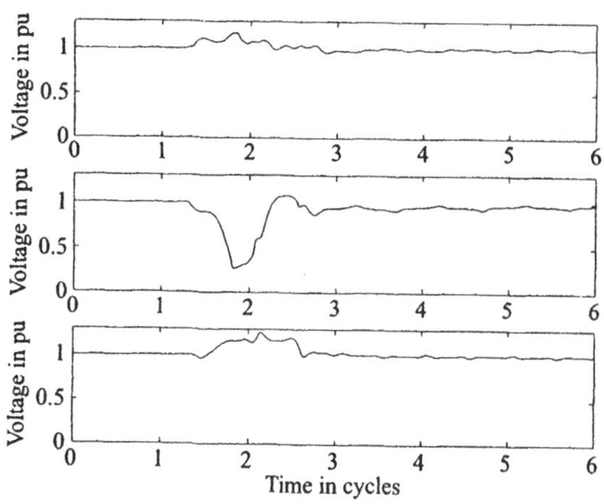

Figure 4.10 Half-cycle rms voltages for the voltage sag shown in Fig. 4.9.

magnitude can be anywhere between 26% and 70% depending on the moment at which the sample is taken. In case a one-cycle window is used to calculate the rms voltage, the situation becomes worse.

The two alternative methods for obtaining the sag magnitude versus time have also been applied to phase b of the event in Fig. 4.9. The half-cycle peak voltage is shown in Fig. 4.11, the half-cycle fundamental voltage component in Fig. 4.12. The shape of the latter is similar to the shape of the half-cycle rms. The half-cycle peak voltage again shows a much sharper transition than the other two methods.

4.2.1.5 Obtaining One Sag Magnitude. Until now, we have calculated the sag magnitude as a function of time: either as the rms voltage, as the peak voltage, or as the fundamental voltage component obtained over a certain window. There are various ways of obtaining one value for the sag magnitude from the magnitude as a function of time. Most monitors take the lowest value. Thinking about equipment sensitivity, this corresponds to the assumption that the equipment trips instantaneously when the voltage drops below a certain value. As most sags have a rather constant rms value during the deep part of the sag, using the lowest value appears an acceptable assumption.

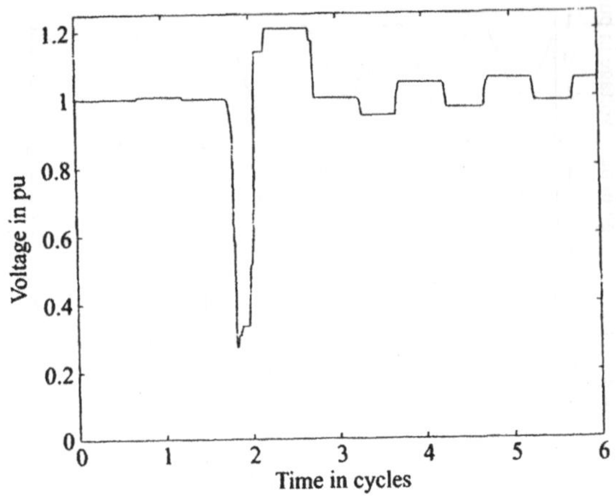

Figure 4.11 Half-cycle peak voltage for phase b of the sag shown in Fig. 4.9.

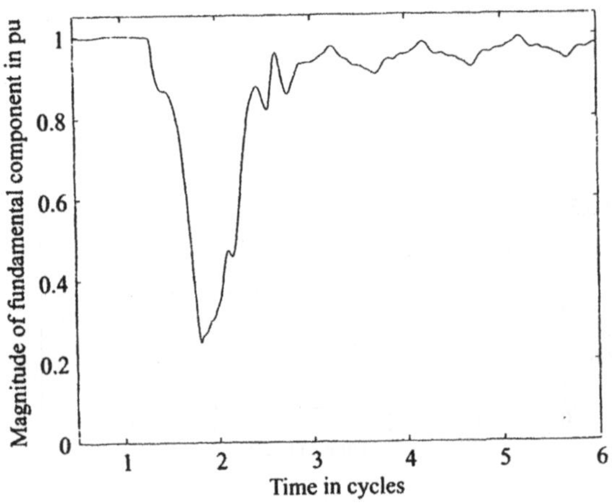

Figure 4.12 Half-cycle fundamental voltage for phase b of the sag shown in Fig. 4.9.

So far there is rather general agreement, both about using the rms value, and about taking the lowest rms value to determine the sag magnitude. But when the sag magnitude needs to be quantified in a number, the agreement is no longer there. One common practice is to characterize the sag through the remaining voltage during the sag. This is then given as a percentage of the nominal voltage. Thus, a 70% sag in a 120 volt system means that the voltage dropped to 84 V. This method of characterizing the sag is recommended in a number of IEEE standards (493–1998, 1159–1995, 1346–1998). The confusion with this terminology is clear. One could be tricked into thinking that a 70% sag refers to a drop of 70%, thus a remaining voltage of 30%. The recommendation is therefore to use the phrase "a sag down to 70%" [3]. The IEC has solved this ambiguity by characterizing the sag through the actual drop in the rms voltage [4]. This has somewhat become common practice in Europe. Characterizing a sag through its drop in voltage does not solve all problems however, because the next question will be: What is the reference voltage? There are arguments in favor of using the pre-fault voltage and there are arguments in favor of using the nominal voltage. The International Union of Producers and Distributors of Electrical Energy (Union International des Producteurs et Distributeurs d'Energie Electrique, UNIPEDE) recommends to use the nominal voltage as a reference [5]. As several definitions are in use, it is important to clearly define the way in which the sag magnitude is defined. In this book sag magnitude is defined as the remaining voltage during the event.

Using the remaining voltage as the sag magnitude, leads to some obvious confusions. The main source of confusion is that a larger sag magnitude indicates a less severe event. In fact, a sag magnitude of 100% corresponds to no sag at all. The use of terms like "large sag" and "small sag" would be extremely confusing. Instead we will talk about a "deep sag" and a "shallow sag." A deep sag is a sag with a low magnitude; a shallow sag has a large magnitude. When referring to equipment behavior we will also use the terms "severe sag" and "mild sag." As far as magnitude is concerned, these terms correspond to "deep sag" and "shallow sag," respectively.

4.2.2 Theoretical Calculations

Consider the power system shown in Fig. 4.13, where the numbers (1 through 5) indicate fault positions and the letters (A through D) loads. A fault in the transmission network, fault position 1, will cause a serious sag for both substations bordering the faulted line. This sag is then transferred down to all customers fed from these two substations. As there is normally no generation connected at lower voltage levels, there is nothing to keep up the voltage. The result is that a deep sag is experienced by all customers A, B, C, and D. The sag experienced by A is likely to be somewhat less deep, as the generators connected to that substation will keep up the voltage. A fault at position 2 will not cause much voltage drop for customer A. The impedance of the transformers between the transmission and the sub-transmission system are large enough to considerably limit the voltage drop at high-voltage side of the transformer. The sag experienced by customer A is further mitigated by the generators feeding in to its local transmission substation. The fault at position 2 will, however, cause a deep sag at both subtransmission substations and thus for all customers fed from here (B, C, and D).

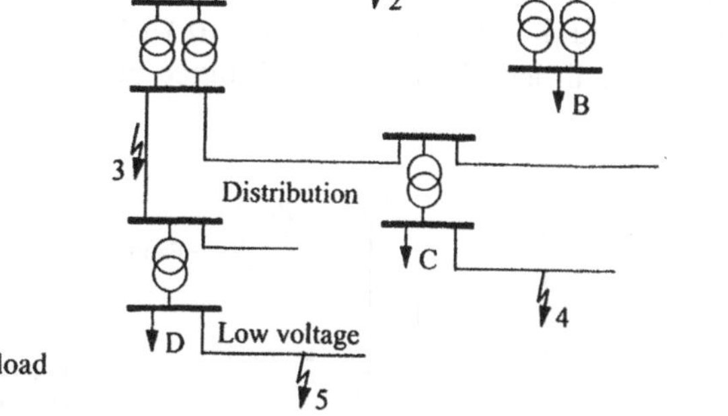

Figure 4.13 Distribution network with load positions and fault positions.

A fault at position 3 will cause a very deep sag for customer D, followed by a short or long interruption when the protection clears the fault. Customer C will only experience a deep sag. If fast reclosure is used in the distribution system, customer C will experience two or more sags shortly after each other for a permanent fault. Customer B will only experience a shallow sag due to the fault at position 3, again due to the transformer impedance. Customer A will probably not notice anything from this fault. Finally, fault 4 will cause a deep sag for customer C and a shallow one for customer D. For fault 5 the result is just the other way around: a deep sag for customer D and a shallow one for customer C. Customers A and B will not be influenced at all by faults 4 and 5.

To quantify sag magnitude in radial systems, the voltage divider model, shown in Fig. 4.14, can be used. This might appear a rather simplified model, especially for transmission systems. But as we will see in the course of this and further chapters, it has turned out to be a rather useful model to predict some of the properties of sags. In Fig. 4.14 we see two impedances: Z_S is the source impedance at the point-of-common coupling; and Z_F is the impedance between the point-of-common coupling and the fault. The point-of-common coupling is the point from which both the fault and the load are fed. In other words: it is the place where the load current branches off from the fault current. We will often abbreviate "point-of-common coupling" as pcc. In the voltage divider model, the load current before as well as during the fault is neglected. There is thus no voltage drop between the load and the pcc. The voltage at the pcc, and thus the voltage at the equipment terminals, can be found from

$$V_{sag} = \frac{Z_F}{Z_S + Z_F} E \tag{4.8}$$

In the remainder of this chapter, we will assume that the pre-event voltage is exactly 1 pu, thus $E = 1$. This results in the following expression for the sag magnitude

$$V_{sag} = \frac{Z_F}{Z_S + Z_F} \tag{4.9}$$

Any fault impedance should be included in the feeder impedance Z_F. We see from (4.9) that the sag becomes deeper for faults electrically closer to the customer (when Z_F becomes smaller), and for systems with a smaller fault level (when Z_S becomes larger).

Note that a single-phase model has been used here, whereas in reality the system is three-phase. That means that this equation strictly speaking only holds for three-phase faults. How the voltage divider model can be used for single-phase and phase-to-phase faults is discussed in Section 4.4.

Equation (4.9) can be used to calculate the sag magnitude as a function of the distance to the fault. Therefore we have to write $Z_F = z \times \mathcal{L}$, with z the impedance of the feeder per unit length and $\mathcal{L}$ the distance between the fault and the pcc, leading to

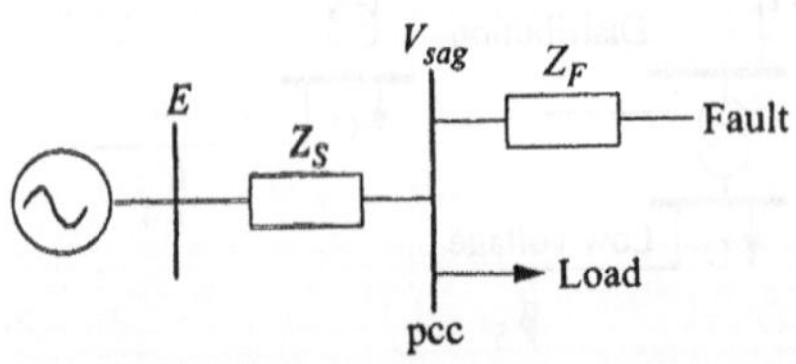

Figure 4.14 Voltage divider model for a voltage sag.

$$V_{sag} = \frac{z\mathcal{L}}{Z_S + z\mathcal{L}} \tag{4.10}$$

The sag magnitude as a function of the distance to the fault has been calculated for a typical 11 kV overhead line, resulting in Fig. 4.15. For the calculations a 150 mm² overhead line was used and fault levels of 750 MVA, 200 MVA, and 75 MVA. The fault level is used to calculate the source impedance at the pcc, the feeder impedance to calculate the impedance between the pcc and the fault. It was assumed that the source impedance is purely reactive, thus $Z_S = j0.161\ \Omega$ for the 750 MVA source. The impedance of the 150 mm² overhead line is $0.117 + j0.315\ \Omega$ per km [10].

As expected, the sag magnitude increases (i.e., the sag becomes less severe) for increasing distance to the fault and for increasing fault level. We also see that faults at tens of kilometers distance may still cause a severe sag.

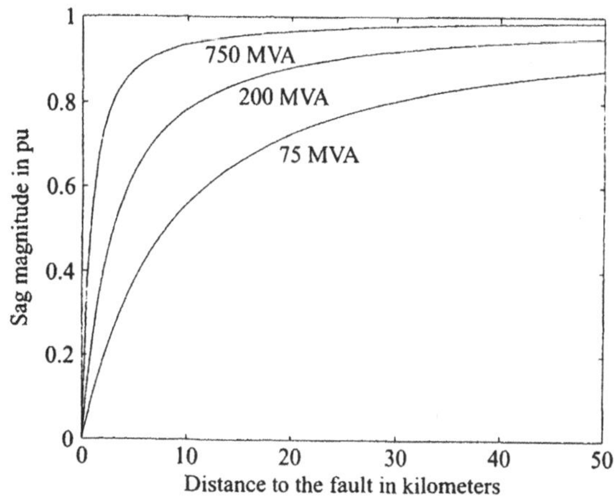

Figure 4.15 Sag magnitude as a function of the distance to the fault, for faults on an 11 kV, 150 mm² overhead line.

4.2.2.1 Influence of Cross Section. Overhead lines of different cross section have different impedance, and lines and cables also have different impedance. It is thus to be expected that the cross section of the line or cable influences the sag magnitude as well. To show this influence, Fig. 4.16 plots the sag magnitude at the pcc

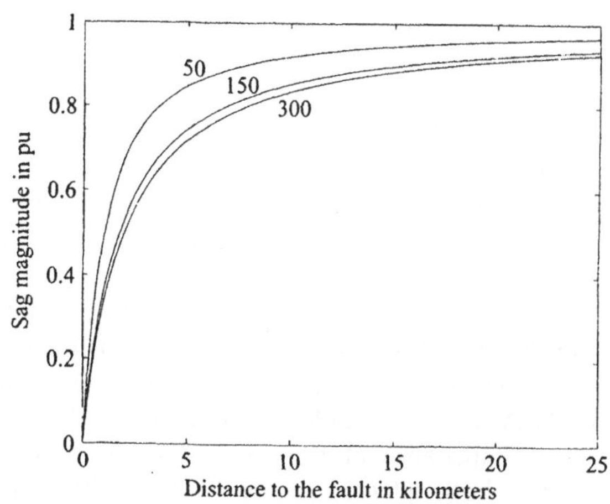

Figure 4.16 Sag magnitude versus distance, for 11 kV overhead lines with different cross sections.

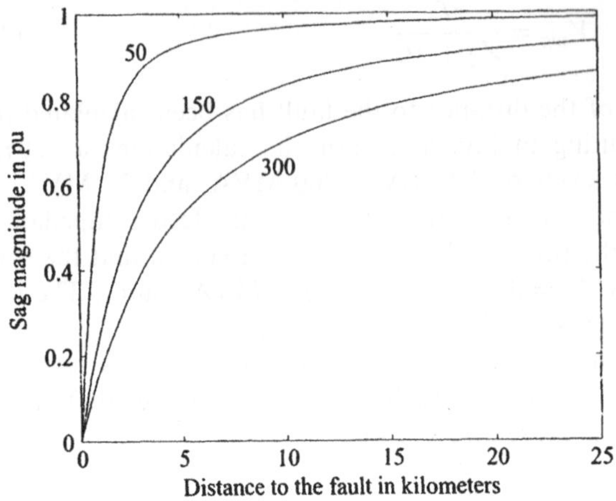

Figure 4.17 Sag magnitude versus distance, for 11 kV underground cables with different cross sections.

as a function of the distance between the fault and the pcc, for 11 kV overhead lines with three different cross sections: 50, 150, and 300 mm^2. A source impedance of 200 MVA has been used. The smaller the cross section, the higher the impedance of the feeder and thus the lower the voltage drop. For overhead lines, the influence is rather small as the reactance dominates the impedance. For underground cables, the influence is much bigger as shown in Fig. 4.17, again for cross sections of 50, 150, and 300 mm^2. The inductance of cables is significantly smaller than for overhead lines, so that the resistance has more influence on the impedance and thus on the sag magnitude. The impedance values used to obtain Fig. 4.16 and Fig. 4.17 are given in Table 4.1. All impedances are for an 11 kV voltage level.

TABLE 4.1 Line and Cable Impedances for 11 kV Feeders Used in Figs. 4.16 and 4.17

	Impedance	
Cross Section	Overhead Line	Cable
50 mm^2	$0.363 + j0.351\ \Omega$	$0.492 + j0.116\ \Omega$
150 mm^2	$0.117 + j0.315\ \Omega$	$0.159 + j0.097\ \Omega$
300 mm^2	$0.061 + j0.298\ \Omega$	$0.079 + j0.087\ \Omega$

Source: Data obtained from [10].

4.2.2.2 Faults behind Transformers. The impedance between the fault and the pcc in Fig. 4.14 not only consists of lines or cables but also of power transformers. As transformers have a rather large impedance, among others to limit the fault level on the low-voltage side, the presence of a transformer between the fault and the pcc will lead to relatively shallow sags.

To show the influence of transformers on the sag magnitude, consider the situation shown in Fig. 4.18: a 132/33 kV transformer is fed from the same bus as a 132 kV line. A 33 kV line is fed from the low-voltage side of the transformer. Fault levels are 3000 MVA at the 132 kV bus, and 900 MVA at the 33 kV bus. In impedance terms, the source impedance at the 132 kV bus is 5.81 Ω, and the transformer impedance is 13.55 Ω, both referred to the 132 kV voltage level. The sensitive load for which we

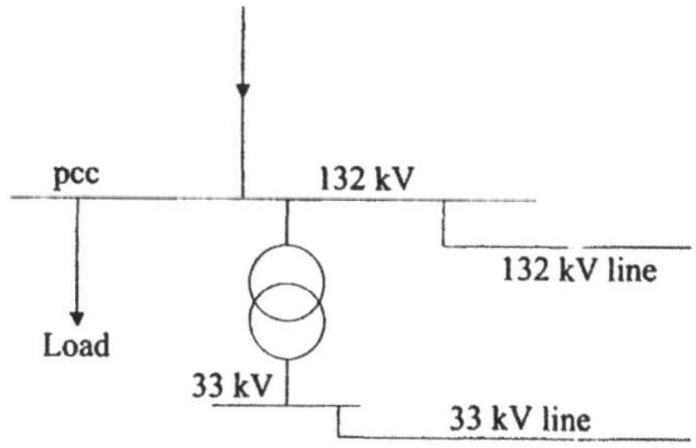

Figure 4.18 Power system with faults at two voltage levels.

want to calculate the sag magnitude is fed from the 132 kV bus via another 132/33 kV transformer. We can again use (4.9), where $Z_S = 5.81\,\Omega$, $Z_F = 13.55\,\Omega + z \times \mathcal{L}$, z is the feeder impedance per unit length, and $\mathcal{L}$ the distance between the fault and the transformer's secondary side terminals. The feeder impedance must also be referred to the 132 kV level: $z = \left(\frac{132\,kV}{33\,kV}\right)^2 \times 0.3\,\Omega/\text{km}$ when the feeder impedance is $0.3\,\Omega/\text{km}$ at 33 kV. The results of the calculations are shown in Fig. 4.19 for faults on the 33 kV line (upper curve) and for faults on the 132 kV line (lower curve). We see that sags due to 33 kV faults are less severe than sags due to 132 kV faults. Not only does the 33 kV curve start off at a higher level (due to the transformer impedance), it also rises much faster. The latter is due to the fact that the feeder impedance seen from the 132 kV level is $(132/33)^2 = 16$ times as high as that seen from the 33 kV level.

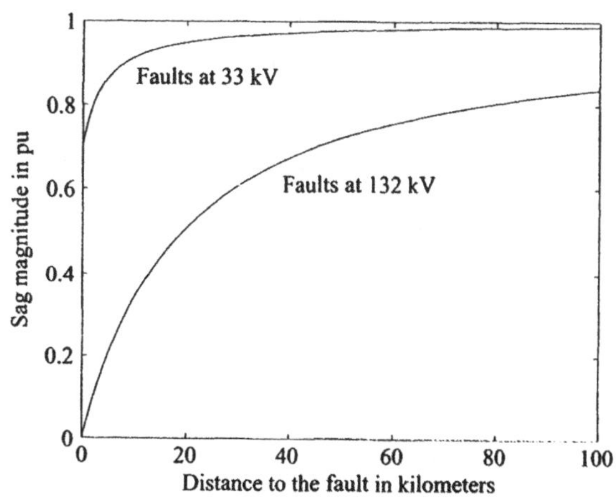

Figure 4.19 Comparison of sag magnitude for 132 kV and 33 kV faults.

4.2.2.3 Fault Levels. Often the source impedance at a certain bus is not immediately available, but instead the fault level is. One can of course translate the fault level into a source impedance and use (4.9) to calculate the sag magnitude. But one may calculate the sag magnitude directly if the fault levels both at the pcc and at the fault position are known. Let S_{FLT} be the fault level at the fault position and S_{pcc} at the point-of-common coupling. For a rated voltage V_n the relations between fault level and source impedance are as follows:

$$S_{FLT} = \frac{V_n^2}{Z_S + Z_F} \tag{4.11}$$

$$S_{pcc} = \frac{V_n^2}{Z_S} \tag{4.12}$$

With (4.9) the voltage at the pcc can be written as

$$V_{sag} = 1 - \frac{S_{FLT}}{S_{pcc}} \tag{4.13}$$

We use (4.13) to calculate the magnitude of sags behind transformers. For this we use typical fault levels in the U.K. power system [13]:

400 V	20 MVA
11 kV	200 MVA
33 kV	900 MVA
132 kV	3000 MVA
400 kV	17000 MVA

Consider a fault at a typical 11 kV bus, i.e., with a fault level of 200 MVA. The voltage sag at the high-voltage side of the 33/11 kV transformer is from (4.13)

$$V_{sag} = 1 - \frac{200\ \text{MVA}}{900\ \text{MVA}} = 78\%$$

In a similar way the whole of Table 4.2 has been filled. The zeros in this table indicate that the fault is at the same or at a higher voltage level. The voltage drops to a low value in such a case. We can see from Table 4.2 that sags are significantly damped when they propagate upwards in the power system. In a sag study we typically only have to take faults one voltage level down into account. And even those are seldom of serious concern. An exception here could be sags due to faults at 33 kV with a pcc at 132 kV. They could lead to sags down to 70%.

TABLE 4.2 Upward Propagation of Sags

	Point-of-Common Coupling			
Fault Point	11 kV	33 kV	132 kV	400 kV
400 V	90%	98%	99%	100%
11 kV	0	78%	93%	99%
33 kV	0	0	70%	95%
132 kV	0	0	0	82%

4.2.2.4 Critical Distance. Equation (4.10) gives the voltage magnitude as a function of the distance to the fault. From this equation we can obtain the distance at which a fault will lead to a sag of a certain magnitude. If we assume equal X/R ratio of source and feeder, we obtain

$$\mathcal{L}_{crit} = \frac{Z_S}{Z} \times \frac{V}{1 - V} \tag{4.14}$$

We refer to this distance as the critical distance for a voltage V. Suppose that a piece of equipment trips when the voltage drops below a certain level (the critical voltage). The definition of critical distance is such that each fault within the critical distance will cause the equipment to trip. This concept will be used in Section 6.5 to estimate the expected number of equipment trips.

If we assume further that the number of faults is proportional to the line length within the critical distance, we would expect that the number of sags below a level V is proportional to $V/(1 - V)$. Another assumption is needed to arrive at this conclusion. Every feeder connected to every pcc needs to be infinitely long without any branching off. Of course this is not the case in reality. Still this equation has been compared with a number of large power quality surveys. The results are shown in Fig. 4.20. Power quality survey results in the United States [11], [12], in the U.K. [13] and in Norway [16] are indicated as dots, the theoretical curve is shown as a solid line. The correspondence is good, despite the obviously serious approximations made.

Even though (4.14) only holds for radial systems, it gives a generally usable relation between the number of voltage sags and the voltage. The expression clearly shows that the majority of sags are shallow, a fact confirmed by most measurements.

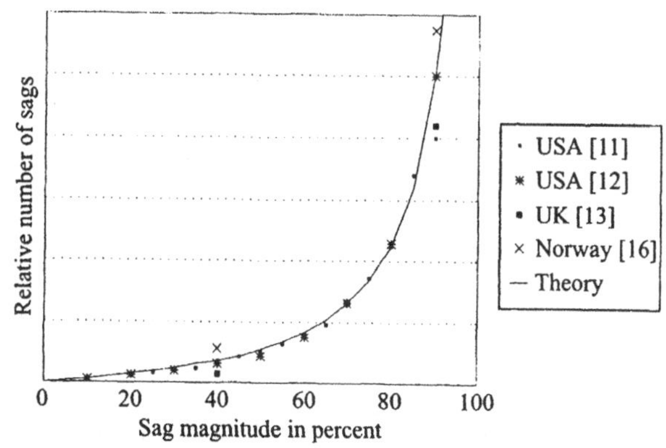

Figure 4.20 Number of sags versus magnitude: theoretical results (solid line) versus monitoring results (dots).

4.2.3 Example of Calculation of Sag Magnitude

We will apply the theoretical concepts developed in the previous sections to the supply shown schematically in Fig. 4.21. This same example will be used again in forthcoming parts of this book. The supply shown in Fig. 4.21 is the existing supply to an industrial customer somewhere in the North of England [15]. The sensitive load consists of several large ac and dc adjustable-speed drives. The dc drives are fed via dedicated transformers at 420 V, the more modern ac drives at 660 V. Most of the data used for the various calculations below have been obtained from the local utility. Where no data was available, data have been used which was considered "as typical as possible." Like often in these kind of studies, the collection of the data requires at least as much effort as the actual calculations. In the rest of this book it will always be assumed that all the required data is readily available.

The first step in a sag analysis is to recognize the possible pcc's. For any fault on one of the 11 kV feeders, the fault current will flow through the STU-11 bus, but not further towards the load. The STU-11 bus is thus the pcc for all faults within the 11 kV network. In the same way, the ROS-33 bus is the pcc for faults on any of the 33 kV feeders. The other possible pcc's are PAD-132 and PAD-400. To calculate the sag magnitude we need the source impedance and the feeder impedance. The source

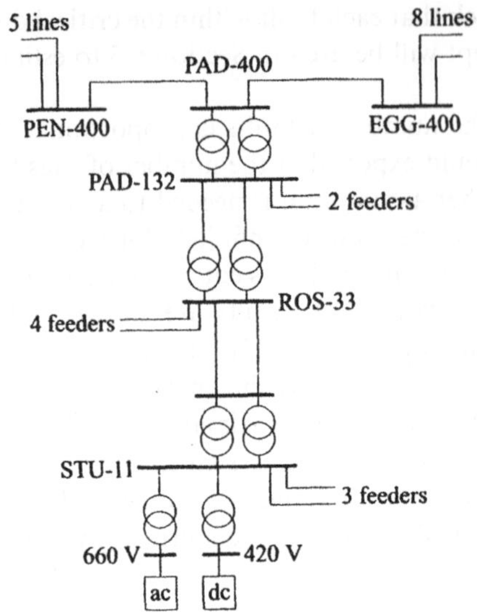

Figure 4.21 Example of power supply to be used for voltage sag calculations.

impedance is given in Table 4.3, the feeder impedance in Table 4.4. All impedances are given for a 100 MVA base. Finally, Table 4.5 gives the transformer connection and neutral grounding. This information is needed in later sections, when unbalanced sags are discussed.

For now we ignore the fact that the impedances are complex and use the absolute values for our calculations. We will come back to the complex impedances in Section 4.5 when phase-angle jumps are discussed. For faults at 11 kV we obtain for the impedances: $z = 27.75\%$ per km and $Z_S = 66.08\%$. The critical distance can be calculated from $\mathcal{L}_{crit} = 2.381 \times \frac{V}{1-V}$.

Calculations for the critical distances at 33 kV and 132 kV proceed in exactly the same way as for the 11 kV system. The results of these calculations are shown in Table 4.6. We see that there are two columns for the 400 kV system in Table 4.3 and in Table 4.6. This has to do with the fact that there are two possible sources for the short-circuit power. If the fault is somewhere between PAD-400 and PEN-400 the fault current will be delivered from the direction of EGG-400. Thus, for such a fault, the impedance Z_S is the source impedance as seen in the direction of EGG-400. The critical distances resulting from this source impedance are shown in Table 4.6 in the column labeled "toward PEN-400." Note that for this the source impedance in the direction of EGG-400 has been used. For faults in the direction of EGG-400, the source impedance in the direction of PEN-400 has been used. Those results are shown in the column labeled "toward EGG-400."

When interpreting Table 4.6 one should realize that these values hold for a radial system with infinitely long lines without any side branches. In reality all feeders have a finite length. In this system the maximum distance from the pcc for a fault at 11 kV is 5 km. The distance to the fault can thus not be more than 5 km and the magnitude of the most shallow sag due to a fault at 11 kV is

$$V_{sag} = \frac{Z_F}{Z_S + Z_F} = \frac{5 \times 0.2727}{5 \times 0.2727 + 0.6608} = 67\% \tag{4.15}$$

Figure 4.22 plots sag magnitude versus distance for faults at all the voltage levels in Fig. 4.21. The horizontal scale is determined by the maximum length of the feeders at that

TABLE 4.3 Source Impedance for the Supply Shown in Fig. 4.21, at a 100 MVA Base

	Zero Sequence	Positive and Negative Sequence
11 kV	787 + j220%	4.94 + j65.9%
33 kV	251%	1.23 + j18.3%
132 kV	0.047 + j2.75%	0.09 + j2.86%
400 kV		
From EGG	0.329 + j2.273%	0.084 + j1.061%
From PEN	0.653 + j5.124%	0.132 + j1.94%

TABLE 4.4 Feeder Data for the Supply Shown in Fig. 4.21

	Positive and Negative Sequence	Zero Sequence	Max Length
11 kV	9.7 + j26%/km	18.4 + j112%/km	5 km
33 kV	1.435 + j3.102%/km	2.795 + j15.256%/km	10 km
132 kV	0.101 + j0.257%/km	0.23 + j0.65%/km	2 km
400 kV	0.001 + j0.018%/km	0.007 + j0.050%/km	> 1000 km

TABLE 4.5 Transformer Connections and Neutral Grounding for the Supply Shown in Fig. 4.21

Voltage Level	Transformer Winding Connection	Neutral Grounding at LV Side
400 kV		solidly grounded
400/132 kV	YY autotransformer	solidly grounded
132/33 kV	Star – Delta	resistance grounded through zig-zag transformer
33/11 kV	Delta – Star	resistance grounded
11 kV/660 V and 11 kV/420 V	Delta – Star	solidly grounded

TABLE 4.6 Critical Distance Calculation for the Network Shown in Fig. 4.21, According to (4.14)

	11 kV	33 kV	132 kV	400 kV Toward PEN-400	400 kV Toward EGG-400
z	27.27%	3.418%	0.276%	0.018%	0.018%
Z_S	66.08%	18.34%	2.861%	1.064%	1.944%
$V = 10\%$	0.3 km	0.6 km	1.2 km	6.6 km	12.0 km
$V = 30\%$	1.0 km	2.3 km	4.4 km	25.3 km	46.3 km
$V = 50\%$	2.4 km	5.4 km	10.4 km	59.1 km	108 km
$V = 70\%$	5.6 km	12.5 km	24.2 km	138 km	252 km
$V = 90\%$	21.4 km	48.3 km	93.3 km	532 km	972 km

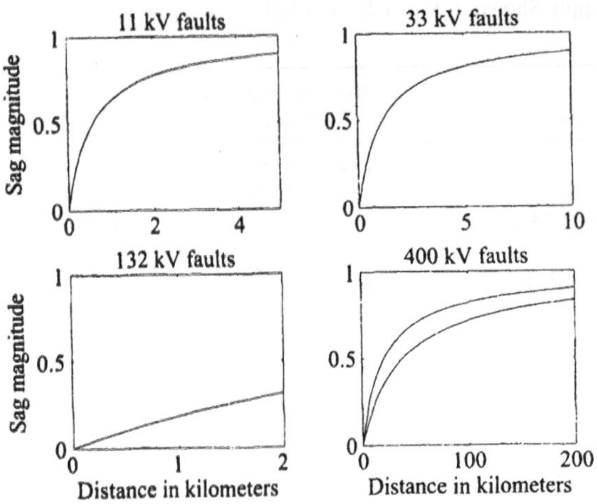

Figure 4.22 Magnitude versus distance for faults at various voltage levels in the supply in Fig. 4.21.

voltage level. For 400 kV a length of 200 km has been taken. The short length of the 132 kV feeders makes that sags due to faults at 132 kV are always very deep.

4.2.4 Sag Magnitude in Non-Radial Systems

In Section 4.2.2 we discussed sag magnitude versus distance in radial systems. Radial systems are common in low-voltage and medium-voltage networks. At higher voltage levels other supply arrangements are common. Some typical cases will be discussed below. We will also present a general way of calculating sag magnitudes in meshed systems.

4.2.4.1 Local Generators. The connection of a local generator to a distribution network, as shown in Fig. 4.23, mitigates voltage sags of the indicated load in two different ways. The generator increases the fault level at the distribution bus, which mitigates voltage sags due to faults on the distribution feeders. This especially holds for a weak system. For a strong system, the fault level cannot be increased much without the risk of exceeding the maximum-allowable short-circuit current of the switchgear. The installation of local generation requires a larger impedance of the feeding transformer.

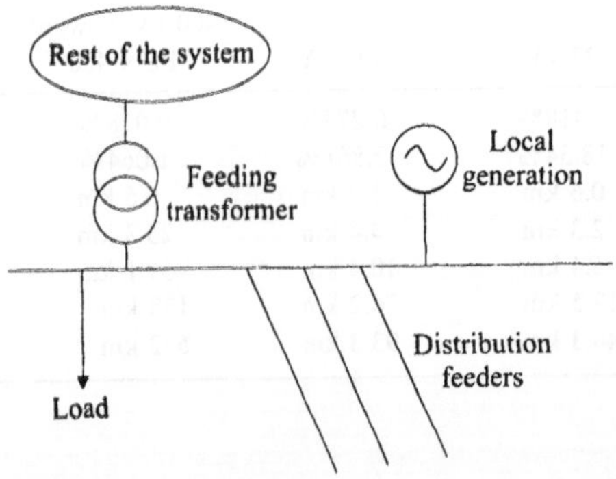

Figure 4.23 Connection of a local generator to a distribution bus.

A local generator also mitigates sags due to faults in the rest of the system. During such a fault the generator keeps up the voltage at its local bus by feeding into the fault. An equivalent circuit to quantify this effect has been drawn in Fig. 4.24: Z_4 is the impedance of the local generator during the fault (typically the transient impedance); Z_1 the source impedance at the pcc; Z_2 the impedance between the fault and the pcc; and Z_3 the impedance between the generator bus and the pcc. Note that the concept of point-of-common coupling strictly speaking no longer holds. This concept, which was introduced for radial networks, assumes one single flow of fault current. By adding a generator close to the load a second flow of fault current is introduced. The pcc as indicated in Fig. 4.24 is the point-of-common coupling before the introduction of the local generator. Without the local generator the voltage at the equipment terminals would be equal to the voltage at the pcc. When a local generator is present, the voltage at the equipment terminals during the sag equals the voltage on the generator bus. This voltage is related to the voltage at the pcc according to the following equation:

$$(1 - V_{sag}) = \frac{Z_4}{Z_3 + Z_4}(1 - V_{pcc}) \tag{4.16}$$

The voltage drop at the generator bus is $\frac{Z_4}{Z_3+Z_4}$ times the voltage drop at the pcc. The voltage drop becomes smaller for larger impedance to the pcc (weaker connection) and for smaller generation impedance (larger generator). The fault contribution of the rest of the system at the generator bus is often mainly determined by the impedance of the feeding transformer. In that case the reduction in voltage drop is approximately equal to the generator contribution to the fault level at the generator bus. Thus, if the generator delivers 50% of the fault current, a sag down to 40% at the pcc (60% voltage drop) will be reduced to a sag down to 70% (30% voltage drop) at the equipment terminals. From (4.16) we can also conclude that there is a non-zero minimum sag magnitude. Even a fault at the pcc will no longer cause a sag down to zero voltage but a sag of magnitude

$$V_{min} = \frac{Z_3}{Z_3 + Z_4} \tag{4.17}$$

For the above-mentioned system, where the local generator is responsible for 50% of the fault level at the generator bus, the lowest sag magnitude due to a fault at a higher voltage level is 50%. During a fault not only local generators contribute to the fault but also induction motors. Using the above reasoning we can conclude that the minimum voltage at the plant bus equals the relative fault level contribution of the induction motors. We will discuss induction motors in more detail in Section 4.8.

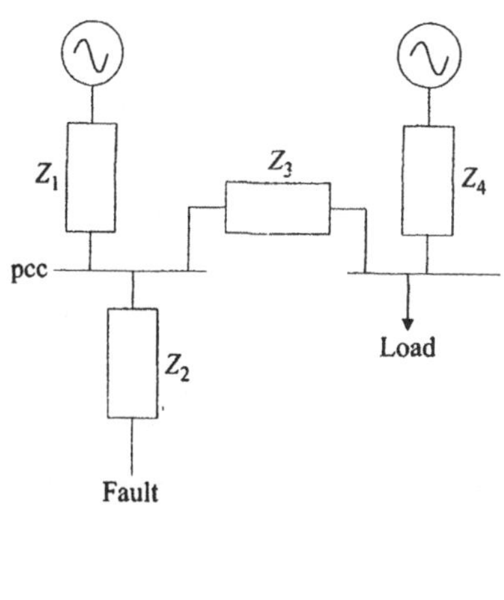

Figure 4.24 Equivalent circuit for system with local generation.

EXAMPLE An example of a system with on-site generation is given in Fig. 4.25: the industrial system is fed from a 66 kV, 1700 MVA substation via two 66/11 kV transformers in parallel. The fault level at the 11 kV bus is 720 MVA, which includes the contribution of two 20 MVA on-site generators with a transient reactance of 17%. The actual industrial load is fed from the 11 kV bus, for which we will calculate the sag magnitude due to faults at 66 kV. The feeder impedance at 66 kV is 0.3 Ω/km.

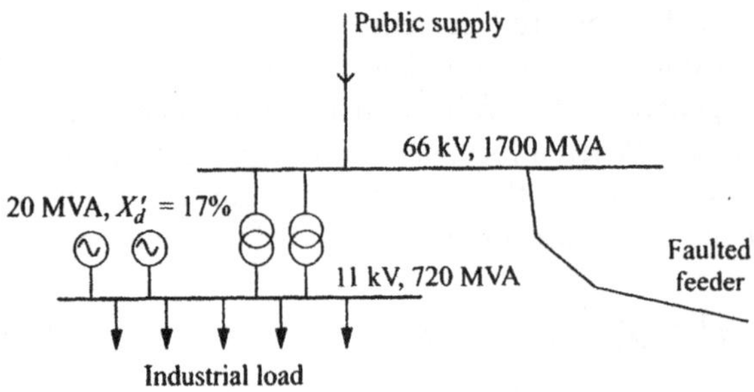

Figure 4.25 Industrial distribution system with on-site generation.

With reference to (4.16) and Fig. 4.24, we get the following impedance values for this system (referred to 66 kV):

$$Z_1 = 2.56\,\Omega$$
$$Z_2 = 0.3\,\Omega/\text{km} \times \mathcal{L}$$
$$Z_3 = 6.42\,\Omega$$
$$Z_4 = 18.5\,\Omega$$

The calculation results are shown in Fig. 4.26. The bottom curve gives the sag magnitude at the 11 kV bus for faults at a 66 kV feeder, when the 11 kV generator is not in operation. In that case the sag magnitude at 11 kV equals the sag magnitude at 66 kV because all load currents have been neglected. The top curve gives the sag magnitude at the 11 kV bus with on-site generator connected. Due to the generator keeping up the voltage at the 11 kV bus, the sag magnitude never drops below 26%. There are two methods to further improve the supply. One can increase the number or size of the generators, which corresponds to decreasing Z_4 in (4.16). Alternatively one can increase Z_3, which leads to a lower fault level at the 11 kV bus.

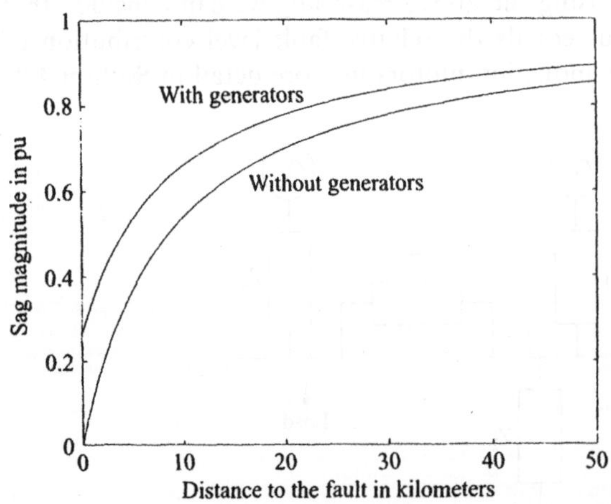

Figure 4.26 Sag magnitude versus distance, with and without on-site generator.

EXAMPLE Another example of the use of (4.16) is given by means of Fig. 4.27. This figure represents half of the transmission system part of the example in Fig. 4.21, containing the substations PAD-400 and EGG-400, plus 30 km of overhead 400 kV line in between them. The impedances have the following values (in % at a 100 MVA base), with $\mathcal{L}$ the distance between EGG-400 and the fault:

$$Z_1 = 1.4\%$$
$$Z_2 = 0.018\%/\text{km} \times \mathcal{L}$$
$$Z_3 = 0.54\%$$
$$Z_4 = 1.94\%$$

The impedance Z_4 represents the source contribution from PEN-400 at PAD-400; Z_3 represents the impedance of 30 km line (0.018%/km); Z_2 the impedance between EGG-400 and the fault, and Z_1 the contribution through the non-faulted lines at EGG-400 (excluding the contribution from PAD-400) during the fault. The latter impedance is likely to be different for faults on different lines. In this study we assumed it to be simply equal to the contribution of all lines at EGG-400 minus the line to PAD-400. As there are a total of nine lines connected to EGG-400 the error made will not be very big.

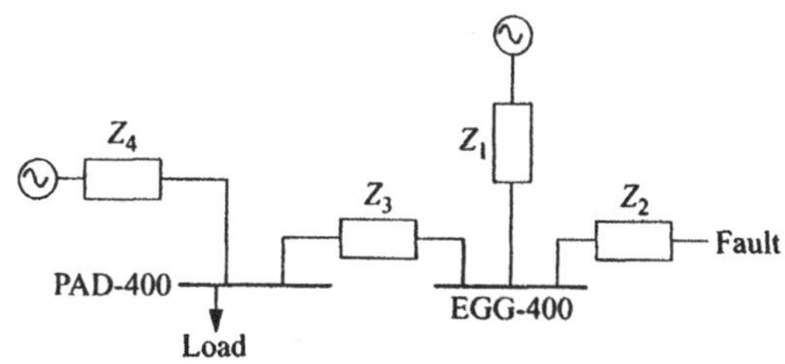

Figure 4.27 Circuit diagram representation of two transmission substations. The sensitive load is fed from the substation on the left.

For faults to the right of EGG-400 we can use (4.16) to calculate the voltage at PAD-400, knowing the voltage at EGG-400. The latter can be obtained from the voltage divider equation with the source impedance formed by the parallel connection of Z_1 and $Z_3 + Z_4$. Note that we still neglect all load currents, so that both source voltages are equal in magnitude and in phase and can be replaced by one source. For faults between PAD-400 and EGG-400 the voltage divider model will give the required voltage directly. The source impedance is now formed by Z_4; the feeder impedance is $0.018\%/\text{km} \times \mathcal{L}$, with $\mathcal{L}$ the distance between PAD-400 and the fault. The resulting sag magnitude as a function of the distance to the fault is shown in Fig. 4.28. For

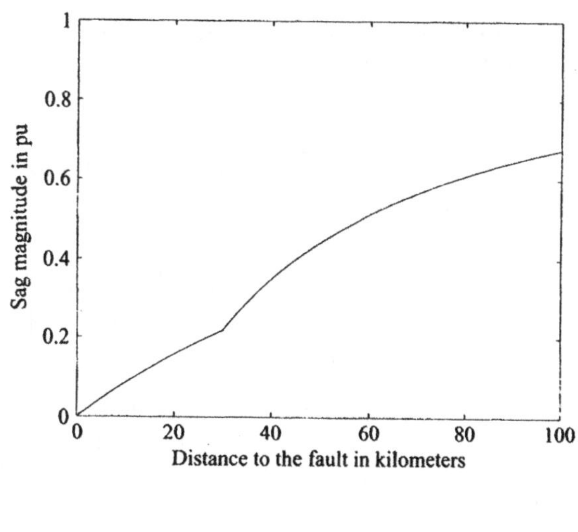

Figure 4.28 Sag magnitude as a function of the distance to the fault, for transmission systems.

distances up to 30 km the sag magnitude changes with distance like in a radial system; for larger distances the magnitude increases faster. Thus, the sag is less severe than for a fault at the same distance in a radial system.

4.2.4.2 Subtransmission Loops. At subtransmission level, the networks often consist of several loops—a typical example is shown in Fig. 4.29. The transmission system is connected to the subtransmission system through two or three transformers. From the busses at the low-voltage side of these transformers a number of substations are fed via a loop. Such a network configuration is also found in industrial power systems. Often the loop only consists of two branches in parallel. The mathematical expressions that will be derived below can also be used to calculate voltage sags due to faults on parallel feeders.

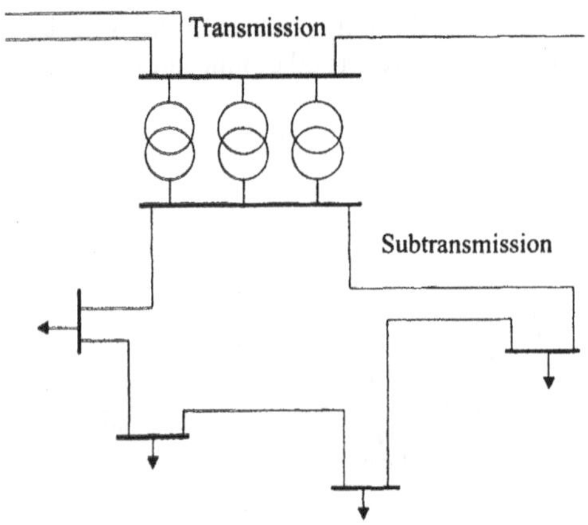

Figure 4.29 Example of subtransmission loop.

To calculate the sag magnitude we need to identify the load bus, the faulted branch, and the non-faulted branch. Knowing these the equivalent scheme in Fig. 4.30 is obtained, where Z_0 is the source impedance at the bus from which the loop is fed; Z_1 is the impedance of the faulted branch of the loop; Z_2 is the impedance of the non-faulted branch; and p is the position of the fault on the faulted branch ($p = 0$ corresponds to a fault at the bus from which the load is fed, $p = 1$ corresponds to a fault at the load bus).

From Fig. 4.30 the voltage at the load bus can be calculated, resulting in the following expression:

$$V_{sag} = \frac{p(1-p)Z_1^2}{Z_0(Z_1+Z_2)+pZ_1Z_2+p(1-p)Z_1^2} \tag{4.18}$$

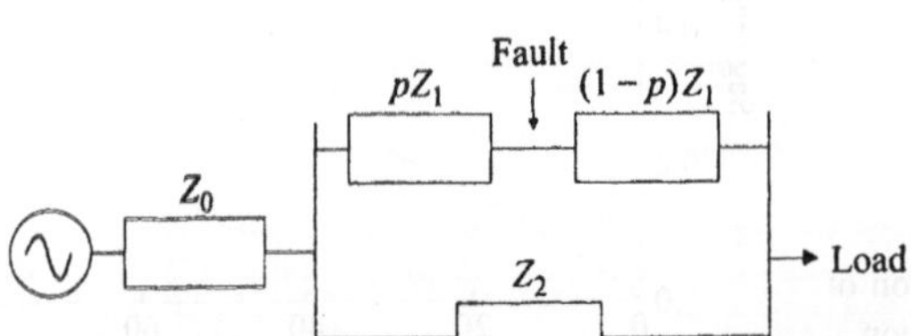

Figure 4.30 Equivalent circuit for subtransmission loop.

The voltage is zero for $p = 0$ (fault at the main subtransmission bus) and for $p = 1$ (fault at the load bus) and has a maximum somewhere in between.

EXAMPLE Consider the system shown in Fig. 4.31: a 125-km 132 kV loop connecting a number of substations. Only the substation feeding the load of interest is shown in the figure. This substation is located at 25 km from the main substation. The fault level at the point-of-supply is 5000 MVA and the feeder impedance 0.3 Ω/km. Faults occur both in the 25 km part and in the 100 km part of the loop, so that both may form the faulted branch. For a fault on the 25 km branch we substitute in (4.18): $Z_1 = 25z$ and $Z_2 = 100z$, with z the feeder impedance per km. For a fault on the 100 km branch, we get $Z_1 = 100z$ and $Z_2 = 25z$.

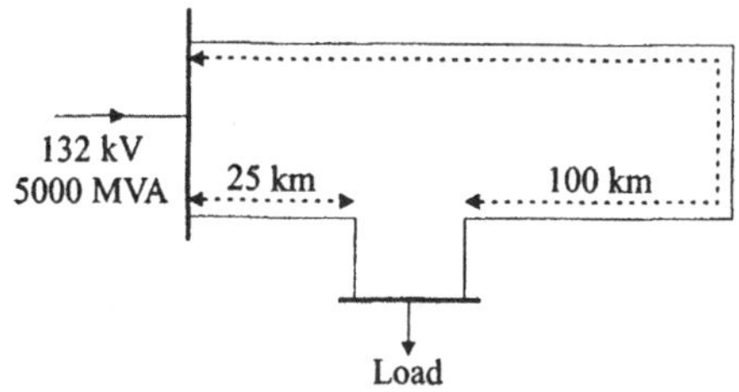

Figure 4.31 Loop system operating at 132 kV.

Figure 4.32 gives the magnitudes of sags due to faults in the 132 kV subtransmission loop. The dashed (top) curve gives the sag magnitude for faults on the 100 km branch, the solid (bottom) curve holds for the 25 km branch. Note that the horizontal scale corresponds to 25 km for the bottom curve and to 100 km for the top curve. Figure 4.33 gives the sag magnitudes for the 100 km and 25 km feeder as a function of the actual distance between the fault and the main 132 kV bus. For comparison, the magnitude is also given for sags due to faults at a radial feeder from the same main 132 kV bus (dotted curve).

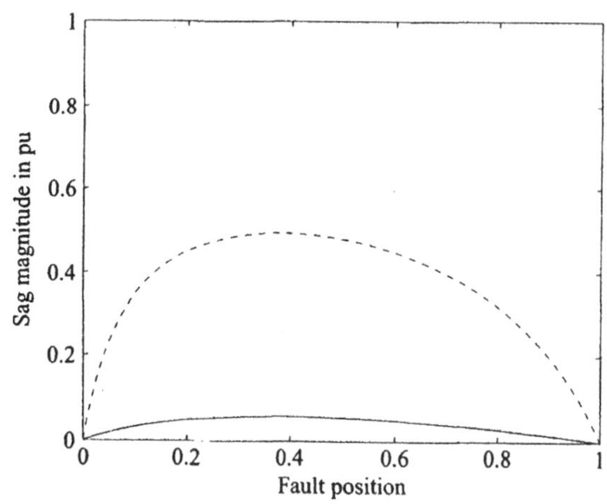

Figure 4.32 Sag magnitudes for faults on a 132 kV loop.

We see from Fig. 4.32 and Fig. 4.33 that each fault on the loop will cause the voltage to drop below 50% of the nominal voltage. A sag due to a fault on a loop is always lower than due to a fault on a radial feeder. Faults close to the point-of-supply will lead to a deep sag. Faults close to the load too. Somewhere in between there is a

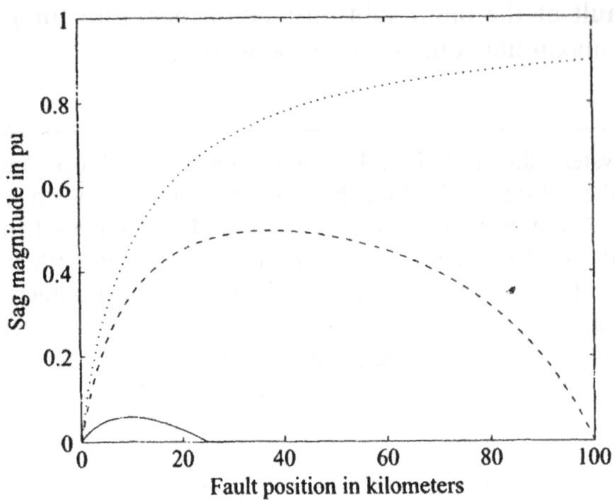

Figure 4.33 Sag magnitude versus distance, for faults on loops (solid and dashed lines) and on a radial feeder (dotted line).

maximum magnitude of the voltage sag due to a fault. The longer the line the higher the maximum. We see from the figure that this maximum is not necessarily in the middle of the branch. The maximum voltage has been calculated as a function of the system parameters. The results are shown in Fig. 4.34 and in Fig. 4.35. To obtain these graphs (4.18) has been rewritten as a function of $z_1 = \frac{Z_1}{Z_0}$ and $z_2 = \frac{Z_2}{Z_0}$; z_1 is the relative impedance of the faulted branch and z_2 of the non-faulted branch. Figure 4.34 gives the maximum voltage as a function of z_2 for various values of z_1 and Fig. 4.35 the other way around. From both figures it follows that the sags become less severe (higher maximum) when the faulted branch becomes longer (higher impedance) and when the non-faulted branch becomes shorter. This can be explained as follows. A longer faulted branch means that the fault can be further away from both busses. A shorter non-faulted branch gives stronger voltage support at the load bus. These relations can easily be understood by considering a fault in the middle of the faulted branch.

The range of values used for both z_1 and z_2 is between 1 and 10. For smaller values of z_1 the sag magnitude becomes very small. Larger values do not give realistic systems. One has to realize that $\frac{1}{Z_0}$ is proportional to the fault level at the point-of-supply. Thus, z_1 and z_2 indicate the variation in fault level for different points in the system. A value of 10 implies that there is at least a factor of six between the highest and the lowest fault level. (Note that the two branches are operated in parallel.) Such a large

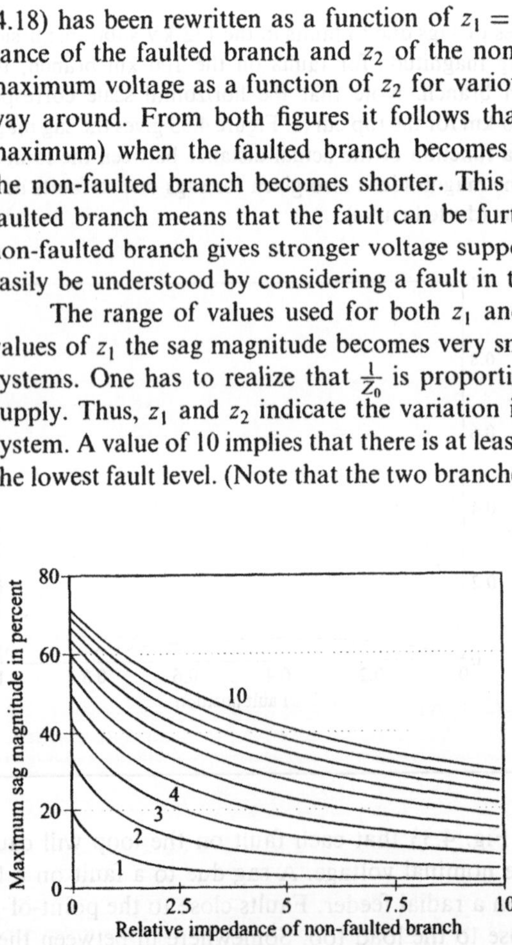

Figure 4.34 Most shallow sag for a fault in a loop, as a function of the impedance of the non-faulted branch for various values of the impedance of the faulted branch.

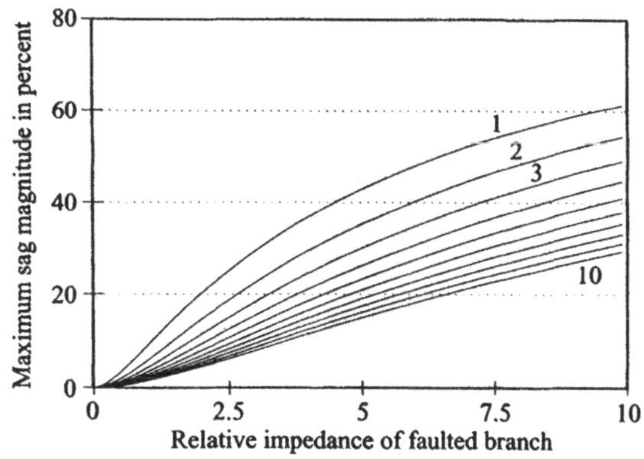

Figure 4.35 Most shallow sag for a fault in a loop, as a function of the impedance of the faulted branch, for various values of the impedance of the non-faulted branch.

range in fault level is rather unlikely in subtransmission systems, as it will lead to large variations in voltage due to load variations.

The general conclusion from Figs. 4.34 and 4.35 is that faults on a loop lead to sags with a magnitude well below 50%, irrespective of the voltage levels. As mentioned before a parallel feeder is a special case of a loop: one in which $z_1 = z_2$. For these we can conclude that the most shallow sag has a magnitude between 20% and 30% for most systems.

4.2.4.3 Branches from Loops. When a load is fed from a loop, like the ones discussed above, a fault on a branch away from that loop will also cause a sag. In that case it is often possible to model the system as shown in Fig. 4.36. The feeder to the fault does not necessarily have to be a single feeder, but could, e.g., represent the effective impedance of another loop. The equivalent circuit for the system in Fig. 4.36 is shown in Fig. 4.37: Z_1 is the source impedance at the main subtransmission bus; Z_2 is the impedance between that bus and the bus from which the load is fed; Z_3 is the impedance between the bus from which the load is fed and the bus from which the fault is fed; Z_4 and Z_5 are the impedances between the latter bus and the main subtransmission bus and the fault, respectively. The voltage at the load bus is found from

$$V_{sag} = \frac{Z_5Z_2 + Z_5Z_3 + Z_5Z_4 + Z_4Z_3}{Z_1Z_2 + Z_1Z_3 + Z_1Z_4 + Z_5Z_2 + Z_5Z_3 + Z_5Z_4 + Z_4Z_2 + Z_4Z_3} \tag{4.19}$$

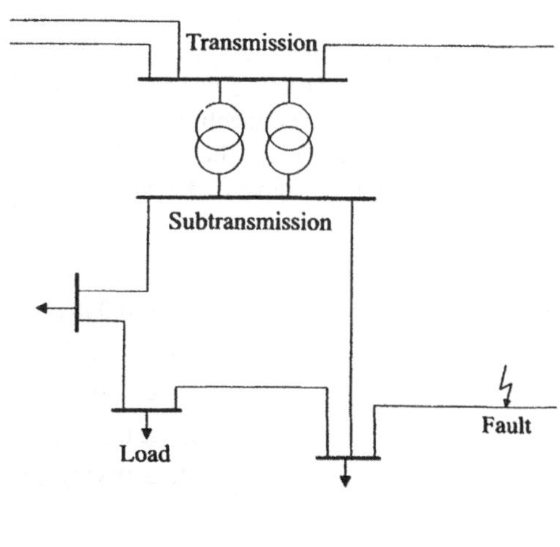

Figure 4.36 System with a branch away from a loop.

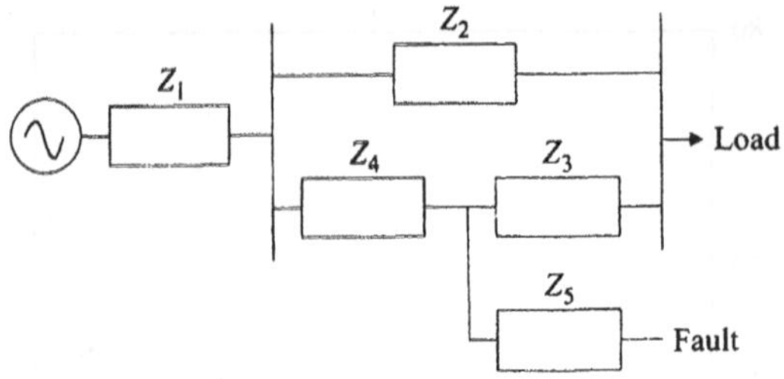

Figure 4.37 Equivalent circuit for system with a branch away from a loop, as in Fig. 4.36.

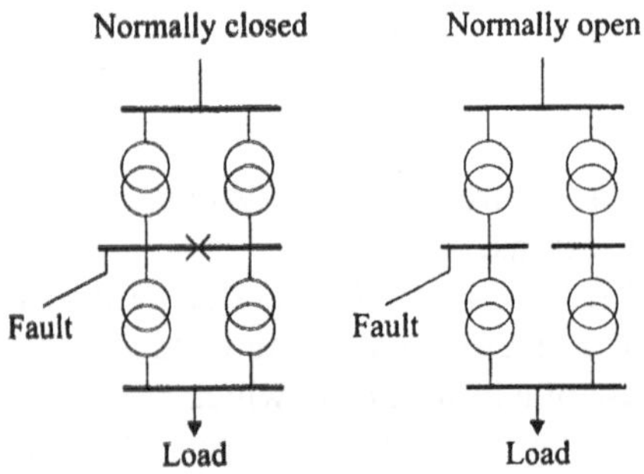

Figure 4.38 Industrial system with breaker at intermediate voltage level closed (left) and open (right).

The same expression can be used to assess an industrial system in which bus splitting is used at an intermediate voltage level. An example of the supply configuration in a large industrial network is shown in Fig. 4.38. In the left example, two transformers are operated in parallel. Typically both transformers feed into a different part of the substation bus, separated through a circuit breaker. This enables an uninterrupted supply after a bus fault. In the network on the right the substation consists of two separate busses, typically with a normally open breaker in between. In case the breaker at an intermediate voltage level is closed, the sag due to a fault at this voltage level will be experienced fully by the load. In case the breaker is open the sag will be mitigated according to (4.19). On the one hand, the source impedance will be less when the breaker is open, leading to a deeper sag at the intermediate voltage level. But on the other hand, the sag at the load bus will be less deep than at the faulted intermediate voltage level.

EXAMPLE Consider the system shown in Fig. 4.38 with the following voltages and fault levels: 2500 MVA at 66 kV, 500 MVA at 11 kV (with the breaker closed), and 50 MVA at 660 V. When the breaker connecting the two 11 kV busses is open, the circuit diagram in Fig. 4.37 can be used to calculate the sag magnitude at the 660 V bus for a fault at an 11 kV feeder. From the fault levels given, the values of various impedances can be calculated (all referred to 11 kV):

$$Z_1 = 0.048\,\Omega$$
$$Z_2 = 4.75\,\Omega$$
$$Z_3 = 4.36\,\Omega$$
$$Z_4 = 0.388\,\Omega$$
$$Z_5 = 0.3\,\Omega/\text{km} \times \mathcal{L}$$

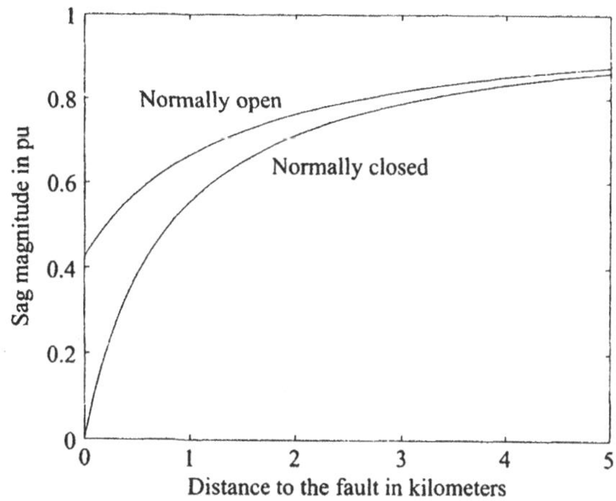

Figure 4.39 Sag magnitude versus distance to the fault, for an industrial system with and without bus-splitting applied to the 11 kV bus.

with $\mathcal{L}$ the distance between the 11 kV bus and the fault, and a feeder impedance of 0.3 Ω/km. When the 11 kV breaker is closed, the system can be treated like a radial system with a source impedance equal to $Z_1 + \frac{1}{2}Z_4$ and a feeder impedance equal to Z_5. A comparison between these two ways of system operation is given in Fig. 4.39. Bus-splitting (operating the system with the 11 kV breaker normally open) clearly limits the influence of 11 kV faults on the load. The improvement is especially large for nearby faults. For faults further away from the 11 kV substation the effect becomes smaller. But industrial medium-voltage systems are seldom larger than a few kilometers. We will come back to this and other ways of mitigating sags through system design and operation in Chapter 7.

4.2.4.4 Parallel Operation across Voltage Levels. In many countries the subtransmission system is not fed from the transmission system at one point but at a number of points, resulting in a system structure similar to the one shown in Fig. 4.40. The number of supply points for the subtransmission system varies from country to country. The 275 kV systems in the U.K. are fed like this; also the 130 kV system in Sweden and the 150 kV system in Belgium [23].

This type of configuration can be treated like a loop that extends over two voltage levels. For a fault within the loop we can apply (4.18), for a fault on a feeder away from the loop (4.19) can be used. The equations remain the same independent of the voltage level at which the fault takes place. The only thing that changes are the impedance values.

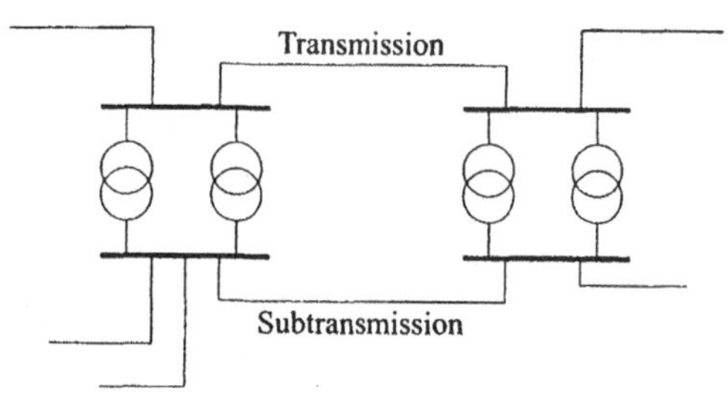

Figure 4.40 Parallel operation of transmission and subtransmission systems.

4.2.5 Voltage Calculations in Meshed Systems

When the system becomes more complicated than the examples discussed previously, closed expressions for the voltage during the sag get very complicated and unfeasible to handle. For meshed systems, matrix calculations have proven to be very efficient for computer-based analysis. The calculation of the voltages during a fault is based on two principles from circuit theory: Thevenin's superposition theorem; and the node impedance matrix. Both are discussed in detail in many books on power systems. Here we will only give a brief description.

- According to Thevenin's superposition theorem voltages and currents in the system during a sag are the sum of two contributions: currents and voltages before the event, and currents and voltages due to the change in voltage at the fault position. Currents and voltages before the fault are due to all generators across the system. Currents and voltages due to the fault originate at a voltage source at the fault position. All other voltage sources are considered short-circuited during the calculation of the latter contribution.
- The node impedance matrix **Z** relates node voltages and node currents:

$$V = \mathbf{Z}I \tag{4.20}$$

  with V the vector of (complex) node voltages and I the vector of (complex) node currents. The node voltage is the voltage between a node and the reference node (typically ground). The node current is equal to the sum of all currents flowing toward a node. For most nodes the node current is zero according to Kirchhoff's current law. The only exception are generator nodes, where the node current is the current flowing from the generator into the system.

Consider a system with N nodes plus a reference node. The voltages before the fault are denoted as $V_k^{(0)}$. A short-circuit fault occurs at node f. According to Thevenin's superposition theorem we can write the voltage during the fault at any node k as

$$V_k = V_k^{(0)} + \Delta V_k \tag{4.21}$$

where ΔV_k is the change in voltage at node k due to the fault. This latter term is due to a voltage source $-V_f^{(0)}$ at the fault position. To calculate ΔV_k all other voltage sources in the system are short-circuited, so that node f is the only node with a non-zero node current. After using the information, (4.20) becomes

$$\Delta V_k = Z_{kf} I_f \tag{4.22}$$

At the fault position ($k = f$) we know that $\Delta V_f = -V_f^{(0)}$ so that

$$I_f = -\frac{V_f^{(0)}}{Z_{ff}} \tag{4.23}$$

and

$$V_k = V_k^{(0)} - \frac{Z_{kf}}{Z_{ff}} V_f^{(0)} \tag{4.24}$$

The pre-fault voltages are normally close to unity, so that (4.24) can be approximated by

$$V_k = 1 - \frac{Z_{kf}}{Z_{ff}} \tag{4.25}$$

The moment the node impedance matrix is known, calculating sag magnitudes becomes very easy. The drawback with this method is that the node impedance matrix needs to be calculated. This can be done through a recursive procedure where the matrix is updated for each new branch added. Alternatively one can first calculate the node admittance matrix from the branch impedances. The node impedance matrix is the inverse of the node admittance matrix.

EXAMPLE Consider the circuit diagram shown in Fig. 4.41. This circuit represents a 275/400 kV system, with nodes 1 and 2 representing 400 kV substations; nodes 3, 4, and 5 representing 275 kV substations; the branches between 1 and 3 and between 2 and 4 representing transformers (the latter two transformers in parallel). The impedance values indicated in the figure are in percent at a 100 MVA base.

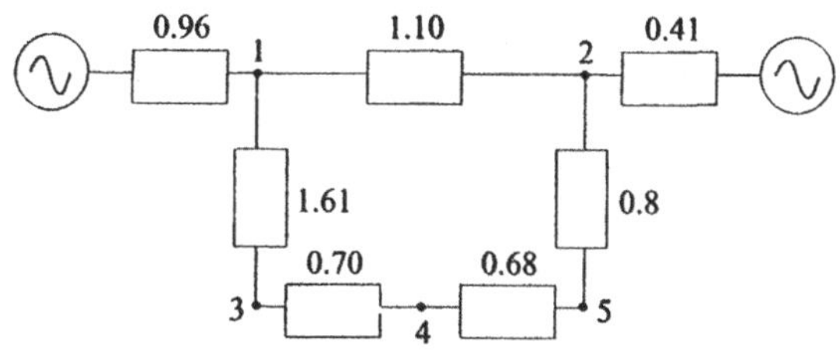

Figure 4.41 Circuit diagram representation of part of a 400/275 kV system.

The node admittance matrix can be built easily from the branch admittances or impedances. An off-diagonal element Y_{kl} of the node admittance matrix is equal to minus the admittance of the branch between nodes k and l. The element is zero if there is no branch between these two nodes. The diagonal element Y_{kk} equals the sum of all admittances of branches to node k including any branch between node k and the reference node. For the circuit in Fig. 4.41 this calculation leads to the node admittance matrix

$$Y = \begin{bmatrix} 2.5719 & -0.9091 & -0.6211 & 0 & 0 \\ -0.9091 & 4.5981 & 0 & -1.25 & 0 \\ -0.6211 & 0 & 2.0497 & 0 & -1.4286 \\ 0 & -1.25 & 0 & 2.7206 & -1.4706 \\ 0 & 0 & -1.4286 & -1.4706 & 2.8992 \end{bmatrix} \tag{4.26}$$

The node impedance matrix is obtained by inverting the node admittance matrix

$$Z = Y^{-1} = \begin{bmatrix} 0.5453 & 0.1771 & 0.3889 & 0.2548 & 0.3209 \\ 0.1771 & 0.3344 & 0.2439 & 0.3012 & 0.2730 \\ 0.3889 & 0.2439 & 1.2534 & 0.6144 & 0.9292 \\ 0.2548 & 0.3012 & 0.6144 & 0.9225 & 0.7707 \\ 0.3209 & 0.2730 & 0.9292 & 0.7707 & 1.1937 \end{bmatrix} \tag{4.27}$$

The voltage at node 5 due to a fault at node 2 is

$$V_5 = 1 - \frac{Z_{52}}{Z_{22}} = 1 - \frac{0.2730}{0.3344} = 0.1836 \tag{4.28}$$

TABLE 4.7 Voltage Sags in the System Shown in Fig. 4.41

	Fault at Node				
Voltage at Node	1	2	3	4	5
1	0	0.4704	0.6897	0.7238	0.7312
2	0.6753	0	0.8054	0.6735	0.7713
3	0.2869	0.2706	0	0.3340	0.2216
4	0.5327	0.0993	0.5098	0	0.3544
5	0.4116	0.1837	0.2586	0.1646	0

Table 4.7 gives the voltage at any node due to a fault at any other node. We see, e.g., that for node 5 a fault at node 2 is more severe than a fault at node 1. This is understandable as the source at node 2 is stronger than the source at node 1.

4.3 VOLTAGE SAG DURATION

4.3.1 Fault-Clearing Time

We have seen in Section 4.2 that the drop in voltage during a sag is due to a short circuit being present in the system. The moment the short-circuit fault is cleared by the protection, the voltage can return to its original value. The duration of a sag is mainly determined by the fault-clearing time, but it may be longer than the fault-clearing time. We will come back to this further on in this section.

Generally speaking faults in transmission systems are cleared faster than faults in distribution systems. In transmission systems the critical fault-clearing time is rather small. Thus, fast protection and fast circuit breakers are essential. Also transmission and subtransmission systems are normally operated as a grid, requiring distance protection or differential protection, both of which are rather fast. The principal form of protection in distribution systems is overcurrent protection. This requires often some time-grading which increases the fault-clearing time. An exception are systems in which current-limiting fuses are used. These have the ability to clear a fault within one half-cycle [6], [7].

An overview of the fault-clearing time of various protective devices is given in reference [8].

- current-limiting fuses: less than one cycle
- expulsion fuses: 10–1000 ms
- distance relay with fast breaker: 50–100 ms
- distance relay in zone 1: 100–200 ms
- distance relay in zone 2: 200–500 ms
- differential relay: 100–300 ms
- overcurrent relay: 200–2000 ms

Some typical fault-clearing times at various voltage levels for a U.S. utility are given in reference [9].

Voltage Level	Best Case	Typical	Worse Case
525 kV	33 ms	50 ms	83 ms
345 kV	50 ms	67 ms	100 ms
230 kV	50 ms	83 ms	133 ms
115 kV	83 ms	83 ms	167 ms
69 kV	50 ms	83 ms	167 ms
34.5 kV	100 ms	2 sec	3 sec
12.47 kV	100 ms	2 sec	3 sec

From this list it becomes clear that the sag duration will be longer when a sag originates at a lower voltage level. Many utilities operate their distribution feeders in such a way that most faults are cleared within a few cycles. Such a way of operation was discussed in detail in Chapter 3. But even for those feeders, a certain percentage of faults will lead to long sags. The difference between the two ways of operation is discussed in more detail in Section 7.1.3.

4.3.2 Magnitude-Duration Plots

Knowing the magnitude and duration of a voltage sag, it can be presented by a point in a magnitude-duration plane. This way of sag characterization has been shown to be extremely useful for various types of studies. We will use it in forthcoming chapters to describe both equipment and system performance. Various types of magnitude-duration plots will be discussed in Section 6.2. The magnitude-duration plot will also be used in Chapter 6 to present the results of power quality surveys. An example of a magnitude-duration plot is shown in Fig. 4.42. The numbers in Fig. 4.42 refer to the following sag origins:

1. Transmission system faults
2. Remote distribution system faults
3. Local distribution system faults
4. Starting of large motors
5. Short interruptions
6. Fuses

Consider the general system configuration shown in Fig. 4.43. A short-circuit fault in the local distribution network will typically lead to a rather deep sag. This is

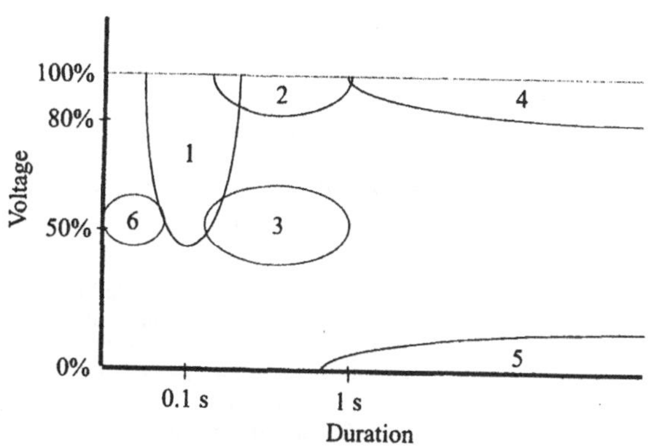

Figure 4.42 Sags of different origin in a magnitude-duration plot.

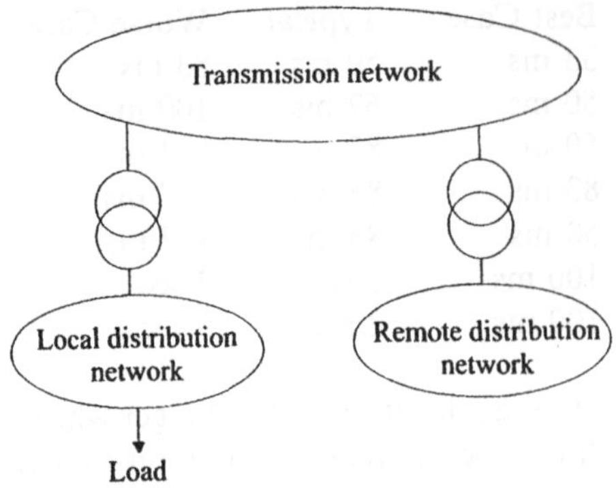

Figure 4.43 General structure of power system, with distribution and transmission networks.

due to the limited length of distribution feeders. When the fault occurs in a remote distribution network, the sag will be much more shallow due to the transformer impedance between the fault and the pcc. For a fault in any distribution network, the sag duration may be up to a few seconds.

Transmission system faults are typically cleared within 50 to 100 ms, thus leading to short-duration sags. Current-limiting fuses lead to sag durations of one cycle or less, and rather deep sags if the fault is in the local distribution or low-voltage network. Faults in remote networks, cleared by current-limiting fuses, lead to short and shallow sags, not indicated in the figure. Finally the figure contains voltage sags due to motor starting, shallow and long duration (see Section 4.9) and short interruptions, deep and long duration (see Chapter 3).

4.3.3 Measurement of Sag Duration

Measurement of sag duration is much less trivial than it might appear from the previous section. For a sag like in Fig. 4.1 it is obvious that the duration is about $2\frac{1}{2}$ cycles. However, to come up with an automatic way for a power quality monitor to obtain the sag duration is no longer straightforward. A commonly used definition of sag duration is the number of cycles during which the rms voltage is below a given threshold. This threshold will be somewhat different for each monitor but typical values are around 90%. A power quality monitor will typically calculate the rms value once every cycle. This gives an overestimation of the sag duration as shown in Fig. 4.44. The

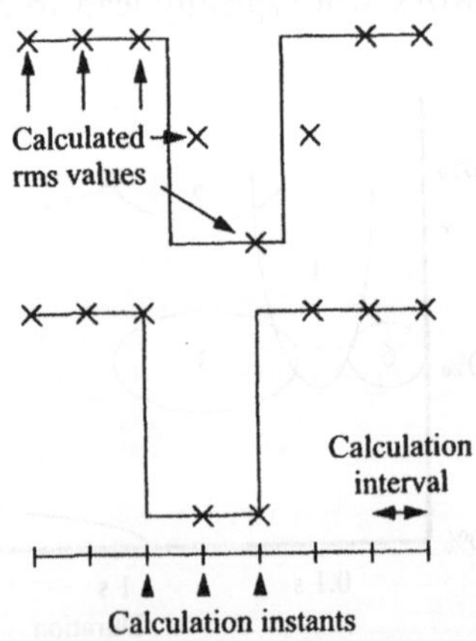

Figure 4.44 Estimation of sag duration by power quality monitor for a two-cycle sag: overestimation by one cycle (upper graph); correct estimation (lower graph).

normal situation is shown in the upper figure. The rms calculation is performed at regular instants in time and the voltage sag starts somewhere in between two of those instants. As there is no correlation between the calculation instants and the sag commencement, this is the most likely situation. We see that the rms value is low for three samples in a row. The sag duration according to the monitor will be three cycles. Here it is assumed that the sag is deep enough for the intermediate rms value to be below the threshold. For shallow sags both intermediate values might be above the threshold and the monitor will record a one-cycle sag. The bottom curve of Fig. 4.44 shows the rare situation where the sag commencement almost coincides with one of the instants on which the rms voltage is calculated. In that case the monitor gives the correct sag duration.

Calculating the rms voltage once a cycle, it is obvious that the resulting sag duration will be an integer number of cycles. For a $2\frac{1}{2}$-cycle sag the computed duration will be either two or three cycles. But even when a sliding window is used to calculate the rms voltage as a function of time, an erroneous sag duration might result. To show this possible error for a measured sag, we have plotted in Fig. 4.45 the half-cycle rms of the sag shown in Fig. 4.1, together with the absolute value of the measured voltage. The "actual sag duration" obtained from the sudden drop and rise in the voltage is 2.4 cycles. For large thresholds the recorded sag duration will be an overestimation. A 90% threshold gives a 2.8 cycle sag duration, and 80% threshold a 2.5 cycles duration. For lower thresholds the recorded sag duration is an underestimation: a 60% threshold gives a 2.1 cycle duration and a 40% threshold a 2.0 cycle duration. In reality, thresholds this low will not be used, but the same effect will be obtained when the depth of the sag is varied and the threshold is kept constant. The duration of deep sags will be overestimated, and the duration of shallow ones underestimated.

As the shortest-duration window for calculating the sag magnitude is one half-cycle, an error up to one half-cycle must be accepted. Several methods have been suggested to measure sag initiation and voltage recovery more accurately. These methods also give a more accurate value of sag duration [134], [201], [202]. Using the fundamental voltage component results in a similar transition between pre-sag and during-sag voltage, thus similar errors in sag duration. Using the half-cycle peak voltage will give a much sharper transition, as long as sag initiation and voltage recovery are close to voltage maximum. Sag initiation and voltage recovery around the voltage zero-crossing will give a smoother transition and a larger uncertainly in sag duration.

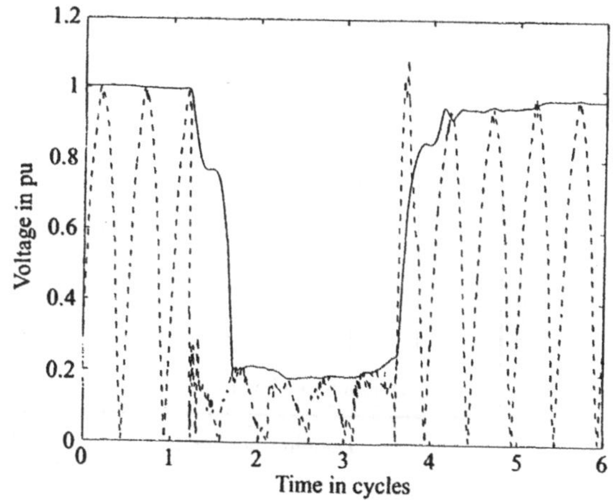

Figure 4.45 Half-cycle rms voltage together with absolute value of the voltage (dashed line) of the sag shown in Fig. 4.1.

The above-mentioned error in sag duration is only significant for short-duration sags. For longer sags it does not really matter. But for longer sags the so-called post-fault sag will give a serious uncertainty in sag duration. When the fault is cleared the voltage does not recover immediately. Some of this effect can be seen in Fig. 4.3 and Fig. 4.4. The rms voltage after the sag is slightly lower than before the sag. The effect can be especially severe for sags due to three-phase faults. The explanation for this effect is as follows [17], [18]. Due to the drop in voltage during the sag, induction motors will slow down. The torque produced by an induction motor is proportional to the square of the voltage, so even a rather small drop in voltage can already produce a large drop in torque and thus in speed. The moment the fault is cleared and the voltage comes back, the induction motors start to draw a large current: up to 10 times their nominal current. Immediately after the sag, the air-gap field will have to be built up again. In other words, the induction motor behaves like a short-circuited transformer. After the flux has come back into the air gap, the motor can start re-accelerating which also requires a rather large current. It is this post-fault inrush current of induction motors which leads to an extended sag. The post-fault sag can last several seconds, much longer than the actual sag.

Such a post-fault sag will cause uncertainty in the sag duration as obtained by a power quality monitor: different monitors can give different results. This is shown schematically in Fig. 4.46. Assume that monitor 1 has a setting as indicated, and monitor 2 a slightly higher setting. Both monitors will record a sag duration much longer than the fault-clearing time. The fault-clearing time can be estimated from the duration of the deep part of the sag. We see that monitor 2 will record a significantly longer duration than monitor 1.

A measured sag with a long post-fault component is shown in Fig. 4.47. The three phases are shown in the same figure to better indicate the post-fault voltage sag. Note that the sag is unbalanced during the fault, but balanced after the fault.

The rms voltage versus time for the sag shown in Fig. 4.47 is plotted in Fig. 4.48. We see a large drop in voltage in two phases and a small one in the third phase. The fault-clearing time is about four cycles; the fault leading to this sag took place at 132 kV, the voltages were measured at 11 kV. The sag duration has been determined as the time during which the rms voltage is below a certain threshold. Figure 4.49 plots this duration as a function of the threshold, for the three phases. One of the phases only drops to 88% so that any threshold setting below 88% will give zero sag duration for that phase. The sag duration obtained for the other two phases is about four cycles for thresholds below 90%, increasing fast for higher threshold settings.

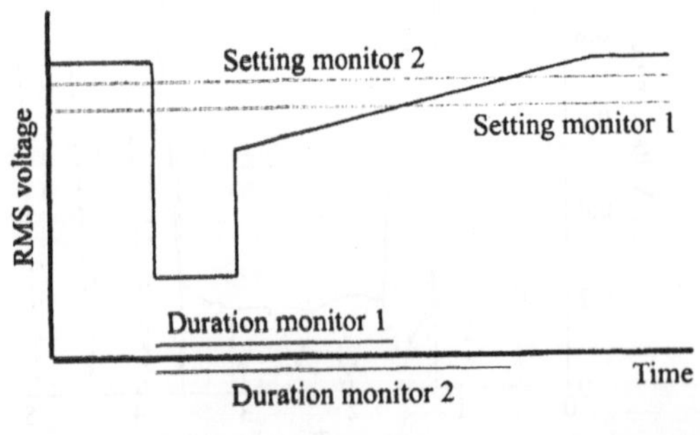

Figure 4.46 Error in sag duration due to post-fault sag.

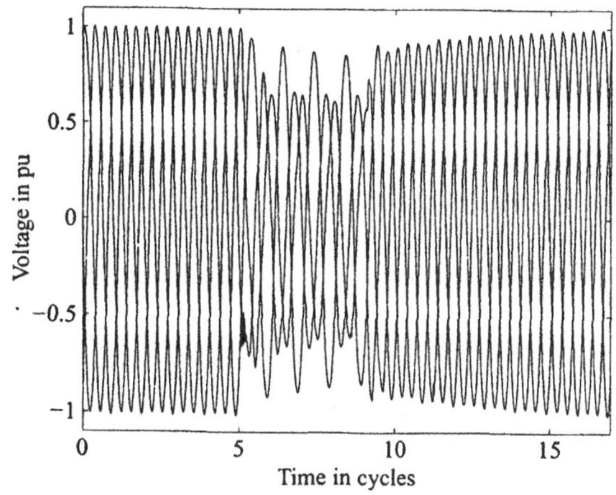

Figure 4.47 Measured sag with a clear post-fault component (Data obtained from Scottish Power.)

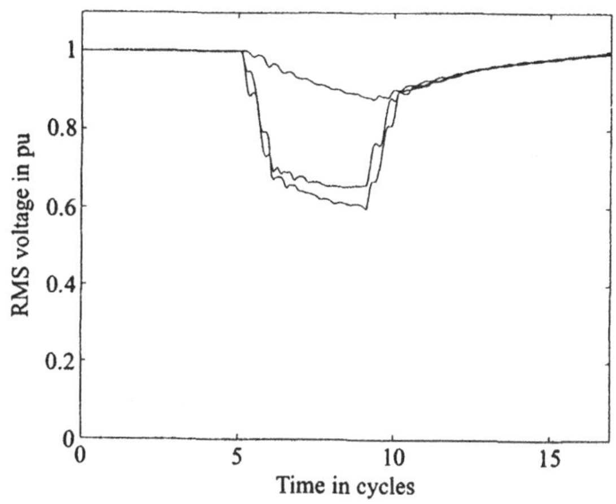

Figure 4.48 The rms voltages versus time for the sag shown in Fig. 4.47.

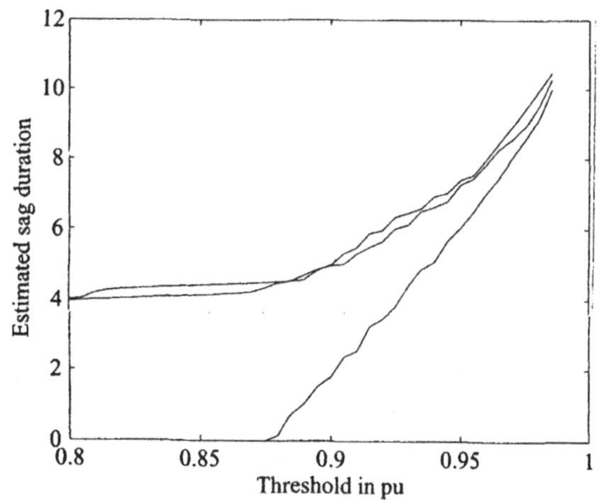

Figure 4.49 Sag duration versus threshold setting for the three phases of the sag shown in Figs. 4.47 and 4.48.

4.4 THREE-PHASE UNBALANCE

The analysis of sag magnitude presented in the previous sections considers only one phase. For example, the voltage divider model in Fig. 4.14 was introduced for three-phase faults: the impedances used in that figure are the positive-sequence values. But most short circuits in power systems are single phase or two phase. In that case we need to take all three phases into account or use the symmetrical component theory. A good and detailed description of the use of symmetrical components theory for the analysis of non-symmetrical faults is given in reference [24] and in several other books on power system analysis and is not repeated here. We will only use the results of the theory to calculate the voltages in the three phases due to a non-symmetrical short circuit.

For non-symmetrical faults the voltage divider in Fig. 4.14 can still be used but it has to be split into its three components: a positive-sequence network, a negative-sequence network, and a zero-sequence network. The three component networks are shown in Fig. 4.50, where V_1, V_2, and V_0 represent positive-, negative-, and zero-sequence voltage, respectively, at the pcc; Z_{S1}, Z_{S2}, and Z_{S0} are the source impedance values and Z_{F1}, Z_{F2}, and Z_{F0} the feeder impedance values in the three components. The three components of the fault current are denoted by I_1, I_2, and I_0. The positive-sequence source is denoted by E. There is no source in the negative and zero-sequence networks. The three component networks have to be connected into one equivalent circuit at the fault position. The connection of the component networks depends on the fault type. For a three-phase fault all three networks are shorted at the fault position. This leads to the standard voltage divider model for the positive sequence, and zero voltage and current for the negative and zero sequences.

4.4.1 Single-Phase Faults

For a single-phase fault, the three networks shown in Fig. 4.50 should be connected in series at the fault position. The resulting circuit for a single-phase fault in

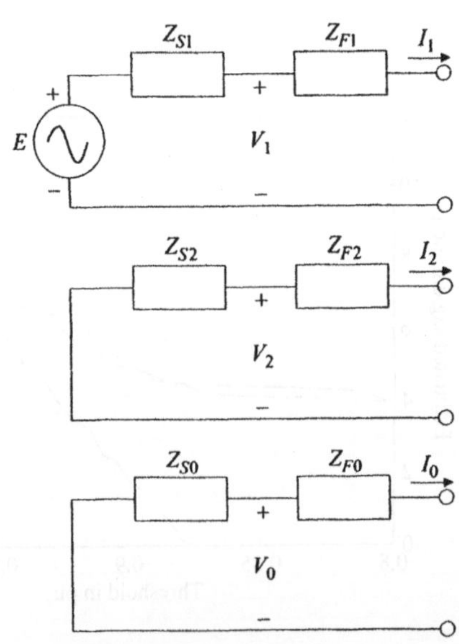

Figure 4.50 Positive- (top), negative- (center), and zero- (bottom) sequence networks for the voltage divider shown in Fig. 4.14.

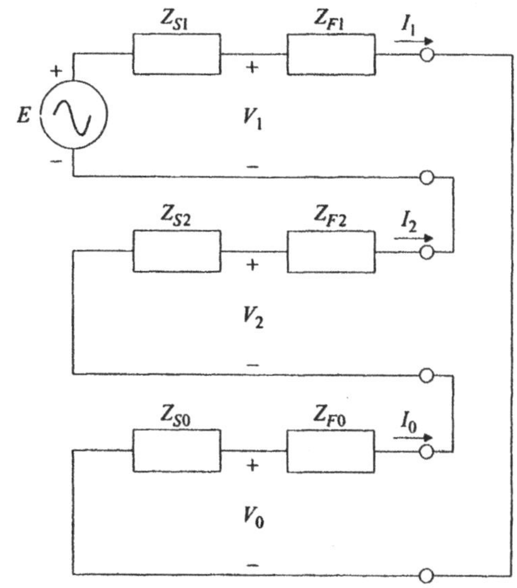

Figure 4.51 Equivalent circuit for a single-phase fault.

phase a, is shown in Fig. 4.51. If we again make $E = 1$, like in the single-phase model in Fig. 4.14, the following expressions are obtained for the component voltages at the pcc:

$$V_1 = \frac{Z_{F1} + Z_{S2} + Z_{F2} + Z_{S0} + Z_{F0}}{(Z_{F1} + Z_{F2} + Z_{F0}) + (Z_{S1} + Z_{S2} + Z_{S0})} \tag{4.29}$$

$$V_2 = \frac{Z_{S2}}{(Z_{F1} + Z_{F2} + Z_{F0}) + (Z_{S1} + Z_{S2} + Z_{S0})} \tag{4.30}$$

$$V_0 = \frac{Z_{S0}}{(Z_{F1} + Z_{F2} + Z_{F0}) + (Z_{S1} + Z_{S2} + Z_{S0})} \tag{4.31}$$

The voltages in the three phases at the pcc during the fault are obtained by transforming back from sequence domain to phase domain:

$$\begin{aligned} V_a &= V_1 + V_2 + V_0 \\ V_b &= a^2 V_1 + aV_2 + V_0 \\ V_c &= aV_1 + a^2 V_2 + V_0 \end{aligned} \tag{4.32}$$

For the faulted phase voltage V_a we get

$$V_a = \frac{Z_{F1} + Z_{F2} + Z_{F0}}{(Z_{F1} + Z_{F2} + Z_{F0}) + (Z_{S1} + Z_{S2} + Z_{S0})} \tag{4.33}$$

We can obtain the original voltage divider equation (4.9) by defining $Z_F = Z_{F1} + Z_{F2} + Z_{F0}$ and $Z_S = Z_{S1} + Z_{S2} + Z_{S0}$. Thus, the voltage divider model of Fig. 4.14 and (4.9) still holds for single-phase faults. The condition thereby is that the resulting voltage is the voltage in the faulted phase, and that the impedance values used are the sum of the positive-, negative-, and zero-sequence impedances. From (4.29) through (4.32) we can calculate the voltages in the non-faulted phases, which results into the following expressions for the three voltages:

$$V_a = 1 - \frac{Z_{S1} + Z_{S2} + Z_{S0}}{(Z_{F1} + Z_{F2} + Z_{F0}) + (Z_{S1} + Z_{S2} + Z_{S0})}$$
$$V_b = a^2 - \frac{a^2 Z_{S1} + a Z_{S2} + Z_{S0}}{(Z_{F1} + Z_{F2} + Z_{F0}) + (Z_{S1} + Z_{S2} + Z_{S0})} \tag{4.34}$$
$$V_c = a - \frac{a Z_{S1} + a^2 Z_{S2} + Z_{S0}}{(Z_{F1} + Z_{F2} + Z_{F0}) + (Z_{S1} + Z_{S2} + Z_{S0})}$$

Note that the expression for V_a has been slightly rewritten to explicitly obtain the voltage drop as a separate term.

These voltages are shown as a phasor diagram in Fig. 4.52. The voltage drop in the non-faulted phases consists of three terms:

- a voltage drop proportional to the positive-sequence source impedance, along the direction of the pre-fault voltage.
- a voltage drop proportional to the negative-sequence source impedance, along the direction of the pre-fault voltage in the other non-faulted phase.
- a voltage drop proportional to the zero-sequence source impedance, along the direction of the pre-fault voltage in the faulted phase.

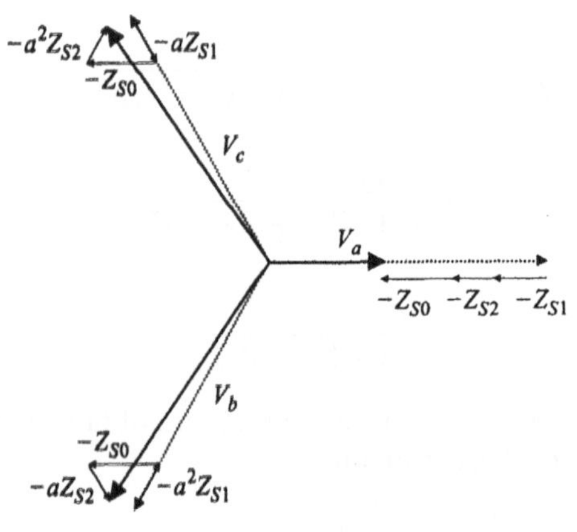

Figure 4.52 Phase-to-ground voltages during a single-phase fault.

The voltage between the two non-faulted phases is

$$V_b - V_c = (a^2 - a)\left[1 - \frac{Z_{S1} - Z_{S2}}{(Z_{F1} + Z_{F2} + Z_{F0}) + (Z_{S1} + Z_{S2} + Z_{S0})}\right] \tag{4.35}$$

We see that the change in this voltage is only due to the difference between positive-sequence and negative-sequence source impedances. As these two are normally about equal, the voltage between the non-faulted phases is normally not influenced by the fault. Below we will simplify the expressions (4.34) and (4.35) for two cases:

- Positive-, negative-, and zero-sequence source impedances are equal.
- Positive- and negative-sequence source and feeder impedances are equal.

4.4.1.1 Solidly-Grounded Systems. In a solidly-grounded system, the source impedances in the three sequence components are often about equal. The three voltage drops in the non-faulted phases now cancel, resulting in the following voltages during the fault:

$$\begin{aligned} V_a &= -\frac{Z_{S1}}{\frac{1}{3}(Z_{F1}+Z_{F2}+Z_{F0})+Z_{S1}} \\ V_b &= a^2 \\ V_c &= a \end{aligned} \tag{4.36}$$

The voltage in the faulted phase is the same as during a three-phase fault, the voltages in the non-faulted phase are not affected.

4.4.1.2 Impedance-Grounded Systems. In a resistance or high-impedance grounded system, the zero-sequence source impedance differs significantly from the positive and negative-sequence source impedances. We can, however, assume that the latter two are equal. Also in systems where the source impedance consists for a large part of line or cable impedances (e.g., in transmission systems) positive- and zero-sequence impedances can be significantly different. The resulting expressions for the voltages at the pcc during a single-phase fault are, when $Z_{S1} = Z_{S2}$ and $Z_{F1} = Z_{F2}$:

$$\begin{aligned} V_a &= 1 - \frac{Z_{S0}+2Z_{S1}}{(2Z_{F1}+Z_{F0})+(2Z_{S1}+Z_{S0})} \\ V_b &= a^2 - \frac{Z_{S0}-2Z_{S1}}{(2Z_{F1}+Z_{F0})+(2Z_{S1}+Z_{S0})} \\ V_c &= a - \frac{Z_{S0}-2Z_{S1}}{(2Z_{F1}+Z_{F0})+(2Z_{S1}+Z_{S0})} \end{aligned} \tag{4.37}$$

The voltage drop in the non-faulted phases only contains a zero-sequence component (it is the same in both phases). We will see later that the zero-sequence component of the voltage is rarely of importance for the voltage sag as experienced at equipment terminals. Sags at the same voltage level as the equipment terminals are rare. During the transfer of the sag down to lower voltage levels, the transformers normally block the zero-sequence component of the voltage. Even if the fault occurs at the same voltage level as the equipment terminals, the equipment is normally connected in delta so it will not notice the zero-sequence component of the voltage. Thus the voltage drop in the non-faulted phases is not of importance from an equipment point of view. We can therefore add a zero-sequence voltage to (4.37) such that the voltage drop in the non-faulted phases disappears. The resulting expressions are

$$\begin{aligned} V'_a &= V_a + \frac{Z_{S0}-Z_{S1}}{(2Z_{F1}+Z_{F0})+(2Z_{S1}+Z_{S0})} = 1 - \frac{3Z_{S1}}{(2Z_{F1}+Z_{F0})+(2Z_{S1}+Z_{S0})} \\ V'_b &= V_b + \frac{Z_{S0}-Z_{S1}}{(2Z_{F1}+Z_{F0})+(2Z_{S1}+Z_{S0})} = a^2 \\ V'_c &= V_c + \frac{Z_{S0}-Z_{S1}}{(2Z_{F1}+Z_{F0})+(2Z_{S1}+Z_{S0})} = a \end{aligned} \tag{4.38}$$

The expression for the voltage in the faulted phase is somewhat rewritten, to enable a comparison with (4.36):

$$V'_a = 1 - \frac{Z_{S1}}{\frac{1}{3}(Z_{F1}+Z_{F2}+Z_{F0})+\frac{1}{3}(Z_{S0}-Z_{S1})+Z_{S1}} \tag{4.39}$$

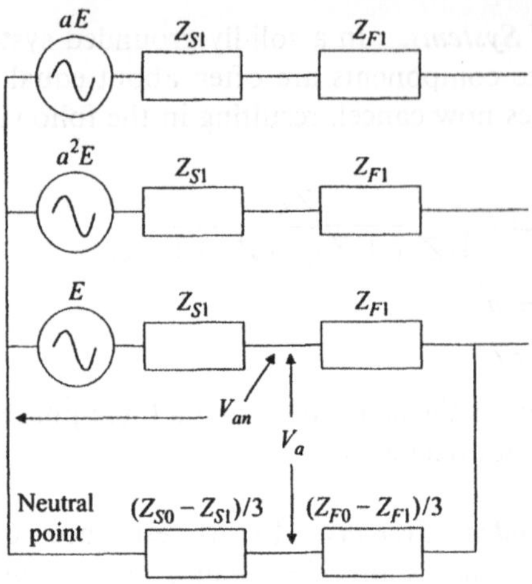

Figure 4.53 Three-phase voltage divider model.

The denominator contains an additional term $\frac{1}{3}(Z_{S0} - Z_{S1})$ compared to (4.36). This can be interpreted as an additional impedance between the pcc and the fault. When this impedance is positive, thus when $Z_{S0} > Z_{S1}$, the sag becomes more shallow. In resistance and reactance-grounded systems, $Z_{S0} \gg Z_{S1}$, so that even a terminal fault, $Z_{F1} + Z_{F2} + Z_{F0} = 0$, will lead to a shallow sag.

Note that in solidly-grounded systems, the zero-sequence source impedance may be less than the positive-sequence one, $Z_{S0} < Z_{S1}$, so that the additional impedance is negative. For nearby faults, we will thus obtain a negative voltage V'_a.

All this might look like a mathematical trick to get rid of the voltage drop in the non-faulted phases. There is, however, some physical significance to this. To show this, the three-phase voltage divider is drawn in a commonly used way [24] in Fig. 4.53. From this model we can calculate the phase-to-neutral voltages at the pcc; with $E = 1$ the calculation results into

$$\begin{aligned} V_{an} &= 1 - \frac{3Z_{S1}}{(2Z_{F1} + Z_{F0}) + (2Z_{S1} + Z_{S0})} \\ V_{bn} &= a^2 \\ V_{cn} &= a \end{aligned} \tag{4.40}$$

The correspondence between (4.40) and (4.38) is obvious. The voltages in (4.38) thus correspond to the phase-to-neutral voltages. Note that the "neutral" in Fig. 4.53 is not a physical neutral but a kind of mathematical neutral. In resistance- or high-impedance grounded systems the physical neutral (i.e., the star point of the transformer) is a good approximation of this "mathematical neutral." The expressions derived not only hold for resistance-grounded systems, but for each system in which we can assume positive- and negative-sequence impedances equal.

EXAMPLE Consider again the system shown in Fig. 4.21, and assume that a single-phase fault occurs on one of the 132 kV feeders. The 132 kV system is solidly grounded, therefore the positive- and zero-sequence source impedances are similar. For the feeders, the zero-sequence impedance is about twice the positive- and negative- sequence impedance. Positive- and negative-sequence impedance are assumed equal.

$$\begin{aligned} Z_{S1} &= Z_{S2} = 0.09 + j2.86\% \\ Z_{S0} &= 0.047 + j2.75\% \\ Z_{F1} &= Z_{F2} = 0.101 + j0.257\%/\text{km} \\ Z_{F0} &= 0.23 + j0.65\%/\text{km} \end{aligned}$$

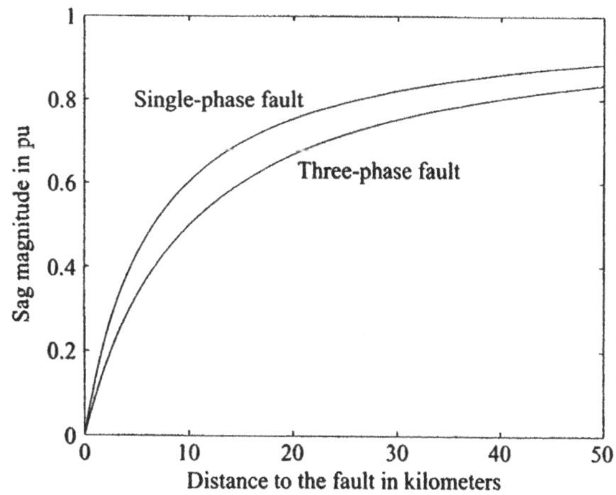

Figure 4.54 Voltage in the faulted phase for single-phase and three-phase faults on a 132 kV feeder in Fig. 4.21.

By using the above-given equations, the voltages in the three phases have been calculated for single-phase as well as for three-phase faults. The results for the faulted phase are shown in Fig. 4.54. The difference is mainly due to the difference in feeder impedance. Note that it is assumed here that the feeders are at least 50 km long, where they are in reality only 2 km long. The zero-sequence feeder impedance increases faster than the positive-sequence impedance, with increasing distance to the fault. Therefore single-phase faults lead to slightly smaller voltage drops than three-phase faults. As we saw from the equations above, it is the average of the three sequence impedances which determines the voltage drop due to single-phase faults. The voltages in the non-faulted phases showed only a very small change due to the single-phase fault.

EXAMPLE The voltages due to single-phase faults have been calculated for the 11 kV system in Fig. 4.21. As this system is resistance grounded, the zero-sequence source impedance is considerably larger than the positive-sequence impedance.

$$Z_{S1} = Z_{S2} = 4.94 + j65.9\%$$
$$Z_{S0} = 787 + j220\%$$
$$Z_{F1} = Z_{F2} = 9.7 + j26\%/\text{km}$$
$$Z_{F0} = 18.4 + j112\%/\text{km}$$

Note the large zero-sequence source impedance, especially its resistive part. The voltage in the faulted phase for three-phase and single-phase faults is shown in Fig. 4.55 as a function of the distance to the fault. The larger source impedance for single-phase faults more than compensates the larger feeder impedance, which makes that single-phase faults cause deeper sags than three-phase faults.

In a solidly-grounded system the voltage in a non-faulted phase stays about the same during a single-phase fault. In a resistance-grounded system the voltage in the non-faulted phases increases. This effect is shown in Figs. 4.56 and 4.57. Figure 4.56 shows the voltage magnitude versus distance to the fault and Fig. 4.57 the path of the voltages in the complex plane. The circles and the arrows indicate the complex voltages during normal operation. The curves indicate the path of the complex voltages with varying distance to the fault. Where the faulted phase shows a drop in voltage, the non-faulted phases show a large increase in voltage, for one phase even increasing 170% of the nominal voltage. From Fig. 4.57 we see that all three voltages are shifted over a

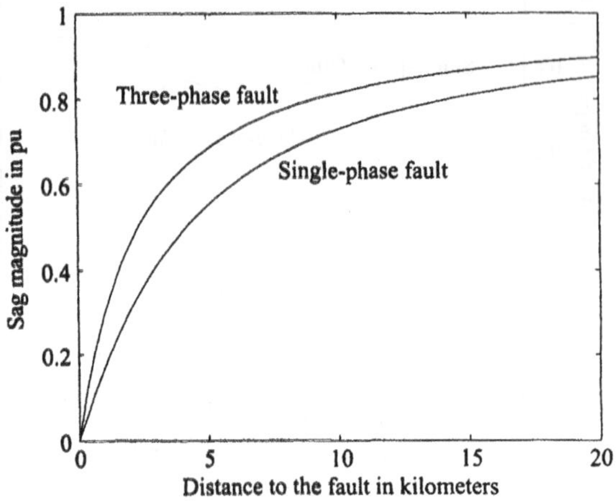

Figure 4.55 Voltage in the faulted phase for single-phase and three-phase faults on an 11 kV feeder in Fig. 4.21.

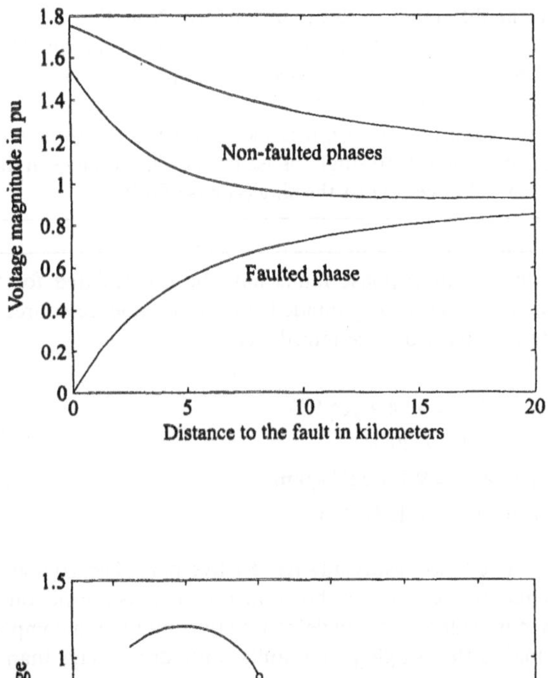

Figure 4.56 Voltage in the faulted and non-faulted phases for a single-phase fault on an 11 kV feeder in Fig. 4.21, as a function of the distance to the fault.

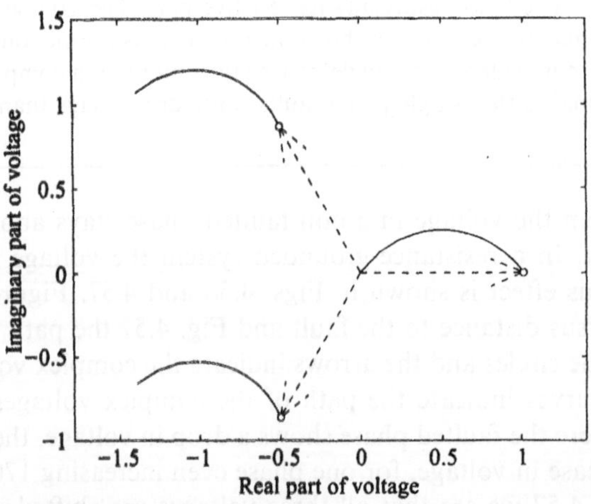

Figure 4.57 Complex voltages due to a fault on an 11 kV feeder in Fig. 4.21.

similar distance in the complex plane. The effect of this common shift (a zero-sequence component) is that the phase-to-phase voltages do not change much.

The phase-to-phase voltages have been calculated from the complex phase voltages by using the following expressions:

$$V_{ab} = \frac{V_a - V_b}{\sqrt{3}}$$
$$V_{bc} = \frac{V_b - V_c}{\sqrt{3}} \tag{4.41}$$
$$V_{ca} = \frac{V_c - V_a}{\sqrt{3}}$$

The factor $\sqrt{3}$ is needed to ensure that the pre-fault phase-to-phase voltages are 1 pu. The resulting voltage magnitudes are shown in Fig. 4.58: note the difference in vertical scale compared to the previous figures. We see that the phase-to-phase voltages are not much influenced by single-phase faults. The lowest voltage magnitude is 89%, the highest 101%.

Figure 4.59 compares phase-to-ground voltage, according to (4.37), and phase-to-neutral voltage, according to (4.40). We see that the drop in phase-to-neutral voltage is

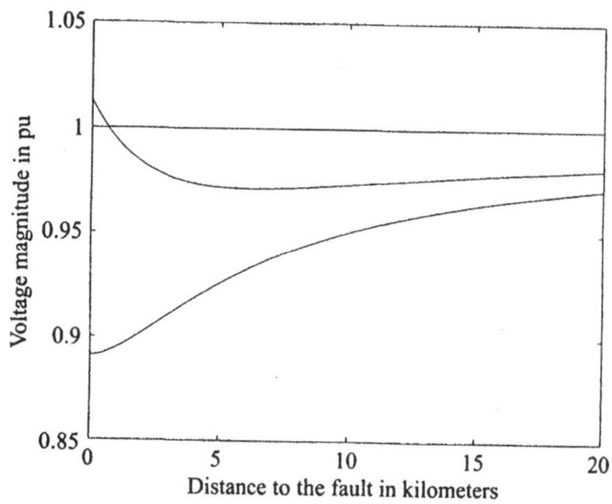

Figure 4.58 Phase-to-phase voltages due to a single-phase fault on an 11 kV feeder in Fig. 4.21, as a function of the distance to the fault.

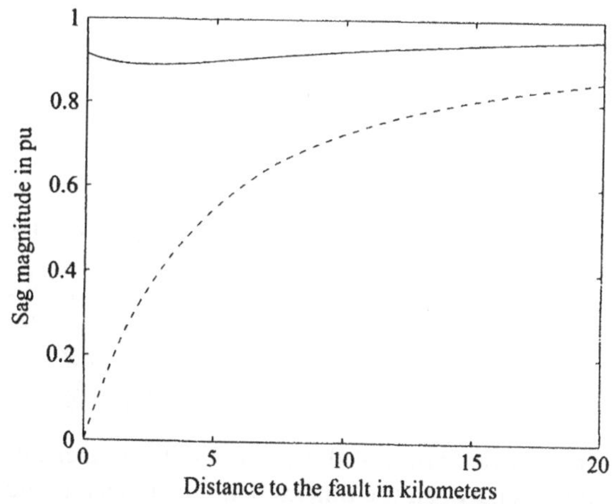

Figure 4.59 Phase-to-ground (dashed) and phase-to-neutral (solid) voltages due to single-phase faults on an 11 kV feeder in Fig. 4.21.

very small. As explained before, this is due to the large zero-sequence source impedance. Also note that the lowest phase-to-neutral voltage occurs for a non-zero distance to the fault.

4.4.2 Phase-to-Phase Faults

For a phase-to-phase fault the positive- and negative-sequence networks are connected in parallel, as shown in Fig. 4.60. The zero-sequence voltages and currents are zero for a phase-to-phase fault.

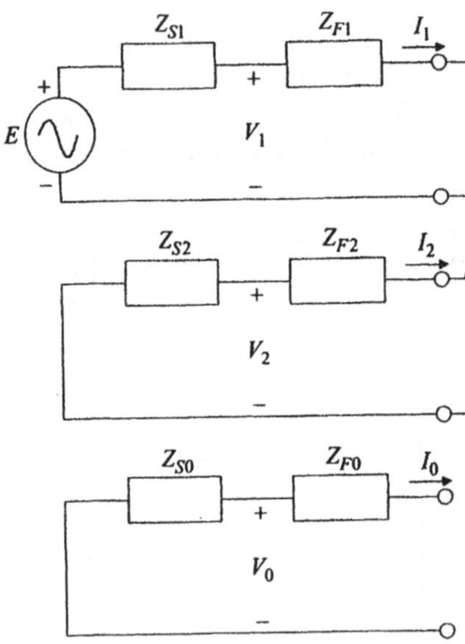

Figure 4.60 Equivalent circuit for a phase-to-phase fault.

The sequence voltages at the pcc are

$$\begin{aligned} V_1 &= E - E\frac{Z_{S1}}{(Z_{S1}+Z_{S2})+(Z_{F1}+Z_{F2})} \\ V_2 &= \frac{Z_{S2}}{(Z_{S1}+Z_{S2})+(Z_{F1}+Z_{F2})} \\ V_0 &= 0 \end{aligned} \tag{4.42}$$

The phase voltages can be found from (4.42) by using (4.32). This results in the following expressions, again with $E = 1$:

$$\begin{aligned} V_a &= 1 - \frac{Z_{S1}-Z_{S2}}{(Z_{S1}+Z_{S2})+(Z_{F1}+Z_{F2})} \\ V_b &= a^2 - \frac{a^2 Z_{S1}-aZ_{S2}}{(Z_{S1}+Z_{S2})+(Z_{F1}+Z_{F2})} \\ V_c &= a - \frac{aZ_{S1}-a^2 Z_{S2}}{(Z_{S1}+Z_{S2})+(Z_{F1}+Z_{F2})} \end{aligned} \tag{4.43}$$

In the calculation of the component voltages and currents, it has been assumed that the fault is between the phases b and c. Thus a is the non-faulted phase, and b and c are the faulted phases. From (4.43) we see that the voltage drop in the non-faulted phase depends on the difference between the positive and negative-sequence source impedances. As these are normally equal, the voltage in the non-faulted phase will not be influenced by the phase-to-phase fault. Under the assumption, $Z_{S1} = Z_{S2}$ (4.43) becomes

$$\begin{aligned} V_a &= 1 \\ V_b &= a^2 - \frac{(a^2 - a)Z_{S1}}{2Z_{S1} + 2Z_{F1}} \\ V_c &= a + \frac{(a^2 - a)Z_{S1}}{2Z_{S1} + 2Z_{F1}} \end{aligned} \tag{4.44}$$

We see that the voltage drop in the faulted phases is equal in magnitude $\frac{Z_{S1}}{2Z_{S1}+2Z_{F1}}$ but opposite in direction. The direction in which the two phase voltages drop is along the pre-fault phase-to-phase voltage between the faulted phases, $V_b - V_c$.

From (4.43) we can derive the following expression for the voltage between the faulted phases

$$V_b - V_c = \frac{Z_{F1} + Z_{F2}}{(Z_{S1} + Z_{S2}) + (Z_{F1} + Z_{F2})}(a^2 - a) \tag{4.45}$$

When we realize that $(a^2 - a)$ is the pre-fault voltage between the two faulted phases, the resemblance with the single-phase voltage divider of Fig. 4.14 and (4.9) becomes immediately clear. The same expressions as for the three-phase fault can be used, but for the voltages between the faulted phases; the impedances in the expression are the sum of positive and negative sequence values.

EXAMPLE Consider phase-to-phase faults on one of the 33 kV feeders in the system shown in Fig. 4.21. The impedance values needed to calculate the voltages during a phase-to-phase fault are as follows:

$$Z_{S1} = Z_{S2} = 1.23 + j18.3\%$$
$$Z_{F1} = Z_{F2} = 1.435 + j3.102\%/\text{km}$$

The resulting complex voltages are shown in Fig. 4.61. The circles and the arrows indicate the pre-fault voltages; the cross indicates the voltages in the faulted phases for a fault at the 33 kV bus.

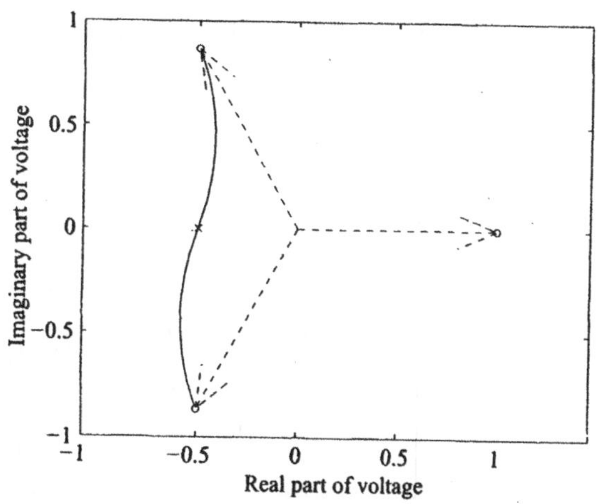

Figure 4.61 Complex voltages due to a phase-to-phase fault (solid line).

We see how the voltages in the two faulted phases move toward each other. The deviation of their path from a straight line is due to the difference in X/R ratio between source and feeder impedance. This is a subject to be discussed in further detail in Section 4.5.

4.4.3 Two-Phase-to-Ground Faults

Single-phase and phase-to-phase faults have been discussed in the two previous sections. The only asymmetrical fault type remaining is the two-phase-to-ground fault. For a two-phase-to-ground fault the three sequence networks are connected in parallel, as shown in Fig. 4.62. It is again possible to calculate component voltages and from these calculate voltages in the three phases in the same way as done for the single-phase and phase-to-phase faults.

The sequence voltages at the pcc for a fault between phases b and c and ground are given by the following expressions:

$$\begin{aligned} V_1 &= 1 - \frac{Z_{S1}(Z_{S0} + Z_{F0} + Z_{S2} + Z_{F2})}{D} \\ V_2 &= \frac{Z_{S2}(Z_{S0} + Z_{F0})}{D} \\ V_0 &= \frac{Z_{S0}(Z_{S2} + Z_{F2})}{D} \end{aligned} \tag{4.46}$$

with

$$D = (Z_{S0} + Z_{F0})(Z_{S1} + Z_{F1} + Z_{S2} + Z_{F2}) + (Z_{S1} + Z_{F1})(Z_{S2} + Z_{F2}) \tag{4.47}$$

From (4.46) it is possible to calculate the phase-to-ground voltages in the three phases

$$\begin{aligned} V_a &= 1 + \frac{(Z_{S2} - Z_{S1})(Z_{S0} + Z_{F0})}{D} + \frac{(Z_{S0} - Z_{S1})(Z_{S2} + Z_{F2})}{D} \\ V_b &= a^2 + \frac{(aZ_{S2} - a^2 Z_{S1})Z_0}{D} + \frac{(Z_{S0} - a^2 Z_{S1})Z_2}{D} \\ V_c &= a + \frac{(a^2 Z_{S2} - aZ_{S1})Z_0}{D} + \frac{(Z_{S0} - aZ_{S1})Z_2}{D} \end{aligned} \tag{4.48}$$

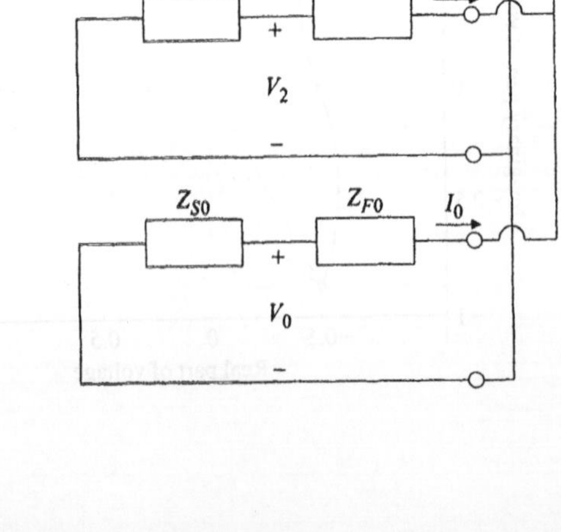

Figure 4.62 Equivalent circuit for a two-phase-to-ground fault.

There are two effects which cause a change in voltage in the non-faulted phase (V_a): the difference between the positive- and the negative-sequence source impedance; and the difference between the positive- and the zero-sequence source impedance. For both effects the non-faulted phase voltage drops when the positive-sequence impedance increases. Negative- and positive-sequence impedance are normally rather close, so that the second term in (4.48) may be neglected. The third term, which depends on the difference between zero- and positive-sequence source impedance, could cause a serious change in voltage. As the zero-sequence source impedance is often larger than the positive-sequence one, we expect a rise in voltage in the non-faulted phase. Like with single-phase faults we can eliminate this term by considering phase-to-neutral voltages instead of phase-to-ground voltages.

Looking at the voltages in the faulted phases and realizing that Z_{S1} is close to Z_{S2} we see that the second term is a voltage drop in the direction of the other faulted phase; $(a - a^2)$ is the pre-fault voltage between the faulted phases. For $Z_{S0} = Z_{S1}$ the third term in (4.48) is a voltage drop towards the non-faulted phase pre-fault voltage, for $Z_{S0} \ll Z_{S1}$ the third term is a drop along the positive real axis, as shown in Fig. 4.63. The voltage drop according to A in Fig. 4.63 is the same drop as for a phase-to-phase fault. The ground-connection causes an additional drop in the voltage in the two faulted phases, somewhere in between directions B and C. It is assumed here that all impedances have the same X/R ratio.

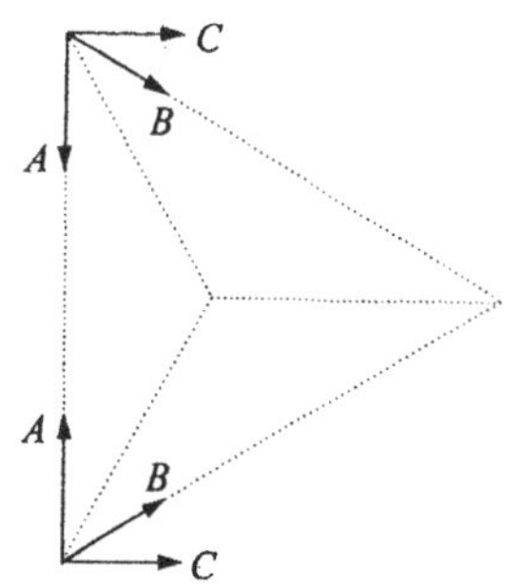

Figure 4.63 Voltage drops in the faulted phase during a two-phase-to-ground fault. A: second term in (4.48); B: third term for $Z_{S1} = Z_{S0}$; C: third term for $Z_{S1} \ll Z_{S0}$.

As said before, positive- and negative-sequence impedances are normally very close. In that case we can simplify the expressions by substituting $Z_{S1} = Z_{S2}$ and $Z_{F1} = Z_{F2}$. But when we are only interested in phase-to-neutral voltages it is easier to use the three-phase voltage divider model introduced in Fig. 4.53 for single-phase faults. For two-phase-to-ground faults the equivalent circuit is redrawn in Fig. 4.64.

Without any further calculation we can see from Fig. 4.64 that the phase-to-neutral voltage in the non-faulted phase is not influenced by the two-phase-to-ground fault. The phase-to-neutral voltage at the fault point, V_{FN}, is found from applying Kirchhoff's current law to the fault point:

$$\frac{a^2 - V_{FN}}{Z_{S1} + Z_{F1}} + \frac{a - V_{FN}}{Z_{S1} - Z_{F1}} = \frac{V_{FN}}{\frac{1}{3}(Z_{S0} - Z_{S1}) + \frac{1}{3}(Z_{F0} - Z_{F1})} \tag{4.49}$$

Solving (4.49) leads to the following expression for the voltage at the fault point:

$$V_{FN} = -\frac{(Z_{S0} + Z_{F0}) - (Z_{S1} + Z_{F1})}{2(Z_{S0} + Z_{F0}) + (Z_{S1} + Z_{F1})} \tag{4.50}$$

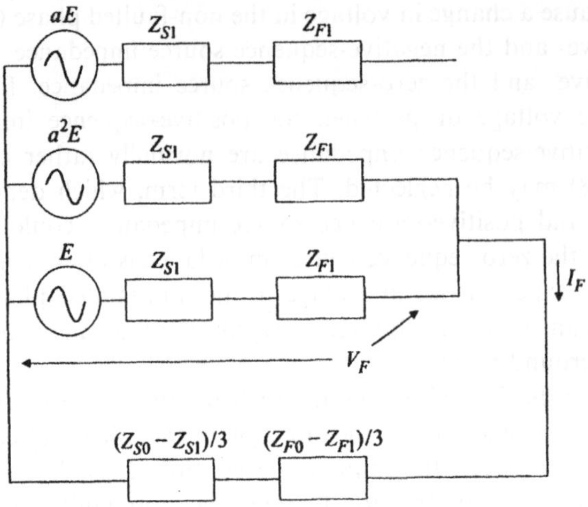

Figure 4.64 Three-phase voltage divider model for a two-phase-to-ground fault.

If zero-sequence and positive-sequence impedances are equal, $Z_{S0} = Z_{S1}$ and $Z_{F0} = Z_{F1}$, we find that

$$V_{FN} = 0 \tag{4.51}$$

If the zero-sequence impedance becomes large, like in a resistance-grounded system, the fault-point voltage is

$$V_{FN} = -\frac{1}{2} \tag{4.52}$$

The latter expression corresponds to the expression obtained for phase-to-phase faults. This is rather obvious if we realize that a large zero-sequence impedance implies that the fault current through the earth return is very small. Thus, the presence of a connection with earth during the fault does not influence the voltages.

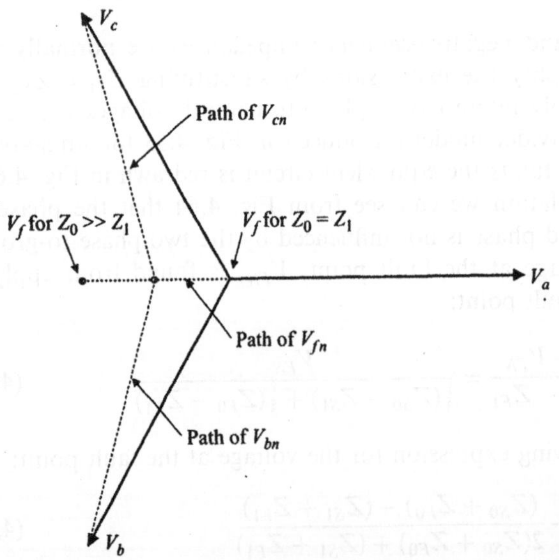

Figure 4.65 Phase-to-neutral voltages in the faulted phases for a two-phase-to-ground fault.

The intermediate case, where $Z_{S1} < Z_{S0} < \infty$, gives a voltage at the fault point somewhere in between these two extremes:

$$-\frac{1}{2} < V_{FN} < 0 \tag{4.53}$$

This voltage and the resulting voltages at the pcc can be obtained from Fig. 4.65. The voltage at the fault point is located between the origin and the point $-\frac{1}{2}$: the former for equal positive- negative-, and zero-sequence impedances, the latter for very large zero-sequence impedance. The voltage at the pcc for a faulted phase is somewhere between the voltage at the fault point and the pre-fault voltage in that phase. This knowledge will later be used for the classification of three-phase unbalanced sags. For calculating sag magnitudes this construction is not of practical use, as the fault-to-neutral voltage V_{FN} depends on the fault position.

4.4.4 Seven Types of Three-Phase Unbalanced Sags

The voltage sags due to the various types of faults have been discussed in the previous sections: three-phase faults in Section 4.2, single-phase faults in Section 4.4.1, phase-to-phase faults in Section 4.4.2, and finally two-phase-to-ground faults in Section 4.4.3. For each type of fault, expressions have been derived for the voltages at the pcc. But as already mentioned, this voltage is not equal to the voltage at the equipment terminals. Equipment is normally connected at a lower voltage level than the level at which the fault occurs. The voltages at the equipment terminals, therefore, not only depend on the voltages at the pcc but also on the winding connection of the transformers between the pcc and the equipment terminals. The voltages at the equipment terminals further depend on the load connection. Three-phase load is normally connected in delta but star-connection is also used. Single-phase load is normally connected in star (i.e., between one phase and neutral) but sometimes in delta (between two phases). Note that we consider here the voltage sag as experienced at the terminals of end-user equipment, not the voltage as measured by monitoring equipment. The latter is typically located at distribution or even at transmission level.

In this section we will derive a classification for three-phase unbalanced voltage sags, based on the following assumptions:

- Positive- and negative-sequence impedances are identical.
- The zero-sequence component of the voltage does not propagate down to the equipment terminals, so that we can consider phase-to-neutral voltages.
- Load currents, before, during, and after the fault, can be neglected.

4.4.4.1 Single-Phase Faults. The phase-to-neutral voltages due to a single-phase-to-ground fault are, under the assumptions mentioned,

$$\begin{aligned} V_a &= V \\ V_b &= -\frac{1}{2} - \frac{1}{2}j\sqrt{3} \\ V_c &= -\frac{1}{2} + \frac{1}{2}j\sqrt{3} \end{aligned} \tag{4.54}$$

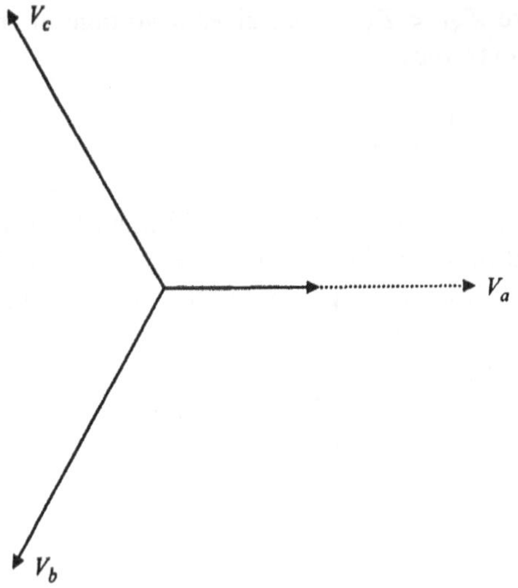

Figure 4.66 Phase-to-neutral voltages before (dashed line) and during (solid line) a phase-to-ground fault.

The resulting phasor diagram is shown in Fig. 4.66. If the load is connected in star, these are the voltages at the equipment terminals. If the load is connected in delta, the equipment terminal voltages are the phase-to-phase voltages. These can be obtained from (4.54) by the following transformation:

$$\begin{aligned} V_a' &= j\frac{V_b - V_c}{\sqrt{3}} \\ V_b' &= j\frac{V_c - V_a}{\sqrt{3}} \\ V_c' &= j\frac{V_a - V_b}{\sqrt{3}} \end{aligned} \tag{4.55}$$

This transformation will be an important part of the classification. The factor $\sqrt{3}$ is aimed at changing the base of the pu values, so that the normal operating voltage remains at 100%. The 90° rotation by using a factor j aims at keeping the axis of symmetry of the sag along the real axis. We will normally omit the primes from (4.55). Applying transformation (4.55) results in the following expression for the three-phase unbalanced voltage sag experienced by a delta-connected load, due to a single-phase fault:

$$\begin{aligned} V_a &= 1 \\ V_b &= -\frac{1}{2} - \left(\frac{1}{6} + \frac{1}{3}V\right)j\sqrt{3} \\ V_c &= -\frac{1}{2} + \left(\frac{1}{6} + \frac{1}{3}V\right)j\sqrt{3} \end{aligned} \tag{4.56}$$

The phasor diagram for the equipment terminal voltages is shown in Fig. 4.67: two voltages show a drop in magnitude and change in phase angle; the third voltage is not influenced at all. Delta-connected equipment experiences a sag in two phases due to a single-phase fault.

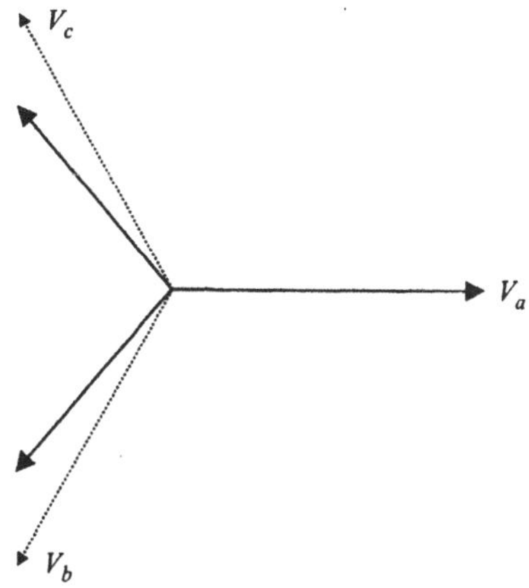

Figure 4.67 Phase-to-phase voltages before (dashed line) and during (solid line) a phase-to-ground fault.

4.4.4.2 Phase-to-Phase Faults. For a phase-to-phase fault the voltages in the two faulted phases move toward each other. The expressions for the phase-to-neutral voltages during a phase-to-phase fault read as follows:

$$\begin{aligned} V_a &= 1 \\ V_b &= -\frac{1}{2} - \frac{1}{2}Vj\sqrt{3} \\ V_c &= -\frac{1}{2} + \frac{1}{2}Vj\sqrt{3} \end{aligned} \tag{4.57}$$

Like before, (4.55) can be used to calculate the voltages experienced by a phase-to-phase connected load, resulting in

$$\begin{aligned} V_a &= V \\ V_b &= -\frac{1}{2}V - \frac{1}{2}j\sqrt{3} \\ V_c &= -\frac{1}{2}V + \frac{1}{2}j\sqrt{3} \end{aligned} \tag{4.58}$$

The corresponding phasor diagrams are shown in Figs. 4.68 and 4.69. Due to a phase-to-phase fault a star-connected load experiences a drop in two phases, a delta-

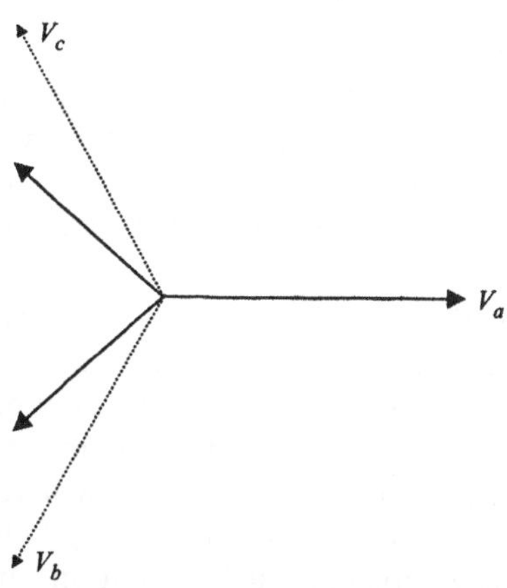

Figure 4.68 Phase-to-neutral voltages before (dashed line) and during (solid line) a phase-to-phase fault.

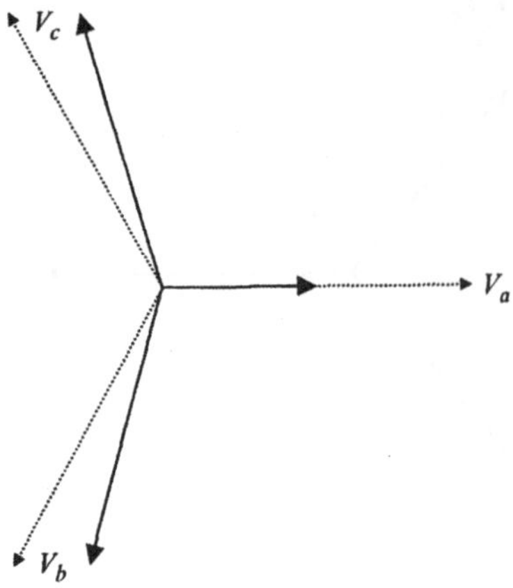

Figure 4.69 Phase-to-phase voltages before (dashed line) and during (solid line) a phase-to-phase fault.

connected load experiences a drop in three phases. For the star-connected load the maximum drop is 50%, for $V = 0$. But for the delta-connected load one phase could drop all the way down to zero. The conclusion that load could therefore best be connected in star is wrong, however. Most sags do not originate at the same voltage level as the equipment terminals. We will see later that the sag at the equipment terminals could be either of the two types shown in Figs. 4.68 and 4.69, depending on the transformer winding connections.

4.4.4.3 Transformer Winding Connections. Transformers come with many different winding connections, but a classification into only three types is sufficient to explain the transfer of three-phase unbalanced sags from one voltage level to another.

1. Transformers that do not change anything to the voltages. For this type of transformer the secondary-side voltages (in pu) are equal to the primary-side voltages (in pu). The only type of transformer for which this holds is the star-star connected one with both star points grounded.
2. Transformers that remove the zero-sequence voltage. The voltages on the secondary side are equal to the voltages on the primary side minus the zero-sequence component. Examples of this transformer type are the star-star connected transformer with one or both star points not grounded, and the delta-delta connected transformer. The delta-zigzag (Dz) transformer also fits into this category.
3. Transformers that swap line and phase voltages. For these transformers each secondary-side voltage equals the difference between two primary-side voltages. Examples are the delta-star (Dy) and the star-delta (Yd) transformer as well as the star-zigzag (Yz) transformer.

Within each of these three categories there will be transformers with different clock number (e.g., Yd1 and Yd11) leading to a different phase shift between primary- and secondary-side voltages. This difference is not of any importance for the voltage sags as experienced by the equipment. All that matters is the change between the pre-fault voltages and the during-fault voltages, in magnitude and in phase-angle. The whole phasor diagram, with pre-fault and during-fault phasors, can be rotated without any influence on the equipment. Such a rotation can be seen as a shift in the zero point on the time axis which of course has no influence on equipment behavior. The three transformer types can be defined mathematically by means of the following transformation matrices:

$$T_1 = \begin{bmatrix} 1 & 0 & 0 \\ 0 & 1 & 0 \\ 0 & 0 & 1 \end{bmatrix} \tag{4.59}$$

$$T_2 = \frac{1}{3}\begin{bmatrix} 2 & -1 & -1 \\ -1 & 2 & -1 \\ -1 & -1 & 2 \end{bmatrix} \tag{4.60}$$

$$T_3 = \frac{j}{\sqrt{3}}\begin{bmatrix} 0 & 1 & -1 \\ -1 & 0 & 1 \\ 1 & -1 & 0 \end{bmatrix} \tag{4.61}$$

Equation (4.59) is straightforward: matrix T_1 is the unity matrix. Equation (4.60) removes the zero-sequence component of the voltage. The matrix T_2 can be understood easily by realizing that the zero-sequence voltage equals $\frac{1}{3}(V_a + V_b + V_c)$. Matrix T_3 in (4.61) describes exactly the same transformation as expression (4.55). The additional advantage of the 90° rotation is that twice applying matrix T_3 gives the same results as once applying matrix T_2. Thus, $T_3^2 = T_2$; in engineering terms: two Dy transformers in cascade have the same effect on the voltage sag as one Dd transformer.

4.4.3.4 Transfer of Voltage Sags across Transformers. The three types of transformers can be applied to the sags due to single-phase and phase-to-phase faults. To get an overview of the resulting sags, the different combinations will be systematically treated below.

- *Single-phase fault, star-connected load, no transformer.*
  This case has been discussed before, resulting in (4.54) and Fig. 4.66. We will refer to this sag as sag X1. Transformer type 1 gives the same results of course.
- *Single-phase fault, delta-connected load, no transformer.*
  The voltage sag for this case is given in (4.56) and shown in Fig. 4.67. This sag will be referred to as sag X2.
- *Single-phase fault, star-connected load, transformer type 2.*
  Transformer type 2 removes the zero-sequence component of the voltage. The zero-sequence component of the phase voltages due to a single-phase fault is found from (4.54) to be equal to $\frac{1}{3}(V - 1)$. This gives the following expressions for the voltages:

$$\begin{aligned} V_a &= \frac{1}{3} + \frac{2}{3}V \\ V_b &= -\frac{1}{6} - \frac{1}{3}V - \frac{1}{2}j\sqrt{3} \\ V_c &= -\frac{1}{6} - \frac{1}{3}V + \frac{1}{2}j\sqrt{3} \end{aligned} \tag{4.62}$$

  This looks like a new type of sag, but we will see later that it is identical to the one experienced by a delta-connected load during a phase-to-phase fault. But for now it will be referred to as sag X3.

- *Single-phase fault, delta-connected load, transformer type 2.*
The phase-to-phase voltages experienced by a delta-connected load do not contain any zero-sequence component. Thus transformer type 2 does not have any influence on the sag voltages. The sag is thus still of type X2.
- *Single-phase fault, star-connected load, transformer type 3.*
Transformer type 3 changes phase voltages into line voltages. Thus star-connected load on secondary side experiences the same sag as delta-connected load on primary side. In this case that is sag X2.
- *Single-phase fault, delta-connected load, transformer type 3.*
There are now two transformations: from star- to delta-connected load, and from primary to secondary side of the transformer. Each of these transformations can be described through matrix T_3 defined in (4.61). Two of those transformations in cascade have the same effect as transformation T_2. Thus, the sag experienced by this delta-connected load is the same as by the star-connected load behind a transformer of type 2; thus, sag type X3.
- *Phase-to-phase fault, star-connected load, no transformer.*
This case was treated before resulting in (4.57) and Fig. 4.68. This will be sag type X4.
- *Phase-to-phase fault, delta-connected load, no transformer.*
The expression for the sag voltages reads as (4.58) and is shown in Fig. 4.69. This type will be referred to as X5.
- *Phase-to-phase fault, star-connected load, transformer type 2.*
As phase-to-phase faults do not result in any zero-sequence voltage, transformer type 2 (which removes the zero-sequence voltage) does not have any effect. The sag thus remains of type X4.
- *Phase-to-phase fault, delta-connected load, transformer type 2.*
Like before, the sag is still of type X5.
- *Phase-to-phase fault, star-connected load, transformer type 3.*
Star-connected load on secondary side of transformer type 2 experiences the same sag as delta-connected load on primary side. This results in type X5.
- *Phase-to-phase fault, delta-connected load, transformer type 3.*
This gives again two identical transformations T_3 in cascade, resulting in one transformation T_2. But that one only removes the zero-sequence component and has thus no influence on sags due to phase-to-phase faults. The result is, thus, again X4.

The effect of a second transformer on sags X1 through X5 is shown in Table 4.8. These results can be obtained by following the same reasoning as above. It becomes clear that

TABLE 4.8 Further Propagation of Sags

	Transformer Type		
Sag Type	1	2	3
X1	X1	X3	X2
X2	X2	X2	X3
X3	X3	X3	X2
X4	X4	X4	X5
X5	X5	X5	X4

the number of combinations is limited: at most five different sag types are possible due to single-phase and phase-to-phase faults.

4.4.4.5 The Basic Types of Sags. We saw that single-phase faults lead to three types of sags, designated sag X1, sag X2, and sag X3. Phase-to-phase faults lead to sag X4 and sag X5. We saw already from the phasor diagrams in Figs. 4.67 and 4.68 that single-phase and phase-to-phase faults lead to similar sags. The sag voltages for sag type X2 are

$$\begin{aligned} V_a &= 1 \\ V_b &= -\frac{1}{2} - \left(\frac{1}{6} + \frac{1}{3}V\right)j\sqrt{3} \\ V_c &= -\frac{1}{2} + \left(\frac{1}{6} + \frac{1}{3}V\right)j\sqrt{3} \end{aligned} \tag{4.63}$$

For sag type X4 the voltages are

$$\begin{aligned} V_a &= 1 \\ V_b &= -\frac{1}{2} - \frac{1}{2}Vj\sqrt{3} \\ V_c &= -\frac{1}{2} + \frac{1}{2}Vj\sqrt{3} \end{aligned} \tag{4.64}$$

Comparing these two sets of equations shows that (4.63) can be obtained by replacing V in (4.64) by $\frac{1}{3} + \frac{2}{3}V$. If we define the magnitude of sag X4 as V, then sag X2 is a sag of type X4 with magnitude $\frac{1}{3} + \frac{2}{3}V$.

In the same way we can compare sag X3:

$$\begin{aligned} V_a &= \frac{1}{3} + \frac{2}{3}V \\ V_b &= -\frac{1}{6} - \frac{1}{3}V - \frac{1}{2}j\sqrt{3} \\ V_c &= -\frac{1}{6} - \frac{1}{3}V + \frac{1}{2}j\sqrt{3} \end{aligned} \tag{4.65}$$

and sag X5:

$$\begin{aligned} V_a &= V \\ V_b &= -\frac{1}{2}V - \frac{1}{2}j\sqrt{3} \\ V_c &= -\frac{1}{2}V + \frac{1}{2}j\sqrt{3} \end{aligned} \tag{4.66}$$

Again we obtain (4.65) by replacing V in (4.66) by $\frac{1}{3} + \frac{2}{3}V$. The result is that only three types remain: X1, X4, and X5. A fourth type of sag is the sag due to three-phase faults, with all three voltages down the same amount. The resulting classification is shown in Table 4.9 in equation form and in Fig. 4.70 in phasor form. All sags in Fig. 4.70 have a magnitude of 50%. From the discussion about sags due to single-phase and phase-to-phase faults, together with the definition of the four types, the origin and the propagation of the sags becomes straightforward. The results are summarized in Table 4.10 for the origin of sags and in Table 4.11 for their propagation to lower voltage levels. The superscript (*) behind the sag type in Tables 4.10 and

TABLE 4.9 Four Types of Sags in Equation Form

Type A	Type B
$V_a = V$ $V_b = -\frac{1}{2}V - \frac{1}{2}jV\sqrt{3}$ $V_c = -\frac{1}{2}V + \frac{1}{2}jV\sqrt{3}$	$V_a = V$ $V_b = -\frac{1}{2} - \frac{1}{2}j\sqrt{3}$ $V_c = -\frac{1}{2} + \frac{1}{2}j\sqrt{3}$
Type C	**Type D**
$V_a = 1$ $V_b = -\frac{1}{2} - \frac{1}{2}jV\sqrt{3}$ $V_c = -\frac{1}{2} + \frac{1}{2}jV\sqrt{3}$	$V_a = V$ $V_b = -\frac{1}{2}V - \frac{1}{2}j\sqrt{3}$ $V_c = -\frac{1}{2}V + \frac{1}{2}j\sqrt{3}$

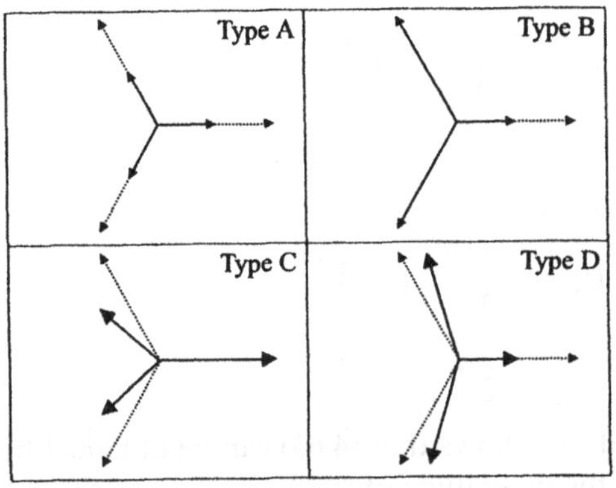

Figure 4.70 Four types of sag in phasor-diagram form.

TABLE 4.10 Fault Type, Sag Type, and Load Connection

Fault Type	Star-connected Load	Delta-connected Load
Three-phase	sag A	sag A
Phase-to-phase	sag C	sag D
Single-phase	sag B	sag C*

TABLE 4.11 Transformation of Sag Type to Lower Voltage Levels

Transformer Connection	Sag Type A	Sag Type B	Sag Type C	Sag Type D
YNyn	type A	type B	type C	type D
Yy, Dd, Dz	type A	type D*	type C	type D
Yd, Dy, Yz	type A	type C*	type D	type C

4.11 indicates that the sag magnitude is not equal to V but equal to $\frac{1}{3} + \frac{2}{3}V$, with V the voltage in the faulted phase or between the faulted phases in Table 4.10 and the magnitude of the sag on primary side in Table 4.11. Note that in effect these two definitions of V are the same.

4.4.4.6 Two-Phase-to-Ground Faults. Two-phase-to-ground faults can be treated in the same way as single-phase and phase-to-phase faults. We will assume that the voltage in the non-faulted phase is not influenced by the fault. As we have seen in Section 4.4.3 this corresponds to the situation in which positive-, negative-, and zero-sequence impedances are equal. This can be seen as an extreme case. A zero-sequence impedance larger than the positive-sequence impedance will shift the resulting voltages toward those for a phase-to-phase fault.

The phase-to-ground voltages at the pcc due to a two-phase-to-ground fault are

$$\begin{aligned} V_a &= 1 \\ V_b &= -\frac{1}{2}V - \frac{1}{2}Vj\sqrt{3} \\ V_c &= -\frac{1}{2}V + \frac{1}{2}Vj\sqrt{3} \end{aligned} \tag{4.67}$$

After a Dy transformer or any other transformer of type 3, the voltages are

$$\begin{aligned} V_a &= V \\ V_b &= -\frac{1}{3}j\sqrt{3} - \frac{1V}{2} - \frac{1}{6}Vj\sqrt{3} \\ V_c &= +\frac{1}{3}j\sqrt{3} - \frac{1}{2}V + \frac{1}{6}Vj\sqrt{3} \end{aligned} \tag{4.68}$$

After two transformers of type 3 or after one transformer of type 2, we get

$$\begin{aligned} V_a &= \frac{2}{3} + \frac{1}{3}V \\ V_b &= -\frac{1}{3} - \frac{1}{6}V - \frac{1}{2}Vj\sqrt{3} \\ V_c &= -\frac{1}{3} - \frac{1}{6}V + \frac{1}{2}Vj\sqrt{3} \end{aligned} \tag{4.69}$$

These three sags are different from the four types found earlier. It is not possible to translate one into the other. Two-phase-to-ground faults lead to three more types of sags, resulting in a total of seven. The three new types are shown in phasor-diagram form in Fig. 4.71 and in equation form in Table 4.12. Sags due to two-phase-to-ground faults and sags due to phase-to-phase faults are compared in Fig. 4.72. For a type C sag the voltages change along the imaginary axis only, for type D along the real axis only.

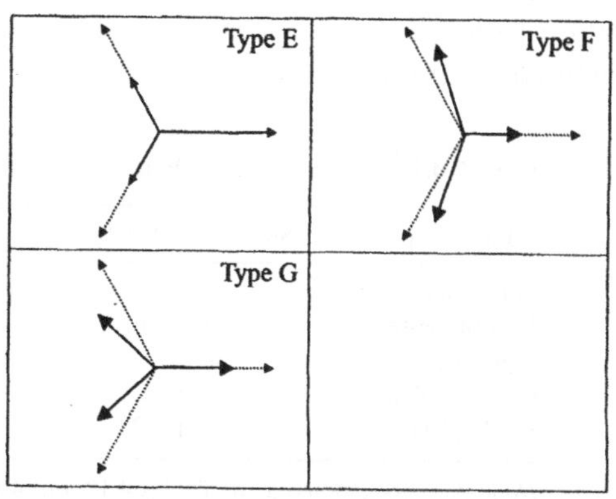

Figure 4.71 Three-phase unbalanced sags due to two-phase-to-ground faults.

TABLE 4.12 Sags Due to Two-Phase-to-Ground Faults

Type E	Type F
$V_a = 1$	$V_a = V$
$V_b = -\frac{1}{2}V - \frac{1}{2}Vj\sqrt{3}$	$V_b = -\frac{1}{3}j\sqrt{3} - \frac{1}{2}V - \frac{1}{6}Vj\sqrt{3}$
$V_c = -\frac{1}{2}V + \frac{1}{2}Vj\sqrt{3}$	$V_c = +\frac{1}{3}j\sqrt{3} - \frac{1}{2}V + \frac{1}{6}Vj\sqrt{3}$

Type G
$V_a = \frac{2}{3} + \frac{1}{3}V$
$V_b = -\frac{1}{3} - \frac{1}{6}V - \frac{1}{2}Vj\sqrt{3}$
$V_c = -\frac{1}{3} - \frac{1}{6}V + \frac{1}{2}Vj\sqrt{3}$

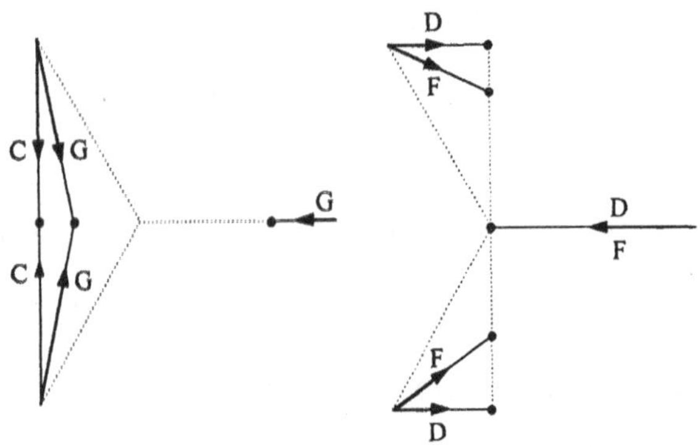

Figure 4.72 Comparison of three-phase unbalanced sags due to two-phase-to-ground faults (F and G) with three-phase unbalanced sags due to phase-to-phase and single-phase-to-ground faults (C and D). The arrows indicate the direction of change in the three complex voltages for the different sag types.

For types F and G the voltages drop along both axis. The resulting voltages at the equipment terminals are lower during a two-phase-to-ground fault. An additional difference is that all three voltages drop in magnitude for a type G sag. Note also that for a type D and type F sag the drop in the worst-affected phase is the same, whereas for a type C and a type G sag the drop in voltage between the two worst-affected phases is the same. This property will be used when defining the magnitude of measured three-phase unbalanced sags.

Sag types F and G have been derived by assuming that positive-, negative-, and zero-sequence impedances are the same. If the zero-sequence impedance is larger than the positive-sequence impedance, the resulting sag will be somewhere in between type C and type G, or in between type D and type F.

4.4.4.7 Seven Types of Three-Phase Unbalanced Sags. Origin of sags and transformation to lower voltage levels for all seven types of three-phase unbalanced sags are summarized in Tables 4.13 and 4.14. An example of the sag transformation to

TABLE 4.13 Origin of Three-Phase Unbalanced Sags

Fault Type	Star-connected Load	Delta-connected Load
Three-phase	Type A	Type A
Two-phase-to-ground	Type E	Type F
Phase-to-phase	Type C	Type D
Single-phase	Type B	Type C*

Note: Asterisk defined as in Tables 4.10 and 4.11.

TABLE 4.14 Transformation of Sag Type to Lower Voltage Levels

Transformer Connection	Sag on Primary Side						
	Type A	Type B	Type C	Type D	Type E	Type F	Type G
YNyn	A	B	C	D	E	F	G
Yy, Dd, Dz	A	D*	C	D	G	F	G
Yd, Dy, Yz	A	C*	D	C	F	G	F

lower voltage levels is shown in Fig. 4.73. A fault at 33 kV causes the voltage at the pcc to drop to 50% of the nominal voltage. For a three-phase fault the situation is easy: at any level and for any load connection the sag is of type A and with a magnitude of 50%. For a phase-to-phase fault the voltage between the faulted phases at the pcc drops to 50%. For star-connected load the resulting sags are type C, 50% at 33 kV; type D, 50% at 11 kV; and again type C, 50% at 660 V. In case the fault is a single-phase one, the voltage in the faulted phase drops to 50% at the pcc. This corresponds to a sag of type B and magnitude 50% at 33 kV. After the first Dy transformer the zero-sequence component of the voltages has been removed. Star-connected load at 11 kV will experience a sag of type C with a magnitude of 67%. Delta-connected load will experience a sag of type D with a magnitude of 67%. For load fed at 660 V the situation is just the other way around: star-connected load experiences a sag of type D; delta-connected load one of type C.

4.4.4.8 Overview. In the beginning of this section we assumed that the zero-sequence component of the voltages did not propagate down to the equipment terminals. We used this assumption to obtain an expression for the voltages during a single-phase-to-ground fault. Under this same assumption we find that three-phase unbalanced sags of type B or type E cannot occur at the equipment terminals. At the equipment terminals we only find the following five types of three-phase unbalanced sags:

- type A due to three-phase faults.
- type C and type D due to single-phase and phase-to-phase faults.
- type F and type G due to two-phase-to-ground faults.

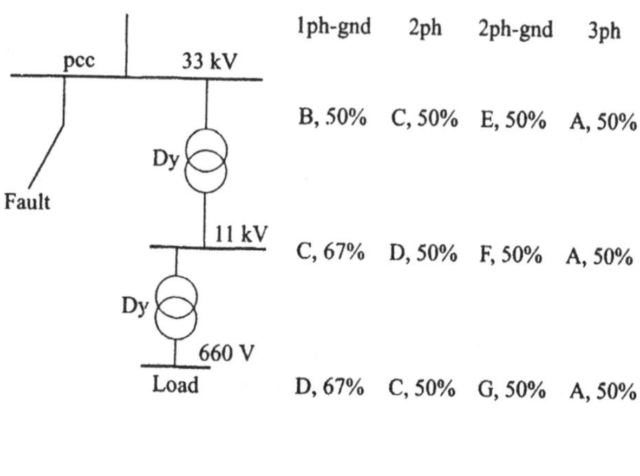

Figure 4.73 Example of sag transformation, for star-connected load.

The latter two types can be considered as distorted versions of type C and D. Sags of type C and D are also distorted by the presence of induction motor load. The presence of induction motor load makes that positive- and negative-sequence source impedances are no longer equal. One of the effects of this is that the voltage in the "non-faulted phase" for a type C sag is no longer equal to 100%. This has been the basis for a classification and characterization of three-phase unbalanced sags into three types, corresponding to our types A, C, and D [203], [204].

4.5 PHASE-ANGLE JUMPS

A short circuit in a power system not only causes a drop in voltage magnitude but also a change in the phase angle of the voltage. In a 50 Hz or 60 Hz system, voltage is a complex quantity (a phasor) which has magnitude and phase angle. A change in the system, like a short circuit, causes a change in voltage. This change is not limited to the magnitude of the phasor but includes a change in phase angle as well. We will refer to the latter as the phase-angle jump associated with the voltage sag. The phase-angle jump manifests itself as a shift in zero crossing of the instantaneous voltage. Phase-angle jumps are not of concern for most equipment. But power electronics converters using phase-angle information for their firing instants may be affected. We will come back to the effect of phase-angle jumps on equipment in Chapter 5.

Figure 4.74 shows a voltage sag with a phase-angle jump of +45°: the during-fault voltage leads the pre-fault voltage. A sag with a phase-angle jump of −45° is shown in Fig. 4.75: the during-fault voltage lags the pre-fault voltage. Both sags have a magnitude of 70%. In both figures, the pre-fault voltages have been continued as a dashed curve. Note that these are synthetic sags, not measurement results.

The origin of phase-angle jumps will be explained for a three-phase fault, as that enables us to use the single-phase model. Phase-angle jumps during three-phase faults are due to the difference in X/R ratio between the source and the feeder. A second cause of phase-angle jumps is the transformation of sags to lower voltage levels. This phenomenon has already been mentioned when unbalanced sags were discussed in Section 4.4.

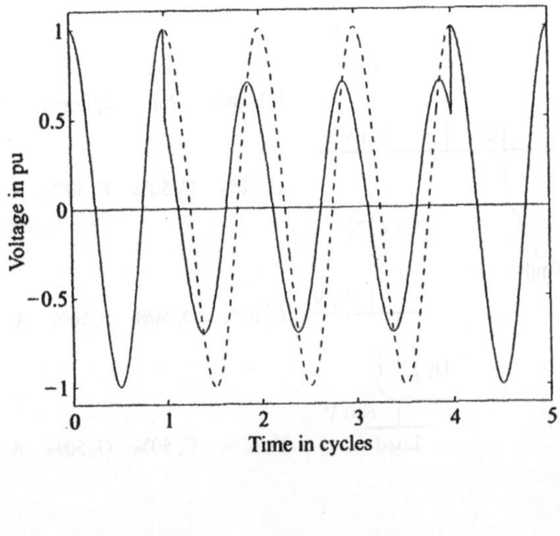

Figure 4.74 Synthetic sag with a magnitude of 70% and a phase-angle jump of +45°.

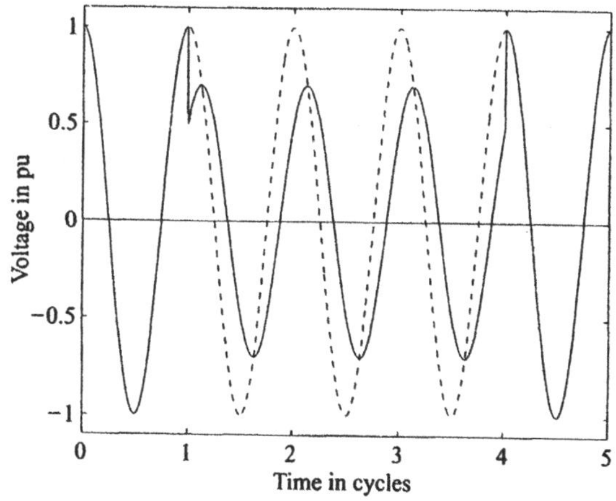

Figure 4.75 Synthetic sag with a magnitude of 70% and a phase-angle jump of −45°.

4.5.1 Monitoring

To obtain the phase-angle jump of a measured sag, the phase-angle of the voltage during the sag must be compared with the phase-angle of the voltage before the sag. The phase-angle of the voltage can be obtained from the voltage zero-crossing or from the phase of the fundamental component of the voltage. The complex fundamental voltage can be obtained by doing a Fourier transform on the signal. This enables the use of Fast-Fourier Transform (FFT) algorithms.

To explain an alternative method, consider the following voltage signal:

$$v(t) = X\cos(\omega_0 t) - Y\sin(\omega_0 t) = Re\{(X + jY)e^{j\omega_0 t}\} \tag{4.70}$$

with ω_0 the fundamental (angular) frequency. Two new signals are obtained from this signal, as follows:

$$v_d(t) = 2v(t) \times \cos(\omega_0 t) \tag{4.71}$$

$$v_q(t) = 2v(t) \times \sin(\omega_0 t) \tag{4.72}$$

which we can write as

$$v_d(t) = X + X\cos(2\omega_0 t) + Y\sin(2\omega_0 t) \tag{4.73}$$

$$v_q(t) = -Y + Y\cos(2\omega_0 t) + X\sin(2\omega_0 t) \tag{4.74}$$

Averaging the two resulting signals over one half-cycle of the fundamental frequency gives the required fundamental voltage.

$$X + jY = \overline{v}_d(t) - j\overline{v}_q(t) \tag{4.75}$$

Knowing the values of X and Y, the sag magnitude can be calculated as $\sqrt{X^2 + Y^2}$ and the phase-angle jump as $\arctan\frac{Y}{X}$.

This algorithm has been applied to the recorded sag in Fig. 4.1. The resulting sag magnitude is shown in Fig. 4.76 and the phase-angle jump in Fig. 4.77. The effect of averaging $v_d(t)$ and $v_q(t)$ over one full cycle of the fundamental frequency is shown in Fig. 4.78 for the sag magnitude and in Fig. 4.79 for the phase-angle jump. The effect of a larger window is that the transition is slower, but the overshoot in phase-angle is less. Which window length needs to be chosen depends on the application.

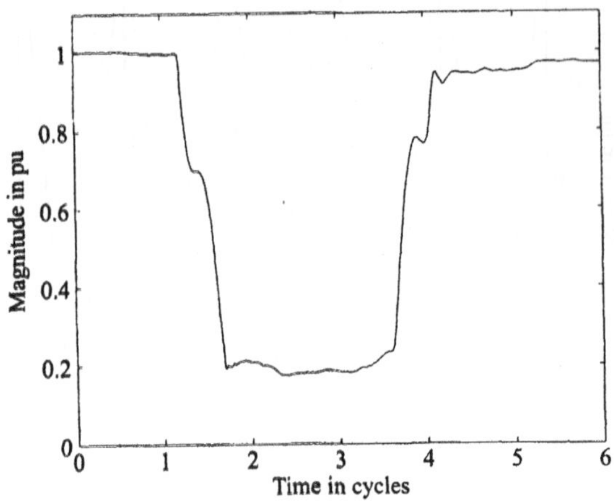

Figure 4.76 Amplitude of the fundamental voltage versus time for the voltage sag shown in Fig. 4.1—a half-cycle window has been used.

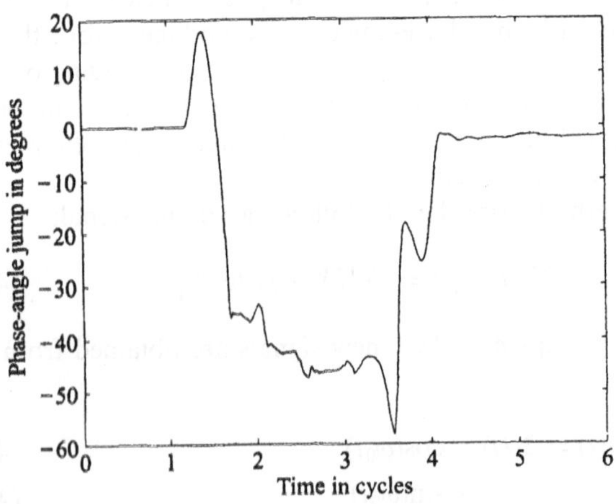

Figure 4.77 Argument of the fundamental voltage versus time for the voltage sag shown in Fig. 4.1—a half-cycle window has been used.

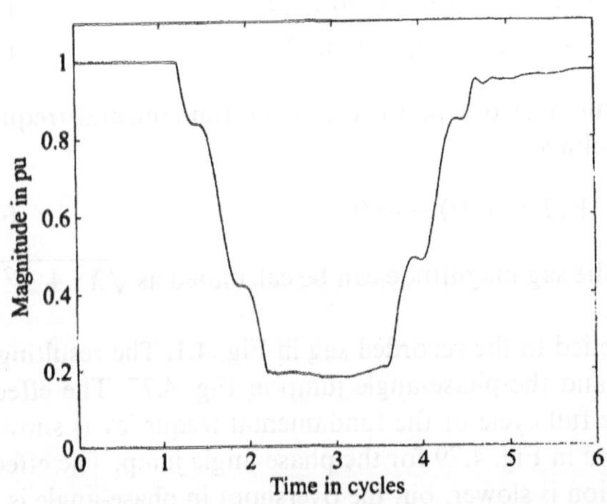

Figure 4.78 Amplitude of the fundamental voltage versus time for the voltage sag shown in Fig. 4.1—a one-cycle window has been used.

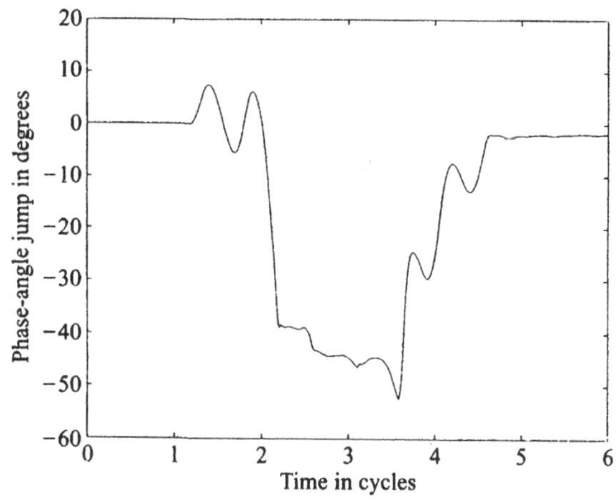

Figure 4.79 Argument of the fundamental voltage versus time for the voltage sag shown in Fig. 4.1—a one-cycle window has been used.

4.5.2 Theoretical Calculations

4.5.2.1 Origin of Phase-Angle Jumps. To understand the origin of phase-angle jumps associated with voltage sags, the single-phase voltage divider model of Fig. 4.14 can be used again, with the difference that Z_S and Z_F are complex quantities which we will denote as $\overline{Z}_S$ and $\overline{Z}_F$. Like before, we neglect all load currents and assume $E = 1$. This gives for the voltage at the point-of-common coupling (pcc):

$$\overline{V}_{sag} = \frac{\overline{Z}_F}{\overline{Z}_S + \overline{Z}_F} \tag{4.76}$$

Let $\overline{Z}_S = R_S + jX_S$ and $\overline{Z}_F = R_F + jX_F$. The argument of $\overline{V}_{sag}$, thus the phase-angle jump in the voltage, is given by the following expression:

$$\Delta\phi = \arg(\overline{V}_{sag}) = \arctan\left(\frac{X_F}{R_F}\right) - \arctan\left(\frac{X_S + X_F}{R_S + R_F}\right) \tag{4.77}$$

If $\frac{X_S}{R_S} = \frac{X_F}{R_F}$, expression (4.77) is zero and there is no phase-angle jump. The phase-angle jump will thus be present if the X/R ratios of the source and the feeder are different.

4.5.2.2 Influence of Source Strength. Consider again the power system used to obtain Fig. 4.15. Instead of the sag magnitude we calculated the phase-angle jump, resulting in Fig. 4.80. We again see that a stronger source makes the sag less severe: less drop in magnitude as well as a smaller phase-angle jump. The only exception is for terminal faults. The phase-angle jump for zero distance to the fault is independent of the source strength. Note that this is only of theoretical value as the phase-angle jump for zero distance to the fault, and thus for zero voltage magnitude, has no physical meaning.

4.5.2.3 Influence of Cross Section. Figure 4.81 plots phase-angle jump versus distance for 11 kV overhead lines of different cross sections. The resistance of the source has been neglected in these calculations: $R_S = 0$. The corresponding sag magnitudes were shown in Fig. 4.16. From the overhead line impedance data shown in Table 4.1 we can calculate the X/R ratio of the feeder impedances: 1.0 for the

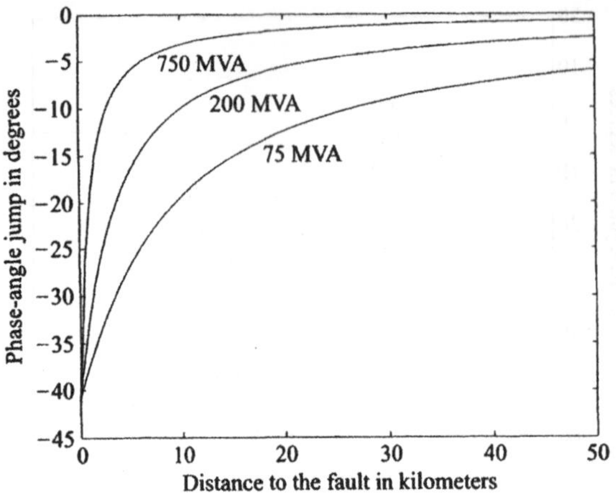

Figure 4.80 Phase-angle jump versus distance, for faults on a 150 mm^2 11 kV overhead feeder, with different source strength.

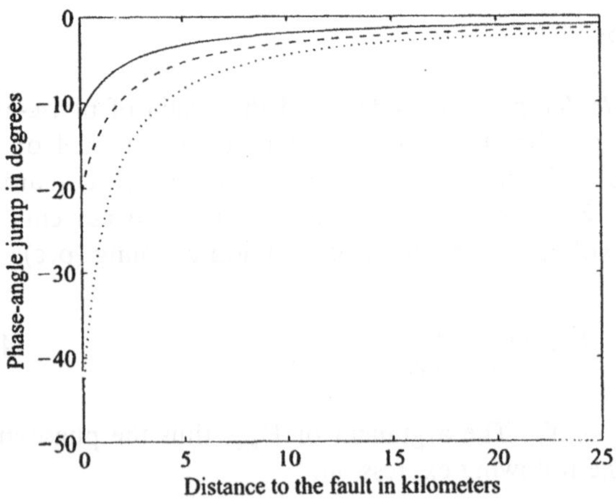

Figure 4.81 Phase-angle jump versus distance, for overhead lines with cross section 300 mm^2 (solid line), 150 mm^2 (dashed line), and 50 mm^2 (dotted line).

50 mm^2 line, 2.7 for the 150 mm^2, and 4.9 for the 300 mm^2; the phase-angle jump decreases for larger X/R ratio of the feeder.

The results for underground cables are shown in Fig. 4.82. Cables with a smaller cross section have a larger phase-angle jump for small distances to the fault, but the phase-angle jump also decays faster for increasing distance. This is due to the (in absolute value) larger impedance per unit length. The corresponding sag magnitudes were shown in Fig. 4.17.

Sag magnitude and phase-angle jump, i.e., magnitude and argument of the complex during-fault voltage, can be plotted in one diagram. Figure 4.83 shows the voltage paths in the complex plane, where the pre-sag voltage is in the direction of the positive real axis. The further the complex voltage is from $1 + j0$, the larger the change in complex voltage due to the fault. The difference between the pre-sag voltage and the actual voltage is referred to as the missing voltage. We will come back to the concept of missing voltage in Section 4.7.2.

Instead of splitting the disturbance into real and imaginary parts one may plot magnitude against phase-angle jump as done in Fig. 4.84. From the figure we can conclude that the phase-angle jump increases (in absolute value) when the drop in voltage increases (thus, when the sag magnitude decreases). Both an increase in

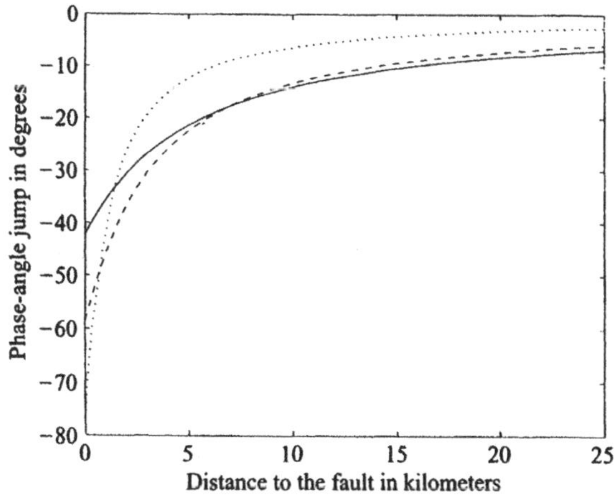

Figure 4.82 Phase-angle jump versus distance, for underground cables with cross section 300 mm^2 (solid line), 150 mm^2 (dashed line), and 50 mm^2 (dotted line).

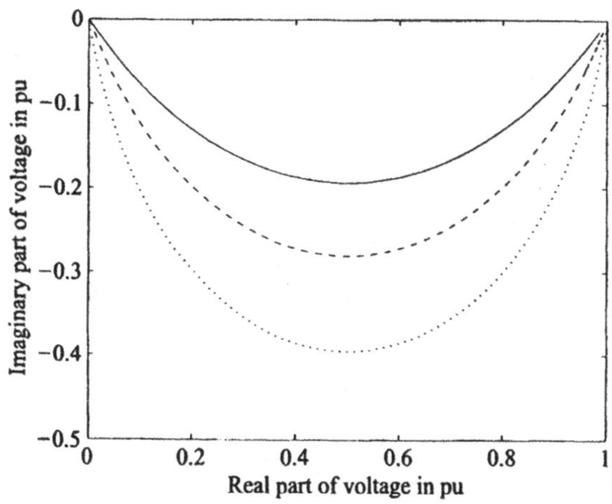

Figure 4.83 Path of the voltage in the complex plane when the distance to the fault changes, for underground cables with cross section 300 mm^2 (solid line), 150 mm^2 (dashed line), and 50 mm^2 (dotted line).

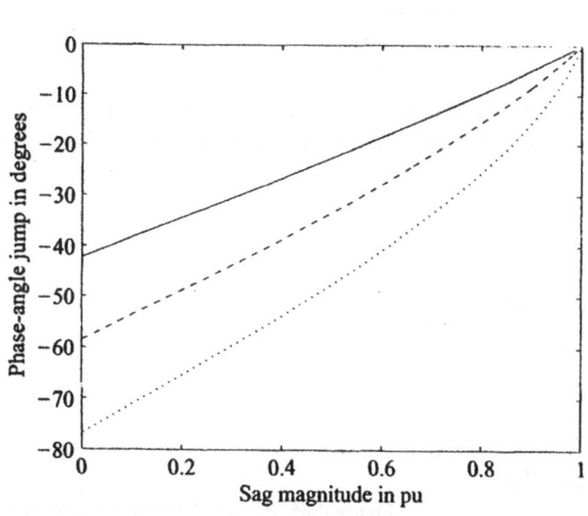

Figure 4.84 Magnitude versus phase-angle jump, for underground cables with cross section 300 mm^2 (solid line), 150 mm^2 (dashed line), and 50 mm^2 (dotted line).

phase-angle jump and a decrease in magnitude can be described as a more severe event. Knowing that both voltage drop and phase-angle jump increase when the distance to the fault increases, we can conclude that a fault leads to a more severe event the closer it is to the point-of-common coupling. We will later see that this only holds for three-phase faults. For single-phase and phase-to-phase faults this is not always the case.

4.5.2.4 Magnitude and Phase-Angle Jump Versus Distance. To obtain expressions for magnitude and phase-angle jump as a function of the distance to the fault we substitute $\overline{Z}_F = \overline{z}\mathcal{L}$ in (4.76) with $\overline{z}$ the complex feeder impedance per unit length, resulting in

$$\overline{V}_{sag} = \frac{\overline{z}\mathcal{L}}{\overline{Z}_S + \overline{z}\mathcal{L}} \tag{4.78}$$

The phase-angle jump is found from

$$\arg(\overline{V}_{sag}) = \arg(\overline{z}\mathcal{L}) - \arg(\overline{Z}_S + \overline{z}\mathcal{L}) \tag{4.79}$$

The phase-angle jump is thus equal to the angle in the complex plane between $\overline{z}\mathcal{L}$ and $\overline{Z}_S + \overline{z}\mathcal{L}$. This is shown in Fig. 4.85, where ϕ is the phase-angle jump and α is the angle between source impedance $\overline{Z}_S$ and feeder impedance $\overline{z}$.

$$\alpha = \arctan\left(\frac{X_F}{R_F}\right) - \arctan\left(\frac{X_S}{R_S}\right) \tag{4.80}$$

We will refer to α as the "impedance angle;" it is positive when the X/R-ratio of the feeder is larger than that of the source. Note that this is a rare situation: the impedance angle is in most cases negative. Using the cosine rule twice in the lower triangle in Fig. 4.85 gives the two expressions

$$|\overline{Z}_S + \overline{z}\mathcal{L}|^2 = |\overline{z}\mathcal{L}|^2 + |\overline{Z}_S|^2 - 2|\overline{z}\mathcal{L}||\overline{Z}_S|\cos(180^\circ + \alpha) \tag{4.81}$$

$$|\overline{Z}_S|^2 = |\overline{Z}_S + \overline{z}\mathcal{L}|^2 + |\overline{z}\mathcal{L}|^2 - 2|\overline{Z}_S + \overline{z}\mathcal{L}||\overline{z}\mathcal{L}|\cos(-\phi) \tag{4.82}$$

Substituting (4.81) into (4.82) and some rewriting gives an expression for the phase-angle jump as a function of distance

$$\cos(\phi) = \frac{\lambda + \cos\alpha}{\sqrt{1 + \lambda^2 + 2\lambda\cos\alpha}} \tag{4.83}$$

where $\lambda = z\mathcal{L}/Z_S$ is a measure of the "electrical" distance to the fault and α the impedance angle. Note that it is not so much the difference in X/R ratio which deter-

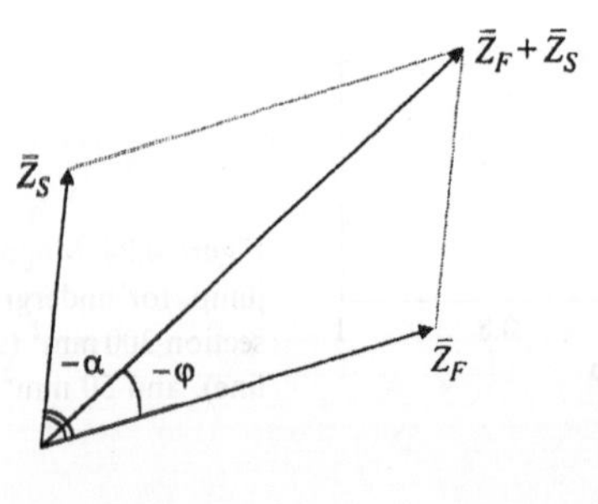

Figure 4.85 Phasor diagram for calculation of magnitude and phase-angle jump.

mines the size of the phase-angle jump but the actual angle between source and feeder impedance. For example, a source with $X_S/R_S = 40$ and a feeder with $X_F/R_F = 2$ gives an impedance angle of

$$\alpha = \arctan(2) - \arctan(40) = 63.4^\circ - 88.6^\circ = -25.2^\circ \tag{4.84}$$

where a source with $X_S/R_S = 3$ and a feeder with $X_F/R_F = 1$ gives an impedance angle of $\alpha = -26.6^\circ$. The latter will result in more severe phase-angle jumps.

The maximum angular difference occurs for underground cables in distribution systems. For a source X/R of 10 and a cable X/R of 0.5 we obtain an impedance angle of about -60°. In the forthcoming sections the value of -60° is used as the worst case. Although this is a rather rare case, it assists in showing the various relationships. Small positive phase-angle jumps may occur in transmission systems where X/R ratio of source and feeder impedance are similar. Impedance angles exceeding $+10^\circ$ are very unlikely. For most of the forthcoming studies we will assume that the impedance angle varies between 0 and -60°.

From (4.83) we can conclude that the maximum phase-angle jump occurs for $\mathcal{L} = 0$, $\lambda = 0$ and that it is equal to the impedance angle α.

The magnitude of the sag is obtained from (4.79) as

$$V_{sag} = \frac{|z\mathcal{L}|}{|z\mathcal{L} + Z_S|} \tag{4.85}$$

With (4.81) the following expression for the sag magnitude as a function of the distance to the fault is obtained:

$$V_{sag} = \frac{\lambda}{(1+\lambda)} \frac{1}{\sqrt{1 - \frac{2\lambda(1-\cos\alpha)}{(1+\lambda)^2}}} \tag{4.86}$$

Note that the first factor in the right-hand side of (4.86) gives the sag magnitude when the difference in X/R ratio is neglected ($\alpha = 0$). This is the same expression as (4.9) in Section 4.2. The error in making this approximation is estimated by approximating the second factor in (4.86) for small values of α:

$$\frac{1}{\sqrt{1 - \frac{2\lambda(1-\cos\alpha)}{(1+\lambda)^2}}} \approx \frac{1}{1 - \frac{\lambda(1-\cos\alpha)}{(1+\lambda)^2}} \approx 1 + \frac{\lambda}{(1+\lambda)^2}(1-\cos\alpha) \approx 1 + \frac{\lambda}{(1+\lambda)^2}\alpha^2 \tag{4.87}$$

The error is proportional to α^2. Thus, for moderate values of α the simple expression without considering phase-angle jumps can be used to calculate the sag magnitude.

4.5.2.5 Range of Magnitude and Phase-Angle Jump. The relation between magnitude and phase-angle jump is plotted for four values of the impedance angle in Fig. 4.86. Magnitude and phase-angle jump have been calculated by using (4.83) and (4.86). During a three-phase fault all three phases will experience the same change in magnitude and phase-angle. The relation shown in Fig. 4.86 thus also holds for single-phase equipment. When testing equipment for sags due to three-phase faults one should consider that magnitude and phase-angle jump can reach the whole range of combinations in Fig. 4.86.

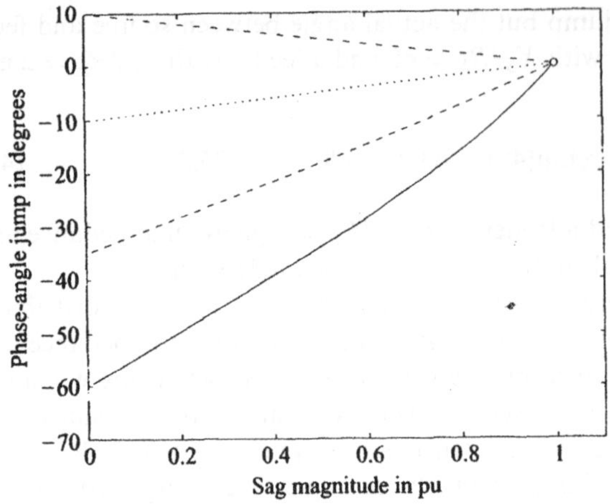

Figure 4.86 Relation between magnitude and phase-angle jump for three-phase faults: impedance angles: −60° (solid curve); −35° (dashed); −10° (dotted); +10° (dash-dot).

EXAMPLE Magnitude and phase-angle jump have been calculated for sags due to three-phase faults at the various voltage levels in the example supply shown in Fig. 4.21. Using the data in Tables 4.3 and 4.4 we can calculate the complex voltage at the pcc for any fault in the system. The absolute value and argument of this complex voltage are shown in Fig. 4.87. The complex voltage has been calculated for distances to the fault less than the maximum feeder length indicated in the last column of Table 4.4. As the maximum feeder length at 132 kV is only 2 km, the sag magnitude due to 132 kV faults does not exceed 20%. We see that distribution system faults give phase-angle jumps up to 20^0, with the largest ones due to 33 kV faults. Transmission system faults only cause very mild phase-angle jumps. These magnitudes and phase-angle jumps hold for single-phase as well as three-phase equipment, connected to any voltage level and irrespective of the load being connected in star or in delta.

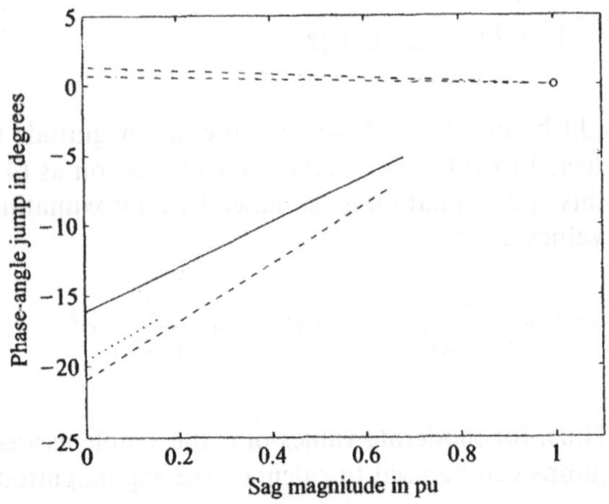

Figure 4.87 Magnitude and phase-angle jump for three-phase sags in the example supply in Fig. 4.21—solid line: 11 kV; dashed line: 33 kV; dotted line: 132 kV; dash-dot line: 400 kV.

4.6 MAGNITUDE AND PHASE-ANGLE JUMPS FOR THREE-PHASE UNBALANCED SAGS

4.6.1 Definition of Magnitude and Phase-Angle Jump

4.6.1.1 Three Different Magnitudes and Phase-Angle Jumps. The magnitude of a voltage sag was defined in Section 4.2 as the rms value of the voltage during the fault. As long as the voltage in only one phase is considered this is an implementable definition, despite the problems with actually obtaining the rms value. For three-phase unbalanced sags the problem becomes more complicated as there are now three rms values to choose from. The most commonly used definition is: *The magnitude of a three-phase unbalanced sag is the rms value of the lowest of the three voltages.* Alternatives suggested earlier are to use the average of the three rms values, or the lowest value but one [205]. Here we will propose a magnitude definition based on the analysis of three-phase unbalanced sags.

First we need to distinguish between three different kinds of magnitude and phase-angle jump. In all cases magnitude and phase-angle jump are absolute value and argument, respectively, of a complex voltage.

- The *initial complex voltage* is the voltage at the point-of-common coupling at the faulted voltage level. For a single-phase-to-ground fault the initial complex voltage is the voltage between the faulted phase and ground at the pcc. For a phase-to-phase fault the initial complex voltage is the voltage between the two faulted phases. For a two-phase-to-ground or a three-phase fault it can be either the voltage in one of the faulted phases or between two faulted phases (as long as pu values are used). The initial sag magnitude is the absolute value of the complex initial voltage; the initial phase-angle jump is the argument of the complex initial voltage.
- The *characteristic complex voltage* of a three-phase unbalanced sag is defined as the value of V in Tables 4.9 and 4.12. We will give an easy interpretation of the characteristic complex voltage later on. The characteristic sag magnitude is the absolute value of the characteristic complex voltage. The characteristic phase-angle jump is the argument of the characteristic complex voltage. These can be viewed as generalized definitions of magnitude and phase-angle jumps for three-phase unbalanced sags.
- The *complex voltages at the equipment terminals* are the values of V_a, V_b, and V_c in Tables 4.9 and 4.12 and in several of the equations around these tables. The sag magnitude and phase-angle jump at the equipment terminals are absolute value and argument, respectively, of the complex voltages at the equipment terminals. For single-phase equipment these are simply sag magnitude and phase-angle jump as previously defined for single-phase voltage sags.

4.6.1.2 Obtaining the Characteristic Magnitude. In Section 4.4 we have introduced seven types of sags together with their characteristic complex voltage V. For type D and type F the magnitude is the rms value of the lowest of the three voltages. For type C and type G it is the rms value of the difference between the two lowest voltages (in pu). From this we obtain the following method of determining the characteristic magnitude of a three-phase sag from the voltages measured at the equipment terminals:

- Determine the rms values of the three voltages.
- Determine the rms values of the three voltage differences.
- The magnitude of the three-phase sag is the lowest of these six values.

It is easy to see from the expressions given earlier, that this will give the value of $|V|$ as used for the definition of the three-phase unbalanced sags. An exception are sags of type B and type E. For sags conforming to (4.54) and (4.67) the method would still give the exact value for the magnitude. But the difference between zero-sequence and positive-sequence source impedance makes that the actual sags can deviate significantly. In that case the method is likely to give a completely wrong picture. Another problem is that for these sags the magnitude changes when they propagate to a lower voltage level. This makes measurements at a medium voltage level not suitable for predicting the sag magnitude at the equipment terminals. This problem can be solved by removing the zero-sequence component from the voltage and applying the method to the remaining voltages. The complete procedure proceeds as follows:

- obtain the three voltages as a function of time: $V_a(t)$, $V_b(t)$, and $V_c(t)$.
- determine the zero-sequence voltage:

$$V_0(t) = \frac{V_a(t) + V_b(t) + V_c(t)}{3} \tag{4.88}$$

- determine the remaining voltages after subtracting the zero-sequence voltage:

$$\begin{aligned} V'_a(t) &= V_a(t) - V_0(t) \\ V'_b(t) &= V_b(t) - V_0(t) \\ V'_c(t) &= V_c(t) - V_0(t) \end{aligned} \tag{4.89}$$

- determine the rms values of the voltages V'_a, V'_b, and V'_c.
- determine the three voltage differences:

$$\begin{aligned} V_{ab}(t) &= \frac{V_a(t) - V_b(t)}{\sqrt{3}} \\ V_{bc}(t) &= \frac{V_b(t) - V_c(t)}{\sqrt{3}} \\ V_{ca}(t) &= \frac{V_c(t) - V_a(t)}{\sqrt{3}} \end{aligned} \tag{4.90}$$

- determine the rms values of the voltages V_{ab}, V_{bc}, and V_{ca}.
- the magnitude of the three-phase sag is the lowest of the six rms values.

In case also phase-angle jump and sag type are needed, it is better to use a more mathematically correct method. A method based on symmetrical components has recently been proposed by Zhang [203], [204].

EXAMPLE This procedure has been applied to the voltage sag shown in Fig. 4.1. At first the rms values have been determined for the three measured phase-to-ground voltages, resulting in Fig. 4.88. The rms value has been determined each half-cycle over the preceeding 128 samples (one half-cycle). We see the behavior typical for a single-phase fault on an overhead feeder: a drop in voltage in one phase and a rise in voltage in the two remaining phases.

After subtraction of the zero-sequence component, all three voltages show a drop in magnitude (see Fig. 4.89). The phase-to-ground voltages minus the zero-sequence are indicated through solid lines, the phase-to-phase voltages through dashed lines. The lowest rms value is reached for a phase-to-ground voltage, which indicates a sag of type D. This is not surprising as the original sag was of type B (albeit with a larger than normal zero-sequence component). After removal of the zero-sequence voltage a sag of type D remains. The characteristic magnitude of this three-phase unbalanced sag is 63%.

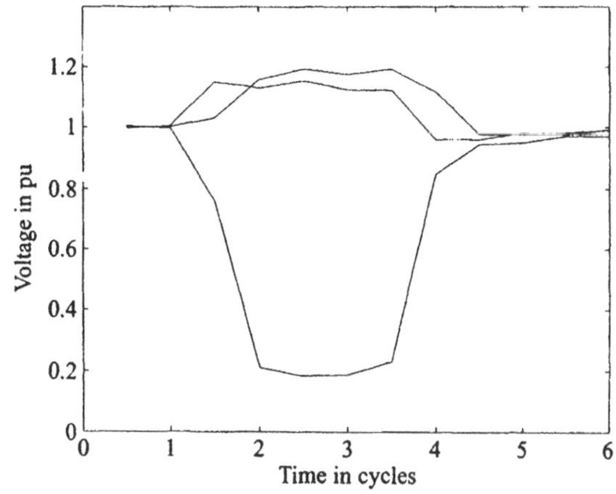

Figure 4.88 The rms values of the phase-to-ground voltages for the sag shown in Fig. 4.1.

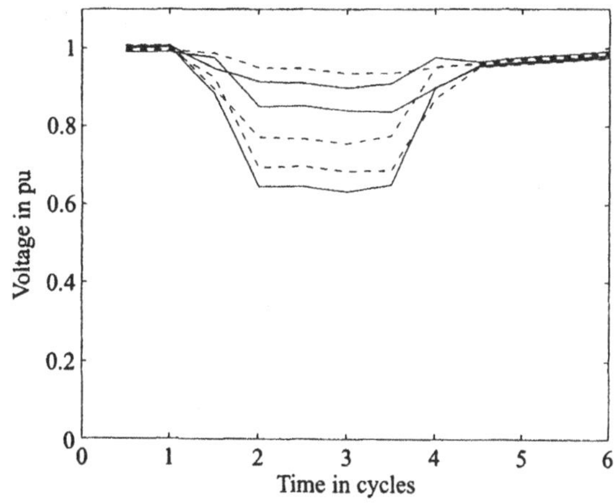

Figure 4.89 The rms values of phase-to-phase (dashed lines) and phase-to-ground voltages after removal of the zero-sequence component (solid lines) for the sag shown in Fig. 4.1.

4.6.2 Phase-to-Phase Faults

The impact of phase-to-phase faults depends on the transformer winding connections between the fault and the equipment. As shown in Section 4.4, the result is a sag either of type C or of type D. It was shown in Section 4.4.2 that the voltage between the faulted phases can be obtained by using the same voltage divider model as for the three-phase sag. The latter has been used to obtain expressions (4.83) and (4.86) for phase-angle jump and magnitude versus distance. These expressions can thus also be used to calculate initial magnitude and initial phase-angle jump: absolute value and argument of the voltage between the faulted phases at the pcc. The three-phase unbalanced sags in Section 4.4 were all derived under the assumption that the initial voltage drops in magnitude without change in phase angle. In case of a phase-angle jump in the initial voltage, the characteristic voltage of the three-phase unbalanced sag at the pcc also becomes complex. The expressions in Tables 4.9 and 4.12 still hold with the exception that the characteristic voltage V has become a complex number. The characteristic voltage for sag types C and D does not change when they are transformed down to lower voltage levels, so that the characteristic complex voltage remains equal to the initial complex voltage.

4.6.2.1 Sags of Type C. The phasor diagram for a sag of type C is shown in Fig. 4.90, where ϕ is the characteristic phase-angle jump and V the characteristic magnitude. Depending on the phase to which it is connected, single-phase equipment will experience a sag with magnitude V_b and phase-angle jump ϕ_b, a sag with magnitude V_c and phase-angle jump ϕ_c, or no sag at all. Due to the initial phase-angle jump ϕ the voltage magnitudes in the two faulted phases are no longer equal. Note that in Fig. 4.90 $\phi < 0$, $\phi_b < 0$, and $\phi_c > 0$.

From Fig. 4.90 expressions can be derived for magnitude and phase-angle jump at the equipment terminals. As a first step the sine rule and the cosine rule are applied to the two triangles indicated in Fig. 4.90 resulting in

$$V_b^2 = \frac{1}{4} + \frac{3}{4}V^2 - 2 \cdot \frac{1}{2} \cdot \frac{1}{2} V\sqrt{3}\cos(90^\circ - \phi) \tag{4.91}$$

$$\frac{\sin(60^\circ + \phi_b)}{\frac{1}{2}V\sqrt{3}} = \frac{\sin(90^\circ - \phi)}{V_b} \tag{4.92}$$

$$V_c^2 = \frac{1}{4} + \frac{3}{4}V^2 - 2 \cdot \frac{1}{2} \cdot \frac{1}{2} V\sqrt{3}\cos(90^\circ + \phi) \tag{4.93}$$

$$\frac{\sin(60^\circ - \phi_c)}{\frac{1}{2}V\sqrt{3}} = \frac{\sin(90^\circ + \phi)}{V_c} \tag{4.94}$$

from which the following desired expressions are obtained:

$$\begin{aligned} V_a &= 1 \\ V_b &= \sqrt{\frac{1}{4} + \frac{3}{4}V^2 - \frac{1}{2}V\sqrt{3}\sin(\phi)} \\ V_c &= \sqrt{\frac{1}{4} + \frac{3}{4}V^2 + \frac{1}{2}V\sqrt{3}\sin(\phi)} \end{aligned} \tag{4.95}$$

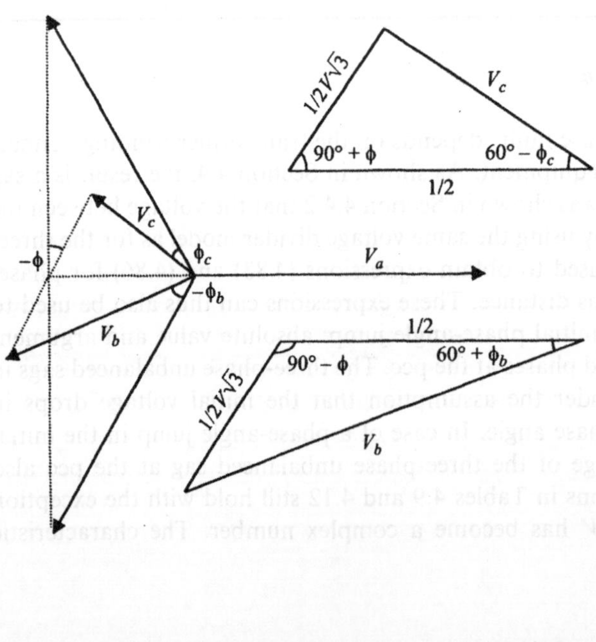

Figure 4.90 Phasor diagram for a sag of type C with characteristic magnitude V and characteristic phase-angle jump ϕ.

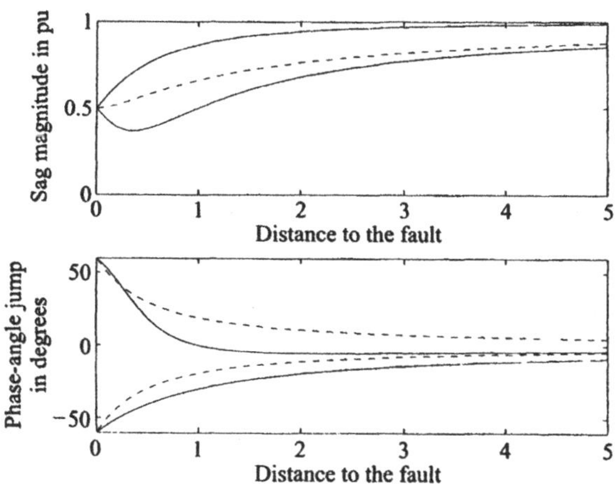

Figure 4.91 Magnitude (top) and phase-angle jump (bottom) for sags of type C due to phase-to-phase faults. Dashed line: zero impedance angle (no characteristic phase-angle jump). Solid line: −60° impedance angle (large characteristic phase-angle jump).

$$
\begin{aligned}
\phi_a &= 0 \\
\phi_b &= -60^\circ + \arcsin\left(\frac{1}{2}\sqrt{3}\frac{V}{V_b}\cos(\phi)\right) \\
\phi_c &= 60^\circ - \arcsin\left(\frac{1}{2}\sqrt{3}\frac{V}{V_c}\cos(\phi)\right)
\end{aligned}
\tag{4.96}
$$

Combining (4.95) and (4.96) with (4.83) and (4.86) gives the magnitude and phase-angle jump in the three phases as a function of the distance to the fault. This is done in Fig. 4.91 for impedance angles equal to 0 and −60°. The horizontal scale corresponds to $\lambda = \frac{z\mathcal{L}}{Z_S}$ as in (4.83). We see that the severity of sags decreases with increasing distance when there is no characteristic phase-angle jump. The introduction of a characteristic phase-angle jump creates asymmetry between the faulted phases. We see, e.g., that the voltage in one of the phases initially decreases with increasing distance to the fault. For one of the phases the phase-angle jump drops to zero rather quickly, whereas for the other phase the phase-angle jump remains high much longer.

Figure 4.92 plots magnitude versus phase-angle jump for four values of the impedance angle. We can see that the characteristic phase-angle jump significantly disturbs the symmetry between the two faulted phases. Also the voltage can drop well below 50%, which is not possible without characteristic phase-angle jump.

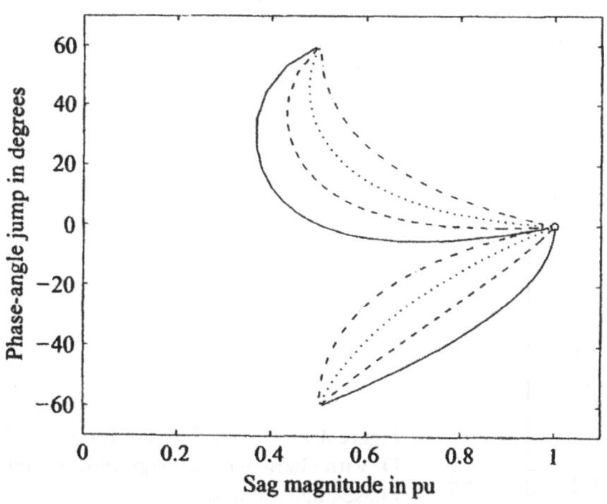

Figure 4.92 Magnitude versus phase-angle jump for sag type C due to phase-to-phase faults for impedance angle −60° (solid line), −40° (dashed), −20° (dotted), 0 (dash-dot).

4.6.2.2 Sags of Type D. The phasor diagram for a type D sag is shown in Fig. 4.93, where ϕ is again the characteristic phase-angle jump. One phase will go down significantly with a phase-angle jump equal to the characteristic value. Equipment connected to one of the two other phases will see a small drop in voltage and a phase-angle jump of up to 30°. Severe characteristic phase-angle jumps can even lead to voltage swells. The two phases with the small voltage drop can experience positive as well as negative phase-angle jumps. The phase with the large voltage drop always experiences a negative phase-angle jump.

From Fig. 4.93 magnitude and phase-angle jump in the three phases can be calculated for a sag of type D. Applying the sine rule and the cosine rule to the two triangles indicated in Fig. 4.93 gives the following expressions:

$$V_b^2 = \frac{1}{4}V^2 + \frac{3}{4} - 2 \cdot \frac{1}{2}V \cdot \frac{1}{2}\sqrt{3}\cos(90^\circ + \phi) \tag{4.97}$$

$$\frac{\sin(30^\circ - \phi_b)}{\frac{1}{2}V} = \frac{\sin(90^\circ + \phi)}{V_b} \tag{4.98}$$

$$V_c^2 = \frac{1}{4}V^2 + \frac{3}{4} - 2 \cdot \frac{1}{2}V \cdot \frac{1}{2}\sqrt{3}\cos(90^\circ - \phi) \tag{4.99}$$

$$\frac{\sin(30^\circ + \phi_c)}{\frac{1}{2}V} = \frac{\sin(90^\circ - \phi)}{V_c} \tag{4.100}$$

Rewriting these expressions results in

$$\begin{aligned} V_a &= V \\ V_b &= \sqrt{\frac{3}{4} + \frac{1}{4}V^2 + \frac{1}{2}V\sqrt{3}\sin(\phi)} \\ V_c &= \sqrt{\frac{3}{4} + \frac{1}{4}V^2 - \frac{1}{2}V\sqrt{3}\sin(\phi)} \end{aligned} \tag{4.101}$$

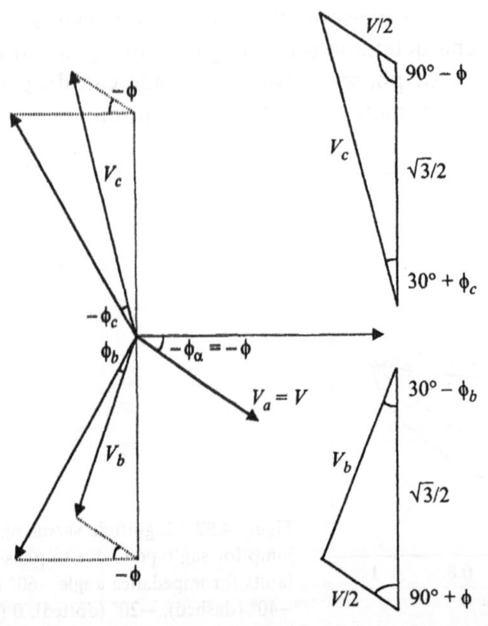

Figure 4.93 Phasor diagram for a sag of type D, with characteristic magnitude V and phase-angle jump ϕ.

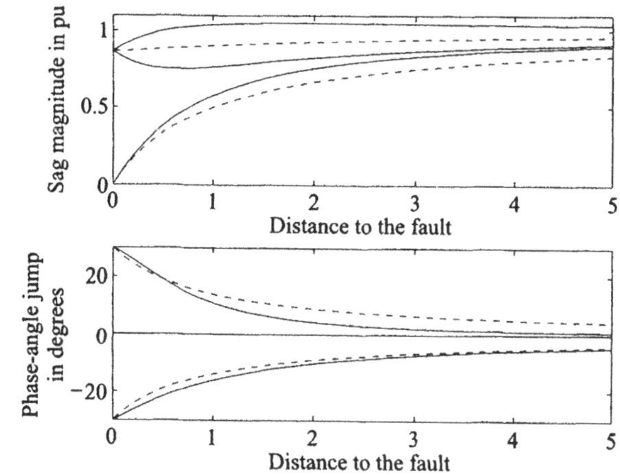

Figure 4.94 Magnitude (top) and phase-angle jump (bottom) for sags of type D due to phase-to-phase faults. Dashed line: zero impedance angle. Solid line: impedance angle of −60°.

$$\begin{aligned}\phi_a &= \phi \\ \phi_b &= 30^\circ - \arcsin\left(\frac{V}{2V_b}\cos(\phi)\right) \\ \phi_c &= -30^\circ + \arcsin\left(\frac{V}{2V_c}\cos(\phi)\right)\end{aligned} \tag{4.102}$$

Again we can plot magnitude and phase-angle jump versus distance and magnitude versus phase-angle jump. Figure 4.94 gives magnitude and phase-angle jump as a function of distance for impedance angles equal to zero and −60°. Here we see that the voltage drop in the non-faulted phases is rather small; the voltage drops to about 75%. The characteristic phase-angle jump causes an additional drop in voltage at the equipment terminals. Magnitude versus phase-angle jump is plotted in Fig. 4.95 for four values of the impedance angle.

4.6.2.3 Range of Magnitude and Phase-Angle Jump. As mentioned before, phase-to-phase faults lead to sags of type C or of type D. Combining the range of magnitude and phase-angle jump due to type C sags (Fig. 4.92) with the range due

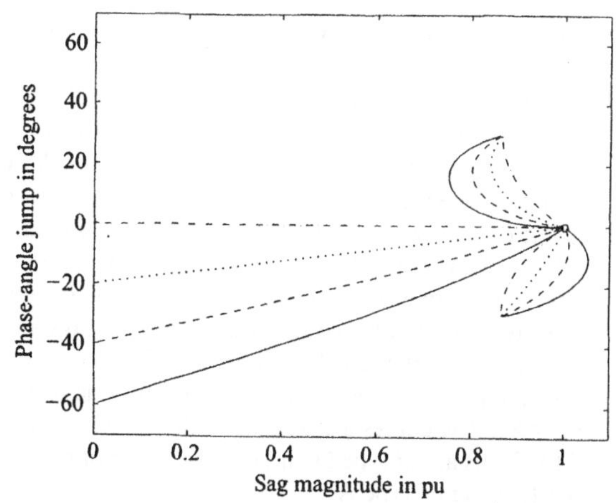

Figure 4.95 Magnitude versus phase-angle jump for sag type D due to phase-to-phase faults: impedance angle −60° (solid line), −40° (dashed), −20° (dotted), 0 (dash-dot).

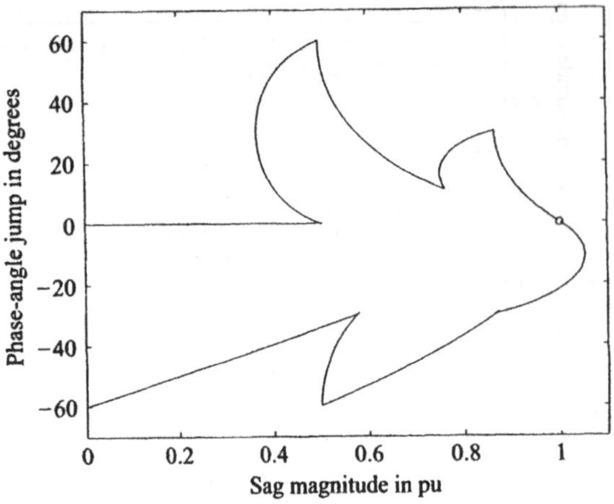

Figure 4.96 Range of sags due to phase-to-phase faults, as experienced by single-phase equipment.

to type D sags (Fig. 4.95) gives the whole range of sags experienced by single-phase equipment during phase-to-phase faults. The merger of the two mentioned figures is shown in Fig. 4.96, where only the outer contour of the area is indicated.

Sags due to three-phase faults are automatically included in Fig. 4.96. A three-phase fault gives a sag with the initial magnitude and the initial phase-angle jump, in all the three phases. Such a sag also appears in one of the phases for a type D sag due to a phase-to-phase fault. This is the large triangular area in Fig. 4.96. Sags due to single-phase and two-phase-to-ground faults have not yet been included. These will be treated below.

EXAMPLE: PHASE-TO-PHASE FAULTS, THREE-PHASE LOAD The magnitude and phase-angle jump due to phase-to-phase faults have been calculated for faults in the example supply in Fig. 4.21. The calculations have been performed for two different types of load:

- three-phase load connected in delta at 660 V.
- single-phase load connected in star (phase-to-neutral) at 420 V.

For a three-phase load, we can use the classification introduced in Section 4.4 to characterize the sag. The magnitude and phase-angle jump of these three-phase unbalanced sags are the same as those of sags due to three-phase faults. The only difference is the type of sag. A phase-to-phase fault at 11 kV will, for delta-connected load at 11 kV, lead to a sag of type D. The Dy transformer between the fault (at 11 kV) and the load (at 660 V) will change this into a type C sag. Thus, the delta-connected load at 660 V will, due to a phase-to-phase fault at 11 kV, experience a sag of type C. The characteristic magnitude and phase-angle jump of this three-phase unbalanced sag will be equal to the magnitude and phase-angle jump of the voltage (in any phase) due to a three-phase fault at the same position as the phase-to-phase fault. Using the same reasoning we find that phase-to-phase faults at 33 kV lead to type D sags and faults at 132 kV and 400 kV to sags of type C. The results of the calculations are shown in Fig. 4.97: characteristic magnitude and phase-angle jump of three-phase unbalanced sags due to phase-to-phase faults. Note the similarity with Fig. 4.87. The curves are at exactly the same position; the only difference is that the ones due to 33 kV faults are of type D and the others are of type C. Three-phase faults at any voltage level will lead to a sag of type A.

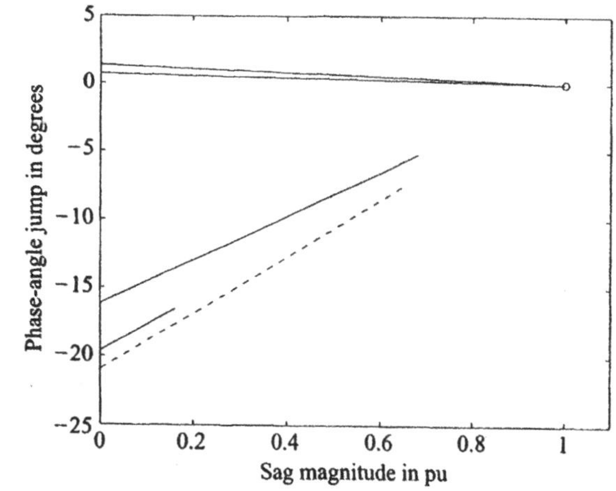

Figure 4.97 Characteristic magnitude and phase-angle jump for sags due to phase-to-phase faults in the example supply in Fig. 4.21—solid line: type C sags, dashed line: type D sags.

EXAMPLE: PHASE-TO-PHASE FAULTS, SINGLE-PHASE LOAD Magnitude and phase-angle jump at the equipment terminals due to phase-to-phase faults have been calculated for a single-phase load connected phase-to-neutral at 420 V. The classification of three-phase sags no longer fully describes the voltage at the equipment terminals. The additional information needed is the phases between which the fault takes place. One can calculate the voltage sag in one phase for three different faults; but it is easier to calculate the voltages in the three phases for one fault. These three voltages are the voltages in one phase for the three different faults. We saw before that we do not need to calculate the whole transfer of the sag from the faulted voltage level to the load terminals. All we need to do is determine whether the equipment terminal voltage corresponds to phase-to-phase or phase-to-neutral voltage at the faulted voltage level. In this example, the equipment terminal voltage corresponds to phase-to-phase voltages at 11 kV, 132 kV, and 400 kV and to phase-to-neutral voltages at 33 kV.

The resulting magnitude and phase-angle jump are plotted in Fig. 4.98. Faults at 11 kV, 132 kV, and 400 kV cause a three-phase unbalanced sag of type D for star-connected equipment. For a type D sag one voltage drops to a low value, and the two remaining voltages show a small drop with a phase-angle jump up to 30°. Note the symmetry in the sags originating at 400 kV, which is not present in the sags originating at 11 kV and 132 kV. This is due to the large initial

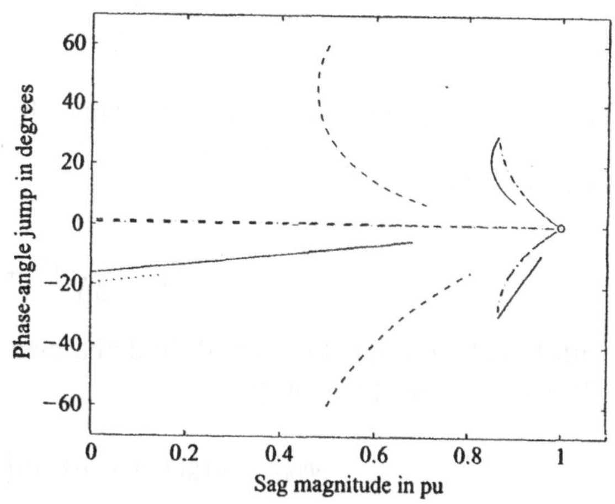

Figure 4.98 Magnitude and phase-angle jump at the equipment terminals due to phase-to-phase faults in the supply in Fig. 4.21, experienced by single-phase load connected phase-to-ground at 420 V—solid line: 11 kV, dashed line: 33 kV, dotted line: 132 kV, dash-dot line: 400 kV.

phase-angle jump for the latter two. Faults at 33 kV cause a sag of type C, with two voltages down to about 50% and phase-angle jumps up to ±60°.

4.6.3 Single-Phase Faults

For single-phase faults the situation becomes slightly more complicated. Expressions (4.83) and (4.86) can still be used to calculate magnitude and phase-angle jump of the voltage in the faulted phase at the pcc (i.e., the initial magnitude and phase-angle jump). Star-connected equipment at the same voltage level as the fault would experience a sag of type B. But as we have seen before, this is a rather rare situation. In almost all cases a sag due to a single-phase fault is of type C or type D. The characteristic magnitude of these three-phase unbalanced sags is no longer equal to the initial magnitude. The same holds for the phase-angle jump.

4.6.3.1 Initial and Characteristic Magnitude. To obtain an expression for the characteristic magnitude and phase-angle jump, we need to go back to the type B sag. The voltages for a type B sag are

$$\begin{aligned} V_a &= V\cos\phi + jV\sin\phi \\ V_b &= -\frac{1}{2} - \frac{1}{2}j\sqrt{3} \\ V_c &= -\frac{1}{2} + \frac{1}{2}j\sqrt{3} \end{aligned} \tag{4.103}$$

with V the initial magnitude and ϕ the initial phase-angle jump. When this three-phase unbalanced sag propagates to lower voltage levels, the zero-sequence voltage is lost. The zero-sequence component for (4.103) is

$$V_0 = \frac{1}{3}(V_a + V_b + V_c) = \frac{1}{3}(V\cos\phi + jV\sin\phi - 1) \tag{4.104}$$

Subtracting the zero-sequence voltage from (4.103) gives a three-phase unbalanced sag of type D. Characteristic magnitude and phase-angle jump for a sag of type D are equal to the absolute value and the argument of the complex voltage in the worst-effected phase, V_a in this case.

$$V_a = \frac{1}{3} + \frac{2}{3}(V\cos\phi + jV\sin\phi) \tag{4.105}$$

Note that this expression can also be obtained by substituting $V = V\cos\phi + jV\sin\phi$ in (4.62). For three-phase unbalanced sags due to single-phase faults the characteristic magnitude becomes

$$V_{char} = |V_a| = \frac{2}{3}\sqrt{V^2 + V\cos\phi + \frac{1}{4}} \tag{4.106}$$

with V and ϕ the initial magnitude and phase-angle jump, and V_a according to (4.105). The characteristic phase-angle jump is

$$\phi_{char} = \arg(V_a) = \arctan\left(\frac{2V\sin\phi}{1 + 2V\cos\phi}\right) \tag{4.107}$$

For small values of ϕ these expressions can be approximated by using

$$\sin\phi \approx \phi$$
$$\cos\phi \approx 1$$
$$\arctan(x\phi) \approx x\phi, \quad x < 1$$

resulting in

$$V'_{char} = \frac{1}{3} + \frac{2}{3}V \tag{4.108}$$

$$\phi'_{char} = \frac{2V\phi}{1+2V} \tag{4.109}$$

Figures 4.99 and 4.100 show the error made by using the approximated expressions (4.108) and (4.109). The error has been defined as $1 - \frac{V_{char}}{V'_{char}}$. The calculations have been performed for impedance angles equal to −60°, −40°, and −20°. Even for a system with large phase-angle jumps, an impedance angle of −60°, the errors are not very big. Only for calculating the characteristic phase-angle jump with deep sags might it be needed to use the exact expression. One should realize, however, that the

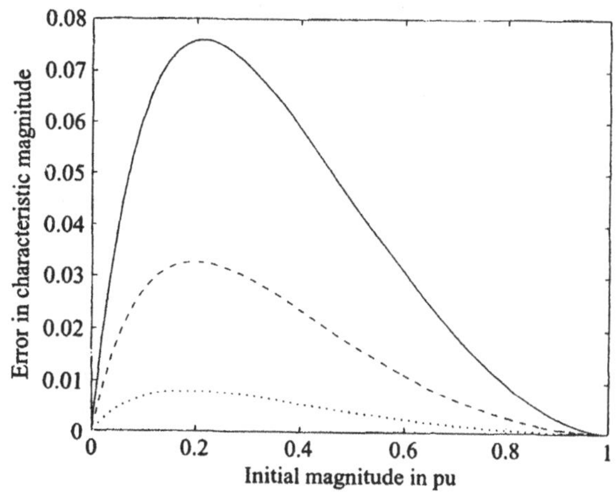

Figure 4.99 Transformation of sags due to single-phase faults—error in approximate expressions for characteristic magnitude. Impedance angle: −60° (solid line); −40° (dashed); −20° (dotted).

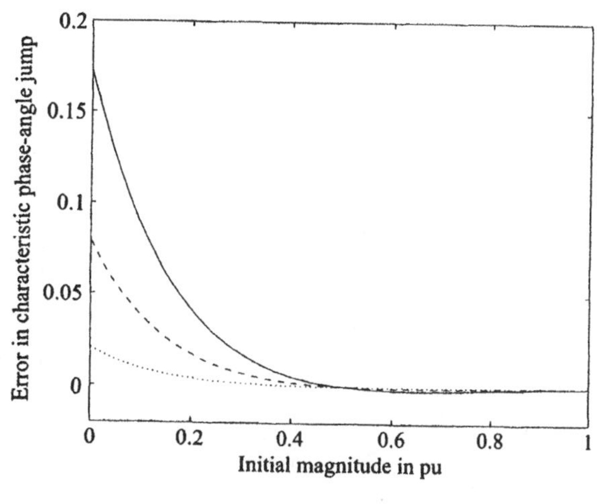

Figure 4.100 Transformation of sags due to single-phase faults—error in approximate expressions for characteristic phase-angle jump. Impedance angle: −60° (solid line); −40° (dashed); −20° (dotted).

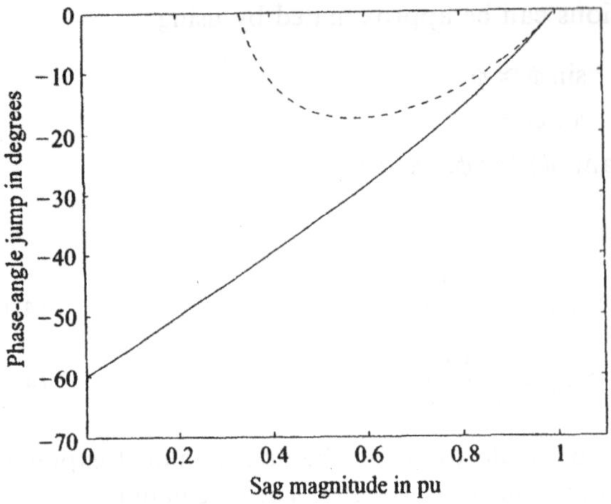

Figure 4.101 Relation between phase-angle jump and magnitude of sags due to single-phase faults: characteristic values (dashed curve) and initial values (solid curve).

characteristic phase-angle jump is close to zero for single-phase faults with a small initial magnitude, as can be seen from (4.107). The absolute error is even for an impedance angle of −60° less than 1°.

Figure 4.101 compares initial magnitude and phase-angle jump with the characteristic values. An impedance angle of −60° has been used. The bottom (solid) curve also gives the relation between characteristic magnitude and phase-angle jump due to phase-to-phase and three-phase faults. Sags due to single-phase faults are clearly less severe: in magnitude as well as in phase-angle jump.

4.6.3.2 Sags of Type C and Type D. Knowing characteristic magnitude and phase-angle jump for the type C or type D sag it is again possible to calculate magnitude and phase-angle jump at the equipment terminals. This results in similar curves as for sags due to phase-to-phase faults. The main difference is that voltage sags due to single-phase faults are less severe than due to phase-to-phase faults. Figure 4.102 plots magnitude versus phase-angle jump for sag type C, for four values of the impedance angle. The lowest sag magnitude at the equipment terminals is about 58%, the largest phase-angle jump is 30°.

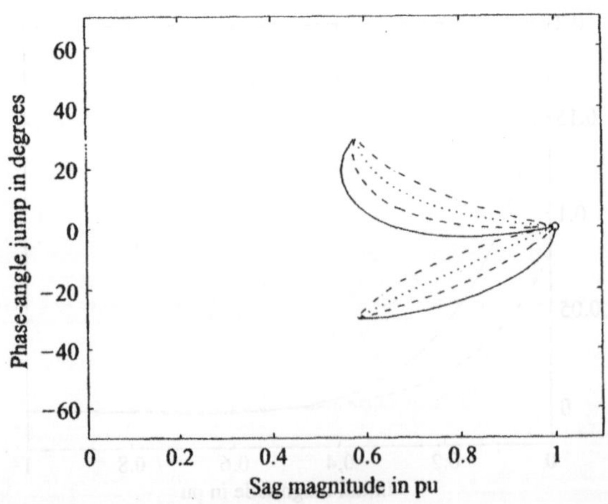

Figure 4.102 Range of sags experienced by single-phase equipment for sag type C and single-phase fault, impedance angle: −60° (solid line), −40° (dashed), −20° (dotted), 0 (dash-dot).

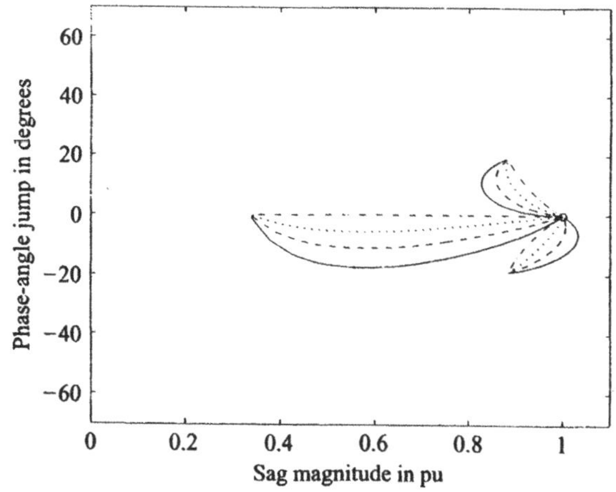

Figure 4.103 Range of sags experienced by single-phase equipment for sag type D and single-phase fault, impedance angle: −60° (solid line), −40° (dashed), −20° (dotted), 0 (dash-dot).

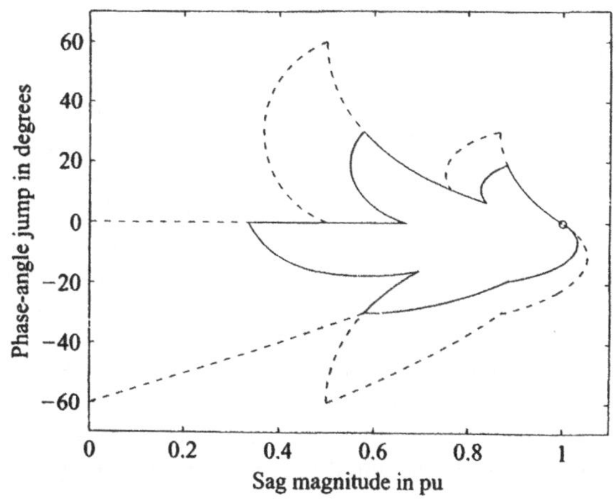

Figure 4.104 Range of sags due to single-phase faults (solid curve) and due to phase-to-phase faults (dashed curve).

Figure 4.103 repeats this for type D sags due to single-phase faults. The lowest sag magnitude is 33% with a maximum phase-angle jump of 19°. Sags due to type C and type D are merged into one plot in Fig. 4.104 which gives the whole range of sags experienced by single-phase equipment due to single-phase faults. This range is smaller than the range due to phase-to-phase faults, indicated by a dashed line in Fig. 4.104.

EXAMPLE: SINGLE-PHASE FAULTS, THREE-PHASE LOAD The calculations for phase-to-phase faults shown in the previous section have been repeated for single-phase faults. For single-phase faults at the various voltage levels in Fig. 4.21, the sag magnitude, phase-angle jump, and type have been calculated for delta-connected (three-phase) load at 660 V. Equations (4.108) and (4.109) have been derived for a system with equal positive, negative and zero-sequence impedance. This is a good approximation for the (solidly grounded) 132 kV system but not for the (resistance-grounded) 11 kV and 33 kV systems. At 400 kV the source impedance is mainly determined by overhead lines, so that the zero-sequence source impedance is larger than the positive-sequence value. To calculate the characteristic magnitude of three-phase unbalanced sags due to single-phase faults, we can first calculate the phase-to-neutral voltage in the faulted phase according to (4.40). Characteristic values are obtained from this by applying (4.108) and (4.109). Alternatively we can calculate the complex phase-to-

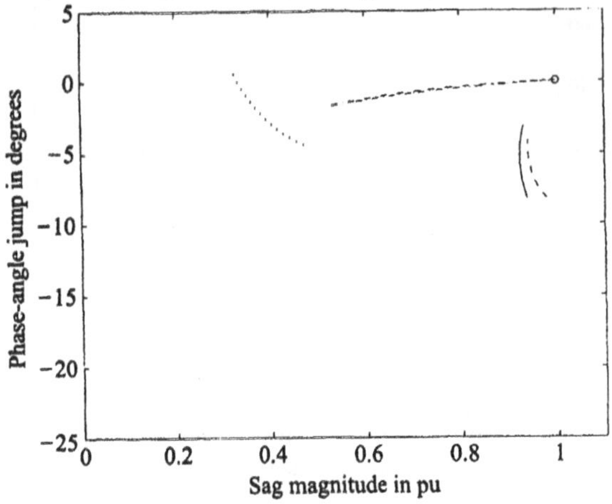

Figure 4.105 Characteristic magnitude and phase-angle jump for sags due to single-phase faults in the example supply in Fig. 4.21, experienced by three-phase load-connected phase-to-phase at 660 V—solid line: 11 kV, dashed line: 33 kV, dotted line: 132 kV, dash-dot line: 400 kV.

ground voltages at the pcc, and apply a type 2 transformer to these. A type 2 transformer removes the zero-sequence voltage and results in a three-phase unbalanced sag of type D. Magnitude and phase-angle jump of the worst-affected phase are equal to the characteristic values. In other words, the characteristic complex voltage can be obtained by subtracting the zero-sequence voltage from the voltage in the faulted phase at the pcc.

The results are shown in Fig. 4.105. We see that single-phase faults at 11 kV and 33 kV cause only a small drop in voltage, but a moderate phase-angle jump. This is due to the resistance grounding applied at these voltage levels. Sags originating in the 132 kV and 400 kV networks show a much larger drop in voltage magnitude but a smaller phase-angle jump. Note that the curves for sags due to 400 kV faults do not start at 33% voltage as expected for solidly-grounded systems. The reason is that the source impedance in PAD-400 mainly consists of overhead lines. Therefore the zero-sequence impedance is larger than the positive-sequence impedance. For faults in the direction of PEN, the source impedances are $Z_{S1} = 0.084 + j1.061$, $Z_{S0} = 0.319 + j2.273$, which gives for the initial phase-to-neutral voltage during a terminal fault:

$$V_{an} = 1 - \frac{3Z_{S1}}{2Z_{S1} + Z_{S0}} = 0.2185 + j0.0243 \tag{4.110}$$

The characteristic magnitude at a lower voltage level is found from

$$V_{char} = \left| \frac{1}{3} + \frac{2}{3} V_{an} \right| = 0.519 \tag{4.111}$$

For single-phase faults in the direction of EGG we find: $V_{an} = 0.3535 - j0.0026$ and $V_{char} = 0.571$. This is a moderate version of the effect which leads to very shallow sags in resistance-grounded systems. Note that we still assume the system to be radial, which gives an erroneous result for single-phase faults at 400 kV. This explains the difference in resulting voltage sags for a terminal fault in the two directions. The actual value is somewhere between 0.519 and 0.571. The difference is small enough to be neglected here.

Figure 4.105 does not plot the sag type: faults at 33 kV lead to a type C sag; faults at 11 kV, 132 kV, and 400 kV cause a sag of type D at the equipment terminals for delta-connected load. At the equipment terminals it is not possible to distinguish between a sag due to a single-phase fault and a sag due to a phase-to-phase fault: they both cause sags of type C or type D. Therefore, we have merged Figs. 4.97 and 4.105 into one figure. The result is displayed in Fig. 4.106, showing characteristic magnitude and phase-angle jump of all three-phase unbalanced sags due to single-phase and phase-to-phase faults, as experienced by a delta-connected three-phase load at 660 V. We see that the equipment experiences the whole range of magnitudes and phase-angle jumps. These have to be considered when specifying the voltage-tolerance requirements of equipment. To

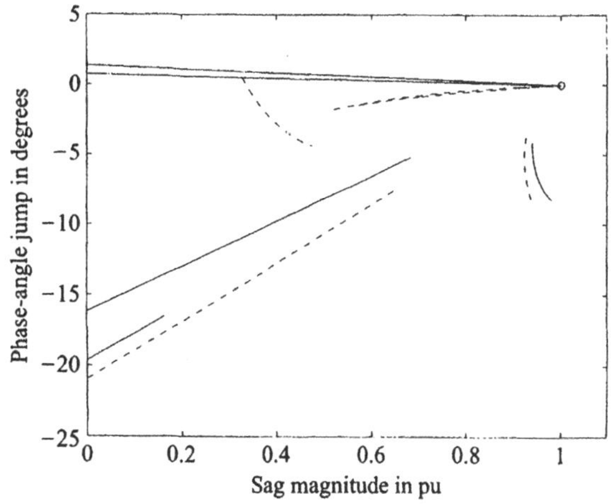

Figure 4.106 Characteristic magnitude and phase-angle jump for three-phase unbalanced sags in Fig. 4.21, experienced by three-phase delta-connected load—solid line: type C, dashed line: type D.

be able to fully interpret these results, two more dimensions are needed. At first, one has to realize that not all sags are of equal duration. Typically sags due to 11 kV and 33 kV faults are of longer duration than those due to 132 kV and 400 kV faults. What is also different for different sags is its likelihood. Roughly speaking one can say that deeper sags are less likely than shallower sags. We will come back to probabilities in detail in Chapter 6. To include magnitude, phase-angle jump, duration, and probability in one, two-dimensional, figure is very difficult if not impossible.

EXAMPLE: SINGLE-PHASE FAULTS, SINGLE-PHASE LOAD The magnitude and phase-angle jump have been calculated for voltage sags due to single-phase faults, experienced by single-phase star-connected load. For this we have calculated either the phase-to-phase voltage, or the phase-to-ground voltage minus the zero-sequence voltage, at the faulted voltage level. For a single-phase fault at 11 kV, star-connected load at 420 V experiences a sag of type C. The complex voltages at the equipment terminals are equal to the phase-to-phase voltages at the pcc. The same calculation method can be used for single-phase faults at 132 kV and at 400 kV. Single-phase faults at 33 kV lead to sags of type D. The complex voltages at the equipment terminals can be calculated as the phase-to-ground voltages at the pcc minus the zero-sequence component. The results of these calculations are shown in Fig. 4.107. We see that the voltage never drops below 50%, and that the phase-angle jumps are between −30° and +30°. Faults at 11 kV and 33 kV again only cause shallow sags due to the system being resistance-grounded. Due to a 33 kV fault, the load can even experience a small voltage swell. Faults at 400 kV are also somewhat damped because the zero-sequence source impedance is about twice the positive-sequence value. Therefore, sags due to single-phase faults are milder than expected for a solidly-grounded system. In the 132 kV system, the zero-sequence source impedance is even a bit smaller than the positive sequence value, thus leading to deep sags. But at 420 V they appear as a type C in which the drop in phase voltages is not below 50%. For this specific system, single-phase faults do not cause very deep sags for star-connected load. Note that this is not a general conclusion. Had the 11 kV/420 V transformer been of type Dd, the equipment would have experienced voltage drops down to 30% (see Fig. 4.105).

To get a complete picture of all sags experienced by the single-phase load, we have merged Fig. 4.87 (three-phase faults), Fig. 4.98 (phase-to-phase faults), and Fig. 4.107 (single-phase faults), resulting in Fig. 4.108. Here we see the whole range of values both in magnitude and in phase-angle jump.

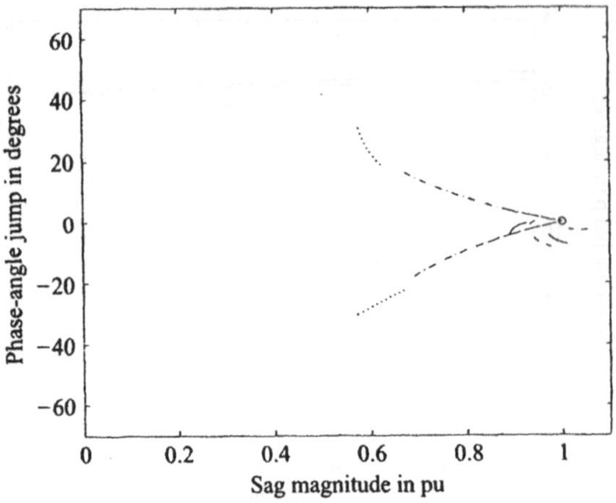

Figure 4.107 Magnitude and phase-angle jump for sags due to single-phase faults in the example supply in Fig. 4.21, experienced by single-phase load-connected phase-to-ground at 420 V—solid line: 11 kV, dashed line: 33 kV, dotted line: 132 kV, dash-dot line: 400 kV.

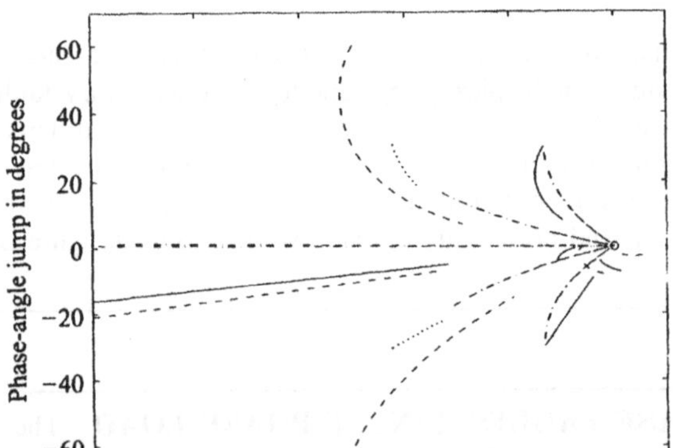

Figure 4.108 Magnitude and phase-angle jump for all sags in the example supply in Fig. 4.21, experienced by single-phase load-connected phase-to-ground at 420 V—solid line: 11 kV, dashed line: 33 kV, dotted line: 132 kV, dash-dot line: 400 kV.

4.6.4 Two-Phase-to-Ground Faults

The analysis of two-phase-to-ground faults does not differ from the treatment of phase-to-phase faults. We saw in Section 4.4.4 that two-phase-to-ground faults lead to three-phase unbalanced sags of type E, type F, or type G. Type E is a rare type which we will not discuss here. Like type B for the single-phase-to-ground fault, the type E contains a zero-sequence component which is normally not transferred to the utility voltage, and never seen by delta-connected equipment.

For type F and type G we can again plot characteristic magnitude against phase-angle jump. The relation between the characteristic magnitude and phase-angle jump of the unbalanced three-phase sag is identical to the relation between the initial magnitude and phase-angle jump, i.e. magnitude and phase-angle jump of the voltage in the faulted phases at the pcc. This relation is described by (4.83) and (4.86) and is shown in Fig. 4.86.

4.6.4.1 Sags of Type F. A detailed phasor diagram of a sag of type F is shown in Fig. 4.109. Like with a type D sag, one phase drops significantly in magnitude, and the other two phases less. The difference with the type D sag is in the latter two

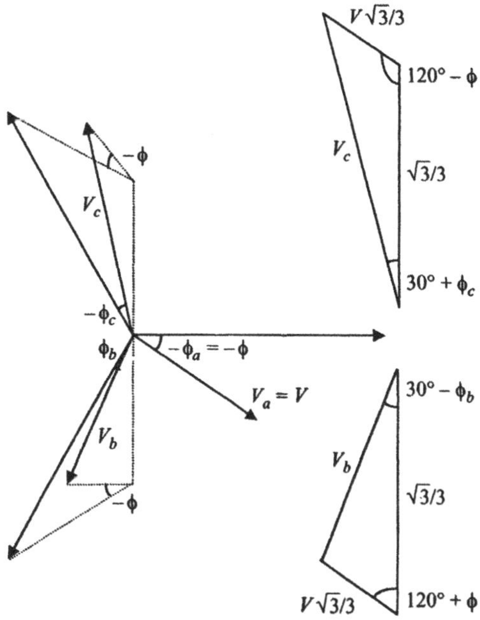

Figure 4.109 Phasor diagram for three-phase unbalanced sag of type F with characteristic magnitude V and characteristic phase-angle jump ϕ.

phases. With a type D sag they drop from $-\frac{1}{2} \pm \frac{1}{2}j\sqrt{3}$ to $\pm\frac{1}{2}j\sqrt{3}$, but with a type F sag they drop significantly more: to $\pm\frac{1}{3}j\sqrt{3}$. The lowest magnitude for a type D sag is 86.6%, whereas it is 57.7% for a type F sag.

In the upper triangle indicated in Fig. 4.109 we can again apply the cosine and sine rule to obtain magnitude and phase-angle jump at the equipment terminals. Note that in Fig. 4.109, $\phi < 0$, $\phi_b > 0$, and $\phi_c < 0$. The cosine rule gives

$$V_c^2 = \left(\frac{1}{3}\sqrt{3}\right)^2 + \left(\frac{1}{3}V\sqrt{3}\right)^2 - 2 \times \frac{1}{3}\sqrt{3} \times \frac{1}{3}V\sqrt{3} \times \cos(120° - \phi) \tag{4.112}$$

which results in an expression for the voltage magnitude V_c:

$$V_c = \sqrt{\frac{1}{3} + \frac{1}{3}V^2 - \frac{2}{3}V\cos(120° - \phi)} \tag{4.113}$$

The sine rule in the same triangle gives

$$\frac{\sin(30° + \phi_c)}{\frac{1}{3}V\sqrt{3}} = \frac{\sin(120° - \phi)}{V_c} \tag{4.114}$$

The phase-angle jump ϕ_c follows as

$$\phi_c = -30° + \arcsin\left\{\frac{V}{V_c\sqrt{3}}\sin(120° - \phi)\right\} \tag{4.115}$$

The same rules can be applied to the lower triangle, which leads to the following expressions for magnitude V_b and phase-angle jump ϕ_b:

$$V_b = \sqrt{\frac{1}{3} + \frac{1}{3}V^2 - \frac{2}{3}V\cos(120° + \phi)} \tag{4.116}$$

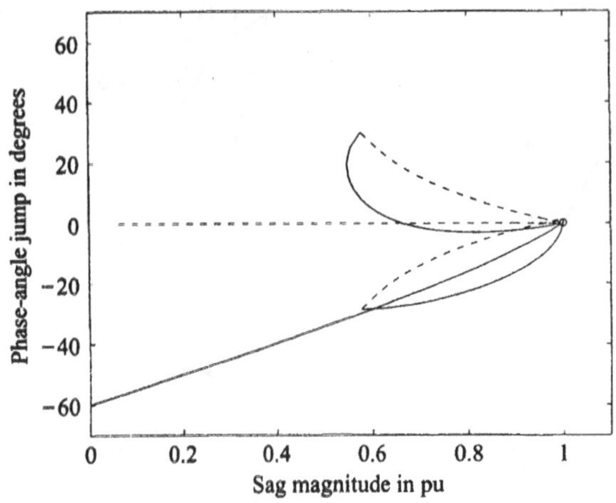

Figure 4.110 Magnitude and phase-angle jump at the equipment terminals for a type F sag, due to a two-phase-to-ground fault. The curves are given for an impedance angle of 0 (dashed line) and −60° (solid line).

$$\phi_b = 30^\circ - \arcsin\left\{\frac{V}{V_b\sqrt{3}}\sin(120^\circ + \phi)\right\} \tag{4.117}$$

From these equations we can again calculate magnitude and phase-angle jump at the equipment terminals, e.g., as a function of the distance to the fault. Figure 4.110 plots magnitude versus phase-angle jump for a type F sag due to a two-phase-to-ground fault. We see that one phase behaves again like the sag due to a three-phase fault. The other two phase are somewhat like the two phases with a shallow sag in the type D sag shown in Fig. 4.95. The difference is that for a type F sag the voltages show a significantly larger drop. The maximum phase-angle jump for these two phases is again 30°.

4.6.4.2 Sags of Type G. A detailed phasor diagram for a type G sag is shown in Fig. 4.111. The complex voltage in phase a drops to a value of $\frac{2}{3}$ (no drop for a sag of type C); the complex voltages in phase b and c drop to a value of $-\frac{1}{3}$ ($-\frac{1}{2}$ for type C).

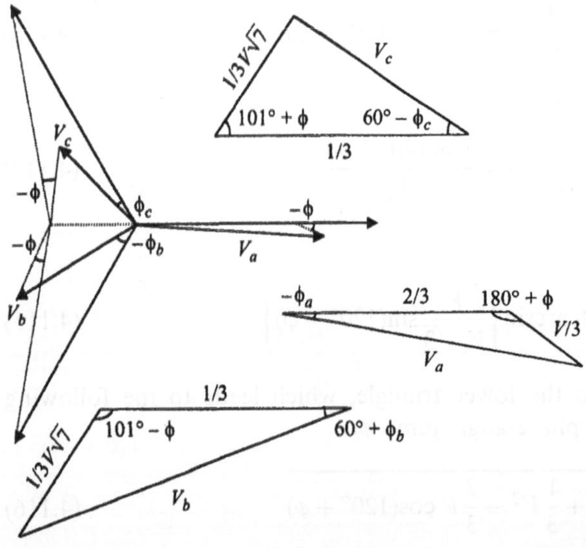

Figure 4.111 Detailed phasor diagram for three-phase unbalanced sag of type G with characteristic magnitude V and characteristic phase-angle jump ϕ.

The cosine rule and the sine rule applied to the triangle on the right give the following expressions:

$$V_a^2 = \frac{4}{9} + \frac{1}{9}V^2 - 2 \times \frac{2}{3} \times \frac{V}{3}\cos(180° + \phi) \tag{4.118}$$

$$\frac{\sin(180° + \phi)}{V_a} = \frac{\sin(-\phi_a)}{\frac{1}{3}V} \tag{4.119}$$

This leads again to expressions for magnitude and phase-angle jump at the equipment terminals.

$$V_a = \sqrt{\frac{4}{9} + \frac{1}{9}V^2 + \frac{4}{9}V\cos\phi} \tag{4.120}$$

$$\phi_a = \arcsin\left(\frac{V}{3V_a}\sin\phi\right) \tag{4.121}$$

Repeating the calculations for the other triangles gives expressions for magnitude and phase-angle jump in the other two phases. Note the angle 101° and the factor $\frac{1}{3}\sqrt{7}$. These originate from the triangle formed by the complex numbers 0, $-\frac{1}{3}$, and $-\frac{1}{2} \pm \frac{1}{2}j\sqrt{3}$.

$$V_b = \frac{1}{3}\sqrt{1 + 7V^2 - 2V\sqrt{7}\cos(101° - \phi)} \tag{4.122}$$

$$\phi_b = -60° + \arcsin\left(\frac{1}{3}\sqrt{7}\frac{V}{V_b}\sin(101° - \phi)\right) \tag{4.123}$$

$$V_c = \frac{1}{3}\sqrt{1 + 7V^2 - 2V\sqrt{7}\cos(101° + \phi)} \tag{4.124}$$

$$\phi_c = 60° - \arcsin\left(\frac{1}{3}\sqrt{7}\frac{V}{V_c}\sin(101° + \phi)\right) \tag{4.125}$$

The results for type G sags are shown in Fig. 4.112. We see that the type G sag is somewhat similar to the type C sag, as shown in Fig. 4.92. Unlike the phase-to-phase

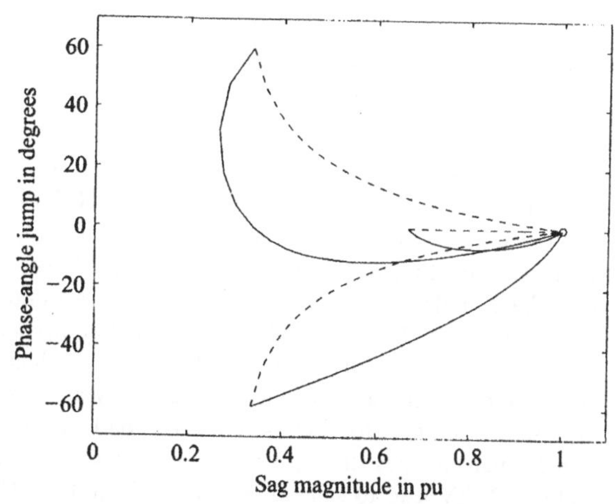

Figure 4.112 Magnitude and phase-angle jump at the equipment terminals for a type G sag, due to a two-phase-to-ground fault. The curves are given for an impedance angle of 0 (dashed line) and −60° (solid line).

fault, two-phase-to-ground faults cause two voltages to drop to 33% instead of 50%. For faults some distance away from the pcc the voltage magnitude can even become a bit less than 33% due to the initial phase-angle jump. Another difference with the phase-to-phase fault is that all three phases drop in magnitude. The third phase, which is not influenced at all by a phase-to-phase fault, may drop to 67% during a two-phase-to-ground fault.

4.6.4.3 Range of Magnitude and Phase-Angle Jump. Merging Fig. 4.110 and Fig. 4.112 gives the whole range of magnitudes and phase-angle jumps experienced by a single-phase load due to two-phase-to-ground faults. In Fig. 4.113 the area due to two-phase-to-ground faults (solid curve) is compared with the area due to phase-to-phase faults (dashed curve). We see that there are certain combinations of magnitude and phase-angle jump which can occur due to phase-to-phase faults but not due to two-phase-to-ground faults, but also the other way around. These curves have been obtained under the assumption that zero-sequence and positive-sequence impedances are equal. For a zero-sequence impedance larger than the positive-sequence source impedance, the resulting sags due to two-phase-to-ground faults are closer toward sags due to phase-to-phase faults. The results are that even a larger range of magnitude and phase-angle jumps can be expected. An increasing zero-sequence impedance will mean that the area enclosed by the solid curve in Fig. 4.113 will shift toward the area enclosed by the dashed curve. The latter is reached for an infinite zero-sequence impedance value.

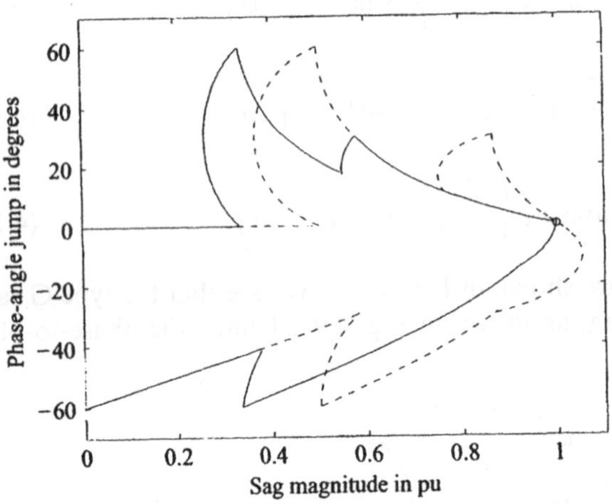

Figure 4.113 Range of magnitude and phase-angle jump at the equipment terminals due to phase-to-phase (dashed curve) and two-phase-to-ground faults (solid curve).

EXAMPLE: TWO-PHASE-TO-GROUND FAULTS, SINGLE-PHASE LOAD

For the same example system as used before (Fig. 4.21) the complex voltages at the equipment terminals due to two-phase-to-ground faults have been calculated. Characteristic magnitude and phase-angle jump due to a two-phase-to-ground fault are the same as due to a phase-to-phase fault. For three-phase delta-connected equipment we can directly use the results obtained for phase-to-phase faults in Fig. 4.97. For two-phase-to-ground faults, the solid lines refer to sags of type G, the dashed lines to sags of type F. A two-phase-to-ground fault at 11 kV leads to a sag of type F for delta-connected load, according to Table 4.13. The Dy 11 kV/660 V transformer changes this into a sag of type G, according to Table 4.14. Two-phase-to-ground faults at 33 kV lead to sags of type F, and faults at 132 kV and 400 kV to type G.

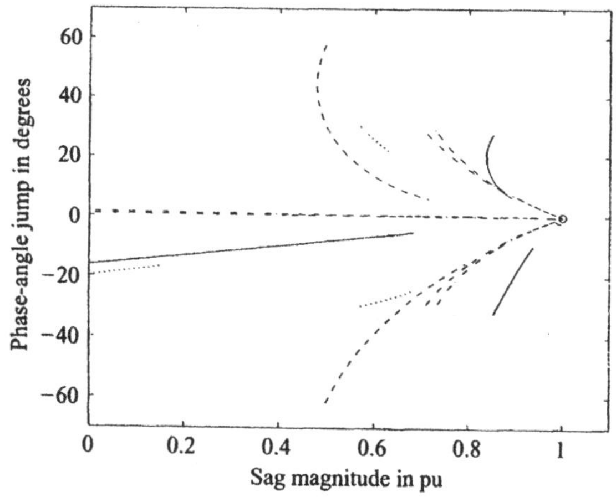

Figure 4.114 Magnitude and phase-angle jump at the equipment terminals due to two-phase-to-ground faults in Fig. 4.21, experienced by single-phase load-connected phase-to-ground at 420 V — solid line: 11 kV, dashed line: 33 kV, dotted line: 132 kV, dash-dot line: 400 kV.

For star-connected single-phase load, the situation is completely different. The zero-sequence source and feeder impedances influence the voltages during a two-phase-to-ground fault, but not during a phase-to-phase fault. The voltage sags experienced by single-phase equipment are shown in Fig. 4.114. Faults at 11 kV, 132 kV, and 400 kV cause sags of type G, in which one phase shows a deep sag and the two other phases a shallow sag. At 11 kV the zero-sequence source impedance is much larger than the positive-sequence one, due to the resistance grounding of this voltage level. The resulting sag is very close to the type D sags due to a phase-to-phase fault. The large zero-sequence impedance makes that the ground connection of a two-phase-to-ground fault does not carry much current. The voltage magnitude in the two phases with shallow sags is thus only down to about 90%. For faults at 132 kV, which is solidly grounded, these voltages are down to about 55%. The 400 kV system is also solidly grounded, but the line impedance dominates the source impedance, making that the zero-sequence impedance is more than twice as large as the positive-sequence impedance. In the phase with the largest voltage drop, the voltage magnitude is about the same for the three voltage levels. Faults at 33 kV will cause a type G sag. As the system is resistance grounded this sag is very close to a type C sag due to a phase-to-phase fault.

4.6.5 High-Impedance Faults

In all the previous calculations in this chapter, we have assumed the fault impedance to be zero. The argumentation for this was that the fault impedance could be incorporated in the feeder impedance, Z_F in (4.9). This argument still holds as long as the magnitude of the sag is concerned, but the phase-angle jump can be significantly affected. We will first address three-phase faults and after that single-phase faults. High-impedance faults are more likely for single-phase-to-ground faults than for three-phase faults.

4.6.5.1 Three-Phase Faults. Consider again the basic voltage divider expression (4.9), but this time with the fault resistance R_{flt} explicitly included:

$$V_{sag} = \frac{\overline{Z}_F + R_{flt}}{\overline{Z}_S + \overline{Z}_F + R_{flt}} \tag{4.126}$$

In many cases the source impedance and the feeder impedance are largely reactive, whereas the fault impedance is mainly resistive. The angle between source impedance and feeder plus fault impedance gets close to 90°, which can lead to very large phase-angle jumps.

The fault resistance only noticeably affects the voltage if $|\overline{Z}_F| \ll R_{flt}$, thus for faults close to the point-of-common coupling with the load. For zero distance to the fault we get for the complex voltage (with $\overline{Z}_S = jX_s$):

$$V_{sag} = \frac{R_{flt}}{jX_S + R_{flt}} \tag{4.127}$$

The fault resistance is normally not more than a fraction of the source reactance, in which case the sag magnitude is the ratio of the fault and the source impedances with a phase-angle jump equal to almost 90°.

To quantify the influence of the fault resistance, the complex voltage during the sag was calculated as a function of the distance to the fault for three-phase faults at 11 kV in Fig. 4.21. The calculations have been performed for a zero fault resistance and for fault resistances equal to 10%, 20%, and 30% of the (absolute value of the) source impedance. The sag magnitude (the absolute value of the complex voltage) is plotted in Fig. 4.115 as a function of the distance to the fault. As expected the influence on the sag magnitude is limited to small distances to the fault. The fault resistance increases the impedance between the pcc and the fault, and thus reduces the voltage drop at the pcc.

The phase-angle jump is much more influenced, as shown in Fig. 4.116. The phase-angle jump reaches values up to 80°. For increasing fault resistance the maximum phase-angle jump does not reduce much.

4.6.5.2 Single-Phase Faults. To assess the effect of high-impedance single-phase faults on the voltage at the equipment terminals, we use the classification of three-phase unbalanced sags again. At first we consider a solidly-grounded system, for which we can assume that the two non-faulted phase voltages remain at their pre-fault values. In other words, we have a clean type B sag. The voltage in the faulted phase is influenced by the fault resistance as shown in Figs. 4.115 and 4.116. At the equipment terminals the sag will be of type C or D. Magnitude and phase-angle jump at the equipment terminals are shown in Fig. 4.117 for a type C sag and in Fig. 4.118 for a type D sag. In Fig. 4.117 we see how an increasing fault resistance increases the unbalance between the two affected phases. Although the characteristic

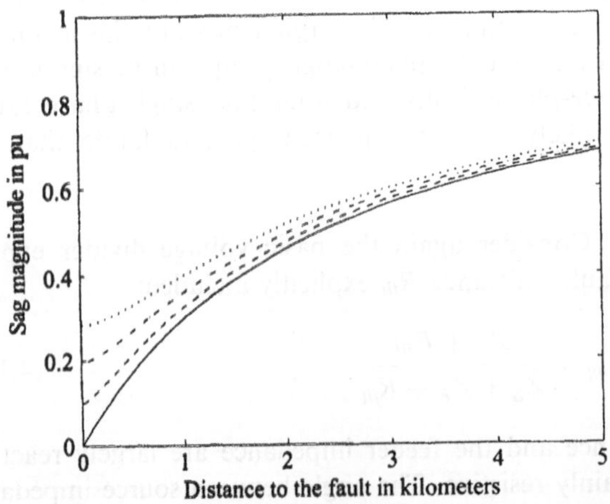

Figure 4.115 Sag magnitude versus distance for three-phase faults with fault resistances equal to zero (solid line), 10% (dashed line), 20% (dash-dot line), and 30% (dotted line) of the source impedance.

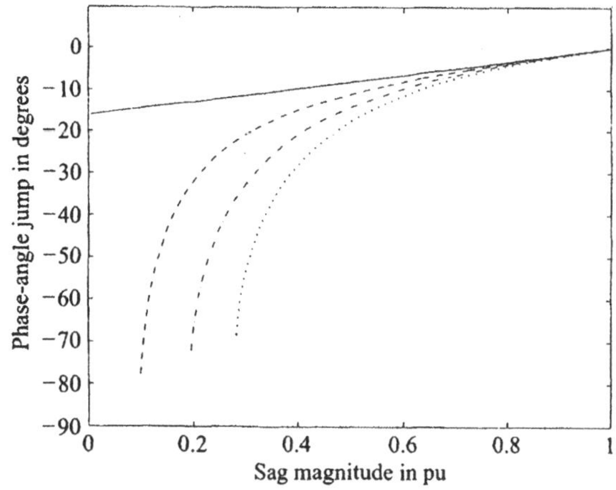

Figure 4.116 Sag magnitude versus phase-angle jump for three-phase faults with fault resistances equal to zero (solid line), 10% (dashed line), 20% (dash-dot line), and 30% (dotted line) of the source impedance.

magnitude increases due to the fault resistance, one of the phases actually drops in voltage. The characteristic magnitude is the difference between the two affected phases in the figure. We also see that the phase-angle jump at the equipment terminals only slightly exceeds 30°, despite the very large initial phase-angle jump. The largest phase-angle jump occurs for a 30% fault resistance at zero distance: −31.9°. In Fig. 4.118 we see that for a type D sag, the fault resistance increases the phase-angle jump in the phase with the large voltage drop, and that it raises one of the other two voltages and reduces the other. Fault resistances above 30% cause a small swell in one of the phases.

For Figs. 4.117 and 4.118, the 11 kV system was assumed to be solidly grounded. Therefore, the zero-sequence source impedance was made equal to the positive-sequence value. In reality this system is resistive grounded: positive- and zero-sequence source impedance are significantly different. The phase-to-neutral voltage is much lower in this case. To calculate the phase-to-neutral voltage a slightly revised version of (4.38) has been used:

$$V_{an} = 1 - \frac{3Z_{S1}}{2Z_{F1} + Z_{F0} + 2Z_{S1} + Z_{S0} + 3R_{flt}} \tag{4.128}$$

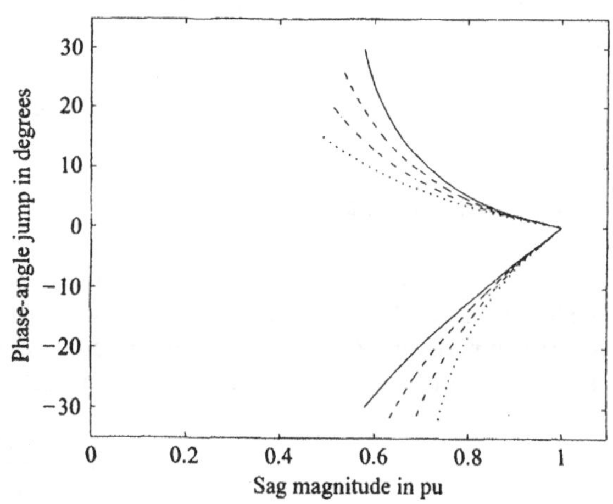

Figure 4.117 Magnitude versus phase-angle jump at the equipment terminals for single-phase faults in a solidly grounded system, sag type C; fault resistances equal to zero (solid line), 10% (dashed line), 20% (dash-dot line), and 30% (dotted line) of the source impedance.

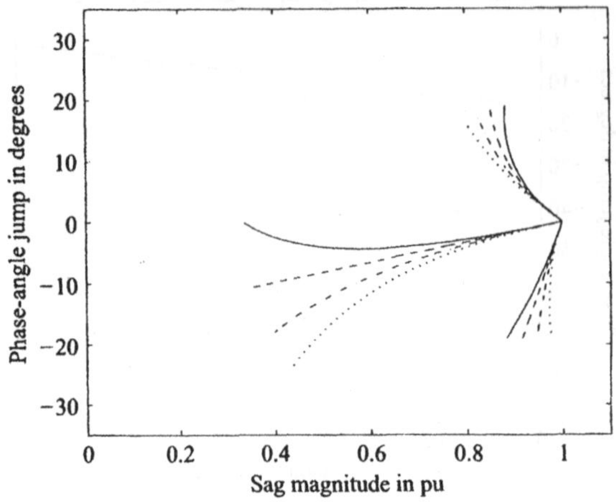

Figure 4.118 Magnitude versus phase-angle jump at the equipment terminals for single-phase faults in a solidly grounded system, sag type D, fault resistances equal to zero (solid line), 10% (dashed line), 20% (dash-dot line), and 30% (dotted line) of the source impedance.

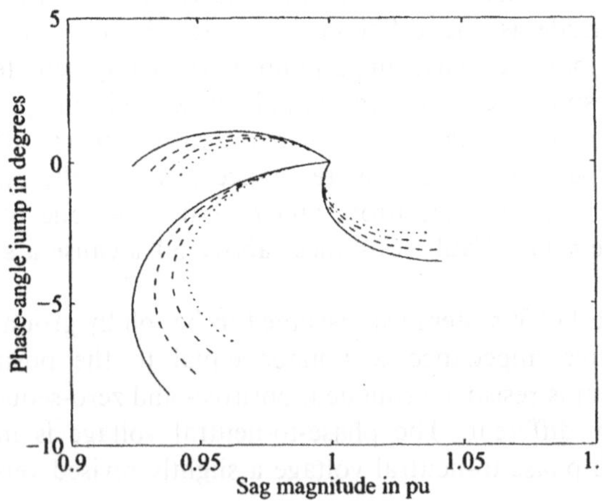

Figure 4.119 Magnitude versus phase-angle jumps at the equipment terminals for single-phase faults in a resistance-grounded system, sag type D; fault resistances equal to zero (solid line), 50% (dashed line), 100% (dash-dot line), and 150% (dotted line) of the source impedance.

The influence of the fault resistance is small in this case, as can be seen in Fig. 4.119. The magnitude and phase-angle jump at the equipment terminals are plotted for a type D sag. Due to the small fault currents arc resistances can reach much higher values in a resistance-grounded system than in a solidly-grounded system. In the calculations leading to Fig. 4.119 fault resistances equal to 50%, 100%, and 150% of the positive-sequence source impedance were used. The main effect of large fault resistances is that the sag becomes less severe in magnitude and in phase-angle jump.

4.6.6 Meshed Systems

All calculations in Sections 4.4 and 4.5 were based on the assumption that the system is radial; thus that we can uniquely identify a point-of-common coupling (pcc), a source impedance Z_S, and a feeder impedance Z_F, as were shown in Fig. 4.14. From Fig. 4.14 we obtained the basic voltage divider equation for the complex sag voltage:

$$V_{sag} = 1 - \frac{Z_S}{Z_S + Z_F} \tag{4.129}$$

In case the system is loaded, we can use Thevenin's superposition theorem which states that the voltage during the fault equals the voltage before the fault plus the change in voltage due to the fault:

$$V_{sag} = V_{pcc}^{(0)} - \frac{Z_S}{Z_S + Z_F} V_f^{(0)} \tag{4.130}$$

with $V_{pcc}^{(0)}$ the pre-fault voltage at the pcc and $V_f^{(0)}$ the pre-fault voltage at the fault position. Note that the source impedance Z_S includes the effect of loads elsewhere in the system.

For a meshed system we need matrix methods to calculate voltage during the fault, as introduced in Section 4.2.5. We obtained the following expression (4.24) for the voltage V_k at node k due to a fault at node f:

$$V_k = V_k^{(0)} - \frac{Z_{kf}}{Z_{ff}} V_f^{(0)} \tag{4.131}$$

with $V_k^{(0)}$ the voltage at node k before the fault and $V_f^{(0)}$ the voltage at the fault position before the fault, and Z_{ij} element ij of the node impedance matrix. Comparing this equation with (4.129) we see that they have the same structure. The voltage divider model can be used for meshed systems, when the following source and feeder impedances are used:

$$Z_S = Z_{kf} \tag{4.132}$$

$$Z_F = Z_{ff} - Z_{kf} \tag{4.133}$$

The main difference is that both Z_S and Z_F are dependent on the fault location. Equivalent source and feeder impedances can be obtained for positive-, negative-, and zero-sequence networks, and all the previously discussed analysis can still be applied.

4.7 OTHER CHARACTERISTICS OF VOLTAGE SAGS

4.7.1 Point-on-Wave Characteristics

The voltage sag characteristics discussed hitherto (magnitude, phase-angle jump, three-phase unbalance) are all related to the fundamental-frequency component of the voltage. They require the calculation of the rms value of the voltage or the complex voltage over a period of one half-cycle or longer. We saw earlier how this leads to an uncertainty in the calculation of sag duration. To obtain a more accurate value for the sag duration one needs to be able to determine "start" and "ending" of the sag with a higher precision. For this one needs to find the so-called "point-on-wave of sag initiation" and the "point-on-wave of voltage recovery" [38], [134]. Both require more advanced analysis techniques, which are still under development. We will see in the next chapter that the point-on-wave characteristics also affect the behavior of some equipment.

4.7.1.1 Point-on-Wave of Sag Initiation. The point-on-wave of sag initiation is the phase angle of the fundamental voltage wave at which the voltage sag starts. This angle corresponds to the angle at which the short-circuit fault occurs. As most faults are associated with a flashover, they are more likely to occur near voltage maximum than near voltage zero. In the sag shown in Fig. 4.1 the point-on-wave of sag initiation is close to voltage maximum. In Fig. 4.9 sag initiation takes place about 35°

after voltage maximum, at least in the phase with the largest voltage drop. In other phases the event starts at another angle compared to the fundamental voltage in that phase.

When quantifying the point-on-wave a reference point is needed. The upward zero crossing of the fundamental voltage is an obvious choice. One is likely to use the last upward zero crossing of the pre-event voltage as reference, as this closely resembles the fundamental voltage. The sag shown in Fig. 4.1 is partly repeated in Fig. 4.120: one cycle (1/60 of a second) starting at the last upward zero crossing before sag initiation. We see that the point-on-wave of sag initiation is about 275°. A closer look at the data learns that this point is between 276° and 280°. The slope at the beginning of the sag actually takes 4°, or about 185 μs. This is probably due to the low-pass character of the measurement circuit.

Figure 4.121 plots all three phases of the sag for which one phase was plotted in Fig. 4.120. For each phase, the zero point of the horizontal axis is the last upward zero crossing before the start of the event in that phase. We see that the point-on-wave is different in the three phases. This is obvious if one realizes that the event starts at the same moment in time in the three phases. As the voltage zero crossings are 120° shifted,

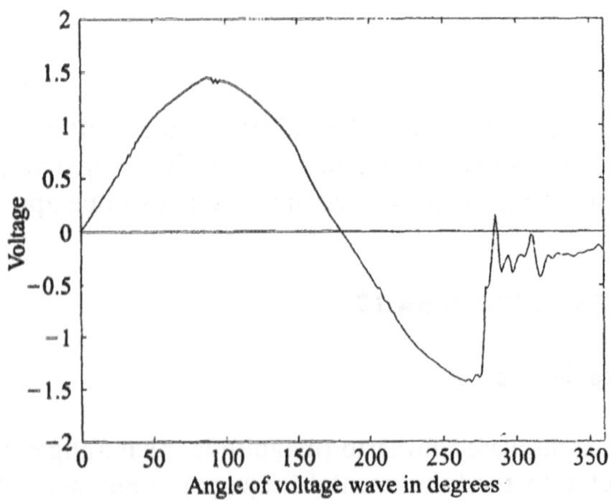

Figure 4.120 Enlargement of the sag shown in Fig. 4.1 indicating the point-on-wave of sag initiation.

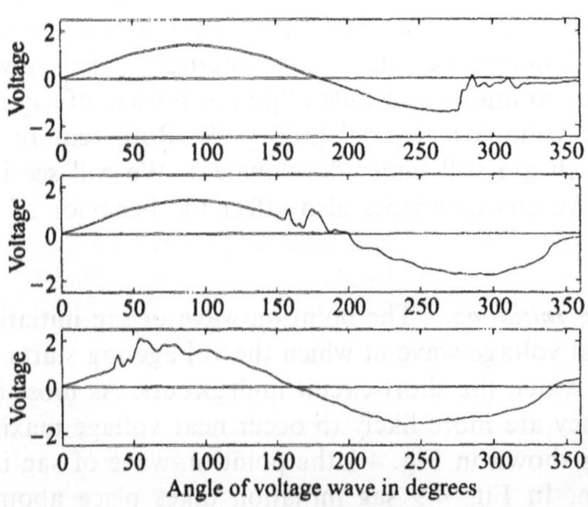

Figure 4.121 Event initiation in the three phases, compared to the last upward voltage zero crossing.

the point-on-wave values differ by 120°. In case phase-to-phase voltages are used, the resulting values are again different. When quantifying point-on-wave it is essential to clearly define the reference.

4.7.1.2 Point-on-Wave of Voltage Recovery. The point-on-wave of voltage recovery is the phase angle of the fundamental voltage wave at which the main recovery takes place. We saw before that most existing power quality monitors look for the point at which the voltage recovers to 90% or 95% of the nominal voltage. Note that there is in many cases no link between these two points. Consider as an example again the sag shown in Fig. 4.1. Voltage recovery in the meaning of this section takes place about 2.5 cycles after sag initiation, even though the voltage does not fully recover for at least another two cycles, as can be seen in Fig. 4.3.

Voltage recovery corresponds to fault clearing, which takes place at current zero crossing. Because the power system is mainly inductive, current zero crossing corresponds to voltage maximum. Thus we expect points-on-wave of voltage recovery to be around 90° and 270°. This assumes that we use the pre-event fundamental voltage as reference, not the during-event voltage. It is the pre-event voltage which drives the fault current and which is thus 90° shifted compared to the fault current. The recovery of the sag in Fig. 4.120 is shown in Fig. 4.122. The recovery is, at least in this case, slower than the sag initiation. The shape of the voltage recovery corresponds to the so-called "transient recovery voltage" well-known in circuit-breaker testing. The smooth sinusoidal curve in Fig. 4.122 is the continuation of the pre-event fundamental voltage. Considering the start of the recovery, we find a point-on-wave of 52°. If we further assume this to be the moment of fault-clearing taking place at current zero, we see that the current lags the voltage by 52°, which gives an X/R ratio at the fault position equal to $\tan^{-1}(52°) = 1.3$.

For a two-phase-to-ground or three-phase fault, fault clearing does not take place in all three phases at the same time. This could make a determination of the point-on-wave of voltage recovery difficult. An unambiguous definition of the reference point and phase is needed to apply this concept to three-phase unbalanced sags.

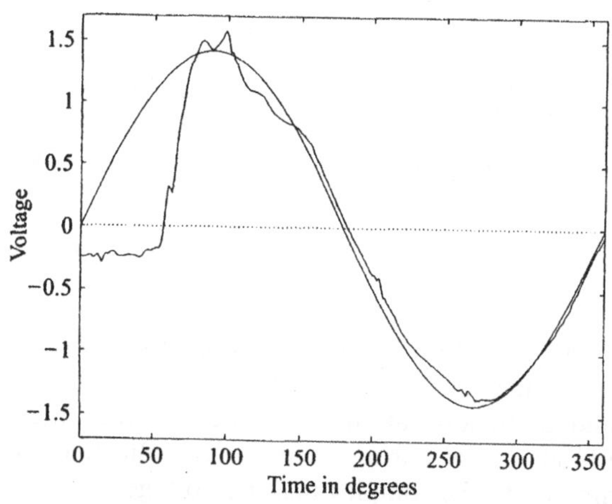

Figure 4.122 Enlargement of Fig. 4.1 showing the point-on-wave of voltage recovery. The smooth curve is the continuation of the pre-sag fundamental voltage.

4.7.2 The Missing Voltage

The missing voltage is another voltage sag characteristic which has been proposed recently [134]. The missing voltage is a way of describing the change in momentary voltage experienced by the equipment. The concept became important with the dimensioning of series-connected voltage-source converters to compensate for the voltage drop due to the fault. We will see in Chapter 7 that the voltage injected by the series compensator is equal to the missing voltage: the difference between the voltage as it would have been without the sag, and the actual voltage during the sag.

4.7.2.1 The Complex Missing Voltage. One can think of the missing voltage as a complex voltage (a phasor), being the difference in the complex plane between the pre-event voltage and the voltage during the sag. The absolute value of this complex missing voltage can be directly read from a plot like shown in Fig. 4.83. In Fig. 4.83 the missing voltage is the distance between the complex voltage during the sag (which is on one of the three curves) and the top-right corner of the diagram (the point $1+j0$).

EXAMPLE Consider a sag on a 50 mm² underground cable, like in Fig. 4.83, with a sag magnitude of 60%. If the pre-event voltage was 100%, the drop in rms value of the voltage is 40%. Having no further information one would be tempted to say that a compensator should inject a voltage with an rms value equal to 40% of nominal.

Looking in the complex plane, we see that a magnitude of 60% corresponds to a complex voltage $\overline{V} = 0.45 - j0.39$. The missing voltage is the difference between the pre-fault voltage and the voltage during the sag, thus $1 - \overline{V} = 0.55 + j0.39$. The absolute value of the missing voltage is 67% in this example. Compare this with the 40% drop in rms voltage.

The complex missing voltage can also be calculated from the magnitude V and the phase-angle jump ϕ of the sag. The complex voltage during the sag is

$$\overline{V} = V\cos\phi + jV\sin\phi \tag{4.134}$$

The missing voltage is simply

$$1 - \overline{V} = 1 - V\cos\phi - jV\sin\phi \tag{4.135}$$

with as absolute value

$$V_{miss} = |1 - \overline{V}| = \sqrt{1 - V^2 - 2V\cos\phi} \tag{4.136}$$

When we neglect the phase-angle jump, thus assume that $\overline{V} = V$, the missing voltage is simply $\tilde{V}_{miss} = 1 - V$. We can assess the error made by writing $1 - V = \sqrt{1 + V^2 - 2V}$. Comparing this with (4.136) gives for the difference between the exact and the approximate expression for the missing voltage:

$$V_{miss}^2 - \tilde{V}_{miss}^2 = 2V(1 - \cos\phi) \tag{4.137}$$

4.7.2.2 The Missing Voltage in Time Domain. The concept of missing voltage can become much more useful by extending it to time domain. A very first step would be to look at the difference between the fundamental pre-event voltage and the fundamental during-event voltage. But that would not give any extra information compared to the complex missing voltage.

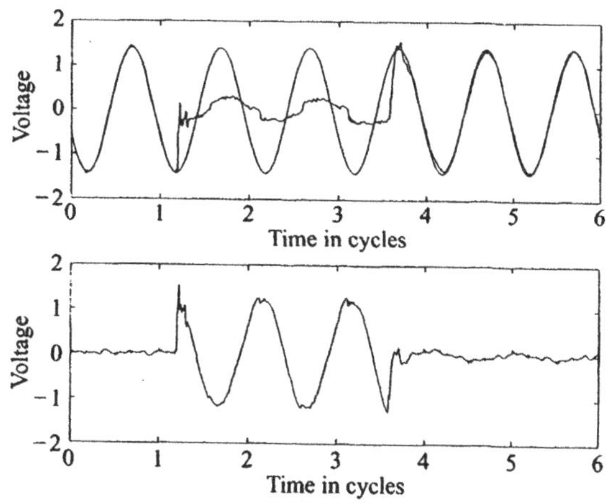

Figure 4.123 Time-domain voltage measurement together with pre-event fundamental voltage (top curve) and the time-domain missing voltage being the difference of those two (bottom curve).

In the top part of Fig. 4.123 the sag from Fig. 4.1 has been plotted again. Together with the actual time-domain voltage wave, the fundamental pre-event voltage has been plotted. The latter is obtained by applying a fast-Fourier-transform algorithm to the first cycle of the voltage wave form. From the complex coefficient for the fundamental term in the Fourier series C_1, the (time-domain) fundamental component of the voltage can be calculated:

$$V_{fund}(t) = \mathrm{Re}\{C_1 e^{-j\omega t}\} \tag{4.138}$$

This fundamental component of the pre-event voltage (pre-event fundamental voltage, for short) is the smooth sinusoidal curve in the top part of Fig. 4.123.

The missing voltage is calculated as the difference between the actual voltage and the pre-event fundamental voltage:

$$V_{miss}(t) = V(t) - V_{fund}(t) \tag{4.139}$$

This missing voltage is plotted in the bottom part of Fig. 4.123. Before the initiation of the sag there is obviously no fundamental component present; during the sag the fundamental component of the missing voltage is large; after the principal sag (after fault clearing) a small fundamental component remains. The reason for this becomes clear from the upper curve: the voltage does not immediately fully recover to its pre-event value.

Figure 4.124 repeats this for the voltage in one of the non-faulted phases, for the same event as in Fig. 4.123 and Fig. 4.1. In the top curve we see that the during-event voltage has a larger rms value than the pre-event voltage. In terms of rms voltages, we would call this an increase in voltage: a voltage swell. But looking at the missing voltage it is not possible to say whether the underlying event is a swell or a sag. This might be seen as a disadvantage of the missing voltage concept. But one should realize that this concept is not meant to replace the other ways of characterizing the sag; instead, it should give additional information.

Finally, Fig. 4.125 plots the missing voltage in all three phases. As expected for a single-phase-to-ground fault, the missing voltage in the two non-faulted phases is the same and in phase with the missing voltage in the faulted phase. After the fault the missing voltages in the three phases form a positive sequence set. This is probably due to the re-acceleration of induction motors fed from the supply.

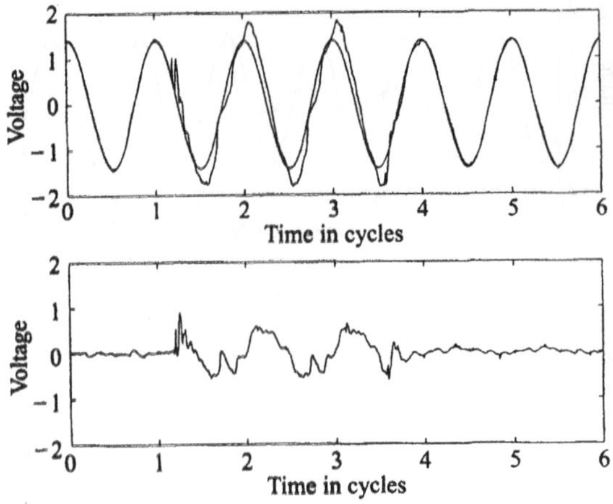

Figure 4.124 Measured voltage with pre-event fundamental voltage (top curve) and missing voltage (bottom curve) during a voltage swell event.

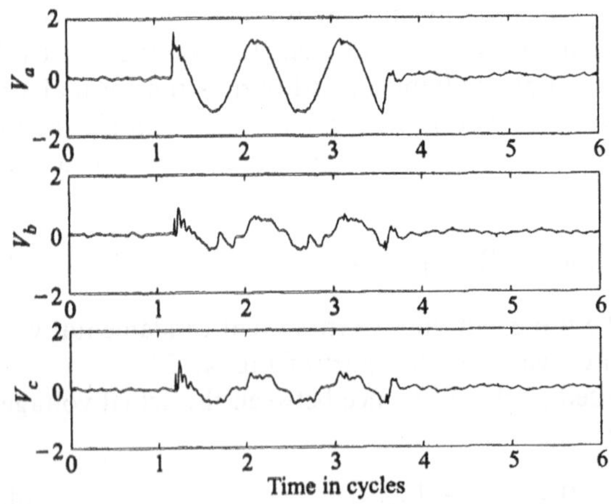

Figure 4.125 Missing voltage for the three phases of a sag due to a single-phase fault.

In Figs. 4.124 and 4.125 we used the fundamental pre-event voltage as a reference to obtain the missing voltage. The concept of missing voltage has been introduced to quantify the deviation of the voltage from its ideal value. In other words: we have used the fundamental pre-event voltage as the ideal voltage. This could become a point of discussion, as there are at least three alternatives:

- Use the full pre-event waveform, including the harmonic distortion, as a reference. One can either take the last cycle before the event or the average over a number of cycles. The latter option is limited in its application because there are normally not more than one or two pre-event cycles available.
- Use the fundamental component of the pre-event waveform as a reference. One can again choose between the fundamental obtained from the last cycle before the event (as was done in Fig. 4.124 and Fig. 4.125) or obtain the fundamental from a number of pre-event cycles.
- Use as a reference, a sinusoidal waveform with the same amplitude and rms value as the system nominal voltage and the same phase angle as the fundamental pre-event waveform. The difference between the last two alternatives is

the same as the discussion between defining the voltage drop with reference to the pre-event rms voltage or with reference to the nominal rms voltage. Both methods have their advantages and can thus be used. But it is important to always indicate which method is used.

4.7.2.3 Distribution of the Missing Voltage. An alternative and potentially very useful way of presenting the missing voltage is through the amount of time that the missing voltage, in absolute value, exceeds given values; in other words, the amount of time during which the deviation from the ideal voltage waveform is larger than a given value.

In the top curve of Fig. 4.126 the missing voltage from Fig. 4.123 is shown again. But this time the absolute value is plotted, instead of the actual waveform. We see, e.g., that this absolute value exceeds the value of 0.5, a total of six times during the event. The cumulative duration of these six periods is 1.75 cycles. The cumulative time during which the missing voltage in absolute value exceeds a given level can be determined for each level. The result of this calculation is shown in the bottom part of Fig. 4.126. This curve can be read as follows: the missing voltage is never larger than 1.53, is during 1 cycle larger than 0.98, during 1.75 cycle larger than 0.5, during two cycles larger than 0.32, etc. The long tail in Fig. 4.126 is due to the post-fault voltage sag as well as to the non-zero pre-event missing voltage. The latter contribution can be removed by either using the full pre-event waveshape as a reference to calculate the missing voltage, or by only considering the missing voltage samples from the instant of sag-initiation onward.

Through the same procedure, distributions of the missing voltage can be obtained for the other two phases, resulting in the curves shown in Fig. 4.127. The missing voltage in the faulted phase (solid curve) is naturally larger than in the non-faulted phases. But still, the missing voltage in the non-faulted phases is significant: during about 1 cycle it exceeds a value of 0.4. We also see a small difference in missing voltage between the two non-faulted phases: the value in phase b is somewhat higher than in phase c.

The missing voltage distribution curve can be used as a generalized way of defining the event duration. The larger the deviation from the ideal voltage one considers, the shorter the "cumulative duration" of the event. The cumulative duration of a

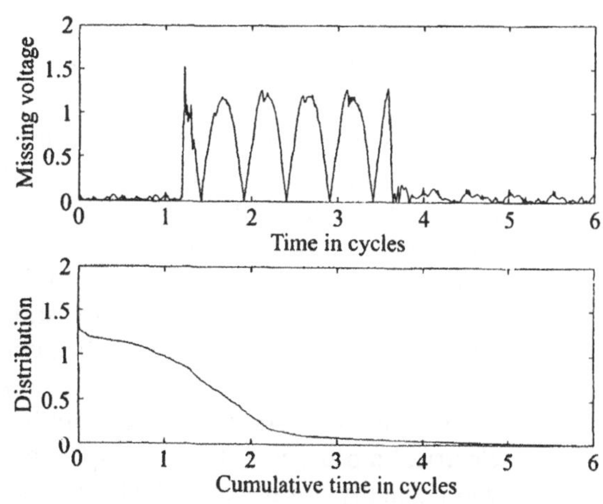

Figure 4.126 Absolute value of the missing voltage (top curve) and the distribution of the missing voltage (bottom curve) for the sag shown in Fig. 4.1.

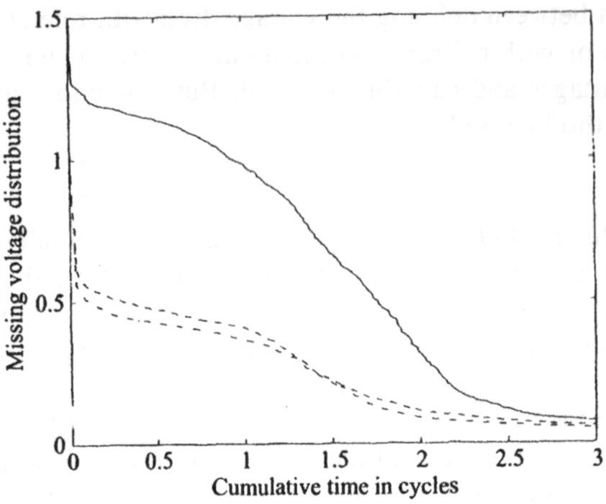

Figure 4.127 Missing voltage distribution for phase a (solid curve), phase b (dashed curve), and phase c (dash-dot curve).

voltage sag for a given deviation would be defined as the total amount of time during which the voltage deviates more than the given value from the ideal voltage waveshape.

4.8 LOAD INFLUENCE ON VOLTAGE SAGS

In the calculation of sag magnitude for various system configurations, in the classification of three-phase sags and in most of the examples, we have assumed that the load currents are zero. In this section we will discuss some situations in which the load currents can have a significant influence on the voltages during a fault. The main load having influence on the voltage during and after a sag is formed by induction and synchronous motors as they have the largest currents during and after a short-circuit fault. But we will also briefly discuss single-phase and three-phase rectifiers as they are a large fraction of the load at many locations.

4.8.1 Induction Motors and Three-Phase Faults

During a three-phase fault the voltages at the motor terminals drop in magnitude. The consequences of this drop are twofold:

- The magnetic flux in the air gap is no longer in balance with the stator voltage. The flux decays with a time constant of up to several cycles. During this decay the induction motor contributes to the fault and somewhat keeps up the voltage at the motor terminals.
- The decay in voltage causes a drop in electrical torque: the electrical torque is proportional to the square of the rms value of the voltage. The mechanical torque in the mean time remains largely unchanged. The result is that the motor slows down. While the motor slows down it will take a larger current with a smaller power factor. This could bring down the voltage even more. For small voltage drops, a new steady state could be reached at a lower speed, depending on the speed-torque behavior of the mechanical load. For deep sags the motor will continue to slow down until it reaches standstill, or until the voltage recovers, whichever comes first. The mechanical time constant of electrical motors is of the order of one second and more. Therefore the motor will normally not have reached zero speed yet upon voltage recovery.

The moment the voltage recovers the opposite phenomena occur. The flux in the air gap will build up again. This causes a large inrush current, which slows down the voltage recovery. After that, the motor will re-accelerate until it reaches its pre-event speed. During the re-acceleration the motor again takes a larger current with a smaller power factor, which causes a post-fault voltage sag sometimes lasting for several seconds.

The contribution of the induction motor load to the fault can be modeled as a voltage source behind reactance. The voltage source has a value of about 1 pu at fault initiation and decays with the subtransient time-constant (between 0.5 and 2 cycles). The reactance is the leakage reactance of the motor, which is between 10% and 20% on the motor base. Note that this is not the leakage reactance which determines the starting current, but the leakage reactance at nominal speed. For double-cage induction machines these two can be significantly different.

EXAMPLE Consider a bolted fault at primary side of a 33/11 kV transformer in the supply shown in Fig. 4.21. The total induction motor load connected to the 11 kV bus is 5% of the fault level. The induction motors have a leakage reactance of 10% on the motor base. We are interested in the voltage at secondary side of the transformer. Consider only the reactive part of the impedances.

The transformer impedance is the difference between the 33 kV and 11 kV fault levels: $Z_T = 47.6\%$ at a 100 MVA base. The fault level at 11 kV is 152 MVA, thus the total motor load is (5% of this): 7.6 MVA. The leakage reactance of the motors is 10% at a 7.6 MVA base, which is $Z_M = 132\%$ at a 100 MVA base. The voltage on secondary side of the transformer is found from the voltage divider equation:

$$V_{load} = \frac{Z_T}{Z_T + Z_M} = 27\% \tag{4.140}$$

To assess the increase in motor current after the fault, we use the common equivalent circuit for the induction motor, consisting of the series connection of the stator resistance R_S, the leakage reactance X_L and the slip-dependent rotor resistance $\frac{R_R}{s}$, with s the motor slip. The motor impedance is

$$Z_M = R_S + jX_L + \frac{R_R}{s} \tag{4.141}$$

The change of motor impedance with slip has been calculated for four induction motors of four different sizes. Motor parameters have been obtained from [135], [136], and the motor impedance has been calculated by using (4.141). The results are shown in Fig. 4.128. For each motor, the impedance at nominal slip is set at 1 pu, and the absolute value of the impedance is plotted between nominal slip and 25% slip. We see for each motor a decrease in motor impedance, and thus an increase in motor current, by a factor of about five. The decrease in impedance is much faster for large machines than for smaller ones.

If we assume the voltage to recover to 1 pu immediately upon fault clearing, the current taken by the motor is the inverse of the impedance (both equal to 1 pu in normal operation). The path of the current in the complex plane is shown in Fig. 4.129. The path is given for an increase in slip from its nominal value to 25%. The positive real axis is in the direction of the motor terminal voltage. For small motors we see predominantly an increase in resistive current, for large motors the main increase is in the inductive part of the current. When the slip increases further, even the resistive part

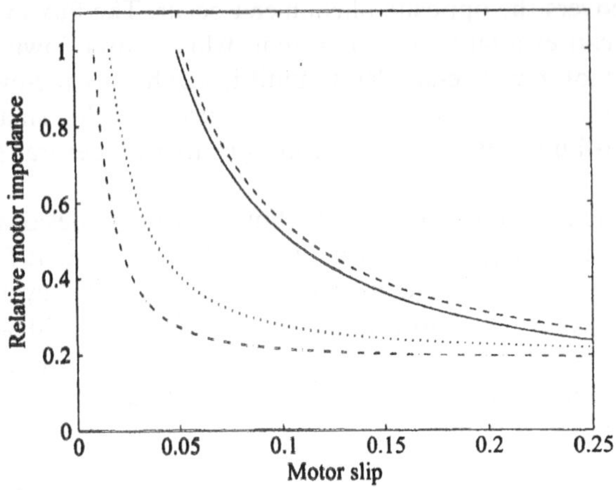

Figure 4.128 Induction motor impedance versus slip; the impedance at nominal slip is 1 pu; 3 hp 220 V (solid line), 50 hp 460 V (dashed line), 250 hp 2300 V (dotted line), 1500 hp 2300 V (dash-dot line).

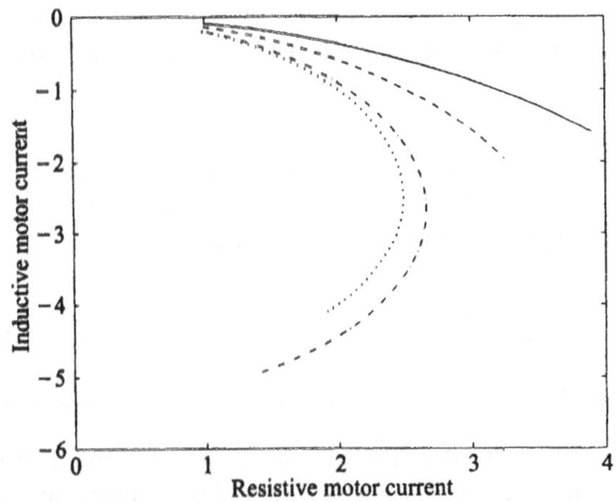

Figure 4.129 Change in induction motor current with increasing slip; the current at nominal slip is 1 pu; 3 hp 220 V (solid line), 50 hp 460 V (dashed line), 250 hp 2300 V (dotted line), 1500 hp 2300 V (dashed line).

of the current starts to decrease. The power factor of the current decreases significantly, especially for large motors.

The influence of large induction motors on voltage sags is described in detail by Yalçinkaya [136]. Fig. 4.130 shows the voltage sags (top curve) and the motor slip (bottom curve) due to a three-phase fault in an industrial system with a large induction motor load. Without induction motor load, the voltage would have been zero during the sag and 1 pu after the sag. The voltage plotted in Fig. 4.130 is the absolute value of a time-dependent phasor, used in a transient-stability program. The effect of the induction motor load is that the voltage during the fault is increased, and after the fault decreased. The slip of all motors increases fast during the sag, and even continues to increase a bit after fault clearing.

The voltage after fault clearing, the so-called post-fault sag, shows an additional decrease about 200 ms after fault clearing. This corresponds to the moment the motor starts to re-accelerate and draws larger currents. The low voltage immediately after fault clearing is due to the large current needed to rebuild the air gap flux.

During the fault the induction motors significantly keep up the voltage. Even toward the end of the sag the voltage at the motor busses is still above 10% of its pre-event value.

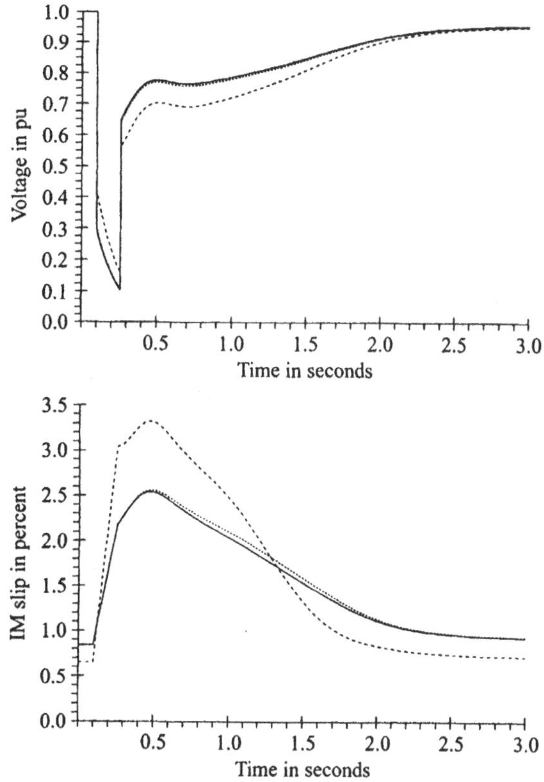

Figure 4.130 Voltage sag (top) and induction motor slip (bottom) for three busses in an industrial power system. (Reproduced from Yalçinkaya [136].)

One should realize that this is a somewhat exceptional case, as the motor load connected to the system is very large. Similar but less severe effects have been noticed in other systems. Another phenomenon which contributes to the post-fault voltage sag is that the fault occurs in one of two parallel transformers. The protection removes the faulted transformer, so that only one transformer is available for the supply after fault clearing. The post-fault fault level is thus significantly less than its pre-fault value. A similar effect occurs for a fault in one of two parallel feeders. The post-fault sag, described here for three-phase faults, has also been observed after single-phase faults.

4.8.2 Induction Motors and Unbalanced Faults

The behavior of an induction motor during an unbalanced fault is rather complicated. Only a network analysis program simulating a large part of the system can give an accurate picture of the quantitative effects. The following phenomena play a part in the interaction between system and induction motor during unbalanced faults.

- During the first one or two cycles after fault initiation the induction motor contributes to the fault. This causes an increase in positive-sequence voltage. Negative- and zero-sequence voltage are not influenced.
- The induction motor slows down, causing a decrease in positive-sequence impedance. This decrease in impedance causes an increase in current and thus a drop in positive-sequence voltage.

- The negative-sequence impedance of the motor is low, typically 10–20% of the nominal positive-sequence impedance. The negative-sequence voltage due to the fault will thus be significantly damped at the motor terminals. The negative-sequence impedance is independent of the slip. The negative-sequence voltage will thus remain constant during the event.
- The induction motor does not take any zero-sequence current. The zero-sequence voltage will thus not be influenced by the induction motor.

4.8.2.1 Simulation Example. Simulations of the influence of induction motor loads on unbalanced sags are shown in [136], [137]. Some of those results are reproduced here. The system studied was a radial one with large induction motor load connected to each of the low-voltage busses. Motor sizes and transformer impedances were chosen such that for each bus the fault level contribution from the source was 15 times the total motor load fed from the bus. Voltages and currents in the system were calculated by using the transient analysis package EMTP. All transformers in the system were connected star-star with both neutral points earthed. Although this is not a very common arrangement, it helped in understanding the phenomena. The voltages at the terminals of one of the motors are shown in Fig. 4.131. Without induction motor influence we would have seen a sag of type B of zero magnitude: zero voltage in phase a, and no change in the voltage in

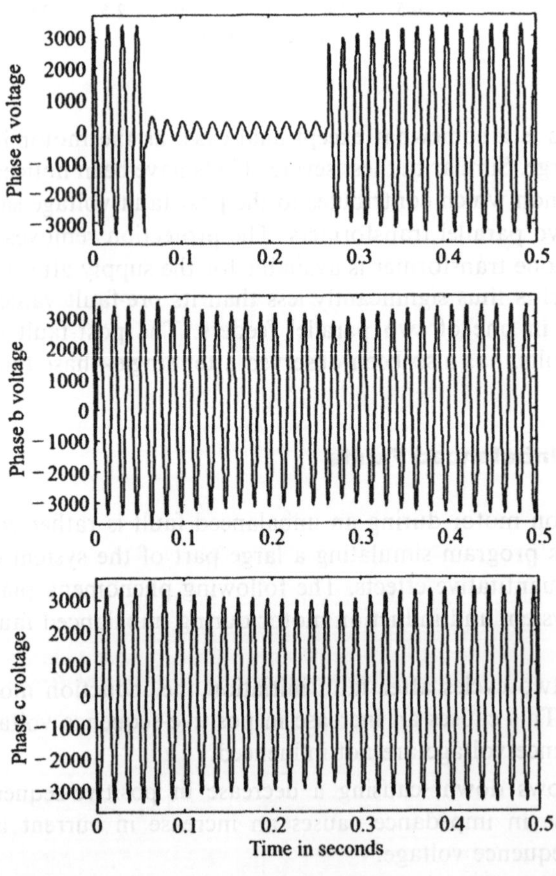

Figure 4.131 Voltages at the motor terminals, due to a single-phase-to-ground fault in the supply. (Reproduced from Yalçinkaya [136].)

phase b and phase c. Instead we see a small non-zero voltage in phase a and in the two non-faulted phases an initial increase followed by a slow decay. After fault clearing the system becomes balanced again, and the three phase voltages thus equal in amplitude. The motor re-acceleration causes a post-fault sag of about 100 ms duration.

The non-zero voltage in the faulted phase is due to the drop in negative-sequence voltage. We saw in (4.32) and (4.34) that the voltage in the faulted phase during a single-phase fault is given as

$$V_a = V_1 + V_2 + V_0 = |V_1| - |V_2| - |V_0| \tag{4.142}$$

The effect of the induction motor is that V_2 drops in absolute value, causing an increase in voltage in the faulted phase.

During the sag, the positive-sequence voltage also drops, which shows up as the slow but steady decrease in voltage in all phases.

The non-faulted phases show an initial increase in voltage. The explanation for this is as follows. The voltage in the non-faulted phases during a single-phase fault is made up of a positive-sequence, a negative-sequence, and a zero-sequence component. For phase c this summation in the complex plane is for the system without induction motor load.

$$V_c = V_{c1} + V_{c0} + V_{c2} = \frac{2}{3}a - \frac{1}{3} - \frac{1}{3}a^2 = a \tag{4.143}$$

Due to the induction motor load, the positive-sequence voltage will not immediately drop from 1 pu to 0.67 pu. The negative-sequence voltage will jump from zero to its new value immediately. The consequence is that the resulting voltage amplitude slightly exceeds its pre-fault value. After a few cycles the induction motor no longer keeps up the positive-sequence voltage. The voltage in the non-faulted phases drops below its pre-event value due to negative- and positive-sequence voltages being less than 33% and 67%, respectively.

The currents taken by the induction motors are shown in Figs. 4.132 and 4.133. Figure 4.132 shows the motor currents for a motor with a small decrease in speed. The slip of this motor increases from 2% to 6% during the sag. The motor shown in Fig. 4.133 experienced a much larger decrease in speed: its slip increased from 3% to 19%. This behavior is difficult to explain without considering symmetrical components. But generally we can observe that the current increases initially in the faulted phase, rises to a higher value in one of the non-faulted phases, and initially drops in the other non-faulted phase. The current in the second non-faulted phase rises again after a certain time, determined by the slowing down of the motor.

For the motor shown in Figs. 4.131 and 4.132 the component voltages and currents have been plotted in Figs. 4.134 and 4.135. From Fig. 4.134 we see that negative and zero-sequence voltage remain constant during the sag, but that the positive-sequence voltage shows a steady decay, due to the decrease in positive-sequence impedance when the motor slows down. Figure 4.135 clearly shows the increase in positive-sequence current when the motor slows down. The zero-sequence current is zero as the motor windings are connected in delta. From Figs. 4.134 and 4.135 the positive- and negative-sequence impedance of the motor load can be calculated, simply through dividing voltage by current. The results are shown in Fig. 4.136, where we see again that the negative-sequence impedance remains constant, whereas the positive-sequence impedance drops. When the motor reaches standstill, it is no longer a dynamic element, and positive- and negative-sequence impedance become equal.

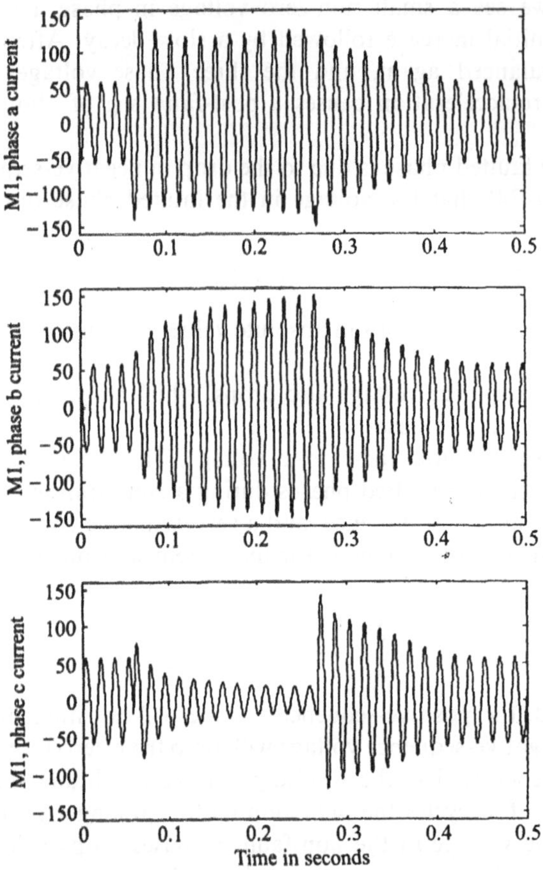

Figure 4.132 Induction motor currents during and after a single-line-to-ground fault in the supply. This motor showed only a small decrease in speed. (Reproduced from Yalçinkaya [136].)

4.8.2.2 Monitoring Example. An example of a three-phase unbalanced sag was shown in Fig. 4.48. The severe post-fault sag indicates the presence of induction motor load. For each of the three sampled waveforms, the complex voltage as a function of time was determined by using the method described in Section 4.5. From the three complex voltages, positive-, negative- and zero-sequence voltages have been calculated. Their absolute values are plotted in Fig. 4.137 as a function of time. The zero-sequence component is very small. The negative-sequence component is zero when the fault is not present and non-zero but constant during the fault. The positive-sequence voltage is 1 pu before the fault, shows a slow decay during the fault, and a slow increase after the fault. This is exactly in correspondence with the above-described theory and simulation results.

4.8.2.3 Simplified Analysis. From the simulation and monitoring results we can extract three stages in the voltage sag:

- The induction motor feeds into the fault, raising the positive-sequence voltage.
- The positive-sequence voltage is the same as it would have been without the induction motor load.
- The induction motor has slowed down, drawing additional positive-sequence current, thus causing the positive-sequence voltage to drop.

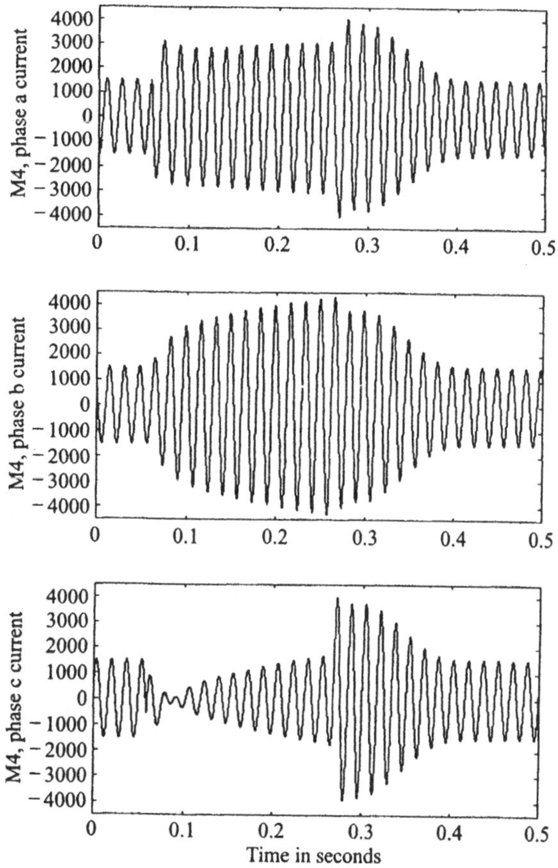

Figure 4.133 Induction motor currents during and after a single-line-to-ground fault in the supply. This motor showed a large decrease in speed. (Reproduced from Yalçinkaya [136].)

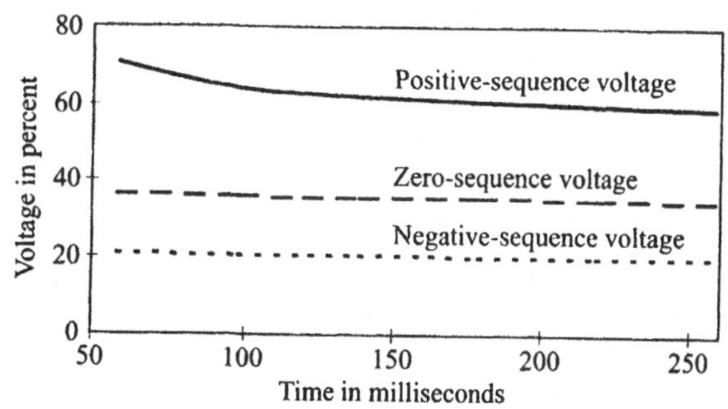

Figure 4.134 Symmetrical components for the voltages shown in Fig. 4.131. (Reproduced from Yalçinkaya [136].)

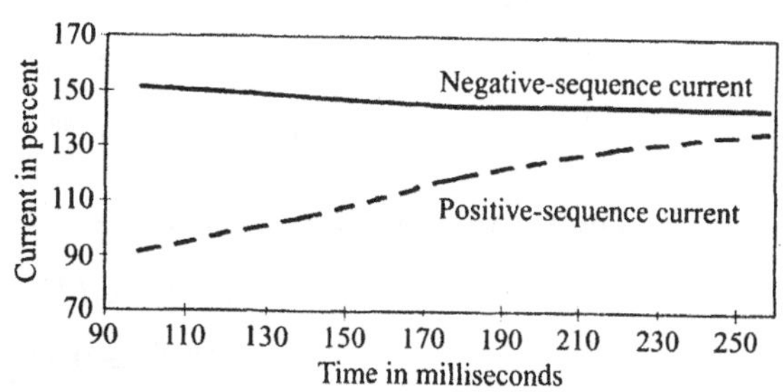

Figure 4.135 Symmetrical components for the currents shown in Fig. 4.132. (Reproduced from Yalçinkaya [136].)

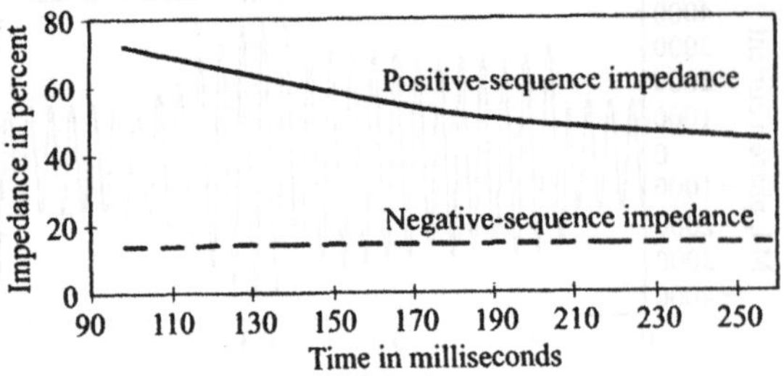

Figure 4.136 Positive- and negative-sequence impedance for an induction motor during a sag. (Reproduced from Yalçinkaya [136].)

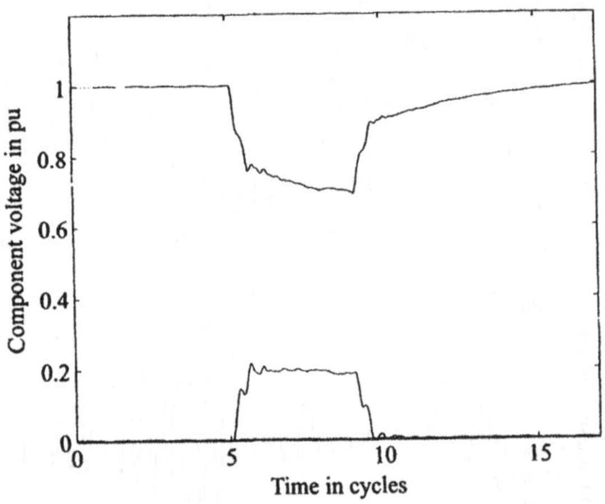

Figure 4.137 Positive-, negative- and zero-sequence voltages for the three-phase unbalanced sag shown in Fig. 4.47.

The negative-sequence voltage is constant during the fault, but lower than without induction motor load. To quantify the effect of induction motors, we use a two-step calculation procedure. At first we calculate positive- and negative-sequence voltage ($V_1^{(no)}, V_2^{(no)}$) for the no-load case. As we saw before this will lead to voltage sags of type C or type D with different characteristic magnitude. We assumed a zero characteristic phase-angle jump. As a second step the influence of the induction motor is incorporated. For this we model the supply as a source generating a type C or type D sag, with a finite source impedance. Note that this is a three-phase Thevenin source representation of the supply during the fault. The effect of the induction motor load is a difference between the source voltages and the voltages at the motor terminals, for positive as well as for negative-sequence components. The voltage at the motor terminals are denoted as $V_1^{(load)}$ and $V_2^{(load)}$. For the three above-mentioned "stages" these relations are assumed to be as follows:

1. The drop in positive-sequence voltage is reduced by 15%, the negative-sequence voltage drops by 30%.

$$V_1^{(load)} = 0.15 + 0.85 V_1^{(no)}$$
$$V_2^{(load)} = 0.7 V_2^{(no)}$$

2. The negative-sequence voltage drops by 30%.

$$V_1^{(load)} = V_1^{(no)}$$
$$V_2^{(load)} = 0.7 V_2^{(no)}$$

3. The positive-sequence voltage drops by 10%, the negative-sequence voltage drops by 30%.

$$V_1^{(load)} = 0.9V_1^{(no)}$$
$$V_2^{(load)} = 0.7V_2^{(no)}$$

The voltages at the motor terminals are calculated from the positive- and negative-sequence voltages $V_1^{(load)}$ and $V_2^{(load)}$. The resulting phase voltages for the three stages are shown in Figs. 4.138 and 4.139. For sag type C the voltages are shown for one of the phases with a deep sag, and for the phase with a shallow sag. The more the motors slow down, the more the voltage in this phase drops. The voltage in the worst-affected phase is initially somewhat higher due to the induction motor influence, but drops when the motor slows down and the positive-sequence voltage drops in value as well. For type D we see that the voltage in the least-affected phases drops during all stages of the sag. The voltage in the worst-affected phase increases initially but decreases later.

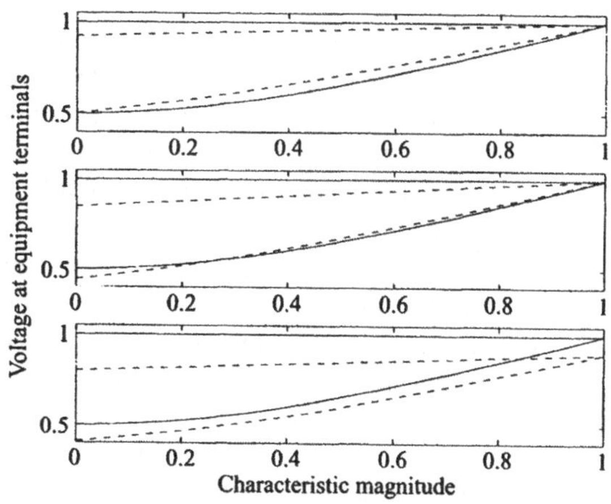

Figure 4.138 Voltages at the equipment terminals, for three stages of induction motor influence for type C sags. The solid lines are without induction motor influence, the dashed lines with.

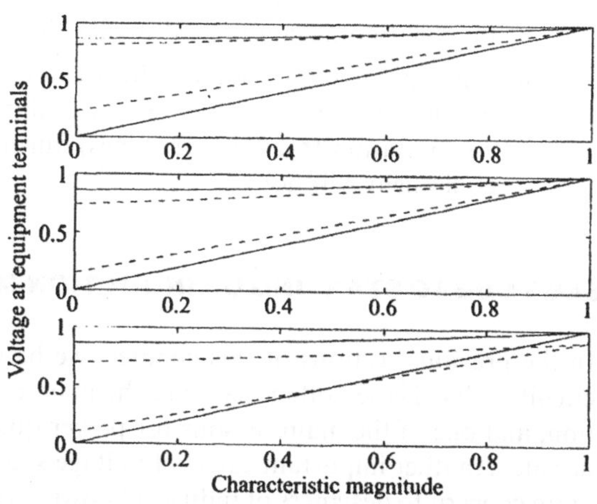

Figure 4.139 Voltages at the equipment terminals, for three stages of induction motor influence for type D sags. The solid lines are without induction motor influence, the dashed lines with.

From the curves in Figs. 4.138 and 4.139 we can see the following two patterns:

- The lowest voltage increases, the highest voltage decreases, thus the unbalance becomes less. This is understandable if we realize that the negative-sequence voltage drops significantly.
- For longer sags all voltages drop. This is due to the drop in positive-sequence voltage.

4.8.3 Power Electronics Load

In systems with a large fraction of the load formed by single-phase or three-phase rectifiers, these can also influence the voltage during and after the voltage sag. Below some qualitative aspects of the effect of rectifiers on the voltage will be discussed briefly. Different aspects will dominate in different systems. The behavior of power electronics equipment during voltage sags is discussed in detail in Chapter 5.

- Especially for longer and deeper sags, a large part of the electronics load will trip. This will reduce the load current and thus increase the voltage, during as well as after the sag.
- Equipment that does not trip will initially take a smaller current from the supply or even no current at all because the dc bus voltage is larger than the peak of the ac voltage. Within a few cycles the dc bus capacitor has discharged sufficiently for the rectifier to start conducting again. Normally the total power taken by the load remains constant so that the ac current will be higher. This current has a high harmonic contents so that the harmonic voltage distortion during the sag will increase.
- Upon voltage recovery, the dc bus capacitors will take a large current pulse from the supply. This can postpone the voltage recovery by up to one cycle.
- For three-phase rectifiers, under unbalanced sags, the largest current flows between the two phases with the largest voltage difference. The effect is that the voltage in these phases drops and increases in the other phase. The three-phase rectifier thus reduces the unbalance between the phases. In this sense they behave similar to induction motor load. For unbalanced sags the current to three-phase rectifiers contains so-called non-characteristic harmonics, noticeably a third harmonic current, so that the voltage during the sag contains a third harmonic component higher than normal.
- Three-phase controlled rectifiers will experience a longer commutation period because the source voltage is lower during the sag. This leads to more severe commutation transients (notches) during the sag. Again this assumes that the equipment will not trip.

4.9 SAGS DUE TO STARTING OF INDUCTION MOTORS

In the previous sections of this chapter, we have discussed voltage sags due to short-circuit faults. These voltage sags are the main cause of equipment failure and malfunction, and one of the main reasons for power quality to become an issue during the last decade. Another important cause of voltage sags, one which has actually been of much more concern to designers of industrial power systems in the past, is the starting of large induction motors. Also the switching on of other loads will cause a voltage sag, just like the switching off of a capacitor bank. But in those latter cases the drop in voltage is rather small, and the voltage only drops but does not recover. Therefore the term "voltage magnitude step" would be more accurate.

During start-up an induction motor takes a larger current than normal, typically five to six times as large. This current remains high until the motor reaches its nominal speed, typically between several seconds and one minute. The drop in voltage depends strongly on the system parameters. Consider the system shown in Fig. 4.140, where Z_S is the source impedance and Z_M the motor impedance during run-up.

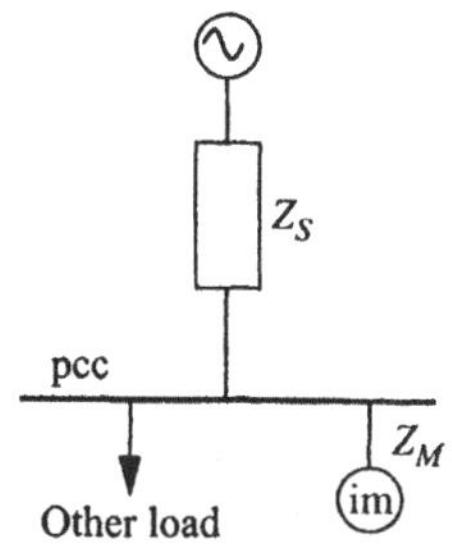

Figure 4.140 Equivalent circuit for voltage sag due to induction motor starting.

The voltage experienced by a load fed from the same bus as the motor is found from the voltage divider equation:

$$V_{sag} = \frac{Z_M}{Z_S + Z_M} \tag{4.144}$$

Like with most previous calculations, a source voltage of 1 pu has been assumed. When a motor of rated power S_{motor} is fed from a source with short-circuit power S_{source}, we can write for the source impedance:

$$Z_S = \frac{V_n^2}{S_{source}} \tag{4.145}$$

and for the motor impedance during starting

$$Z_M = \frac{V_n^2}{\beta S_{motor}} \tag{4.146}$$

with β the ratio between the starting current and the nominal current.

Equation (4.144) can now be written as

$$V_{sag} = \frac{S_{source}}{S_{source} + \beta S_{motor}} \tag{4.147}$$

Of course one needs to realize that this is only an approximation. The value can be used to estimate the sag due to induction motor starting, but for an accurate result one needs a power system analysis package. The latter will also enable the user to incorporate the effect of other motors during starting of the concerned motor. The drop in voltage at the other motor's terminals will slow them down and cause an additional increase in load current and thus an additional drop in voltage.

EXAMPLE Suppose that a 5 MVA motor is started from a 100 MVA, 11 kV supply. The starting current is six times the nominal current. This is a rather large motor for a supply of this strength, as we will see soon. The voltage at the motor terminals during motor starting can be estimated as

$$V_{sag} = \frac{100\,\text{MVA}}{100\,\text{MVA} + 6 \times 5\,\text{MVA}} = 77\% \tag{4.148}$$

In case the voltage during motor starting is too low for equipment connected to the same bus, one can decide to use a dedicated transformer. This leads to the network shown in Fig. 4.141.

Let again Z_S be the source impedance at the pcc, Z_M the motor impedance during run-up, and Z_T the transformer impedance. The magnitude of the voltage sag experienced by the sensitive load is

$$V_{sag} = \frac{Z_T + Z_M}{Z_S + Z_T + Z_M} \tag{4.149}$$

Introducing, like before, the short-circuit power of the source S_{source}, the rated power of the motor S_{motor} and assuming that the transformer has the same rated power of the motor and an impedance ϵ, we get from (4.149):

$$V_{sag} = \frac{(1 + 6\epsilon)S_{Source}}{(1 + 6\epsilon)S_{source} + 6S_{motor}} \tag{4.150}$$

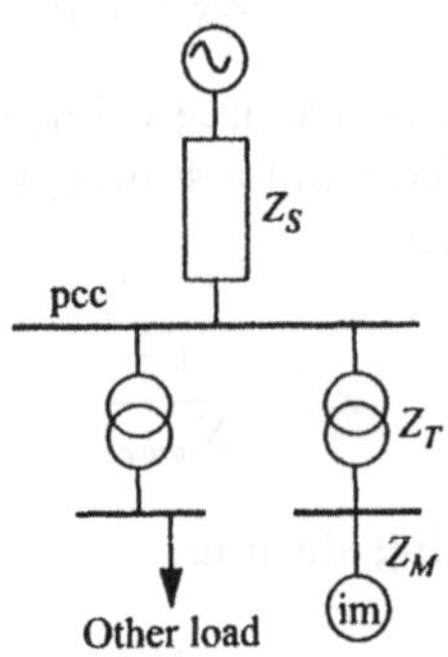

Figure 4.141 Induction motor starting with dedicated transformer for the sensitive load.

EXAMPLE Consider a dedicated supply for the motor in the previous example. The motor is fed through a 5 MVA, 5% 33/11 kV transformer from a 300 MVA, 33 kV supply. Note that the fault current at the 33 kV bus is identical to the fault current at the 11 kV in the previous example. That gives the following parameter values: $S_{source} = 300$ MVA, $S_{motor} = 5$ MVA, and $\epsilon = 0.05$, giving, from (4.150), a sag magnitude of 93%. Most loads will be able to withstand such a voltage reduction. Note that the reduction in sag magnitude is mainly due to the increased fault level at the pcc, not so much due to the transformer impedance. Neglecting the transformer impedance ($\epsilon = 0$ in (4.150)) gives $V_{sag} = 91\%$.

The duration of the voltage sag due to motor starting depends on a number of motor parameters, of which the motor inertia is the main one. When determining the run-up time, it is also important to determine the sag magnitude at the motor terminals.

The torque produced by the motor is proportional to the square of the terminal voltage. That makes that a sag down to 90% causes a drop in torque down to 81%. It is the difference between mechanical load torque and electrical torque which determines the acceleration of the motor, and thus the run-up time. Assume that the mechanical torque is half the electrical torque during most of the run-up if the terminal voltage is nominal. This assumption is based on the general design criterion that the pull-out torque of an induction motor is about twice the torque at nominal operation. When the voltage drops to 90% of nominal the electrical torque drops to 81% of nominal which is 162% of the mechanical torque. The accelerating torque, the difference between electrical and mechanical torque drops from 100% to 62%, a drop of 38%.

EXAMPLE Consider again the 5 MVA induction motor started from a 100 MVA 11 kV supply. The voltage at the motor terminals during run-up drops to 77% as we saw before. The electrical torque drops to 59% of nominal which is 118% of the mechanical torque. The accelerating torque thus drops from 100% to only 18%, and the run-up time will increase by a factor of 6.

A dedicated transformer alone cannot solve this problem, as the voltage at the motor terminals remains low. What is needed here is a stronger supply. To limit the voltage drop at the motor terminals to V_{min}, the source strength, from (4.147), needs to be

$$S_{source} = \frac{6S_{motor}}{1 - V_{min}} \tag{4.151}$$

A 5 MVA motor, with a minimum-permissable voltage of 85% during starting, needs a source strength of at least $\frac{6 \times 5\,\text{MVA}}{0.15} = 200\,\text{MVA}$. To keep the voltage above 90%, the source strength needs to be 300 MVA.

From these examples it will be clear that large voltage drops are not only a problem for sensitive load, but that they also lead to unacceptably long run-up times. The situation becomes even worse if more motors are connected to the same bus, as they will further pull down the voltage. Voltage drops due to induction motor starting are seldom deeper than 85%.

5

Voltage Sags—Equipment Behavior

In this chapter we will study the impact of voltage sags on electrical equipment. After the introduction of some general terminology, we will discuss three types of equipment which are perceived as most sensitive to voltage sags.

1. Computers, consumer electronics, and process-control equipment which will be modeled as a single-phase diode rectifier. Undervoltage at the dc bus is the main cause of tripping.
2. Adjustable-speed ac drives which are normally fed through a three-phase rectifier. Apart from the undervoltage at the dc bus, current unbalance, dc voltage ripple, and motor speed are discussed.
3. Adjustable-speed dc drives which are fed through a three-phase controlled rectifier. The firing-angle control will cause additional problems due to phase-angle jumps. Also the effect of the separate supply to the field winding is discussed.

This chapter closes with a brief discussion of other equipment sensitive to voltage sags: induction and synchronous motors, contactors, and lighting.

5.1 INTRODUCTION

5.1.1 Voltage Tolerance and Voltage-Tolerance Curves

Generally speaking electrical equipment operates best when the rms voltage is constant and equal to the nominal value. In case the voltage is zero for a certain period of time, it will simply stop operating completely. No piece of electrical equipment can operate indefinitely without electricity. Some equipment will stop within one second like most desktop computers. Other equipment can withstand a supply interruption much longer; like a lap-top computer which is designed to withstand (intentional) power interruptions. But even a lap-top computer's battery only contains enough energy for typically a few hours. For each piece of equipment it is possible to determine how long it will continue to operate after the supply becomes interrupted. A rather simple test would give the answer. The same test can be done for a voltage of 10% (of nominal), for a voltage of 20%, etc. If the voltage becomes high enough, the equipment will be able to operate on it indefinitely. Connecting the points obtained by performing these tests results in the so-called "voltage-tolerance curve." An example of a voltage-tolerance curve is shown in Fig. 5.1. In this case information is provided for the voltage tolerance of power stations connected to the Nordic transmission system [149]. The auxiliary supply should be able to tolerate a voltage drop down to 25% for 250 ms. It should be able to operate on a voltage of 95% of nominal. No requirements are given for voltages below 25% of nominal as these are very unlikely for the infeed to the auxiliary supply of a power station. One may claim that this is not a voltage-tolerance curve, but a requirement for the voltage tolerance. One could refer to this as a voltage-tolerance requirement and to the result of equipment tests as a voltage-tolerance performance. We will refer to both the measured curve, as well as to the requirement, as a voltage-tolerance curve. It will be clear from the context whether one refers to the voltage-tolerance requirement or the voltage-tolerance performance.

The concept of voltage-tolerance curve for sensitive electronic equipment was introduced in 1978 by Thomas Key [1]. When studying the reliability of the power supply to military installations, he realized that voltage sags and their resulting tripping of mainframe computers could be a greater threat to national security than complete interruptions of the supply. He therefore contacted some manufacturers for their design criteria and performed some tests himself. The resulting voltage-tolerance curve became known as the "CBEMA curve" several years later. We will come back to the CBEMA curve when discussing computing equipment further on. Note that curves plotting minimum voltage against maximum duration have been used for synchronous machines for many years already, but not for electronic equipment. We will come back to the voltage tolerance of synchronous machines in Section 5.5.

The voltage-tolerance curve is also an important part of IEEE standard 1346 [22]. This standard recommends a method of comparing equipment performance with the supply power quality. The voltage-tolerance curve is the recommended way of presenting the equipment performance. The concept of "voltage sag coordination chart" [20], which is at the heart of IEEE standard 1346, will be presented in detail in Section 6.2.

While describing equipment behavior through the voltage-tolerance curve, a number of assumptions are made. The basic assumption is that a sag can be uniquely characterized through its magnitude and duration. We already saw in the previous

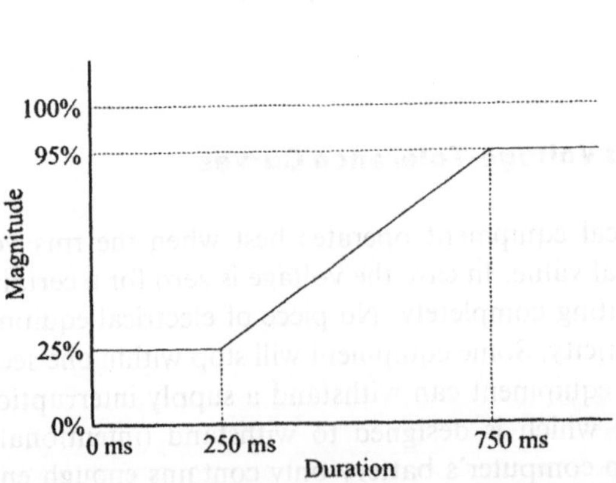

Figure 5.1 Voltage-tolerance requirement for power stations. (Data obtained from [149].)

chapter that this is only an approximation. From an equipment point of view the basic assumption behind the voltage-tolerance curve is: *if two sags have the same magnitude and duration then they will both lead to tripping of the equipment or both not lead to tripping of the equipment*. As we have seen in the previous chapter, the definitions of magnitude and duration of a sag currently in use are far from unique. Further, phase-angle jumps and three-phase voltage unbalance can significantly influence the behavior of equipment. The two-dimensional voltage-tolerance curve clearly has its limitations, especially for three-phase equipment. We will present some extensions to the concept in the next chapter.

An overview of the voltage tolerance of currently available equipment is presented in Table 5.1. The range in voltage tolerance is partly due to the difference between equipment, partly due to the uncertainties mentioned before. With these data, as well as with the voltage-tolerance data presented in the rest of this chapter, one should realize that the values not necessarily apply to a specific piece of equipment. As an example, Table 5.1 gives for motor starters a voltage tolerance between 20 ms, 60% and 80 ms, 40%. Using this range to design an installation could be rather unreliable; using the average value even more. These values are only meant to give the reader an impression of the sensitivity of equipment to voltage sags, not to serve as a database for those designing installations. For the time being it is still necessary to determine the voltage tolerance of each critical part of an installation or to subject the whole installation to a test. In future, voltage-tolerance requirements might make the job easier.

The values in Table 5.1 should be read as follows. A voltage tolerance of a ms, b% implies that the equipment can tolerate a zero voltage of a ms and a voltage of b% of nominal indefinitely. Any sag longer than a ms and deeper than b% will lead to tripping or malfunction of the equipment. In other words: the equipment voltage-tolerance curve is rectangular with a "knee" at a ms, b%.

TABLE 5.1 Voltage-Tolerance Ranges of Various Equipment Presently in Use

	Voltage Tolerance		
Equipment	Upper Range	Average	Lower Range
PLC	20 ms, 75%	260 ms, 60%	620 ms, 45%
PLC input card	20 ms, 80%	40 ms, 55%	40 ms, 30%
5 h.p. ac drive	30 ms, 80%	50 ms, 75%	80 ms, 60%
ac control relay	10 ms, 75%	20 ms, 65%	30 ms, 60%
Motor starter	20 ms, 60%	50 ms, 50%	80 ms, 40%
Personal computer	30 ms, 80%	50 ms, 60%	70 ms, 50%

Source: As given data obtained from IEEE Std.1346 [22]. This data should not be used as a basis for design of installations.

5.1.2 Voltage-Tolerance Tests

The only standard that currently describes how to obtain voltage tolerance of equipment is IEC 61000-4-11 [25]. This standard, however, does not mention the term voltage-tolerance curve. Instead it defines a number of preferred magnitudes and durations of sags for which the equipment has to be tested. (Note: The standard uses the term "test levels," which refers to the remaining voltage during the sag.) The equipment does not need to be tested for all these values, but one or more of the magnitudes and

TABLE 5.2 Preferred Magnitudes and Duration for Equipment Immunity Testing According to IEC-61000-4-11 [25]

	Duration in Cycles of 50 Hz					
Magnitude	0.5	1	5	10	25	50
70%						
40%						
0%						

durations may be chosen. The preferred combinations of magnitude and duration are the (empty) elements of the matrix shown in Table 5.2.

The standard in its current form does not set any voltage-tolerance requirements. It only defines the way in which the voltage tolerance of equipment shall be obtained. An informative appendix to the standard mentions two examples of test setups:

- Use a transformer with two output voltages. Make one output voltage equal to 100% and the other to the required during-sag magnitude value. Switch very fast between the two outputs, e.g., by using thyristor switches.
- Generate the sag by using a waveform generator in cascade with a power amplifier.

The IEEE standard 1346 [22] refers to IEC 61000-4-11 for obtaining the equipment voltage tolerance, and specifically mentions the switching between two supply voltages as a way of generating sags. Both methods are only aimed at testing one piece of equipment at a time. To make a whole installation experience a certain voltage sag, each piece needs to be tested hoping that their interconnection does not cause any unexpected deterioration in performance. A method for testing a whole installation is presented in [56]. A three-phase diesel generator is used to power the installation under test. A voltage sag is made by reducing the field voltage. It takes about two cycles for the ac voltage to settle down after a sudden change in field voltage, so that this method can only be used for sags of five cycles and longer.

5.2 COMPUTERS AND CONSUMER ELECTRONICS

The power supply of a computer, and of most consumer electronics equipment normally consists of a diode rectifier along with an electronic voltage regulator (dc/dc converter). The power supply of all these low-power electronic devices is similar and so is their sensitivity to voltage sags. What is different are the consequences of a sag-induced trip. A television will show a black screen for up to a few seconds; a compact disc player will reset itself and start from the beginning of the disc, or just wait for a new command. Televisions and video recorders normally have a small battery to maintain power to the memory containing the channel settings. This is to prevent loss of memory when the television is moved or unplugged for some reason. If this battery no longer contains enough energy, a sag or interruption could lead to the loss of these settings. The same could happen to the settings of a microwave oven, which is often not equipped with a battery.

The process-control computer of a chemical plant is rather similar in power supply to any desktop computer. Thus, they will both trip on voltage sags and interruptions, within one second. But the desktop computer's trip might lead to the loss of 1 hour of work (typically less), where the process-control computer's trip easily leads to a restarting procedure of 48 hours plus sometimes a very dangerous situation. It is clear that the first is merely an inconvenience, whereas the latter should be avoided at any cost.

5.2.1 Typical Configuration of Power Supply

A simplified configuration of the power supply to a computer is shown in Fig. 5.2. The capacitor connected to the non-regulated dc bus reduces the voltage ripple at the input of the voltage regulator. The voltage regulator converts the non-regulated dc voltage of a few hundred volts into a regulated dc voltage of the order of 10 V. If the ac voltage drops, the voltage on the dc side of the rectifier (the non-regulated dc voltage) drops. The voltage regulator is able to keep its output voltage constant over a certain range of input voltage. If the voltage at the dc bus becomes too low the regulated dc voltage will also start to drop and ultimately errors will occur in the digital electronics. Some computers detect an undervoltage at the input of the controller and give a signal for a "controlled" shutdown of the computer, e.g., by parking the hard drive. Those computers will trip earlier but in a more controlled way.

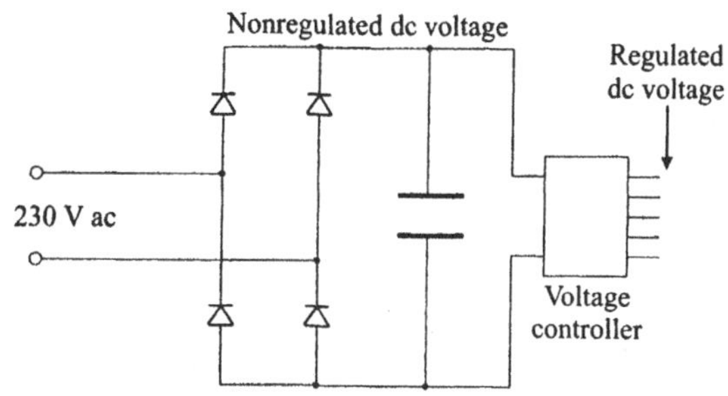

Figure 5.2 Computer power supply.

5.2.2 Estimation of Computer Voltage Tolerance

5.2.2.1 DC Bus Voltages. As shown in Fig. 5.2, a single-phase rectifier consists of four diodes and a capacitor. Twice every cycle the capacitor is charged to the amplitude of the supply voltage. In between the charging pulses the capacitor discharges via the load. The diodes only conduct when the supply voltage exceeds the dc voltage. When the supply voltage drops the diodes no longer conduct and the capacitor continues to discharge until the dc voltage reaches the reduced supply voltage again. In normal operation the capacitor is charged during two small periods each cycle, and discharges during the rest of the cycle. In steady state, the amount of charging and discharging of the capacitor are equal.

To study the effect of voltage sags on the voltage at the (non-regulated) dc bus, the power supply has been modeled as follows:

- The diodes conduct when the absolute value of the supply voltage is larger than the dc bus voltage. While the diodes conduct, the dc bus voltage is equal to the supply voltage.
- The supply voltage is a 1 pu sinewave before the event and a constant-amplitude sinewave during the event but with an amplitude less than 1 pu. The

voltage only shows a drop in magnitude, no phase-angle jump. The supply voltage is not affected by the load current.

- While the diodes do not conduct, the capacitor is discharged by the voltage regulator. The power taken by the voltage regulator is constant and independent of the dc bus voltage.

This model has been used to calculate the dc bus voltages before, during, and after a voltage sag with a magnitude of 50% (without phase-angle jump). The result is shown in Fig. 5.3. As a reference, the absolute value of the ac voltage has been plotted as a dashed line.

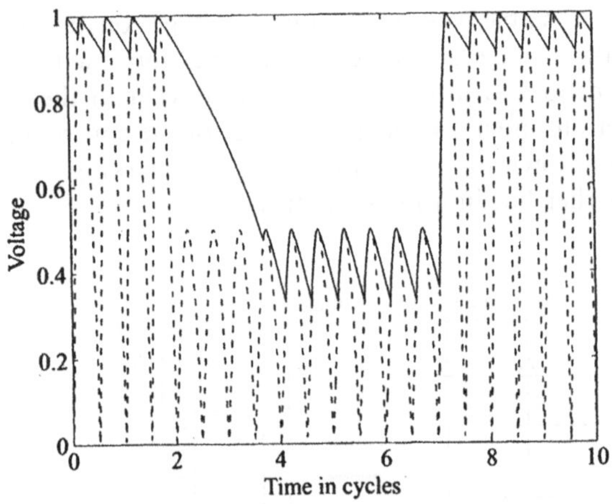

Figure 5.3 Effect of a voltage sag on dc bus voltage for a single-phase rectifier: absolute value of the ac voltage (dashed line) and dc bus voltage (solid line).

Due to the voltage drop, the maximum ac voltage becomes less than the dc voltage. The resulting discharging of the capacitor continues until the capacitor voltage drops below the maximum of the ac voltage. After that, a new equilibrium will be reached. Because a constant power load has been assumed the capacitor discharges faster when the dc bus voltage is lower. This explains the larger dc voltage ripple during the sag.

It is important to realize that the discharging of the capacitor is only determined by the load connected to the dc bus, not by the ac voltage. Thus all sags will cause the same initial decay in dc voltage. But the duration of the decay is determined by the magnitude of the sag. The deeper the sag the longer it takes before the capacitor has discharged enough to enable charging from the supply. In Fig. 5.4 the sags in ac and dc voltage are plotted for voltage sags of different magnitude. The top curves have been calculated for a sag in ac voltage down to 50%, the bottom ones for a sag in ac voltage down to 70%. The dotted lines give the rms voltage at ac side (the sag in ac voltage). We see that the initial decay in dc bus voltage is the same for both sags.

5.2.2.2 Decay of the DC Bus Voltage. Within a certain range of the input voltage, the voltage regulator will keep its output voltage constant, independent of the input voltage. Thus, the output power of the voltage regulator is independent of the input voltage. If we assume the regulator to be lossless the input power is independent of the dc voltage. Thus, the load connected to the dc bus can be considered as a constant power load.

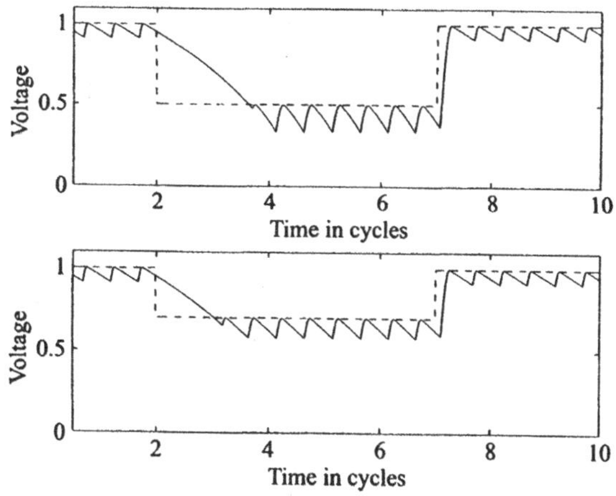

Figure 5.4 Voltage sag at ac side (dashed line) and at the dc bus (solid line) for a sag down to 50% (top) and for a sag down to 70% (bottom).

As long as the absolute value of the ac voltage is less than the dc bus voltage, all electrical energy for the load comes from the energy stored in the capacitor. Assume that the capacitor has capacitance C. The energy a time t after sag initiation is $\frac{1}{2}C\{V(t)\}^2$, with $V(t)$ the dc bus voltage. This energy is equal to the energy at sag initiation minus the energy consumed by the load:

$$\frac{1}{2}CV^2 = \frac{1}{2}CV_0^2 - Pt \tag{5.1}$$

where V_0 is the dc bus voltage at sag initiation and P the loading of the dc bus. Expression (5.1) holds as long as the dc bus voltage is higher than the absolute value of the ac voltage, thus during the initial decay period in Figs. 5.3 and 5.4. Solving (5.1) gives an expression for the voltage during this initial decay period:

$$V = \sqrt{V_0^2 - \frac{2P}{C}t} \tag{5.2}$$

During normal operation, before the sag, the variation in dc bus voltage is small, so that we can linearize (5.2) around $V = V_0$, resulting in

$$V = V_0\sqrt{1 - \frac{2P}{V_0^2C}t} \approx V_0\left(1 - \frac{P}{V_0^2C}t\right) \tag{5.3}$$

where t is the time elapsed since the last recharge of the capacitor. The voltage ripple is defined as the difference between the maximum and the minimum value of the dc bus voltage. The maximum is reached for $t = 0$, the minimum for $t = \frac{T}{2}$, with T one cycle of the fundamental frequency. The resulting expression for the voltage ripple is

$$\epsilon = \frac{PT}{2V_0^2C} \tag{5.4}$$

The voltage ripple is often used as a design criterion for single-phase diode rectifiers.

Inserting the expression for the dc voltage ripple (5.4) in (5.2) gives an expression for the dc voltage during the discharge period, thus during the initial cycles of a voltage sag:

$$V(t) = V_0\sqrt{1 - 4\epsilon\frac{t}{T}} \tag{5.5}$$

where $\frac{t}{T}$ is the number of cycles elapsed since sag initiation. The larger the dc voltage ripple in normal operation, the faster the dc voltage drops during a sag.

5.2.2.3 Voltage Tolerance. Tripping of a computer during a voltage sag is attributed to the dc bus voltage dropping below the minimum input voltage for which the voltage controller can operate correctly. We will refer to this voltage as V_{min}. We will further assume that in normal operation, before the sag, both ac and dc bus voltage are equal to 1 pu.

A sag with a magnitude V will result in a new steady-state dc voltage which is also equal to V, if we neglect the dc voltage ripple. From this we can conclude that the computer will not trip for $V > V_{min}$. For $V < V_{min}$, the dc bus voltage only drops below V_{min} if the sag duration exceeds a certain value t_{max}. The time t_{max} it takes for the voltage to reach a level V_{min} can be found by solving t from (5.5) with $V_0 = 1$:

$$t_{max} = \frac{1 - V_{min}^2}{4\epsilon}T \tag{5.6}$$

When the minimum dc bus voltage is known, (5.6) can be used to calculate how long it will take before tripping. Or in other words: what is the maximum sag duration that the equipment can tolerate. The dc bus voltage at which the equipment actually trips depends on the design of the voltage controller: varying between 50% and 90% dc voltage, sometimes with additional time delay. Table 5.3 gives some values of voltage tolerance, calculated by using (5.6).

Thus, if a computer trips at 50% dc bus voltage, and as the normal operation dc voltage ripple is 5%, a sag of less than four cycles in duration will not cause a maltrip. Any sag below 50% for more than four cycles will trip the computer. A voltage above 50% can be withstood permanently by this computer. This results in what is called a "rectangular voltage-tolerance curve," as shown in Fig. 5.5. Each voltage regulator will have a non-zero minimum operating voltage. The row for zero minimum dc bus voltage is only inserted as a reference. We can see from Table 5.3 that the performance does not improve much by reducing the minimum operating voltage of the voltage controller beyond 50%. When the dc voltage has dropped to 50%, the capacitor has already lost 75% of its energy.

TABLE 5.3 Voltage Tolerance of Computers and Consumer Electronics Equipment: Maximum-Allowable Duration of a Voltage Sag for a Given Minimum Value of the DC Bus Voltage, for Two Values of the DC Voltage Ripple

	Maximum Sag Duration	
Minimum dc Bus Voltage	5% ripple	1% ripple
0	5 cycles	25 cycles
50%	4 cycles	19 cycles
70%	2.5 cycles	13 cycles
90%	1 cycle	5 cycles

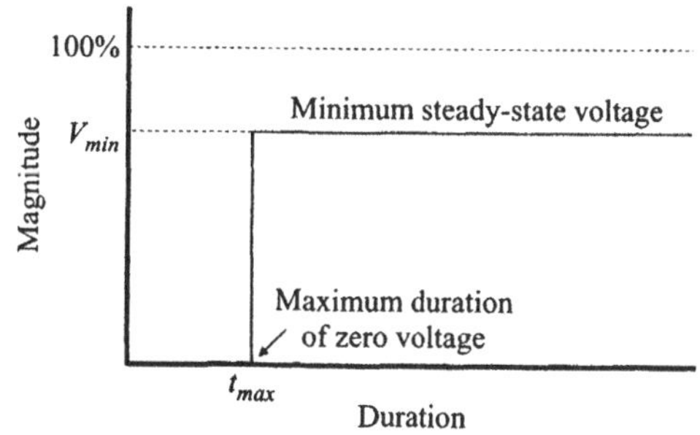

Figure 5.5 Voltage-tolerance curve of a computer: an example of a rectangular voltage-tolerance curve.

5.2.3 Measurements of PC Voltage Tolerance

The voltage tolerance of personal computers has been measured by a number of authors [28], [29], [41], [49], [50]. The voltage-tolerance curves they present are in the same range as found from the simplified model presented in the previous section. Figure 5.6 shows measured voltages and currents for a personal computer. The applied voltage sag was one of the most severe the computer could tolerate.

In Fig. 5.6 we see the dc bus voltage starting to drop the moment the ac voltage drops. During the decay in dc bus voltage, the input current to the rectifier is very small. The output of the voltage controller remains constant at first. But when the dc bus voltage has dropped below a certain value, the dc voltage regulator no longer operates properly and its output also starts to drop. In this case a new steady state is reached where the regulated dc voltage is apparently still sufficient for the digital electronics to operate correctly. During the new steady state, the input current is no longer zero. Upon ac voltage recovery, the dc bus voltage also recovers quickly. This is associated

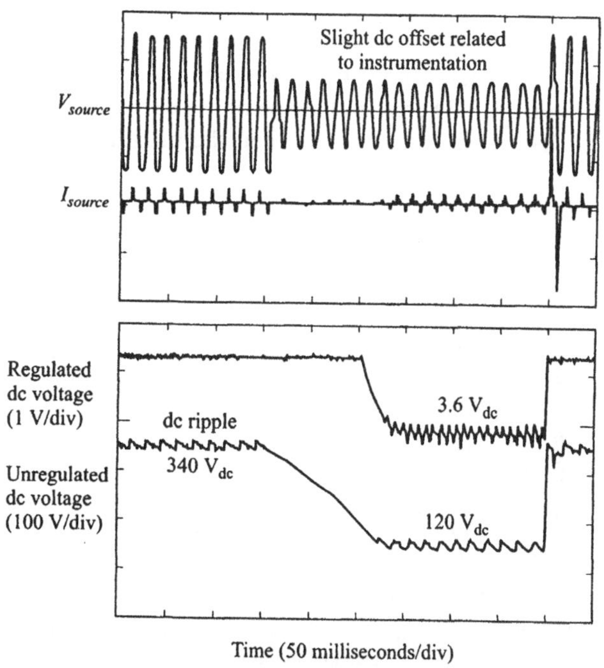

Figure 5.6 Regulated and non-regulated dc voltages for a personal computer, during a 200 ms sag down to 50%: (top-to-bottom) ac voltages; ac current; regulated dc voltage; non-regulated dc voltage. (Reproduced from EPRI Power Quality Database [28].)

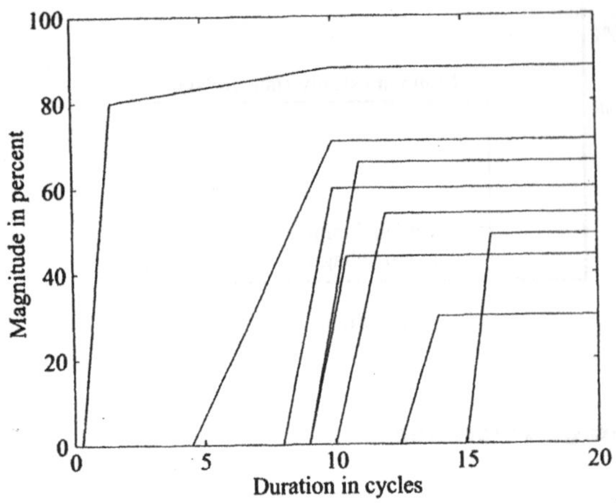

Figure 5.7 Voltage-tolerance curves for personal computers. (Data obtained from EPRI Power Quality Database [29].)

with a very large current peak charging the dc bus capacitor. This current could cause an equipment trip or even a long interruption if fast-acting overcurrent protection devices are used.

The voltage-tolerance curves obtained from various tests are shown in Fig. 5.7 and Fig. 5.8. Figure 5.7 shows the result of a U.S. study [29]. For each personal computer, the tolerance for zero voltage was determined, as well as the lowest steady-state voltage for which the computer would operate indefinitely. For one computer the tolerance for 80% voltage was determined; all other computers could tolerate this voltage indefinitely. We see that there is a large range in voltage tolerance for different computers. The age or the price of the computer did not have any influence. The experiments were repeated for various operating states of the computer: idle; calculating; reading; or writing. It turned out that the operating state did not have any significant influence on the voltage tolerance or on the power consumption. Figure 5.7 confirms that the voltage-tolerance curve has an almost rectangular shape.

Figure 5.8 shows voltage-tolerance curves for personal computers obtained from a Japanese study [49], in the same format and scale as the American measurements in Fig. 5.7. The general shape of the curves is identical, but the curves in Fig. 5.7 indicate less sensitive computers than the ones in Fig. 5.8.

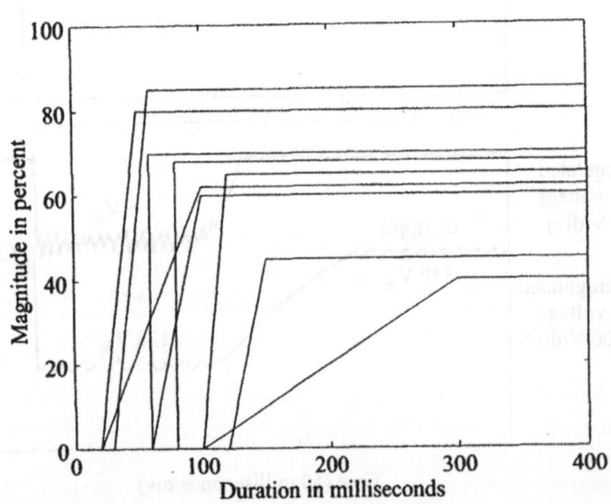

Figure 5.8 Voltage-tolerance curves for personal computers—Japanese tests. (Data obtained from [49].)

Summarizing we can say that the voltage tolerance of personal computers varies over a rather wide range: 30–170 ms, 50–70% being the range containing half of the models. The extreme values found are 8 ms, 88% and 210 ms, 30%.

5.2.4 Voltage-Tolerance Requirements: CBEMA and ITIC

As mentioned before, the first modern voltage-tolerance curve was introduced for mainframe computers [1]. This curve is shown as a solid line in Fig. 5.9. We see that its shape does not correspond with the shape of the curves shown in Figs. 5.5, 5.7, and 5.8. This can be understood if one realizes that these figures give the voltage-tolerance performance for one piece of equipment at a time, whereas Fig. 5.9 is a voltage-tolerance requirement for a whole range of equipment. The requirement for the voltage-tolerance curves of equipment is that they should all be above the voltage-tolerance requirement in Fig. 5.9. The curve shown in Fig. 5.9 became well-known when the Computer Business Equipment Manufacturers Association (CBEMA) started to use the curve as a recommendation for its members. The curve was subsequently taken up in an IEEE standard [26] and became a kind of reference for equipment voltage tolerance as well as for severity of voltage sags. A number of software packages for analyzing power quality data plot magnitude and duration of the sags against the CBEMA curve. The CBEMA curve also contains a voltage-tolerance part for overvoltages, which is not reproduced in Fig. 5.9. Recently a "revised CBEMA curve" has been adopted by the Information Technology Industry Council (ITIC), which is the successor of CBEMA. The new curve is therefore referred to as the ITIC curve; it is shown as a dashed line in Fig. 5.9.

The ITIC curve gives somewhat stronger requirements than the CBEMA curve. This is because power quality monitoring has shown that there are an alarming number of sags just below the CBEMA curve [54].

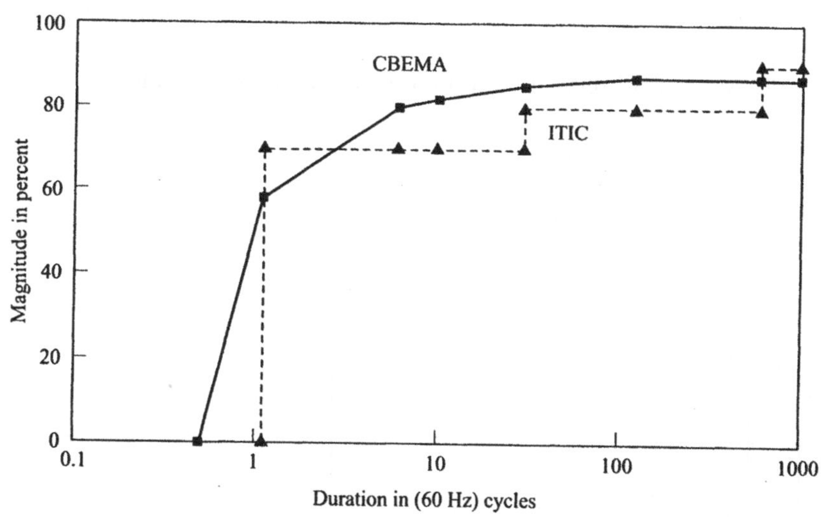

Figure 5.9 Voltage-tolerance requirements for computing equipment: CBEMA curve (solid line) and ITIC curve (dashed line).

5.2.5 Process Control Equipment

Process control equipment is often extremely sensitive to voltage sags; equipment has been reported to trip when the voltage drops below 80% for a few cycles [31], [37], [39], [41]. The consequences of the tripping of process control equipment can be enormous. For example, the tripping of a small relay can cause the shutdown of a large chemical plant, leading to perhaps $100,000 in lost production. Fortunately all this is low-power equipment which can be fed from a UPS, or for which the voltage tolerance can be improved easily by adding extra capacitors, or some backup battery.

Tests of the voltage tolerance of programmable logic controllers (PLC's) have been performed in the same way as the PC tests described before [39]. The resulting voltage-tolerance curves for some controllers are shown in Fig. 5.10. It clearly shows that this equipment is extremely sensitive to voltage sags. As most sags are between 4 and 10 cycles in duration, we can reasonably assume that a PLC trips for each sag below a given threshold, varying between 85% and 35%.

Even more worrying is that some controllers may send out incorrect control signals before actually tripping. This has to do with the different voltage tolerance of the various parts of the controller. The incorrect signals could lead to dangerous process malfunctions.

Additional voltage-tolerance curves for process control equipment, obtained from another study [41], are shown in Fig. 5.11. The numbers with the curves refer to the following devices:

1. Fairly common process controller used for process heating applications such as controlling water temperature.
2. More complicated process controller which can be used to provide many control strategies such as pressure/temperature compensation of flow.
3. Process logic controller.
4. Process logic controller, newer and more advanced version of 3.
5. AC control relay, used to power important equipment.
6. AC control relay, used to power important equipment; same manufacturer as 5.
7. AC control relay used to power motors; motor contactor.

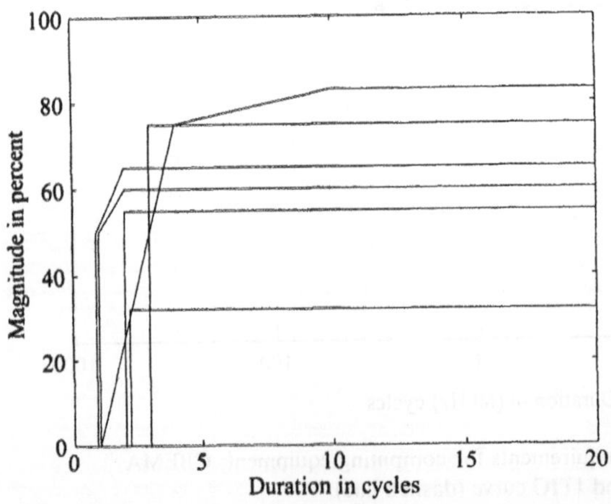

Figure 5.10 Voltage-tolerance curves for programmable logic controllers (PLCs). (Data obtained from [39].)

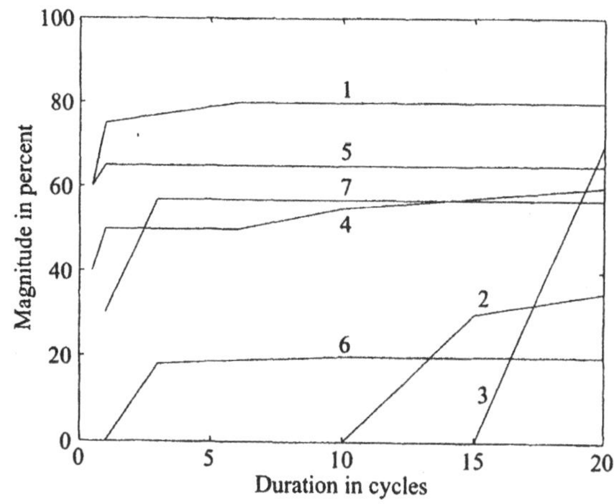

Figure 5.11 Voltage-tolerance curves for various process control equipment [41].

This study confirms that process control equipment is extremely sensitive to voltage disturbances, but also that it is possible to build equipment capable of tolerating long and deep sags. The fact that some equipment already trips for half-a-cycle sags suggests a serious sensitivity to voltage transients as well. The main steps taken to prevent tripping of process control equipment is to power all essential process control equipment via a UPS or to ensure in another way that the equipment can withstand at least short and shallow sags. Devices 2 and 3 in Fig. 5.11 show that it is possible to make process control equipment resilient to voltage sags. But even here the costs of installing a UPS will in almost all cases be justified.

Here are some other interesting observations from Fig. 5.11:

- Device 2 is the more complicated version of device 1. Despite the higher complexity, device 2 is clearly less sensitive to voltage sags than device 1.
- Device 4 is a newer and more advanced version of device 3. Note the enormous deterioration in voltage tolerance.
- Devices 5 and 6 come from the same manufacturer, but show completely different voltage tolerances.

5.3 ADJUSTABLE-SPEED AC DRIVES

Many adjustable-speed drives are equally sensitive to voltage sags as process control equipment discussed in the previous section. Tripping of adjustable-speed drives can occur due to several phenomena:

- The drive controller or protection will detect the sudden change in operating conditions and trip the drive to prevent damage to the power electronic components.
- The drop in dc bus voltage which results from the sag will cause maloperation or tripping of the drive controller or of the PWM inverter.
- The increased ac currents during the sag or the post-sag overcurrents charging the dc capacitor will cause an overcurrent trip or blowing of fuses protecting the power electronics components.

- The process driven by the motor will not be able to tolerate the drop in speed or the torque variations due to the sag.

After a trip some drives restart immediately when the voltage comes back; some restart after a certain delay time and others only after a manual restart. The various automatic restart options are only relevant when the process tolerates a certain level of speed and torque variations. In the rest of this section we will first look at the results of equipment testing. This will give an impression of the voltage tolerance of drives. The effect of the voltage sag on the dc bus voltage, the main cause of equipment tripping, will be discussed next. Requirements for the size of the dc bus capacitor will be formulated. The effect of the voltage sag on the ac current and on the motor terminal voltage will also be discussed, as well as some aspects of automatic restart. Finally, a short overview of mitigation methods will be given.

5.3.1 Operation of AC Drives

Adjustable-speed drives (ASD's) are fed either through a three-phase diode rectifier, or through a three-phase controlled rectifier. Generally speaking, the first type is found in ac motor drives, the second in dc drives and in large ac drives. We will discuss small and medium size ac drives fed through a three-phase diode rectifier in this section, and dc drives fed through controlled rectifiers in the next section.

The configuration of most ac drives is as shown in Fig. 5.12. The three ac voltages are fed to a three-phase diode rectifier. The output voltage of the rectifier is smoothened by means of a capacitor connected to the dc bus. The inductance present in some drives aims at smoothening the dc link current and so reducing the harmonic distortion in the current taken from the supply.

The dc voltage is inverted to an ac voltage of variable frequency and magnitude, by means of a so-called voltage-source converter (VSC). The most commonly used method for this is pulse-width modulation (PWM). Pulse-width modulation will be discussed briefly when we describe the effect of voltage sags on the motor terminal voltages.

The motor speed is controlled through the magnitude and frequency of the output voltage of the VSC. For ac motors, the rotational speed is mainly determined by the frequency of the stator voltages. Thus, by changing the frequency an easy method of speed control is obtained. The frequency and magnitude of the stator voltage are plotted in Fig. 5.13 as a function of the rotor speed. For speeds up to the nominal speed, both frequency and magnitude are proportional to the rotational speed. The

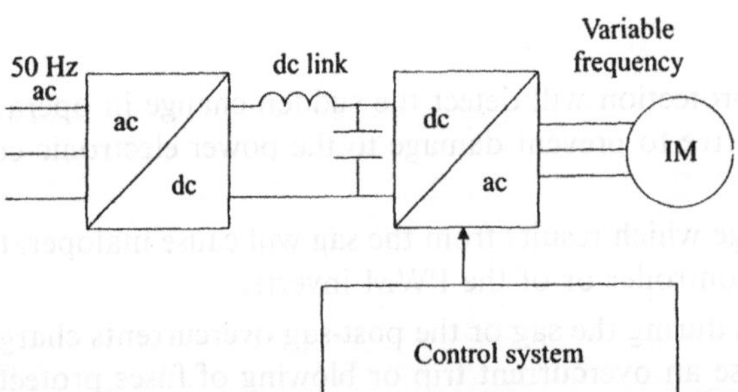

Figure 5.12 Typical ac drive configuration.

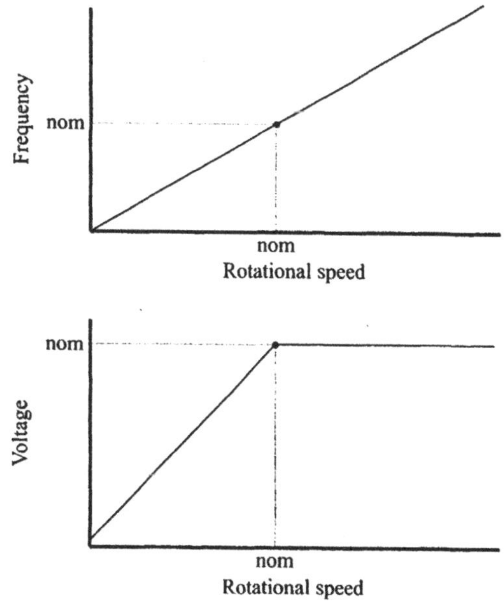

Figure 5.13 Voltage and frequency as a function of speed for an ac adjustable-speed drive.

maximum torque of an induction motor is proportional to the square of the voltage magnitude and inversely proportional to the square of the frequency [53], [206]:

$$T_{max} \approx \frac{V^2}{f^2} \tag{5.7}$$

By increasing both voltage magnitude and frequency, the maximum torque remains constant. It is not possible to increase the voltage magnitude above its nominal value. Further increase in speed will lead to a fast drop in maximum torque.

5.3.2 Results of Drive Testing

The performance of a number of adjustable-speed drives in relation to voltage sag monitoring in an industrial plant is presented in Fig. 5.14 [40]: the circles indicate magnitude and duration of voltage sags for which the drives trip; for the voltage sags indicated by the crosses, the drives did not trip. We see that the drives used in this plant were very sensitive to sags. The voltage tolerance of these drives is 80% of voltage for less than six cycles. The exact duration for which the drives tripped could not be determined as the resolution of the monitors was only six cycles. Similar high sensitivities of adjustable-speed drives to voltage sags have been reported in other studies [2], [35], [42], [48]. Using these data as typical for adjustable-speed drives carries a certain risk. If the drives had not been sensitive to sags, the study would never have been performed. This warning holds for many publications that mention a high sensitivity of equipment to sags. It would thus be very well possible that a large fraction of the adjustable-speed drives are not sensitive to sags at all. To determine the performance of typical drives, one needs to apply tests to randomly selected drives.

Studies after the voltage tolerance of adjustable-speed drives, selected at random are presented in [32], [47]. In one of the studies [47] tests were performed for 20 h.p. and 3 h.p. drives, from several different manufacturers. Each manufacturer provided a 20 h.p. and a 3 h.p. drive. Each drive was tested for the following three voltage magnitude events:

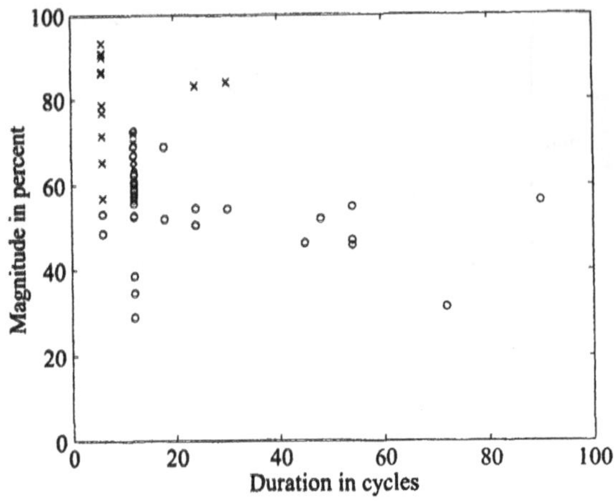

Figure 5.14 Voltage sags which led to drive tripping (o) and voltage sags which did not lead to drive tripping (x). (Data obtained from Sarmiento [40].)

- zero voltage for 33 ms.
- 50% voltage for 100 ms.
- 70% voltage for 1 sec.

The drive performance during the event was classified based on the three types of speed curves shown in Fig. 5.15:

- I: The speed of the motor shows a decrease followed by a recovery.
- II: The speed of the motor reduces to zero after which the drive restarts automatically and accelerates the motor load back to nominal speed.
- III: The motor speed becomes zero, and the drive is unable to restart the motor.

The test results are summarized in Tables 5.4 and 5.5. Each of the columns in the tables gives the number of drives with the indicated performance. For a 50%, 100 ms sag, four of the 20 h.p. drives showed a performance according to curve II in Fig. 5.15 and seven of the drives according to curve III. Table 5.4 gives the results for drives at full load; a distinction is made between 3 h.p. and 20 h.p. drives. Table 5.5 compares the drive behavior at full load with the drive behavior at half-load. These results include 20 h.p. as well as 3 h.p. drives.

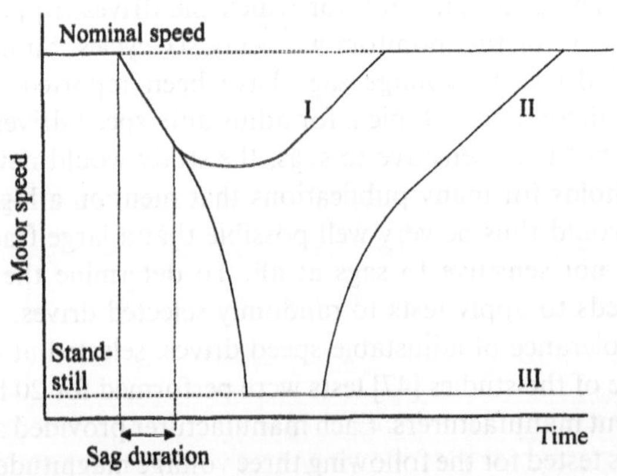

Figure 5.15 Three types of motor speed behavior for an adjustable-speed drive due to a sag.

TABLE 5.4 Results of Voltage-Tolerance Testing of Adjustable-Speed Drives: Number of Drives with the Indicated Performance. I: Only Drop in Speed; II: Automatic Restart; III: Manual Restart

Applied Sag	Drive Performance					
	20 h.p. drives			3 h.p. drives		
	I	II	III	I	II	III
0% 33 ms	4	2	5	12	–	–
50% 100 ms	–	4	7	3	5	4
70% 1000 ms	–	5	6	1	7	4

Source: Data obtained from [47].

TABLE 5.5 Influence of Loading on Drive Voltage Tolerance: Number of Drives with the Indicated Performance. I: Only Drop in Speed; II: Automatic Restart; III: Manual Restart

Applied Sag	Drive Performance					
	Full Load			Half-Load		
	I	II	III	I	II	III
0% 33 ms	7	1	2	8	1	1
50% 100 ms	2	4	4	3	4	3
70% 1000 ms	1	5	4	1	4	5

Source: Data obtained from [47].

From the results in Tables 5.4 and 5.5 one can draw the following conclusions:

- 3 h.p. drives are less sensitive than 20 h.p. drives. This does not necessarily hold in all cases, although a comparison of 3 h.p. versus 20 h.p. drives for the same manufacturer, the same voltage sag, and the same drive loading gives in 25 of the cases a better performance for the 3 h.p. drive; in 20 cases the performance is the same (i.e., in the same class according to the classification above); and only in three cases does the 20 h.p. drive perform better.
- There is no significant difference between the full load and the half-load voltage tolerance. For some loads the performance improves, for others it deteriorates, but for most it does not appear to have any influence. Doing the same comparison as before shows that in two cases performance is better at full load, in four cases it is better at half-load, and in 24 cases the performance falls in the same performance class. For drives falling in performance class I it may be that at full load the drop in speed is more severe than at half-load, but the study did not report this amount of detail.
- Very short interruptions (0%, 33 ms) can be handled by all 3 h.p. drives and by a large part of the 20 h.p. drives.
- Adjustable-speed drives have severe difficulties with sags of 100 ms and longer, especially as one considers that even response I could mean a serious disruption of sensitive mechanical processes.

- The tests confirm that adjustable-speed drives are very sensitive to sags; however, the extreme sensitivity (85%, 8 ms) mentioned by some is not found in this test.

The results of a similar set of tests are reported in [32]: two different voltage sags were applied to 17 drives:

- voltage down to 50% of nominal for 100 ms (6 cycles);
- voltage down to 70% of nominal for 167 ms (10 cycles).

Their results are shown in Table 5.6. The classification used is fairly similar to the one used in Tables 5.4 and 5.5, with the exception that a class "drive kept motor speed constant" is included. This drive performance is indicated as class 0 in Table 5.6. Response classes I, II, and III correspond to the ones used before.

From these studies, it is possible to obtain a kind of "average voltage-tolerance curve" for adjustable-speed drives. The resulting curve is shown in Fig. 5.16, with the measurement points indicated as circles. Tolerance is defined here as performance 0 or I. Note that the actual drives show a large spread in voltage tolerance: some drives could not tolerate any of the applied sags, where one of the drives tolerated all sags. It has further been assumed that the drives could operate indefinitely on 85% voltage.

Conrad et al. [48] obtained voltage tolerance data for adjustable-speed drives through a survey of drive manufacturers. The voltage tolerance stated by the manufacturers is shown in Fig. 5.17. The circles indicate manufacturers which gave minimum voltage as well as maximum sag duration. The other manufacturers, indicated by triangles in Fig. 5.17, only gave a value for the maximum sag duration. Note that 10 out of 13 manufacturers indicate that their drives trip for sags of three cycles or less in duration.

TABLE 5.6 Results of Voltage-Tolerance Tests on Adjustable-Speed Drives

	Response of the Drive			
Sag Applied	0	I	II	III
50% 100 ms	2	1	9	5
70% 170 ms	11	1	5	–

Source: Data obtained from [32].

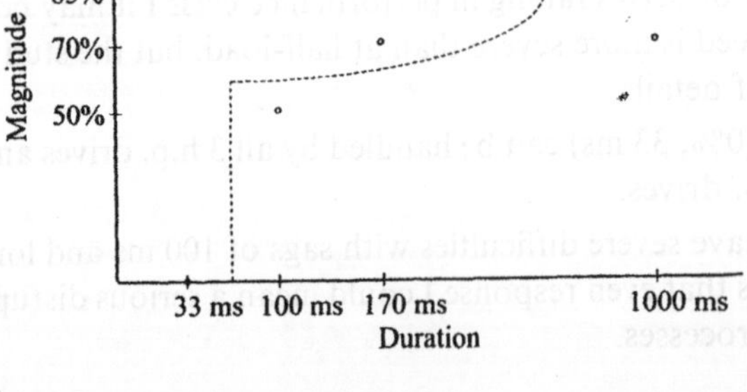

Figure 5.16 Average voltage-tolerance curve for adjustable-speed drives. Note the nonlinear horizontal scale.

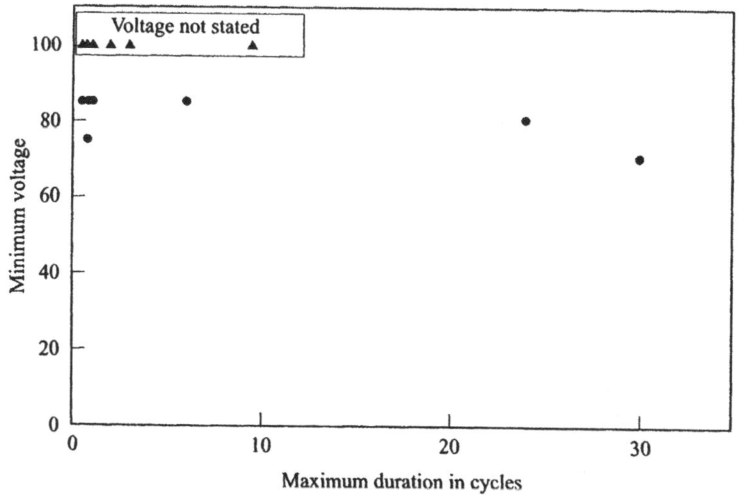

Figure 5.17 Adjustable-speed drive voltage tolerance, according to the drive manufacturer. ● = Magnitude and duration; ▲ = duration only. (Data obtained from [48].)

5.3.2.1 Acceptance Criterion. When testing an adjustable-speed drive, without detailed knowledge of the load driven by the drive, a well-defined criterion is needed to distinguish successful from unsuccessful behavior. The IEC standard 61800-3 [52] gives criteria to assess the performance of adjustable-speed drives for EMC testing. These criteria are given in Table 5.7; they should also be used for voltage sag testing of adjustable-speed drives. The IEC performance criteria can be summarized as follows:

- A: the drive operates as intended;
- B: the drive temporarily operates outside of its intended operating range but recovers automatically;
- C: the drive shuts down safely.

TABLE 5.7 Acceptance Criteria for Drives According to IEC 61800-3 [52]

	Acceptance Criterion		
	A	B	C
Specific performance	No change within the specified tolerance	Noticeable changes, self-recoverable	Shutdown, big changes, not self-recoverable
Torque-generating behavior	Torque within tolerances	Temporary deviation outside of tolerances	Loss of torque
Operations of power electronics and driving circuits	No maloperation of a power semiconductor	Temporary maloperation which cannot cause shutdown	Shutdown, triggering of protection
Information processing and sensing functions	Undisturbed communication and data exchange	Temporary disturbed communication	Errors in communication, loss of data and information
Operation of display and control panel	No change of visible display information	Visible temporary changes of information	Shutdown, obviously wrong display information

5.3.3 Balanced Sags

Many trips of ac drives are due to a low voltage at the dc bus. The trip or maloperation can be due to the controller or PWM inverter not operating properly when the voltage gets too low. But it can also be due to the intervention of undervoltage protection connected to the dc bus. Most likely, the protection will intervene before any equipment malfunction occurs.

The dc bus voltage is normally obtained from the three ac voltages through a diode rectifier. When the voltage at ac side drops, the rectifier will stop conducting and the PWM inverter will be powered from the capacitor connected to the dc bus. This capacitor has only limited energy content (relative to the power consumption of the motor) and will not be able to supply the load much longer than a few cycles. An improved voltage tolerance of adjustable-speed drives can be achieved by lowering the setting of the undervoltage protection of the dc bus. One should thereby always keep in mind that the protection should trip before any malfunction occurs and before components are damaged. Not only is the undervoltage a potential source of damage but also the overcurrent when the ac voltage recovers. If the drive is not equipped with additional overcurrent protection, the dc bus undervoltage should also protect against these overcurrents. Many drives are equipped with fuses in series with the diodes, against large overcurrents. These should not be used to protect against the overcurrent after a sag. Having to replace the fuses after a voltage sag only causes additional inconvenience.

5.3.3.1 Decay of the DC Bus Voltage. The dc bus voltage for an adjustable-speed drive during a sag in three phases behaves the same as the dc bus voltage of a personal computer, as discussed in Section 5.2. When we consider a drive with a motor load P, a nominal dc bus voltage V_0, and capacitance C connected to the dc bus, we can use (5.2) to calculate the initial decay of the dc bus voltage during the sag:

$$V(t) = \sqrt{V_0^2 - \frac{2P}{C}t} \tag{5.8}$$

It has been assumed that the dc bus voltage at sag initiation equals the nominal voltage. We further assumed a constant power load. For the standard PWM inverters this is probably not the case. But one can translate the constant-power assumption into the assumption that the load on ac side of the inverter, i.e., the ac motor, does not notice anything from the sag. Thus, the output power of the inverter is independent of the dc bus voltage. If we neglect the increase in inverter loss for lower dc bus voltage (due to the higher currents) we arrive at the constant-power assumption. The constant-power assumption thus corresponds to assuming an ideal inverter: no drop in voltage at the motor terminals, and no increase in losses during the sag.

5.3.3.2 Voltage Tolerance. The adjustable-speed drive will trip either due to an active intervention by the undervoltage protection (which is the most common situation), or by a maloperation of the inverter or the controller. In both cases the trip will occur when the dc bus voltage reaches a certain value V_{min}. As long as the ac voltage does not drop below this value, the drive will not trip. For sags below this value, (5.8) can be used to calculate the time it takes for the dc bus voltage to reach the value V_{min}:

$$t = \frac{C}{2P}\{V_0^2 - V_{min}^2\} \tag{5.9}$$

EXAMPLE Consider the example discussed in [42]: a drive with nominal dc bus voltage $V_0 = 620\,\text{V}$ and dc bus capacitance $C = 4400\,\mu\text{F}$ powers an ac motor taking an active power $P = 86\,\text{kW}$. The drive trips when the dc bus voltage drops below $V_{min} = 560\,\text{V}$. The time-to-trip obtained from (5.9) is

$$t = \frac{4400\,\mu\text{F}}{2 \times 86\,\text{kW}} \times \left((620\,\text{V})^2 - (560\,\text{V})^2\right) = 1.81\,\text{ms} \tag{5.10}$$

The minimum ac bus voltage for which the drive will not trip is 560/620 = 90%. This drive will thus trip within 2 ms when the ac bus voltage drops below 90%.

Suppose that it would be possible to reduce the setting of the undervoltage protection of the dc bus, to 310 V (50%). That would enormously reduce the number of spurious trips of the drive, because the number of sags below 50% is only a small fraction of the number of sags below 90%. But the time-to-trip for sags below 50% remains very short. Filling in $V_{min} = 310\,\text{V}$ in (5.9) gives $t = 7.38\,\text{ms}$. In fact, by substituting $V_{min} = 0$ we can see that the capacitance is completely empty 9.83 ms after sag initiation, assuming that the load power remains constant. We can conclude that no matter how good the inverter, the drive will trip for any voltage interruption longer than 10 ms.

The amount of capacitance connected to the dc bus of an adjustable-speed drive can be expressed in μF/kW. If we express the dc bus voltage in kV and the time in ms, (5.9) can be written as

$$t = 0.5\left(\frac{C}{P}\right)(V_0^2 - V_{min}^2) \tag{5.11}$$

with (C/P) in μF/kW. With (C/P) in μF/h.p. (5.11) becomes

$$t = 0.67\left(\frac{C}{P}\right)(V_0^2 - V_{min}^2) \tag{5.12}$$

The amount of capacitance connected to the dc bus of modern adjustable-speed drives is between 75 and 360 μF/kW [138]. Figure 5.18 plots the relation between the undervoltage setting for the dc bus (vertical) and the time-to-trip (horizontal scale), for three values of the ratio between dc bus capacitance and motor size according to (5.11). The voltage tolerance of the drive, for balanced sags, can be obtained as follows:

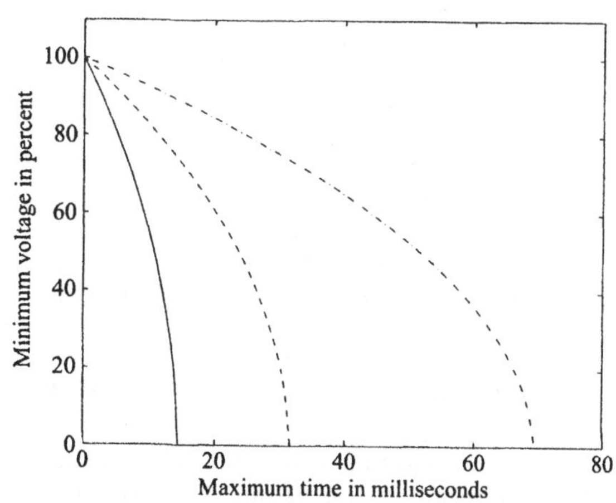

Figure 5.18 Voltage tolerance of adjustable-speed drives for different capacitor sizes. Solid line: 75 μF/kW; dashed line: 165 μF/kW; dotted line: 360 μF/kW.

- The setting of the dc bus undervoltage protection determines the minimum voltage for which the drive is able to operate.
- From the appropriate curve, determined by the capacitor size, the maximum sag duration is found.

We see that even for very low values of the setting of the dc bus undervoltage, the drive will trip within a few cycles.

5.3.3.3 Capacitor Size. It is obvious from the above examples that the amount of capacitance connected to the dc bus of an adjustable-speed drive, is not enough to offer any serious immunity against voltage sags. The immunity can be improved by adding more capacitance to the dc bus. To calculate the amount of capacitance needed for a given voltage tolerance, we go back to (5.8) and assume $V(t_{max}) = V_{min}$, leading to

$$C = \frac{2Pt_{max}}{V_0^2 - V_{min}^2} \tag{5.13}$$

This expression gives the amount of dc bus capacitance needed to obtain a voltage tolerance of V_{min}, t_{max} (i.e., the drive trips when the voltage drops below V_{min} for longer than t_{max}).

EXAMPLE Consider the same drive as in the previous example We want the drive to be able to tolerate sags with durations up to 500 ms. The undervoltage setting remains at 560 V (90% of nominal). The capacitance needed to achieve this is obtained from (5.13) with $t_{max} = 500\,\text{ms}$ and $V_{min} = 560\,\text{V}$:

$$C = \frac{286\,\text{kW} \times 500\,\text{ms}}{(620\,\text{V})^2 - (560\,\text{V})^2} = 1.12\,\text{F} \tag{5.14}$$

This example is used in [42] to compare different ways of improving the drive's voltage tolerance, including the costs of the various options. The total costs of 1.12 F capacitance, with enclosures, fuses, bars, and fans, would be about $200,000 and to place these capacitors would require a space $2.5 \times 18\,\text{m}^2$ and 60 cm high. A battery backup would cost "only" $15,000 and require a space of $2.5 \times 4 \times 0.6\,\text{m}^3$. However the battery block would require more maintenance than the capacitors.

Assume that an undervoltage protection setting of 310 V (50%) is feasible, and that the drive should be able to tolerate voltage sags up to 200 ms in duration. Equation (5.13) can again be used to give the required capacitance, which is 119 mF.

This is only one-tenth of the required capacitance for the original inverter. The costs of installing capacitance would still be higher than for the battery block but the lower maintenance requirements of the capacitors might well tip the balance toward them. Making an inverter that can operate for even lower voltages would not gain much ridethrough time or save capacitors. This is because the stored energy in a capacitor is proportional to the square of the voltage. It would, however, increase the current through the inverter significantly. Bringing the minimum operating voltage down to 25% would double the required current rating of the inverter but still require 95 mF of capacitance; a reduction of only 20%.

5.3.4 DC Voltage for Three-Phase Unbalanced Sags

In normal operation, the dc bus voltage is somewhat smoothened by the capacitance connected to the dc bus. The larger the capacitance, the smaller the voltage ripple.

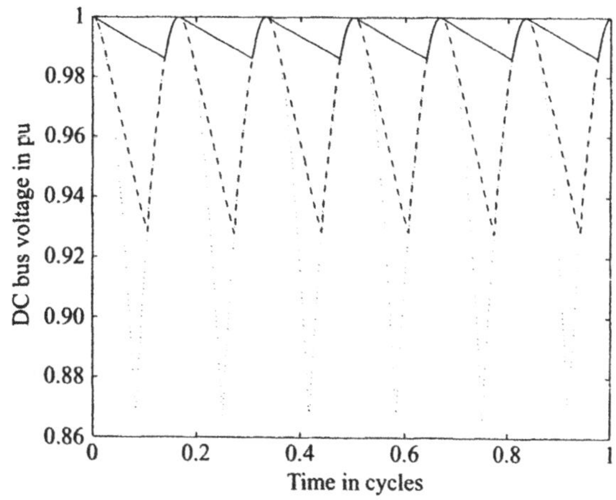

Figure 5.19 DC bus voltage behind a three-phase rectifier during normal operation, for large capacitor (solid line), small capacitor (dashed line), and no capacitor connected to the dc bus (dotted line).

Where with a single-phase rectifier the capacitor is only charged twice a cycle, it is charged six times every cycle for a three-phase rectifier. Figure 5.19 shows the dc bus voltage behind a three-phase rectifier, for various capacitor size. The load fed from the dc bus was assumed to be of the constant-power type. The size of the capacitances was chosen as follows: for the large capacitance and a dc bus voltage of 100%, the initial rate of decay of the voltage is 10% per cycle when the ac side voltage drops; for the small capacitance the initial rate of decay is 75% per cycle. We will relate this to the drive parameters further on.

We saw in Section 4.4 that the most common sags experienced by a three-phase load are type A, type C, and type D. For a type A sag all three phases drop in magnitude the same amount. All six voltage pulses in Fig. 5.19 will drop in magnitude and the load will empty the capacitor connected to the dc bus, until the dc bus voltage drops below the peak of the ac voltage again. The voltage tolerance for this case has been discussed in the previous section.

5.3.4.1 Sags of Type C. For a three-phase unbalanced sag of type C or type D, different phases have different voltage drops. Some phase voltages also show a jump in phase angle. The behavior of the dc bus voltage, and thus of the drive, is completely different than for a balanced voltage sag. The upper plot in Fig. 5.20 shows the voltages at the drive terminals for a sag of type C. Note that these are the line-to-line voltages, as the drive is connected in delta. We see how the voltage drops in two phases, while the sine waves move toward each other. The third phase does not drop in magnitude. A sag with a characteristic magnitude of 50% and zero characteristic phase-angle jump is shown. The voltage magnitudes at the drive terminals are 66.1% (in two phases) and 100% in the third phase; phase-angle jumps are −19.1°, +19.1°, and zero.

The effect of this three-phase unbalanced sag on the dc bus voltage is shown in the lower plot of Fig. 5.20. The capacitor sizes used are the same as in Fig. 5.19. We see that even for the small capacitance, the dc bus voltage does not drop below 70%. For the large capacitance, the dc bus voltage hardly deviates from its normal operating value. In the latter case, the drive will never trip during a sag of type C, no matter how low the characteristic magnitude of the sag. As one phase remains at its pre-event value, the three-phase rectifier simply operates as a single-phase rectifier during the voltage sag. The drop in dc bus voltage (actually: the increase in voltage ripple) is only moderate.

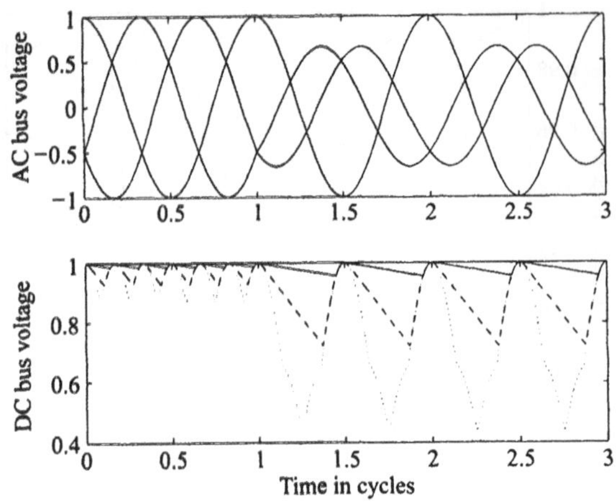

Figure 5.20 Voltage during a three-phase unbalanced sag of type C: ac side voltage (top) and dc side voltages (bottom) for large capacitor (solid line), small capacitor (dashed line), and no capacitor connected to the dc bus (dotted line).

The initial behavior remains identical to the one discussed before for the balanced sag (due to a three-phase fault). The main difference is that the dc bus voltage recovers after one half-cycle. This is due to the one phase that remains at nominal voltage for a sag of type C.

5.3.4.2 Sags of Type D. The voltages on ac side and dc side of the rectifier are shown in Fig. 5.21 for a three-phase unbalanced sag of type D with characteristic magnitude 50% and no characteristic phase-angle jump. The magnitude of the voltages at the drive terminals is 50%, 90.14%, and 90.14%, with phase-angle jumps zero, $-13.9°$ and $+13.9°$.

For a sag of type D, all three phases drop in voltage, thus there is no longer one phase which can keep up the dc bus voltage. Fortunately the drop in voltage is moderate for two of the three phases. Even for a terminal fault, where the voltage in one phase drops to zero, the voltage in the other two phases does not drop below $\frac{1}{2}\sqrt{3} = 86\%$. The top curve in Fig. 5.21 shows how one phase drops significantly in voltage. The other two phases drop less in voltage magnitude and their maxima move away from each other. In the bottom curve of Fig. 5.21 the effect of this on the dc bus

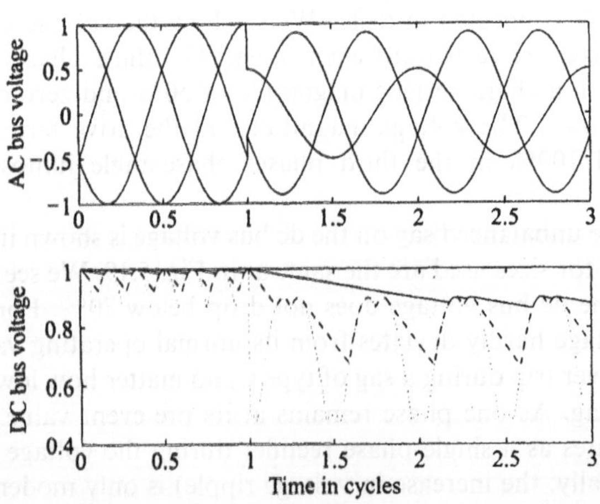

Figure 5.21 Voltage during a three-phase unbalanced sag of type D: ac side voltage (top) and dc side voltages (bottom) for large capacitor (solid line), small capacitor (dashed line), and no capacitor connected to the dc bus (dotted line).

voltage is shown. For not too small values of the dc bus capacitance, the dc bus voltage reaches a value slightly below the peak value of the voltage in the two phases with the moderate drop. Again the effect of the sag on the dc bus voltage, and thus on the motor speed and torque, is much less than for a balanced sag.

5.3.4.3 Phase-Angle Jumps. In Figs. 5.20 and 5.21 it is assumed that the characteristic phase-angle jump is zero. This makes that two of the phase voltages have the same peak value: the highest phases for a sag of type D (Fig. 5.21); the lowest phases for a sag of type C (Fig. 5.20). A non-zero characteristic phase-angle jump makes that one of these two voltages gets lower, and the other higher. The effect of this is shown in Fig. 5.22 for a three-phase unbalanced sag of type D, with a characteristic magnitude of 50%. All phase-angle jumps are assumed negative; positive phase-angle jumps would give exactly the same effect. When there is no capacitance connected to the dc bus (dotted line) the minimum dc bus voltage is determined by the lowest ac side voltage. The effect of the phase-angle jump is that the minimum dc bus voltage gets lower. But for a drive with a large capacitance connected to the dc bus, it is the highest peak voltage which determines the dc bus voltage. For such a drive, the dc bus voltage will increase for increasing phase-angle jump. For a phase-angle jump of −30° the dc bus voltage is even higher than during normal operation. Note that a −30° phase-angle jump is an extreme situation for a sag with a characteristic magnitude of 50%.

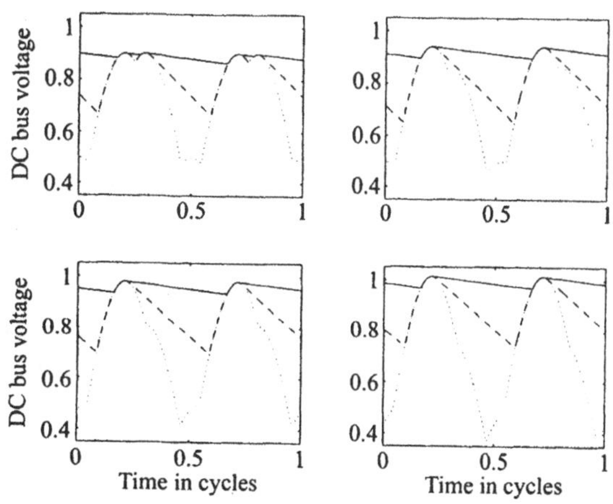

Figure 5.22 DC bus voltage during a three-phase unbalanced sag of type D, with characteristic magnitude 50% and characteristic phase-angle jump zero (top left), 10° (top right), 20° (bottom left), and 30° (bottom right). Solid line: large capacitance; dashed line: small capacitance; dotted line: no capacitance connected to the dc bus.

For three-phase unbalanced sags of type C, the dc bus voltage is determined by the voltage in the phase which does not drop in magnitude. The phase-angle jump has no influence on this value: it simply remains at 100%. Thus for sags of type C the dc bus voltage is not influenced by the phase-angle jump, assuming the capacitance connected to the dc bus is large enough.

5.3.4.4 Effect of Capacitor Size and Sag Magnitude. Some of the effects of the size of the dc bus capacitance on the dc bus voltage during unbalanced sags are summarized in Figs. 5.23 through 5.30. In all the figures, the horizontal axis gives the characteristic magnitude of the sag, the solid line corresponds to a large capacitance connected to the dc bus, the dashed line holds for small capacitance, the dotted

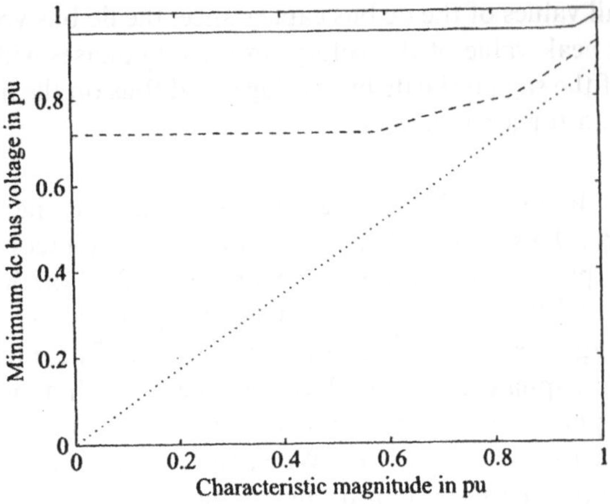

Figure 5.23 Minimum dc bus voltage as a function of the characteristic magnitude of three-phase unbalanced sags of type C. Solid line: large capacitance; dashed line: small capacitance; dotted line: no capacitance connected to the dc bus.

line for no capacitance at all. Figures 5.23 through 5.26 are for three-phase unbalanced sags of type C. Figures 5.27 through 5.30 are the corresponding figures for type D.

Figure 5.23 shows the influence on the minimum dc bus voltage. The dc bus undervoltage protection normally uses this value as a trip criterion. There is thus a direct relation between the minimum dc bus voltage and the voltage tolerance of the drive. We see from the figure that the presence of sufficient capacitance makes that the dc bus voltage never drops below a certain value, no matter how deep the sag at ac side is. This is obviously due to the one phase of the ac voltage which stays at its normal value. For a large capacitance, the drop in dc bus voltage is very small. The smaller the capacitance, the more the drop in dc bus voltage.

Figure 5.24 shows the influence of sag magnitude and capacitor size on the voltage ripple at the dc bus. The larger the capacitance and the larger the characteristic magnitude, the smaller the voltage ripple. Again a large capacitance mitigates the voltage disturbance at the dc bus. Some drives use the voltage ripple to detect malfunctioning of the rectifier. This is more used in controlled rectifiers where a large voltage ripple could indicate an error in one of the firing circuits. The figure is some-

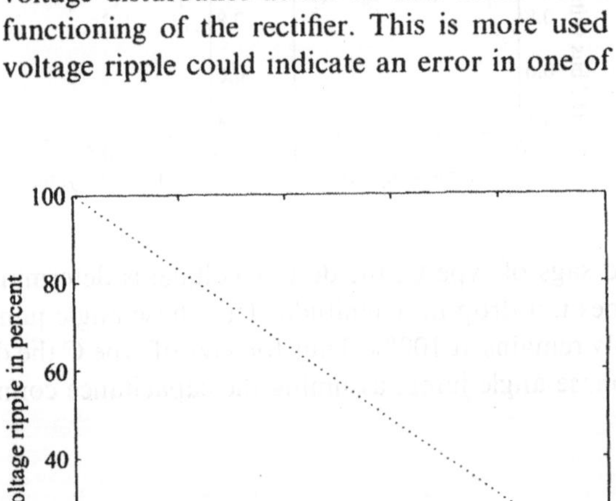

Figure 5.24 Voltage ripple at the dc bus as a function of the characteristic magnitude of three-phase unbalanced sags of type C. Solid line: large capacitance; dashed line: small capacitance; dotted line: no capacitance connected to the dc bus.

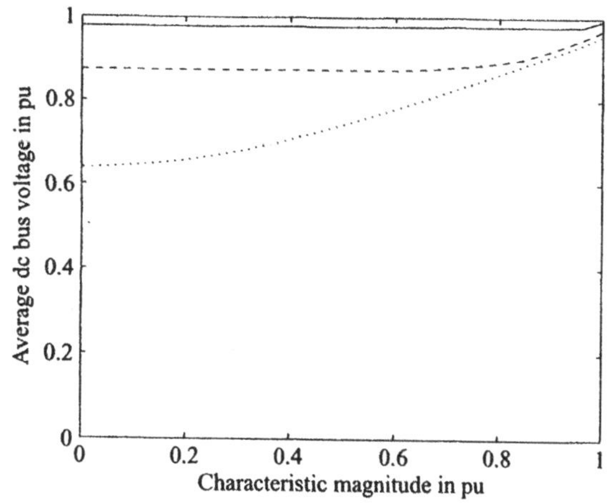

Figure 5.25 Average dc bus voltage as a function of the characteristic magnitude of three-phase unbalanced sags of type C. Solid line: large capacitance; dashed line: small capacitance; dotted line: no capacitance connected to the dc bus.

what misleading in this sense, as a large capacitance would also make it more difficult to detect unbalances in the rectifier (like errors in the thyristor firing). In that case, either a more sensitive setting of the voltage ripple detection should be used (which would overrule the gain in voltage tolerance) or the rectifier currents should be used as a detection criterion (which might introduce more sensitivity to unbalanced sags).

The average dc bus voltage is shown in Fig. 5.25, the rms value in Fig. 5.26. These determine how the motor driven by the drive slows down in speed. We see that the drop in average or rms voltage is not as dramatic as the drop in minimum voltage: although also here, the larger the size of the capacitance, the less the drop in speed. Especially for longer voltage sags, or low-inertia loads, this could be a decisive difference. Of course one needs to assume that the inverter is able to operate during the voltage sag. That is more likely for large capacitance, where the dc bus voltage remains high, than for small capacitance, where the dc bus voltage drops to a low value twice a cycle.

The results for a three-phase unbalanced sag of type D are shown in Figs. 5.27 through 5.30. We saw in Fig. 5.21 that for large capacitance, the new steady state does not settle in immediately. All values for the type D sag have been calculated for the third cycle during the sag. The minimum dc bus voltage for a sag of type D is shown in

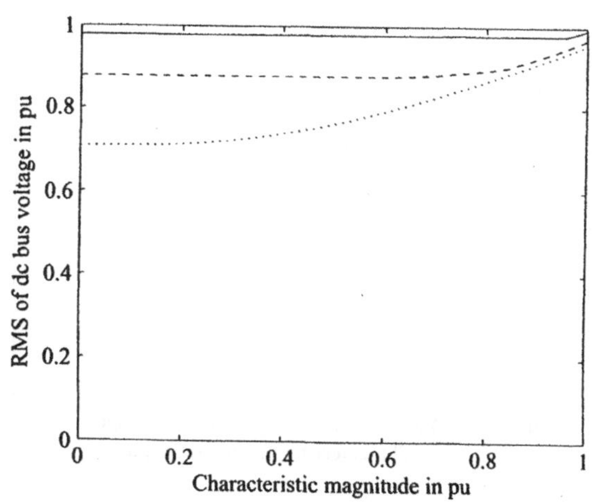

Figure 5.26 The rms of the dc bus voltage as a function of the characteristic magnitude of three-phase unbalanced sags of type C. Solid line: large capacitance; dashed line: small capacitance; dotted line: no capacitance connected to the dc bus.

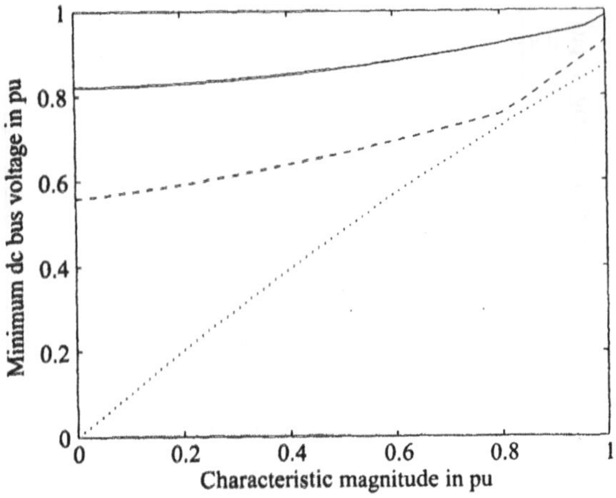

Figure 5.27 Minimum dc bus voltage as a function of the characteristic magnitude of three-phase unbalanced sags of type D. Solid line: large capacitance; dashed line: small capacitance; dotted line: no capacitance connected to the dc bus.

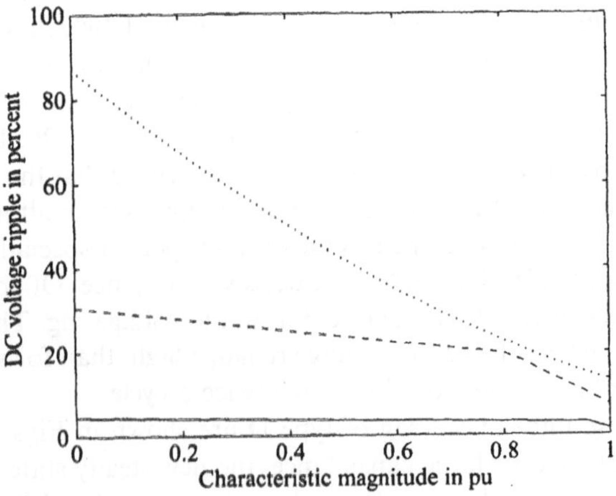

Figure 5.28 Voltage ripple at the dc bus as a function of the characteristic magnitude of three-phase unbalanced sags of type D. Solid line: large capacitance; dashed line: small capacitance; dotted line: no capacitance connected to the dc bus.

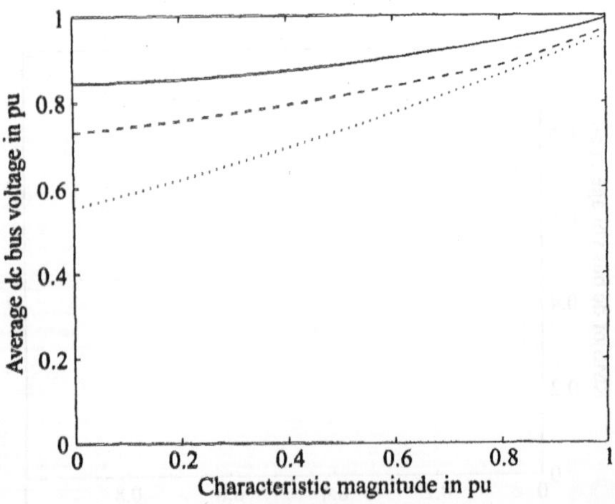

Figure 5.29 Average dc bus voltage as a function of the characteristic magnitude of three-phase unbalanced sags of type D. Solid line: large capacitance; dashed line: small capacitance; dotted line: no capacitance connected to the dc bus.

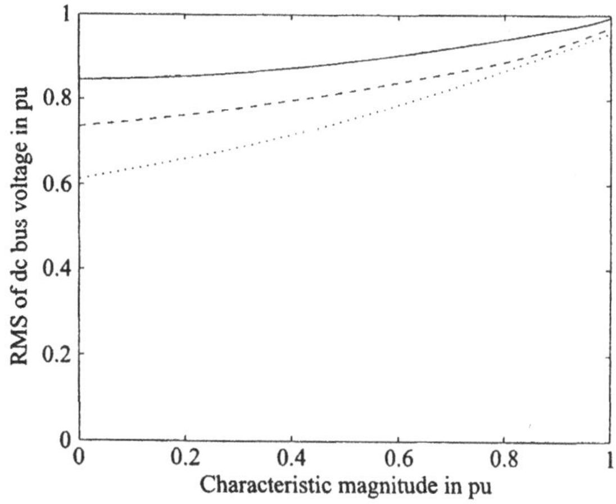

Figure 5.30 The rms of the dc bus voltage as a function of the characteristic magnitude of three-phase unbalanced sags of type D. Solid line: large capacitance; dashed line: small capacitance; dotted line: no capacitance connected to the dc bus.

Fig. 5.27. Comparison with Fig. 5.23 for type C reveals that for a type D sag the minimum dc bus voltage continues to drop with lower characteristic magnitude, even with large capacitor size. But again an increase in capacitance can significantly reduce the voltage drop at the dc bus. For the drive with the large capacitance the dc bus voltage does not drop below 80%, even for the deepest unbalanced sag.

Figure 5.28 plots the voltage ripple for type D sags, which shows a similar behavior as for type C sags. The voltage ripple is calculated as the peak-to-peak ripple related to the normal value. Therefore, the voltage ripple for the drive without capacitance does not reach 100% for a sag of zero characteristic magnitude.

In Figs. 5.29 and 5.30, showing average and rms value of the dc bus voltage, we see similar values as for sags of type C. Again the difference is that the dc bus voltage continues to drop for decreasing characteristic magnitude. Deep sags of type D will cause more drop in motor speed than sags of the same magnitude of type C. For shallow sags the effect on the motor speed will be about the same.

5.3.4.5 Size of the DC Bus Capacitance. In the previous figures, the dc bus voltage was calculated for three values of the size of the capacitance connected to the dc bus. Those were referred to as "large capacitance," "small capacitance," and "no capacitance." Large and small were quantified through the initial decay of the dc bus voltage: 10% per cycle for the large capacitance, 75% per cycle for the small capacitance. Here we will quantify the amount of μF to which this corresponds.

The dc bus voltage $V(t)$ during the sag is governed by the law of conservation of energy: the electric load P is equal to the change in energy stored in the dc bus capacitor C. In equation form this reads as

$$\frac{d}{dt}\left\{\frac{1}{2}CV^2\right\} = P \tag{5.15}$$

Let V_0 be the dc bus voltage at sag initiation. This gives at sag initiation

$$CV_0\frac{dV}{dt} = P \tag{5.16}$$

from which the initial rate of decay of the dc bus voltage can be calculated:

$$\frac{dV}{dt} = \frac{P}{CV_0} \tag{5.17}$$

From (5.16) we can derive an expression for the capacitor size needed to get a certain initial rate of decay of dc bus voltage:

$$C = \frac{P}{V_0 \frac{dV}{dt}} \tag{5.18}$$

EXAMPLE For the same drive parameters as before (620 V, 86 kW) we can use (5.18) to calculate the required size of the capacitance. As a first step we have to translate percent per cycle into volts per second:

$$\begin{aligned} 75\% \text{ per cycle} &= 27{,}900 \text{ V/s} \\ 10\% \text{ per cycle} &= 3730 \text{ V/s} \end{aligned}$$

To obtain a rate of decay of 75% per cycle, we need a capacitance of

$$C = \frac{86\,\text{kW}}{620\,\text{V} \times 27{,}900\,\text{V/s}} = 4970\,\mu\text{F} \tag{5.19}$$

or 57.8 μF/kW. Similarly we find that 37.3 mF or 433 μF/kW corresponds to 10% per cycle. These values need to be compared to the amount of capacitance present in modern drives, which is between 75 and 360 μF/kW, according to [138]. We see that the "large capacitance" curves are feasible with modern adjustable-speed drives.

5.3.4.6 Load Influence. The main load influence on voltage sags is the reduction in negative-sequence voltage due to induction motor load, as explained in Section 4.8. To see what the effect is on adjustable-speed drives, we reproduced type C and type D sags with reduced negative-sequence voltage and calculated dc bus voltage behind a non-controlled rectifier. The three-phase unbalanced sags with reduced negative-sequence voltage were calculated in the same way as for Figs. 4.138 and 4.139. The analysis was performed for a three-phase unbalanced sag with a characteristic magnitude of 50% and zero phase-angle jump. The voltages at the equipment terminals are for a 50% sag of type C:

$$\begin{aligned} V_a &= 1 \\ V_b &= -\frac{1}{2} - \frac{1}{4}j\sqrt{3} \\ V_c &= -\frac{1}{2} + \frac{1}{4}j\sqrt{3} \end{aligned} \tag{5.20}$$

and for a sag of type D:

$$\begin{aligned} V_a &= \frac{1}{2} \\ V_b &= -\frac{1}{4} - \frac{1}{2}j\sqrt{3} \\ V_c &= -\frac{1}{4} + \frac{1}{2}j\sqrt{3} \end{aligned} \tag{5.21}$$

Splitting the phase voltages in sequence components gives

$$V_1 = \frac{3}{4} \qquad V_2 = \frac{1}{4} \tag{5.22}$$

for a sag of type C, and

$$V_1 = \frac{3}{4} \qquad V_2 = -\frac{1}{4} \tag{5.23}$$

for a sag of type D. A "distorted type C" sag is created by keeping the positive-sequence voltage constant, while reducing the negative-sequence voltage. This is to simulate the effect of induction motor load. If we assume that the negative-sequence voltage drops by a factor of β, thus from V_2 to $(1-\beta)V_2$, we obtain the phase voltages from

$$\begin{aligned} V_a &= V_1 + (1-\beta)V_2 \\ V_b &= V_1 + a^2(1-\beta)V_2 \\ V_c &= V_1 + a(1-\beta)V_2 \end{aligned} \tag{5.24}$$

where $a = -\frac{1}{2} + \frac{1}{2}j\sqrt{3}$. The resulting phase voltages are used to calculate the dc bus voltages during the sag, in the same way as for the "nondistorted" sag. The results are shown in Figs. 5.31 through 5.34. Figure 5.31 plots the average dc bus voltage as a function of the drop in negative-sequence voltage. Note that a drop of 50% in negative-sequence voltage requires a very large induction motor load. We see from Fig. 5.31 that the motor load drops the minimum dc bus voltage in case a capacitor is used. For a drive without dc bus capacitor, the minimum dc bus voltage increases. The drop in negative-sequence voltage makes that the three voltages get closer in magnitude, so that the effect of a capacitor becomes less. The same effect is seen in Fig. 5.33 for type D sags. Figs. 5.32 and 5.34 show that also the average dc bus voltage drops for increasing motor load.

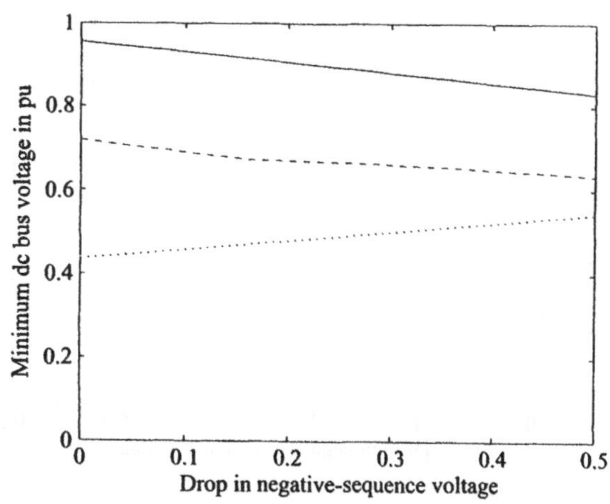

Figure 5.31 Induction motor influence on minimum dc bus voltage for sags of type C. Solid line: large capacitor; dashed line: small capacitor; dotted line: no capacitor connected to the dc bus.

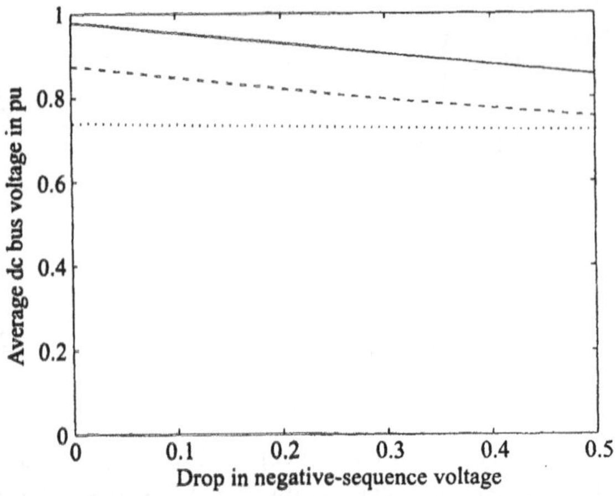

Figure 5.32 Induction motor influence on average dc bus voltage for sags of type C. Solid line: large capacitor; dashed line: small capacitor; dotted line: no capacitor connected to the dc bus.

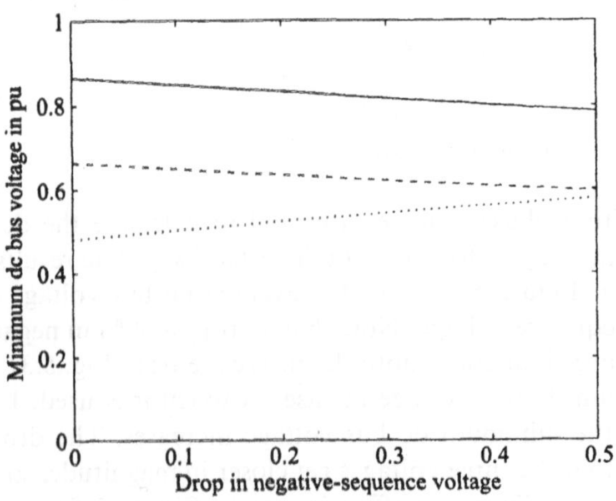

Figure 5.33 Induction motor influence on minimum dc bus voltage for sags of type D. Solid line: large capacitor; dashed line: small capacitor; dotted line: no capacitor connected to the dc bus.

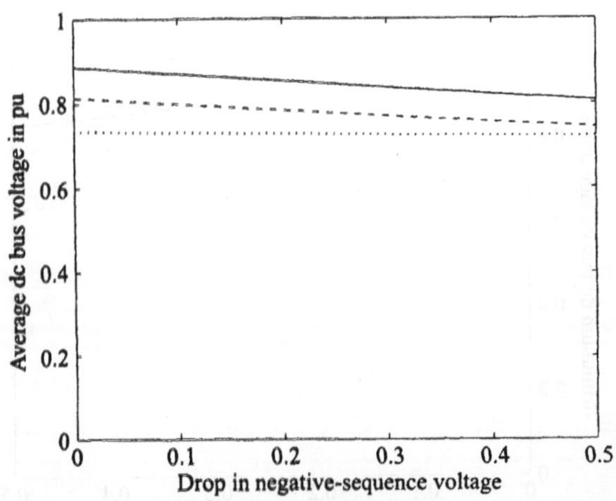

Figure 5.34 Induction motor influence on average dc bus voltage for sags of type D. Solid line: large capacitor; dashed line: small capacitor; dotted line: no capacitor connected to the dc bus.

5.3.4.7 Powering the Controllers. In older drives the control electronics for the PWM inverter was powered from the supply. This made the drive very sensitive to disturbances in the supply. In modern drives the control electronics is powered from the dc bus which can be more constant due to the presence of capacitors. But even here the same reasoning can be used as for process control equipment. Controllers are essentially low-power equipment which only require a small amount of stored energy to ride through sags. The design of the power supply to the drive controller should be such that the controller stays active at least as long as the power electronics or the motor do not require a permanent trip. It should not be that the controller becomes the weak part of the drive. Figure 5.35 shows the typical configuration for powering the controller. The capacitance connected to the dc bus between the rectifier and the inverter is normally not big enough to supply the motor load and the controller during a balanced sag longer than a few cycles. The power supply to the controller can be guaranteed in a number of ways:

- By inhibiting firing of the inverter so that the motor no longer discharges the dc bus capacitance. The power taken by the controller is so much smaller than the motor load, that the capacitor can easily power the controller even for long voltage sags. When the supply voltage recovers, the controller can automatically restart the load.
- Additional capacitance can be installed on low-voltage side of the dc-dc switched mode power supply between the dc bus and the control circuitry. As this capacitance only needs to power the controller, a relatively small amount of capacitance is needed. Also a battery block would do the job.
- Some drives use the rotational energy from the motor load to power the controllers during a voltage sag or short interruption. This causes small additional drop in motor speed, small enough to be negligible. A special control technique for the inverter is needed, as well as a method to detect the sag [33].

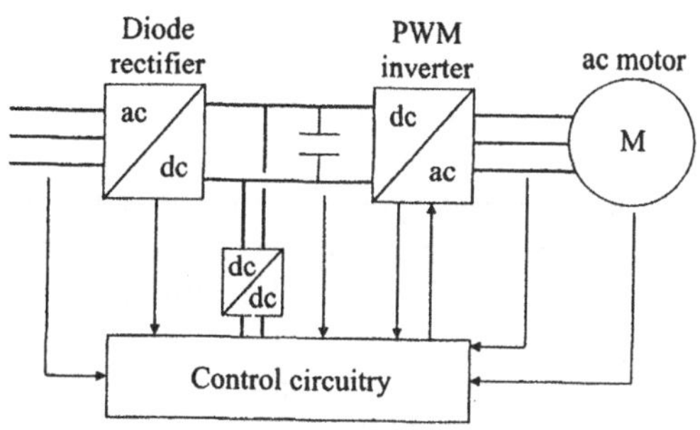

Figure 5.35 Configuration of the power supply to the control circuitry in an adjustable-speed drive.

5.3.5 Current Unbalance

5.3.5.1 Simulations. Unbalance of the ac voltages not only causes an increased ripple in the dc voltage but also a large unbalance in ac currents. The unbalance in current depends on the type of sag. Consider first a sag of type D, where one voltage is much lower than the other two. The upper plot in Fig. 5.36 shows the ac side voltages (in absolute value) compared with the dc bus voltage (solid line near the top) during one cycle, for a sag of type D with characteristic magnitude equal to 50%. Here it is assumed that the dc bus voltage does not change at all during the sag. The

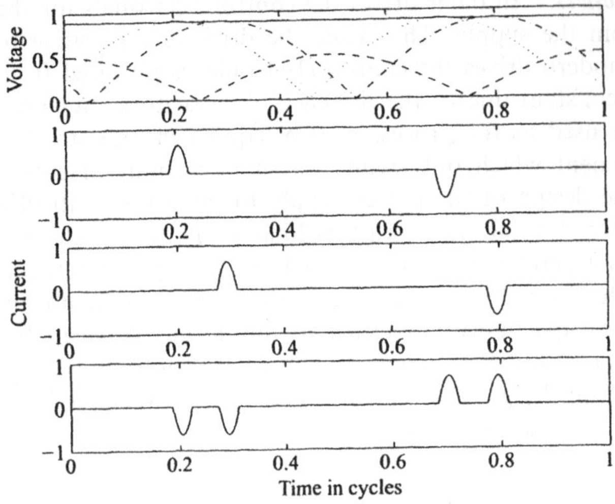

Figure 5.36 AC sideline voltages (top) and currents (phase a, b, and c from top to bottom) for a three-phase unbalanced sag of type D.

rectifier only delivers current when the ac voltage (in absolute value) is larger than the dc voltage. We have assumed that this current is proportional to the difference between the absolute value of the ac voltage and the dc voltage. This results in the line currents as shown in the three remaining plots in Fig. 5.36.

The three voltages in the top plot of Fig. 5.36 are the voltage difference between phase a and phase b (dashed), between phase b and phase c (dash-dot), and between phase c and phase a (dotted). The first pulse occurs when the voltage between a and c exceeds the dc voltage (around $t = 0.2$ cycle). This results in a current pulse in the phases a and c. Around $t = 0.3$ cycle the voltage between b and c exceeds the dc voltage leading to a current pulse in the phases b and c. The pattern repeats itself around $t = 0.7$ cycle and $t = 0.8$ cycle. The currents flow in opposite direction because the ac voltages are opposite now. Whereas at $t = 0.2$ cycle the voltage between c and a was negative resulting in a current from a to c, the voltage is positive now resulting in a current from c to a. The voltage between a and b has dropped so much that there are no current pulses between a and b. This results in two missing pulses per cycle for phase a as well as for phase b.

Whereas in normal operation the capacitor is charged 6 times per cycle, this now only takes place four times per cycle. These four pulses must carry the same amount of charge as the original six pulses. The consequence is that the pulses will be up to 50% higher in magnitude.

For a type C sag the situation is even worse, as shown in the top plot of Fig. 5.37. One line voltage is much higher than the other two, so that only this voltage leads to current pulses. The resulting current pulses in the three phases are shown in the three bottom plots of Fig. 5.37.

Due to a sag of type C the number of current pulses is reduced from 6 per cycle to 2 per cycle, leading to up to 200% overcurrent. Note that a large overcurrent would already arise for a shallow sag. The moment one or two voltages drop below the dc bus voltage, pulses will be missing and the remaining current pulses will have to be higher to compensate for this.

5.3.5.2 Measurements. Figures 5.38, 5.39, and 5.40 show measurements of the input currents of an adjustable-speed drive [27], [30]. Figure 5.38 shows the input

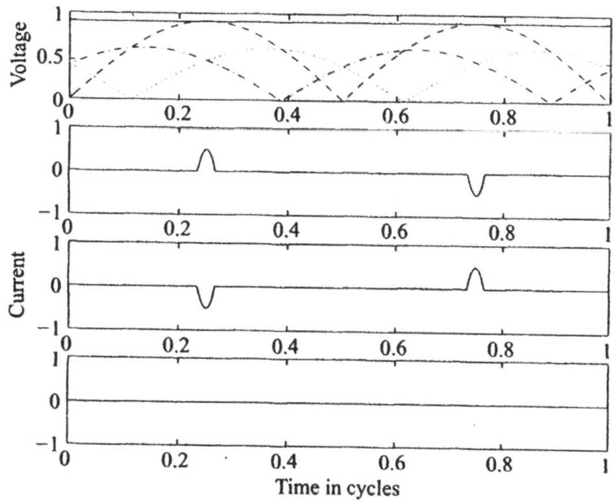

Figure 5.37 AC side voltage (top) and currents (phase a, b, and c from top to bottom) for a three-phase unbalanced sag of type C.

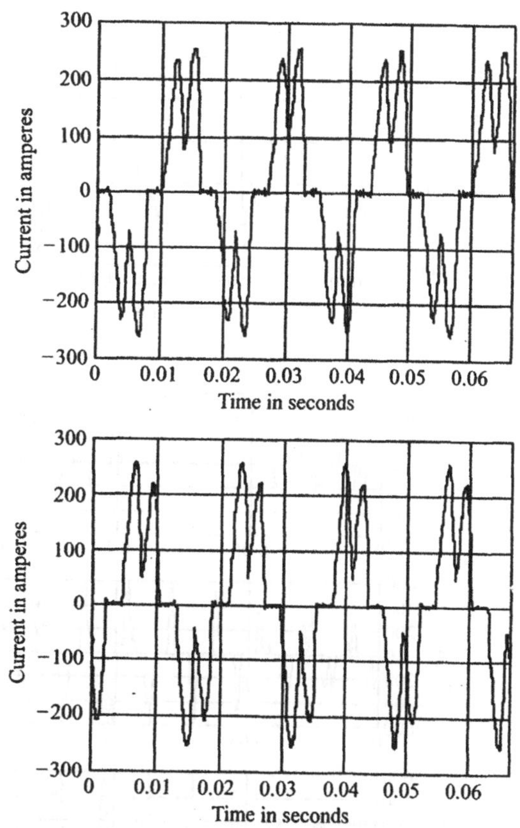

Figure 5.38 Input current for an ac drive in normal operation. (Reproduced from Mansoor [27].)

currents for the drive under normal operating conditions. Only two currents are shown, the third one is similar to one of the other two. The drive is connected in delta, so that each current pulse shows up in two phases. A total of four pulses in each of the three phases implies 6 pulses per cycle charging the capacitor. There was a small unbalance in the supply voltage leading to the difference between the current pulses. We see that the magnitude of the current pulses is between 200 and 250 A.

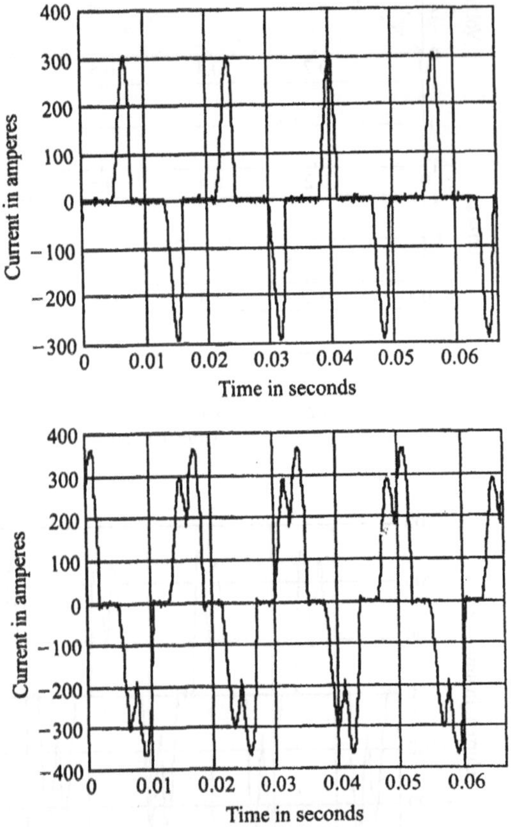

Figure 5.39 Input current for an ac drive with voltage unbalance. (Reproduced from Mansoor [27].)

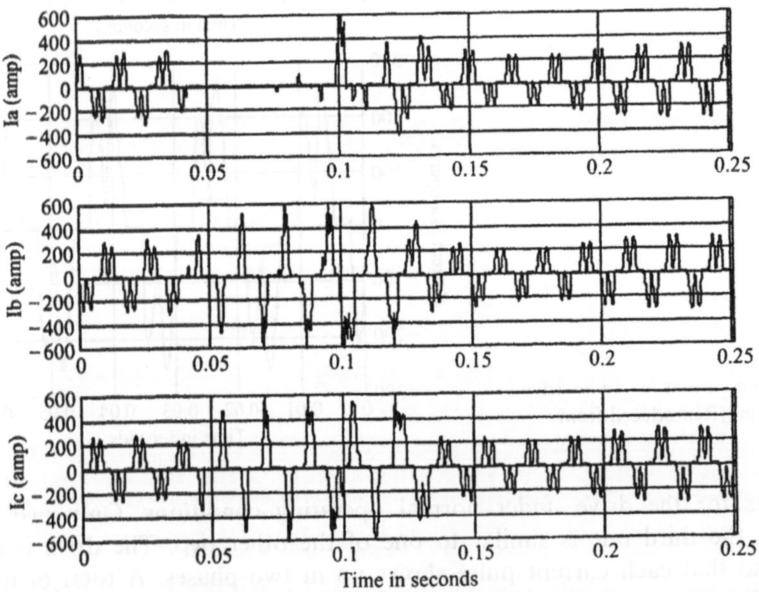

Figure 5.40 Input current for an ac drive during a single-phase fault. (Reproduced from Mansoor [27].)

Figure 5.39 shows the same currents, for an unbalance in the supply voltage. The highest voltage magnitude was 3.6% higher than the lowest one. This small unbalance already leads to two missing pulses both related to the same line voltage. There are now only four pulses left, with a magnitude between 300 and 350 A, confirming the 50% overcurrent predicted above.

Figure 5.40 shows the rectifier input current for a single-phase sag at the rectifier terminals. A measured sag is reproduced by means of three power amplifiers. As explained in Section 4.4.4, a single-phase fault will cause a type D sag on the terminals of delta-connected load. The two remaining pulses per cycle and the peak current of 500 to 600 A confirm the 200% overcurrent predicted above.

5.3.6 Unbalanced Motor Voltages

The dc bus voltage is converted into an ac voltage of the required magnitude and frequency, by using a voltage-source converter (VSC) with pulse-width modulation. The principle of PWM can be explained through Fig. 5.41. A carrier signal V_{cr} with a frequency of typically a few hundred Hertz, is generated and compared with the reference signal V_{ref} (dashed curve in the upper figure). The reference signal is the required motor terminal voltage, with a certain magnitude, frequency, and phase angle. If the reference signal is larger than the carrier signal, the output of the inverter is equal to the positive input signal V_+ and the other way around:

$$\begin{aligned} V_{out} &= V_+, \qquad V_{ref} > V_{cr} \\ V_{out} &= V_-, \qquad V_{ref} < V_{cr} \end{aligned} \tag{5.25}$$

The resulting output voltage V_{out} is shown in the lower plot of Fig. 5.41. It can be shown that the output voltage consists of a fundamental frequency sine wave plus harmonics of the switching frequency [43]. The latter can be removed by a low-pass filter after which the required sinusoidal voltage remains. If the dc bus voltage varies, both the positive and the negative output voltage V_+ and V_- will change proportionally. These variations will thus appear as an amplitude modulation of the output voltage. Let the required motor voltages be

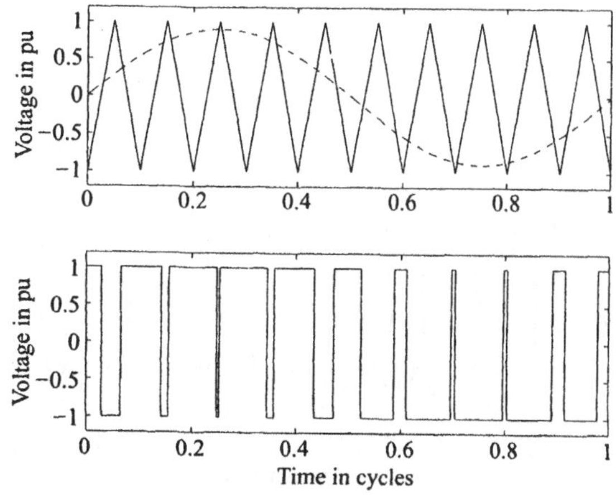

Figure 5.41 Principle of pulse-width modulation: carrier signal with reference signal (dashed) in the top figure; the pulse-width modulated signal in the bottom figure.

$$\begin{aligned} V_a &= V_m \cos(2\pi f_m t) \\ V_b &= V_m \cos(2\pi f_m t - 120^\circ) \\ V_c &= V_m \cos(2\pi f_m t + 120^\circ) \end{aligned} \tag{5.26}$$

We assume that the high-frequency harmonics due to the PWM switching are all removed by the low-pass filter, but that the variations in dc bus voltage are not removed by the filter. The motor voltages for a dc bus voltage $V_{dc}(t)$ are the product of the required voltage and the p.u. dc bus voltage:

$$\begin{aligned} V_a &= V_{dc}(t) \times V_m \cos(2\pi f_m t) \\ V_b &= V_{dc}(t) \times V_m \cos(2\pi f_m t - 120^\circ) \\ V_c &= V_{dc}(t) \times V_m \cos(2\pi f_m t + 120^\circ) \end{aligned} \tag{5.27}$$

Normally the motor frequency will not be equal to the system frequency, thus the ripple in the dc voltage is not synchronized with the motor voltages. This may lead to unbalances and interharmonics in the motor voltages.

The motor terminal voltages have been calculated for sags of type C and D, for various characteristic magnitudes and motor frequencies. A small capacitor was connected to the dc bus. Figure 5.42 shows the results for a 50% sag of type C (see Fig. 5.20) and a motor frequency equal to the fundamental frequency. We see that the motor terminal voltages are seriously distorted by the ripple in the dc bus voltage. One phase drops to 75% while another remains at 100%. The dc bus voltage is shown as a dashed line in the figure. Figure 5.43 shows the result for a 50% sag of type D and a motor frequency of 50 Hz. The effect is similar but less severe than for the type C sag.

Figure 5.44 plots the three motor terminal voltages for a motor frequency of 40 Hz and a supply frequency of 50 Hz. The motor frequency is now no longer an integer fraction of twice the power system frequency (the dc ripple frequency). But two periods of the motor frequency (50 ms) correspond to five half-cycles of the power system frequency. The motor terminal voltage is thus periodic with a period of 50 ms. This subharmonic is clearly visible in Fig. 5.44.

Figure 5.45 shows the unbalance of the voltages at the motor terminals, as a function of the motor speed. The unbalance is indicated by showing both the positive and the negative-sequence component of the voltages. The larger the negative-sequence component, the larger the unbalance. We see that the unbalance is largest for motor

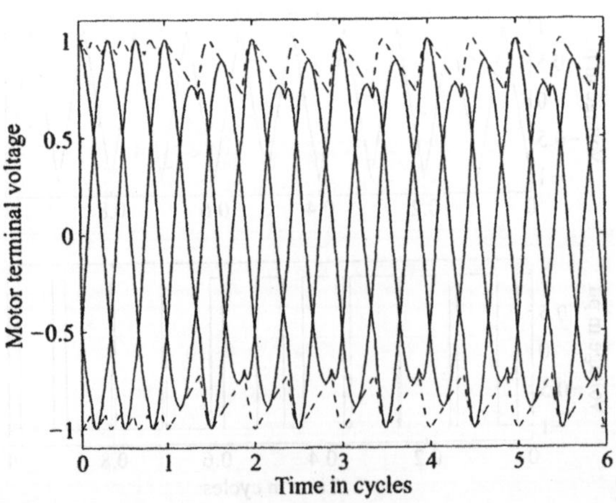

Figure 5.42 Motor terminal voltage due to a three-phase unbalanced sag of type C with a characteristic magnitude of 50%, for a motor frequency of 50 Hz. The dc bus voltage is shown as a dashed curve for reference.

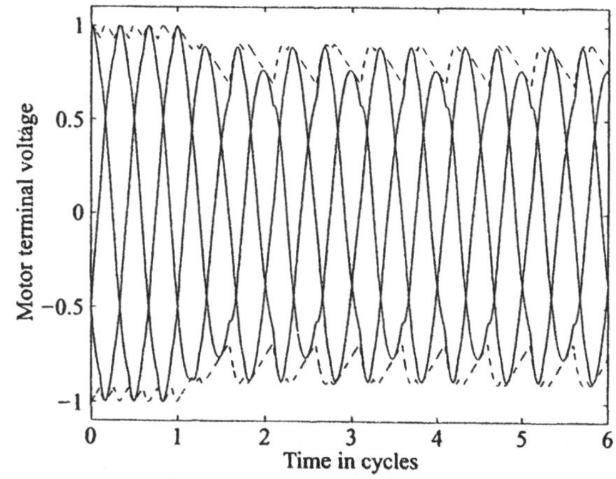

Figure 5.43 Motor terminal voltage due to a three-phase unbalanced sag of type D with a characteristic magnitude of 50%, for a motor frequency of 50 Hz. The dc bus voltage is shown as a dashed curve for reference.

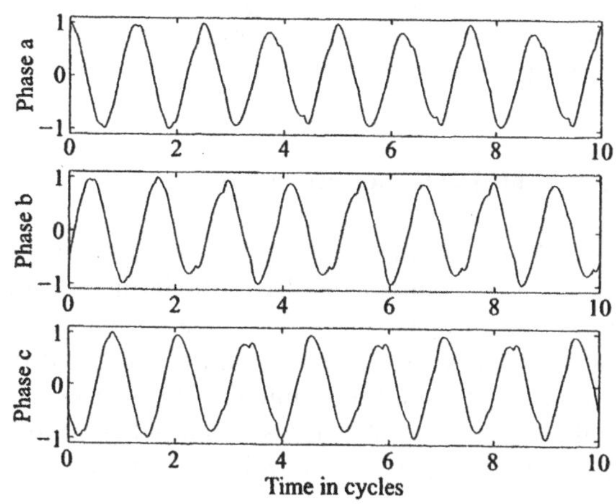

Figure 5.44 Motor terminal voltages due to a three-phase unbalanced sag of type C with a characteristic magnitude of 50%, for a motor speed of 40 Hz.

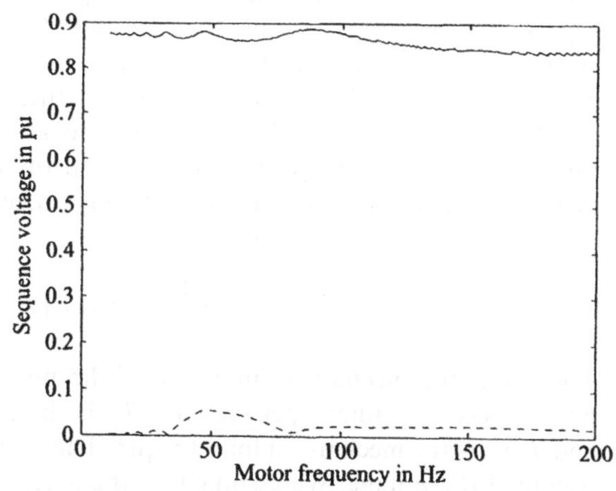

Figure 5.45 Positive- (solid) and negative-sequence component (dashed) of the motor terminal voltages as a function of the motor speed. A sag of type C with a characteristic magnitude of 50% was applied at the supply terminals of the adjustable-speed drive.

TABLE 5.8 Motor Terminal and DC Bus Voltages for AC Drives Due to a 50% Type C Sag

	Positive-sequence voltage		Negative-sequence voltage	dc bus voltage	
	max	min	max	avg.	rms
Small capacitance	88.88%	83.44%	5.56%	87.38%	87.80%
Large capacitance	98.25%	96.91%	0.81%	97.83%	97.84%

speeds around 50 Hz. For low speed the unbalance is very small. Note that the voltage at the supply terminals of the drive (i.e., the type C sag) contains 25% of negative-sequence and 75% of positive-sequence voltage. Even for a small dc bus capacitor the unbalance at the motor terminals is significantly less than at the supply terminals.

The results of the calculations are summarized in Table 5.8. Maximum and minimum positive and negative-sequence voltages have been obtained as in Fig. 5.45. (The lowest negative-sequence voltage was less than 0.01% in both cases.) The average dc bus voltage was obtained as in Fig. 5.25; the rms of the dc bus voltage as in Fig. 5.26. For a large dc bus capacitor, the ripple in the dc bus voltage becomes very small, so that the motor terminal voltages remain balanced, no matter how big the unbalance in the supply.

5.3.7 Motor Deacceleration

Most ac adjustable-speed drives trip on one of the characteristics discussed before. After the tripping of the drive, the induction motor will simply continue to slow down until its speed gets out of the range acceptable for the process. In case the electrical part of the drive is able to withstand the sag, the drop in system voltage will cause a drop in voltage at the motor terminals. We will estimate the motor speed for balanced and unbalanced sags. We will use a simplified motor model: the electrical torque is proportional to the square of the voltage, but independent of the motor speed; the mechanical torque is constant.

5.3.7.1 Balanced Sags. For balanced sags all three phase voltages drop the same amount. We assume that the voltages at the motor terminals are equal to the supply voltages (in p.u.), thus that the sag at the motor terminals is exactly the same as the sag at the rectifier terminals. The dc bus capacitor will somewhat delay the drop in voltage at the dc bus and thus at the motor terminals; but we saw that this effect is relatively small. The voltage drop at the motor terminals causes a drop in torque and thus a drop in speed. This drop in speed can disrupt the production process requiring an intervention by the process control. The speed of a motor is governed by the energy balance:

$$\frac{d}{dt}\left(\frac{1}{2}J\omega^2\right) = \omega(T_{el} - T_{mech}) \tag{5.28}$$

where J is the mechanical moment of the motor plus the mechanical load, ω is the motor speed (in radians per second), T_{el} is the electrical torque supplied to the motor, and T_{mech} is the mechanical load torque. The electrical torque T_{el} is proportional to the square of the voltage. We assume that the motor is running at steady state for a voltage of 1 pu, so that

$$T_{el} = V^2 T_{mech} \tag{5.29}$$

For $V = 1$ electrical and mechanical torque are equal. The resulting expression for the drop in motor speed is

$$\frac{d\omega}{dt} = \frac{(V^2 - 1)T_{mech}}{J} \tag{5.30}$$

Introduce the inertia constant H of the motor-load combination as the ratio of the kinetic energy and the mechanical output power:

$$H = \frac{\frac{1}{2} J\omega_0^2}{\omega_0 T_{mech}} \tag{5.31}$$

with ω_0 the angular frequency at nominal speed; and the slip:

$$s = \frac{\omega_0 - \omega}{\omega_0} \tag{5.32}$$

Combining (5.31) and (5.32) with (5.30) gives an expression for the rate of change of motor slip during a voltage sag (for $\omega \approx \omega_0$):

$$\frac{ds}{dt} = \frac{1 - V^2}{2H} \tag{5.33}$$

Thus for a sag of duration Δt and magnitude V the increase in slip is

$$\Delta s = \frac{ds}{dt}\Delta t = \frac{1 - V^2}{2H}\Delta t \tag{5.34}$$

The larger the inertia constant H, the less the increase in slip. For processes sensitive to speed variations, the voltage tolerance can be improved by adding inertia to the load. Figure 5.46 shows the increase in slip as a function of the sag magnitude and duration, for an inertia constant $H = 0.96$ sec. Note that an increase in slip corresponds to a drop in speed. The increase in slip is given for four different sag durations, corresponding to 2.5, 5, 7.5, and 10 cycles in a 50 Hz system. As expected the speed will drop more for deeper and for longer sags. But even for zero voltage (PWM disabled) the drop in speed is only a few percent during the sag.

If the maximum-allowable slip increase (slip tolerance) is equal to Δs_{max}, the minimum-allowable sag magnitude V_{min} for a sag duration T is found from

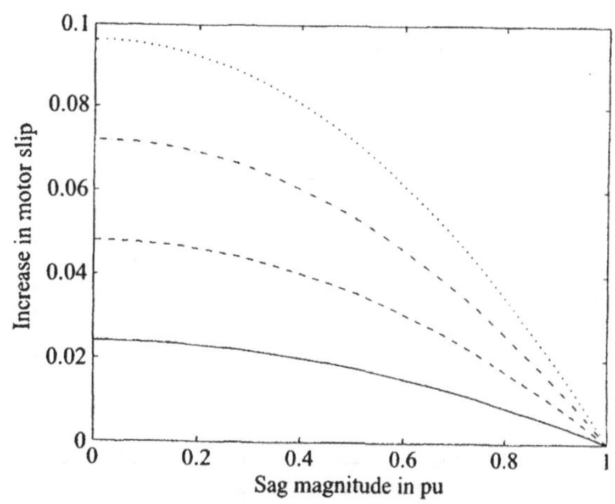

Figure 5.46 Increase in motor slip as a function of the sag magnitude for different sag duration: 50 ms (solid curve), 100 ms (dashed), 150 ms (dash-dot), 200 ms (dotted).

$$V_{min} = \sqrt{1 - \frac{2H\Delta s_{max}}{T}} \tag{5.35}$$

A zero voltage, $V_{min} = 0$, can be tolerated for a duration $2H\Delta s_{max}$. The resulting voltage-tolerance curves have been plotted in Fig. 5.47 for $H = 0.96$ sec and various values of the slip tolerance Δs_{max}. These are the voltage-tolerance curves for an adjustable-speed drive where the drop in speed of the mechanical load is the limiting factor.

Note that some of the earlier quoted tolerances of adjustable-speed drives are even above the 1% or 2% curves. This is mainly due to the sensitivity of the power-electronics part of the drive. Note also that it has been assumed here that the drive stays on-line. Temporary tripping of the drive corresponds to zero voltage at the drive terminals. This will obviously lead to a larger drop in speed.

5.3.7.2 Unbalanced Sags. The curves in Figs. 5.46 and 5.47 have been calculated assuming that the voltages at the motor terminals form a balanced three-phase set. For a balanced sag this will obviously be the case. But as we have seen in the previous section, for an unbalanced sag the motor terminal voltages are also rather balanced. The larger the dc bus capacitance, the more balanced the motor terminal voltages. The above calculations of the motor slip are still applicable. When the motor terminal voltage show a serious unbalance, the positive-sequence voltage should be used.

The effect of three-phase unbalanced sags on the motor speed has been calculated under the assumption that the positive-sequence voltage at the motor terminals is equal to the rms voltage at the dc bus. This is somewhat an approximation, but we have seen that the motor terminal voltage is only slightly unbalanced even for a large unbalance in the supply voltage. This holds especially for a drive with a large dc bus capacitance. The dc bus rms voltages have been calculated in the same way as for Figs. 5.26 and 5.30. These were used to calculate the drop in motor speed according to (5.34) and voltage-tolerance curves were obtained, as in Fig. 5.47. The results for type C sags are shown in Figs. 5.48, 5.49, and 5.50. Figures 5.48 and 5.49 present voltage-tolerance curves for different values of the maximum drop in speed which the load can tolerate, for no capacitance and for a small capacitance, respectively, present at the dc bus. Even the small capacitor clearly improves the drive's voltage tolerance. Below a certain characteristic magnitude of the sag, the rms value of the dc bus voltage remains constant. This

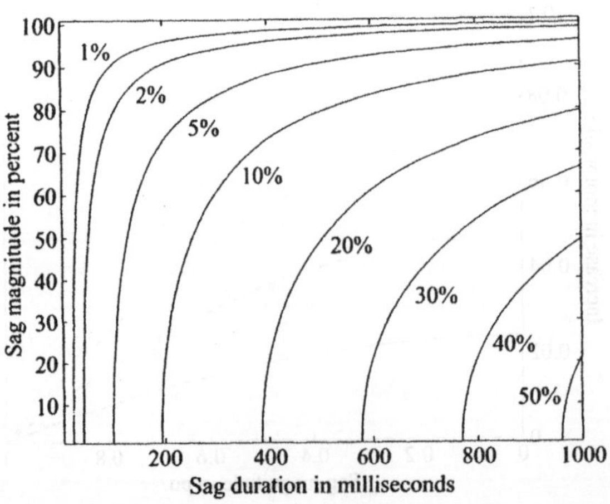

Figure 5.47 Voltage-tolerance curves for adjustable-speed drives, for three-phase balanced sags, for different values of the slip tolerance.

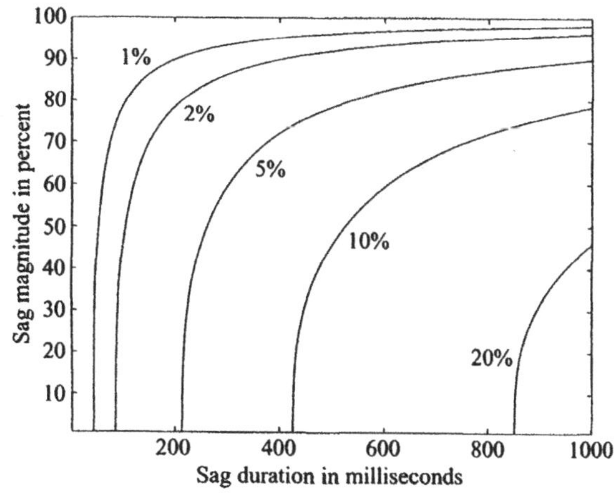

Figure 5.48 Voltage-tolerance curves for sag type C, no capacitance connected to the dc bus, for different values of the slip tolerance.

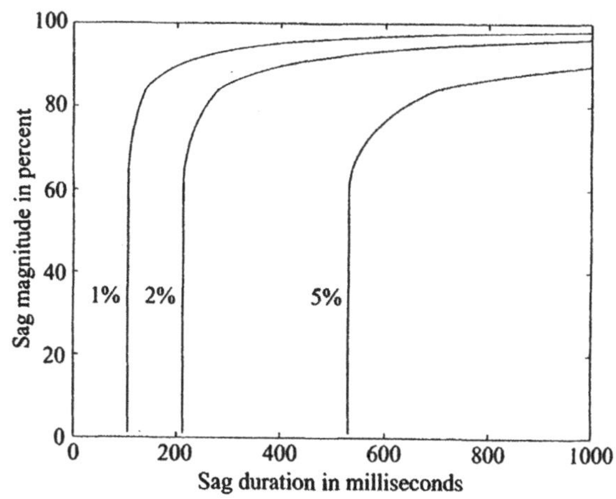

Figure 5.49 Voltage-tolerance curves for sag type C, small capacitance connected to the dc bus, for different values of the slip tolerance.

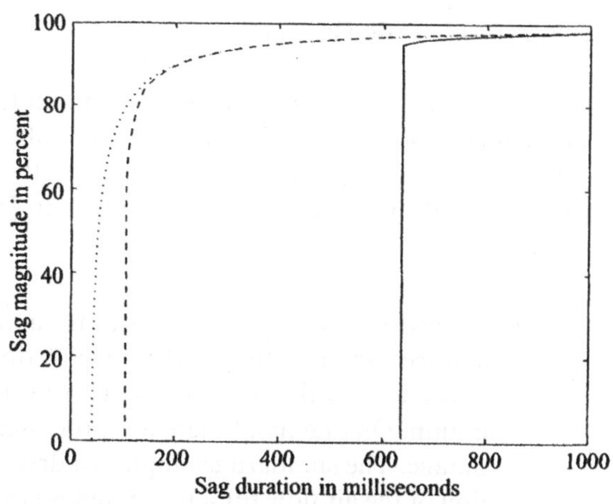

Figure 5.50 Voltage-tolerance curves for sag type C, large (solid line), small (dashed), and no (dotted) capacitance connected to the dc bus.

shows up as a vertical line in Fig. 5.49. Figure 5.50 compares drives with large, small, and no dc bus capacitance for a load with a slip tolerance of 1%. The capacitor size has a very significant influence on the drive performance.

The large improvement in drive performance with capacitor size for type C sags is obviously related to the one phase of the ac supply which does not drop in voltage. For a large capacitance, this phase keeps up the supply voltage as if almost nothing happened. For type D sags, this effect is smaller, as even the least-affected phases drop in voltage magnitude. Figure 5.51 shows the influence of the capacitor size on the voltage tolerance for type D sags. The three curves on the left are for a slip tolerance of 1%, the ones on the right for 10% slip tolerance. The improvement for the 1% case might look marginal, but one should realize that the majority of deep voltage sags have a duration around 100 ms. The large capacitance increases the voltage tolerance from 50 to 95 ms for a 50% sag magnitude. This could imply a serious reduction in the number of equipment trips.

From Figs. 5.48 through 5.51 it becomes clear that the effect of unbalanced sags on the motor speed is small. The best way to prevent speed variations is by using a large dc bus capacitor and by keeping the drive online. The small speed variations which would result may be compensated by a control system in case they cannot be tolerated by the load.

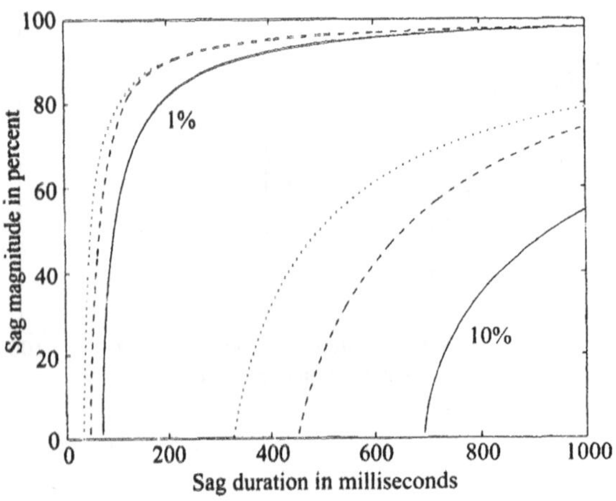

Figure 5.51 Voltage-tolerance curves for sag type D, for two values of the slip tolerance, large (solid line), small (dashed), and no (dotted) capacitance connected to the dc bus.

5.3.8 Automatic Restart

As we saw before many drives trip on undervoltage, for a sag of only a few cycles. This tripping of the drive does however not always imply a process interruption. What happens after the tripping depends on how the motor reacts when the voltage comes back. A good overview of options is given in [51], which served as a basis for the list below.

- Some drives simply trip and wait for a manual restart. This will certainly lead to a process interruption. A drive which does not automatically recover after a trip looks like a rather bad choice. However there are cases in which this is the best option. On one hand there are processes which are not very sensitive to a drive outage. The standard example is a drive used for air-conditioning. An interruption of the air flow for a few minutes is seldom any concern. On the other side of the spectrum one finds processes which are extremely sensitive to speed variation. If a very small speed variation already severely disrupts the process, it is best to not restart the drive. Restarting the drive certainly leads to a speed and torque transient, which could make the situation worse. Safety considerations could dictate that a total stoppage is preferable above an automatic restart.

- Some drives wait a few minutes before the automatic restart. This ensures that the motor load has come to a complete stop. The control system simply starts the motor in the same way it would do for a normal start. With a delayed automatic restart, safety measures have to be taken to ensure that nobody can be injured by the restart of the motor.
- The control system of the drive can apply electrical or mechanical braking to bring the load to a forced stop, after which a normal restart takes place. Without special control measures, it is very hard to restart the drive successfully before it has come to a standstill. Thus forced braking can reduce the time to recovery. The requirement is that the process driven by the drive is able to tolerate the variations in speed and torque due to braking and reacceleration.
- Most drives are able to start under full load, which also implies that they should be able to pick up the already spinning load. The danger of already spinning load is that it might still contain some air-gap flux causing an open-circuit voltage on the motor terminals. When the drive is restarted without any synchronization severe electrical transients are likely to occur due to the residual flux. The solution is to delay the restart for about one second to allow this residual flux to decay. This option will imply that the motor load will be without powering for one or two seconds. In this time the motor speed decays to a typical value of 50% of the nominal speed, depending on the intertia of the load. Also at the moment of restart the inverter frequency will not be equal to the motor speed, the mechanical transient this causes might not be tolerated by the process.
- A speed identification technique can be used to ensure that the inverter picks up the load at the right speed. This reduces the mechanical transient on restarts and makes the motor recover faster. The speed-identification process should be able to determine the motor speed within a few cycles to enable a fast restart of the drive.
- To seriously limit the drop in speed and the time to recovery, the drive needs to restart very soon after the voltage recovers. For this the inverter should be able to resynchronize on the residual stator voltages. This requires extra voltage sensors, thus increasing the price of the drive.
- Instead of resynchronizing the drive after the sag, it is possible to maintain synchronization between inverter and motor during the sag. This requires a more complicated measurement and control mechanism.

Figures 5.52 and 5.53 show the response of a drive with automatic restart. In Fig. 5.52 the drive restarts synchronously which leads to a drop in speed well within 10%. The motor current drops to zero during the sag. This indicates that the operation of the inverter was disabled (by inhibiting the firing of the inverter transistors). The moment the voltage recovered, inverter operation was enabled leading to the large peak in motor current. As the air-gap field in the motor is low and not synchronized with the inverter voltage, it takes another hundred milliseconds before the motor is actually able to

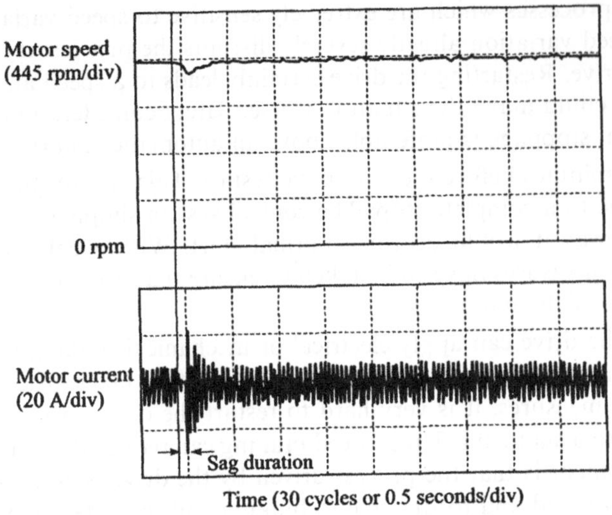

Figure 5.52 Drive response with synchronous restart. (Reproduced from Mansoor [32].)

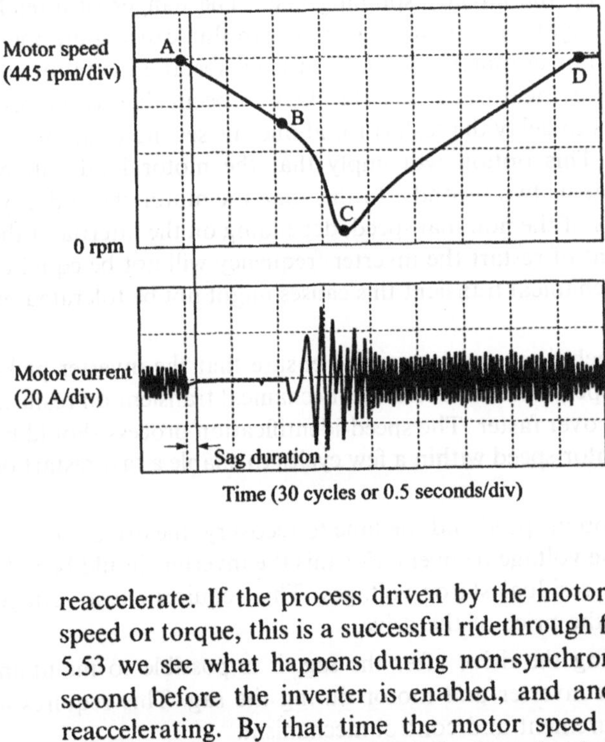

Figure 5.53 Drive response with non-synchronous restart. (Reproduced from Mansoor [32].)

reaccelerate. If the process driven by the motor is able to withstand the variation in speed or torque, this is a successful ridethrough from the process point of view. In Fig. 5.53 we see what happens during non-synchronous restart. It now takes about one second before the inverter is enabled, and another 500 ms for the motor to start reaccelerating. By that time the motor speed has dropped to almost zero. If the motor is used to power any kind of production process this would almost certainly not be acceptable. However, if the motor is used for air-conditioning the temporary drop in speed would not be of any concern.

5.3.9 Overview of Mitigation Methods for AC Drives

5.3.9.1 Automatic Restart. The most commonly used mitigation method is to disable the operation of the inverter, so that the motor no longer loads the drive.

This prevents damage due to overcurrents, overvoltages, and torque oscillations. After the voltage recovers the drive is automatically restarted. The disadvantage of this method is that the motor load slows down more than needed. When synchronous restart is used the drop in speed can be somewhat limited, but non-synchronous restart leads to very large drops in speed or even standstill of the motor. An important requirement for this type of drive is that the controller remain online. Powering of the controllers during the sag can be from the dc bus capacitor or from separate capacitors or batteries. Alternatively, one can use the kinetic energy of the mechanical load to power the dc bus capacitor during a sag or interruption [33], [35], [150].

5.3.9.2 Installing Additional Energy Storage. The voltage-tolerance problem of drives is ultimately an energy problem. In many applications the motor will slow down too much to maintain the process. This can be solved by adding additional capacitors or a battery block to the dc bus. Also the installation of a motor generator set feeding into the dc bus will give the required energy. A large amount of stored energy is needed to ensure tolerance against three-phase sags and short interruptions. For sags due to single-phase and phase-to-phase faults, which are the most common ones, only a limited amount of stored energy is needed as at least one phase of the supply voltage remains at a high value. This appears to be the easiest way of improving the voltage tolerance for the majority of sags.

5.3.9.3 Improving the Rectifier. The use of a diode rectifier is cheap but makes control of the dc bus voltage difficult. The moment the ac voltage maximum drops below the dc bus voltage, the rectifier stops supplying energy and the motor is powered from the capacitor. Using a controlled rectifier consisting of thyristors, like used in dc drives, gives some control of the dc bus voltage. When the ac bus voltage drops the firing angle of the thyristors can be decreased to maintain the dc bus voltage. For unbalanced sags different firing angles are needed for the three phases which could make the control rather complicated. Additional disadvantages are that the control system takes a few cycles to react and that the firing-angle control makes the drive sensitive to phase-angle jumps.

Another option is to use some additional power electronics to draw more current from the supply during the sag. A kind of power electronic current source is installed between the diode rectifier and the dc bus capacitor. This current can be controlled in such a way that it keeps the voltage at the dc bus constant during a voltage sag [150], [151].

By using a rectifier consisting of self-commutating devices (e.g., IGBTs), complete control of the dc voltage is possible. Algorithms have been proposed to keep the dc voltage constant for any unbalance, drop, or change in phase angle in the ac voltages [44], [45], [46]. An additional advantage is that these IGBT inverters enable a sinusoidal input current, solving a lot of the harmonic problems caused by adjustable-speed drives.

The main limitation of all these methods is that they have a minimum operating voltage and will certainly not operate for an interruption.

5.3.9.4 Improving the Inverter. Instead of controlling the dc bus voltage, it is also possible to control the motor terminal voltage. Normally the speed controller assumes a constant dc bus voltage and calculates the switching instants of the inverter from this. We saw earlier that the effect of this is that the dc bus voltage is amplitude modulated on the desired motor terminal voltages. This effect can be compensated by considering the dc bus voltage in the algorithms used to calculate the switching instants. For this (5.25) should be revised as follows, with V_{dc} the dc bus voltage:

$$\begin{aligned} V_{out} &= V_+, \quad \frac{V_{ref}}{V_{dc}} > V_{cr} \\ V_{out} &= V_-, \quad \frac{V_{ref}}{V_{dc}} < V_{cr} \end{aligned} \tag{5.36}$$

This in effect increases the reference voltage when the dc bus voltage drops (instead of pulse-width modulation this results in a kind of "pulse-area modulation"). The drawback of this method is that it will result in additional harmonic distortion, especially when the drive is operated close to nominal speed. Again this method has a minimum voltage below which it will no longer work properly.

5.4 ADJUSTABLE-SPEED DC DRIVES

DC drives have traditionally been much better suited for adjustable-speed operation than ac drives. The speed of ac motors is, in first approximation, proportional to the frequency of the voltage. The speed of dc motors is proportional to the magnitude of the voltage. Voltage magnitude is much easier to vary than frequency. Only with the introduction of power transistors have variable-frequency inverters and thus ac adjustable-speed drives become feasible. In this section we will discuss some aspects of the behavior of dc drives during voltage sags. Modern dc drives come in many different configurations, with different protection and control strategies. A discussion of all these is well beyond the scope of this book. The behavior described below does not cover all types of dc drives and should be viewed as an example of the kind of phenomena that occur when a voltage sag appears at the terminals of a dc drive.

5.4.1 Operation of DC Drives

5.4.1.1 Configuration. A typical configuration of a dc drive is presented in Fig. 5.54. The armature winding, which uses most of the power, is fed via a three-phase controlled rectifier. The armature voltage is controlled through the firing angle of the thyristors. The more the delay in firing angle, the lower the armature voltage. There is normally no capacitor connected to the dc bus. The torque produced by the dc motor is determined by the armature current, which shows almost no ripple due to

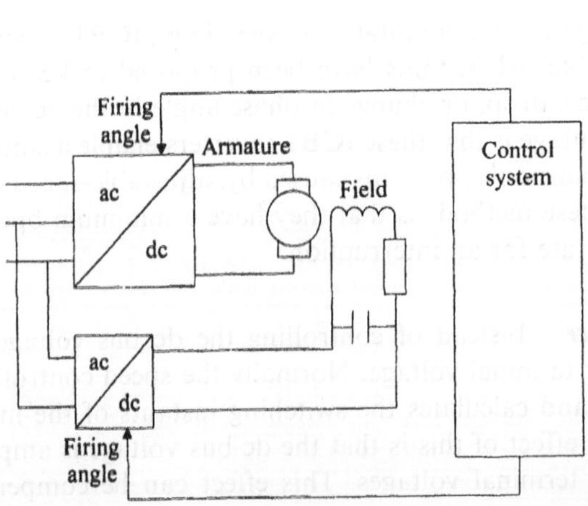

Figure 5.54 Modern dc drive with separately excited armature and field winding.

the large inductance of the armature winding. The field winding takes only a small amount of power; thus a single-phase rectifier is sufficient. The field winding is powered from one of the phase-to-phase voltages of the supply. In case field-weakening is used to extend the speed range of the dc motor, a controlled single-phase rectifier is needed. Otherwise a simple diode rectifier is sufficient. To limit the field current, a resistance is placed in series with the field winding. The resulting field circuit is therefore mainly resistive, so that voltage fluctuations result in current fluctuations and thus in torque fluctuations. A capacitor is used to limit the voltage (and torque) ripple. To limit these torque fluctuations a capacitor is used like the one used to limit the voltage ripple in single-phase rectifiers.

5.4.1.2 DC Motor Speed Control. The standard equivalent circuit for a dc motor is shown in Fig. 5.55. This circuit can only be used for normal operation, because it only considers the dc component of voltages and currents. A model including the inductance of the windings will be discussed further on.

The voltage V_f over the field winding causes a current I_f according to

$$V_f = R_f I_f \tag{5.37}$$

where R_f is the resistance in the field circuit (the resistance of the winding plus any external series resistance). This field current creates the air-gap field

$$\phi_f = kI_f \tag{5.38}$$

which rotates with a speed ω_m thus inducing a voltage E, the so-called "back-EMF" in the armature winding:

$$E = k\omega_m I_f \tag{5.39}$$

This induced voltage limits the armature current I_a:

$$V_a = E + R_a I_a \tag{5.40}$$

where V_a is the voltage over the armature winding and R_a the resistance of the armature winding. Field current and armature current together produce a torque

$$T_m = kI_a I_f \tag{5.41}$$

which accelerates the motor up to the speed at which motor torque and load torque balance.

The design of the motor is typically such that the armature resistance is low and the field resistance relatively high. Neglecting the armature resistance gives the following expression for the armature voltage:

$$V_a \approx E = k\omega_m I_f \tag{5.42}$$

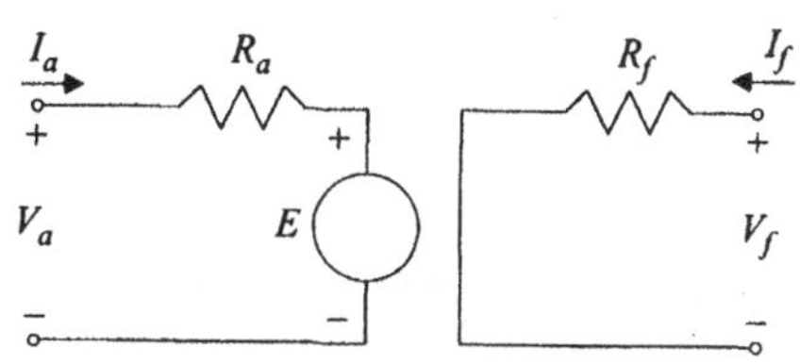

Figure 5.55 Equivalent scheme for dc motor during normal operation.

Rewriting this, and using field voltage as an independent variable, gives the basic expression for the speed control of dc motors:

$$\omega_m = \frac{R_f}{k}\frac{V_a}{V_f} \tag{5.43}$$

The speed of a dc motor is increased by increasing the armature voltage or by decreasing the field voltage. Speed control of a dc drive takes place in two ranges:

1. **Armature voltage control range.** The field voltage is kept at its maximum value and the speed is controlled by the armature voltage. This is the preferred range. The field current is high, thus the armature current has its minimum value for a given torque. This limits the armature losses and the wear on the brushes.
2. **Field weakening range.** Above a certain value the armature voltage can no longer be increased. It is kept constant and the speed is further increased by reducing the field voltage. As there is a maximum value for the armature current, the maximum torque decreases with increasing speed.

5.4.1.3 Firing-Angle Control. The dc component of the output voltage of a thyristor rectifier is varied by means of firing-angle control. The firing angle determines during which part of the cycle the rectifier conducts, and thus the average output voltage. The output voltage of a non-controlled three-phase rectifier was shown in Fig. 5.19 in Section 5.3. A diode starts conducting the moment its forward voltage becomes positive; a thyristor conducts only when the forward voltage is positive and a pulse is applied to its gate. By firing the thyristor at the instant a diode would start conducting, the output voltage of a controlled rectifier is the same as that of a non-controlled one. This is called free-firing. The firing angle of a thyristor is the delay compared to the free-firing point. Figure 5.56 shows the output voltage of a three-phase thyristor rectifier with a firing angle of 50°. For a controlled rectifier the dc bus voltage still consists of six pulses but shifted compared to the output voltage of a non-controlled rectifier. As the conduction period is shifted away from the voltage maximum, the average voltage becomes lower.

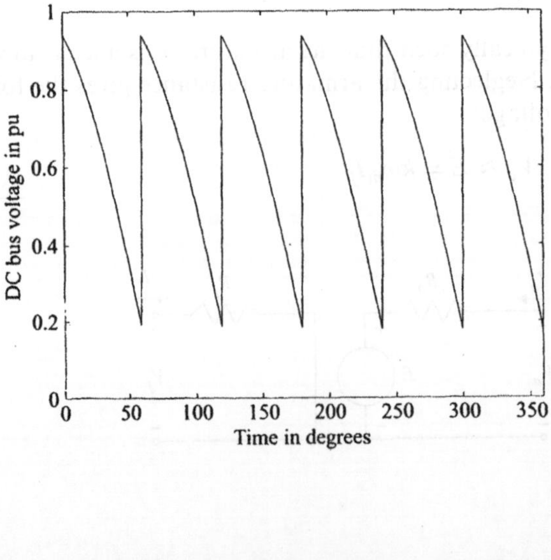

Figure 5.56 Output voltage of controlled rectifier with a firing angle of 50°. No capacitance is connected to the dc bus. Note the difference in vertical scale compared to Fig. 5.19.

A firing angle α delays conduction over a period $\frac{\alpha}{2\pi} \times T$, with T one cycle of the fundamental frequency. The average output voltage (i.e., the dc component) for a firing angle α is

$$V_{dc} = V_{max} \cos \alpha \tag{5.44}$$

with V_{max} the output voltage of a non-controlled rectifier. The voltage also contains an alternating component, with a frequency of six times the power system frequency: 300 Hz in a 50 Hz system; 360 Hz in a 60 Hz system. This voltage component will not lead to large fluctuations in the current and in torque due to the large inductance of the armature winding.

The firing of the thyristors takes place at a certain point of the supply voltage sine wave. For this the control system needs information about the supply voltage. There are different methods of obtaining the correct firing instant:

1. The thyristors are fired with a certain delay compared to the zero-crossing of the actual supply voltage. In normal operation the three voltages are shifted 120° compared to each other. Therefore, the zero-crossing of one voltage is used as a reference and all firing instants are obtained from this reference point. This method of control is extremely sensitive to distortion of the supply voltage. Any change in zero-crossing would lead to a change in firing angle and thus to a change in armature voltage. The problem is especially serious as thyristor rectifiers are the main source of notching, creating large distortion of the supply voltage sine wave [53], [55]. One could end up with a situation where the drive is not immune to its own emission.
2. The output voltage of a phase-locked loop (PLL) is used as a reference. A phase-locked loop generates an output signal exactly in phase with the fundamental component of the input signal. The reference signal is no longer sensitive to short-time variations in the supply voltage. This slow response will turn out to be a serious potential problem during voltage sags associated with phase-angle jumps.
3. A more sophisticated solution is to analyze the voltage in the so-called synchronously rotating dq-frame. In the forwardly rotating frame the voltage consists of a dc component proportional to the positive-sequence supply voltage and a component with twice the fundamental frequency proportional to the negative-sequence supply voltage. In the backwardly rotating frame the dc component is proportional to the negative-sequence voltage. Using a low-pass filter will give complex positive and negative-sequence voltage and thus all required information about the system voltages. The choice of the low-pass filter's cut-off frequency is again a compromise between speed and sensitivity to disturbances [152], [153].

5.4.2 Balanced Sags

A balanced voltage sag leads to a rather complicated transient in the dc motor, with a new steady state at the same speed as the original one. The new steady state will, however, rarely be reached. Most existing drives will trip long before, mainly through the intervention of some kind of protection in the power electronic converters. But even if the drive does not trip, the voltage sag will typically be over well within one second. The new steady state will only be reached for long shallow sags.

According to (5.43), the motor speed is proportional to the ratio of armature voltage and field voltage. The voltage sag in all three phases makes that armature and field voltage drop the same amount; the speed should thus remain the same. The model behind (5.43), however, neglects the transient effects, which are mainly due to the inductance of the motor winding and the inertia of the load. A model of the dc motor, which is valid for transients as well, is shown in Fig. 5.57, where L_a and L_f are the inductance of armature and field winding, respectively.

5.4.2.1 Theoretical Analysis. The qualitative behavior of the motor can be summarized as follows, where it is assumed that neither the control system nor the protection intervenes.

- Because of the voltage sag, the voltage on ac side of the field-winding rectifier will drop. This will lead to a decay in field current. The speed of decay is determined by the amount of energy stored in the inductance and in the capacitance. Typically the capacitor will give the dominant time constant so that the decay in field current can be expressed as follows:

$$I_f(t) = I_{f_0}(1 - e^{-\frac{t}{\tau}}) \tag{5.45}$$

  where I_{f_0} is the initial current and τ is the time constant of the decay in field current. The field current will not decay to zero, as suggested by (5.45), but the decay will stop the moment the field voltage reaches the ac voltage amplitude again. For a voltage drop of 20% the field current will also drop 20%. This is a similar situation as discussed in Section 5.2. The only difference is that the load is a constant impedance instead of constant power. For small dc voltage ripple it may take 10 cycles or more for the capacitor voltage, and thus for the field current, to decay. Note that the ripple in the field current directly translates into a torque ripple. As the latter is often not acceptable, a large capacitance is generally used. Some drives use a constant-voltage transformer to supply the field windings. The effect is again that the field current drops slowly.
- The voltage sag leads to a direct drop in armature voltage, which leads to a decay in armature current. The decay is somewhat different from the decay in field current. The armature current is driven by the difference between the armature voltage and the induced back-EMF. As this difference is normally only a few percent, the change in armature current can be very large. The current quickly becomes zero, but not negative because the rectifier blocks that. From Fig. 5.57 we obtain the following differential equation for the armature current I_a:

$$V_a = L_a \frac{dI_a}{dt} + R_a I_a + E \tag{5.46}$$

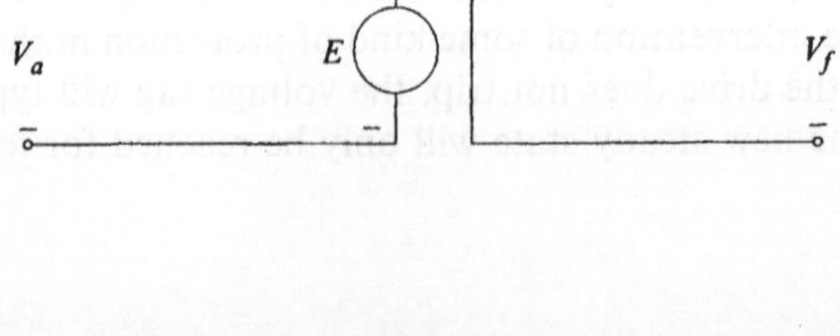

Figure 5.57 Equivalent circuit for a dc motor during transients.

The solution, with I_0 the armature current at time zero, is

$$I_a = \frac{V_a - E}{R_a} + \left(I_0 - \frac{V_a - E}{R_a}\right)e^{-\frac{t}{\tau}} \tag{5.47}$$

where V_a is the armature voltage during the sag, and $\tau = \frac{L_a}{R_a}$. As we saw before, the field current remains close to its pre-event value for at least a few cycles. Because the motor speed does not immediately drop, the back-emf E remains the same. The effect of a drop in armature voltage is thus that the current drops toward a large negative value $(V_a - E)/R_a$.

We will estimate how fast the armature current reaches zero by approximating (5.47) for $t \ll \tau$. Using $e^{-\frac{t}{\tau}} \approx 1 - \frac{t}{\tau}$ gives

$$I_a \approx I_0 - \frac{E - V_a}{L_a} t \tag{5.48}$$

The pre-sag steady-state current I_0 may be obtained from

$$I_0 = \frac{1 - E}{R_a} \tag{5.49}$$

where the steady-state armature voltage is chosen equal to 1 pu. The time for the current to reach zero is, in cycles of the fundamental frequency:

$$t = \frac{1}{2\pi}\left(\frac{X_a}{R_a}\right)\frac{1 - E}{1 - V} \tag{5.50}$$

where X_a is the armature reactance at the fundamental frequency. For $X_a/R_a = 31.4$ and $1 - E = 0.05$ we obtain

$$t = \frac{0.25}{1 - V} \text{ (cycles)} \tag{5.51}$$

For a sag down to 75% the current drops to zero in one cycle; for a 90% sag it takes 2.5 cycles which is still very fast. Thus for the majority of sags the armature current and the torque will drop to zero within a few cycles.

- The drop in armature and in field current leads to a drop in torque which causes a drop in speed. The drop in speed and the drop in field current cause a reduction in back-EMF.
- Sooner or later the back-EMF will become smaller than the armature voltage, reversing the drop in armature current. Because speed as well as field current have dropped the new armature current is higher than the pre-event value.
- The more the speed drops, the more the back-EMF drops, the more the armature current increases, the more the torque increases. In other words, the dc motor has a built-in speed control mechanism via the back-EMF.
- The torque becomes higher than the load torque and the load reaccelerates.
- The load stabilizes at the original speed and torque, but for a lower field current and a higher armature current. The drop in field current equals the drop in voltage; the armature current increases as much as the field current drops, because their product (the torque) remains constant.

5.4.2.2 Simulation of Balanced Sags. Some simulations have been performed to quantify the behavior described above. The results are shown in Figs. 5.58 through 5.61. The simulated drive was configured as shown in Fig. 5.54, with a three-phase rectifier to power the armature winding and a single-phase rectifier for the field winding. The drive was operating at nominal speed, thus with zero firing angle for the rectifiers. In this system the time constant was 100 ms, both for the armature winding and for the field winding. A supply voltage of 660 V was used resulting in a pre-sag motor power of 10 kW and a speed of 500 rpm. The moment of inertia of the load driven by the motor was 3.65 kgm/s^2. The load torque was proportional to the speed. The simulations were performed by solving the differential equations with a step-by-step approximation [154]. The voltage dropped to 80% in all three phases during 500 ms (30 cycles). The plots show two cycles pre-sag, 30 cycles during-sag, and 88 cycles post-sag.

The armature current is shown in Fig. 5.58. The armature current drops to zero in a very short time due to the phenomenon described before. As a direct consequence the torque becomes zero also, as shown in Fig. 5.60. This in turn leads to a fast drop in speed, as shown in Fig. 5.61. After a few cycles the field current (Fig. 5.59) and the speed have dropped sufficiently for the back-EMF to become lower that the armature

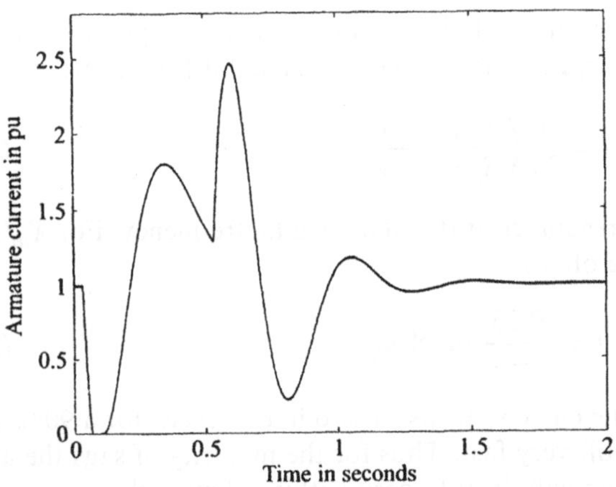

Figure 5.58 DC motor armature current during balanced sag.

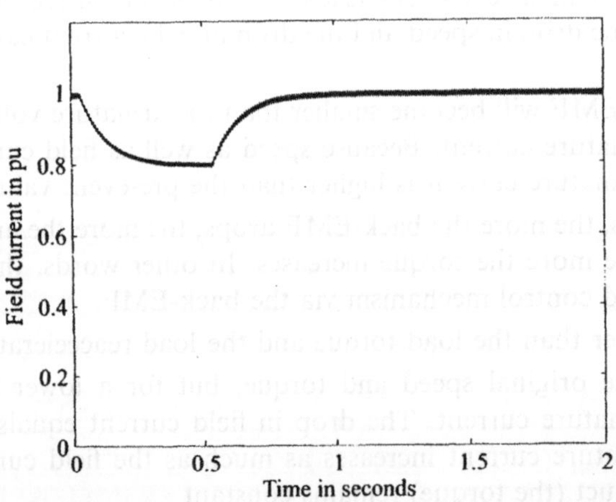

Figure 5.59 DC motor field current during balanced sag.

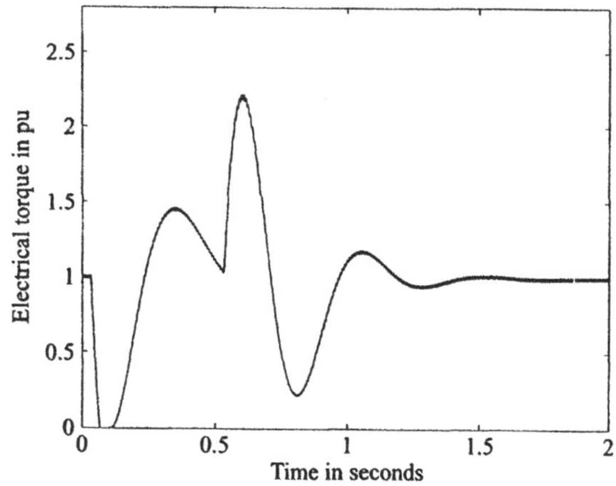

Figure 5.60 Torque produced by dc motor during balanced sag.

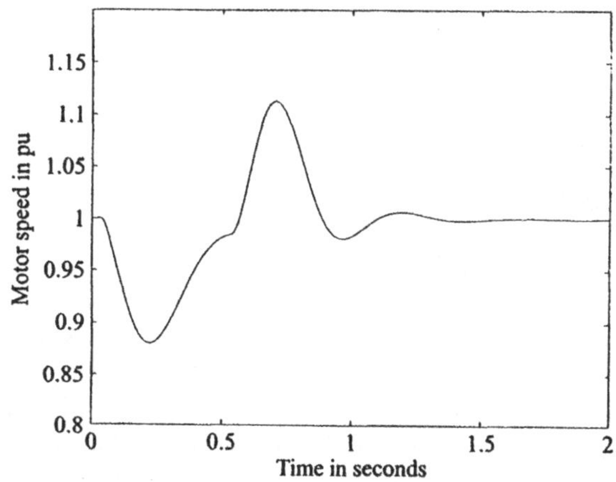

Figure 5.61 Speed of dc motor during balanced sag.

voltage. From this moment on the armature current and the torque recover and a few hundred milliseconds later even exceed their pre-sag value. The result is that the motor picks up speed again.

Upon voltage recovery, around $t = 0.5$ in the figures, the opposite effect occurs. The armature voltage becomes much larger than the back-EMF leading to a large overcurrent, a large torque, and even a significant overspeed. The post-sag transient is over after about one second. Note that the simulated behavior was due to a sag down to 80%, a rather shallow sag. Due to the fast drop in armature current even such a shallow sag will already lead to a serious transient in torque and speed.

5.4.2.3 Intervention by the Control System. The control system of a dc drive can control a number of parameters: armature voltage, armature current, torque, or speed. In case the control system is able to keep armature and field voltage constant, the drive will not experience the sag. However, the control system will typically take a few cycles to react, so that the motor will still experience the fast drop in armature current. The use of such a control system may also lead to an even more severe transient at voltage recovery. The armature voltage will suddenly become much higher than the back-emf leading to a very fast rise in armature current, torque, and speed.

If the motor aims at keeping the motor speed constant, the drop in speed (as shown in Fig. 5.61) will be counteracted through a decrease in firing angle of the thyristor rectifier. For a deep sag the firing angle will quickly reach its minimum value. Further compensation of the drop in armature voltage would require control of the field voltage. But as we saw above, the field voltage is kept intentionally constant so that control is difficult.

5.4.2.4 Intervention by the Protection. The typical reason for the tripping of a dc drive during a voltage sag is that one of the settings of the protection is exceeded. As shown in Figs. 5.58 through 5.61, voltage, current, speed, and torque experience a large transient. The protection could trip on any of these parameters, but more often than not, the protection simply trips on dc bus undervoltage.

DC drives are often used for processes in which very precise speed and positioning are required, e.g., in robotics. Even small deviations in speed cannot be tolerated in such a case. We saw before that the motor torque drops very fast, even for shallow sags, so that the drop in speed will become more severe than for an ac drive. A shallow sag will already have the same effect on a dc drive as a zero voltage on an ac drive: in both cases the torque produced by the motor drops to zero.

5.4.3 Unbalanced Sags

One of the effects of unbalanced sags on dc drives is that armature and field voltage do not drop the same amount. The armature voltage is obtained from a three-phase rectifier, the field voltage from a single-phase rectifier. During an unbalanced sag, the single-phase rectifier is likely to give a different output voltage than the three-phase rectifier. If the field voltage drops more than the armature voltage, the new steady-state speed could be higher than the original speed. However, initially both armature and field current decrease, leading to a decrease in torque and thus in speed. The slowest speed recovery takes place when the field voltage remains constant. The back-EMF only starts to drop when the motor slows down. The armature current will remain zero longer when the field voltage stays constant.

- If the field voltage drops more than the armature voltage, the back-emf will quickly be less than the armature voltage, leading to an increase in armature current. Also the new steady-state speed is higher than the pre-event speed. Overcurrent in the armature winding and overspeed are the main risk.
- If the field voltage drops less than the armature voltage, the armature current's decay will only be limited by the drop in motor speed. It will take a long time before the motor torque recovers. As the new steady-state speed is lower than the pre-event speed, underspeed becomes the main risk.

Simulations have been performed for the same drive configuration as before. But instead of a balanced sag, a number of unbalanced sags were applied to the drive. The results of two sags of type D and one sag of type C are shown here. All three sags had a duration of 10 cycles, a characteristic magnitude of 50%, and zero characteristic phase-angle jump. Note that in this case the sag type refers to the line-to-line voltages, not the the line-to-neutral voltage. The rectifier is delta-connected; thus the line-to-line voltages more directly influence the drive behavior.

- *SAG I*: a sag of type D with the large voltage drop in the phase from which the field winding is powered. The field voltage thus drops to 50%. The results for sag I are shown in Figs. 5.62 through 5.65.
- *SAG II*: a sag of type D with a small voltage drop in the phase from which the field winding is powered, making the field voltage drop to about 90%. The results for sag II are shown in Figs. 5.66 through 5.69.
- *SAG III*: a sag of type C with the field winding powered from the phase without voltage drop. The field voltage thus remains at 100%. The results for sag III are similar to those for sag II and therefore not reproduced in detail.

All plots show two cycles before the sag, 10 cycles during the sag, and 48 cycles after the sag. From the figures we can see that a deep sag in the field voltage (sag I) causes a high overshoot in the armature current (Fig. 5.63), in the torque (Fig. 5.64), and in the speed (Fig. 5.65). For a shallow sag in the field voltage (sag II) the armature current and torque are zero for a long time, but with a smaller overshoot (Figs. 5.67 and 5.68); the speed shows a large drop but only a small overshoot (Fig. 5.69). Note the ripple in the armature current during the sag. The unbalance in the ac voltage leads to a much larger

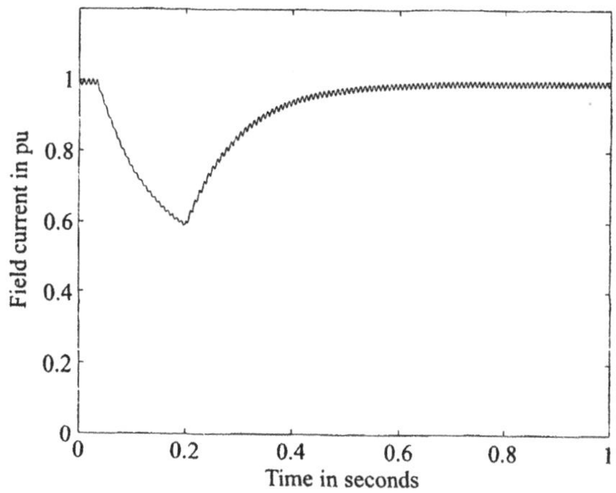

Figure 5.62 Field current for sag type D, with large drop in field voltage.

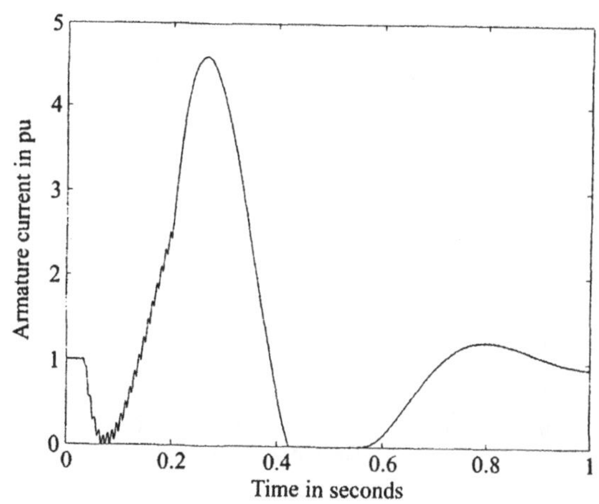

Figure 5.63 Armature current for sag type D, with large drop in field voltage.

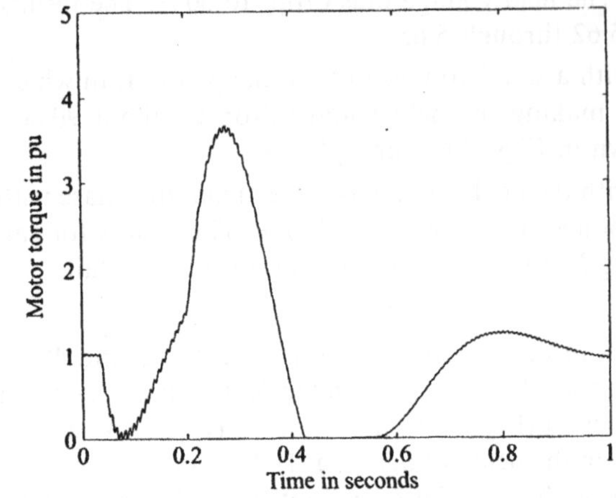

Figure 5.64 Motor torque for sag type D, with large drop in field voltage.

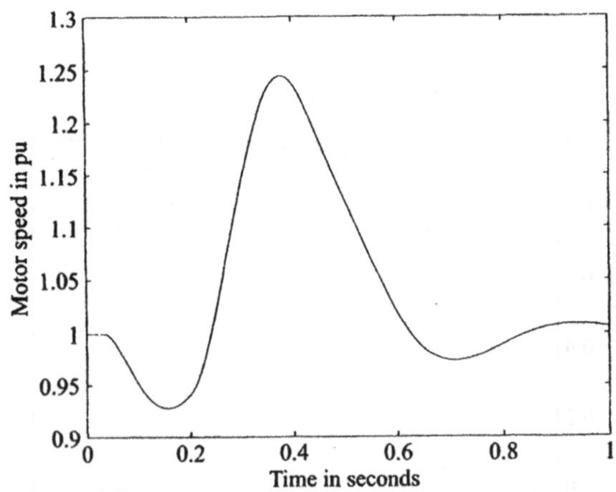

Figure 5.65 Motor speed for sag type D, with large drop in field voltage.

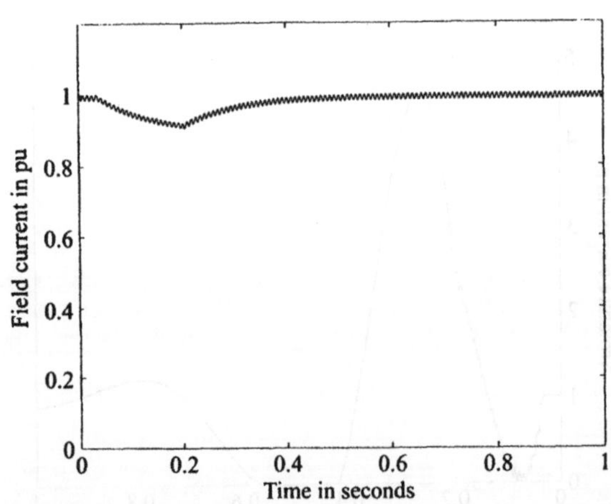

Figure 5.66 Field current for sag type D, with small drop in field voltage.

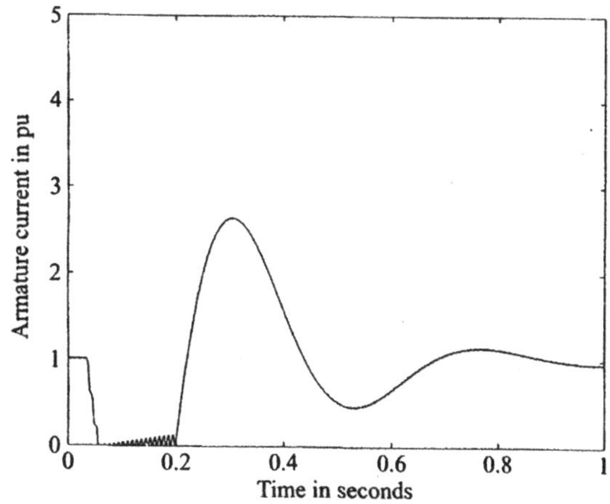

Figure 5.67 Armature current for sag type D, with small drop in field voltage.

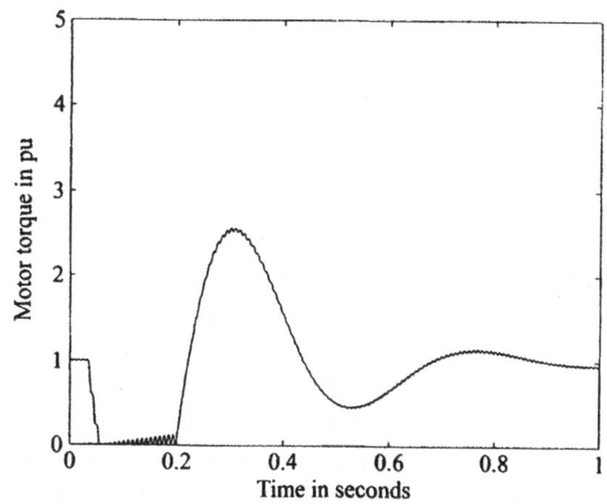

Figure 5.68 Motor torque for sag type D, with small drop in field voltage.

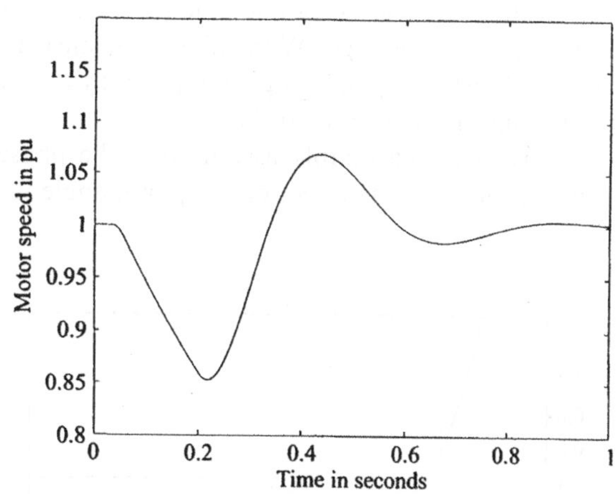

Figure 5.69 Motor speed for sag type D, with small drop in field voltage.

ripple in armature voltage than during normal operation. This ripple disappears upon voltage recovery and is also not present during a balanced sag (Fig. 5.58).

The maximum and minimum values for current, torque, and speed are shown in Table 5.9. All values are given as a percentage of the average pre-event value. Tripping of the drive can be due to undervoltage or overcurrent. The undervoltage is similar for the three sags; thus sag I is the most severe one for the electrical part of the drive because of the large armature current. The mechanical process can, however, get disrupted due to torque variations and variations in speed. For a process sensitive to underspeed, sags II and III are most severe; for a process sensitive to torque variations, sag I is the most severe one. The main conclusion is that unbalanced sags require testing for all phases; it is hard to predict beforehand which sag will be most severe to the drive.

TABLE 5.9 DC Drive Performance During Unbalanced Sags in Different Phases

			Field Current		Armature Current		Motor Torque		Motor Speed	
Sag	Type	Field Voltage	min	max	min	max	min	max	min	max
I	D	50%	59%	100%	0	460%	0	367%	93%	124%
II	D	90%	90%	100%	0	264%	0	256%	85%	107%
III	C	100%	100%	100%	0	229%	0	229%	85%	114%

5.4.4 Phase-Angle Jumps

Phase-angle jumps affect the angle at which the thyristors are fired. The firing instant is normally determined from the phase-locked loop (PLL) output, which takes at least several cycles to react to the phase-angle jump.

A calculated step response of a conventional digital phase-locked loop to a phase-angle jump is shown by Wang [57]. His results are reproduced in Fig. 5.70, where we can see that it takes about 400 ms for the PLL to recover. The error gets smaller than 10% after about 250 ms, which is still longer than the duration of most sags. Thus for our initial analysis we can assume that the firing instants remain fixed to the pre-event voltage zero-crossings. With additional measures it is possible to make PLLs which respond faster to phase-angle jumps, but those will be more sensitive to harmonics and other high-frequency disturbances.

We can reasonably assume that the phase-locked-loop output does not change during the sag. The effect of the phase-angle jump is that the actual voltage is shifted

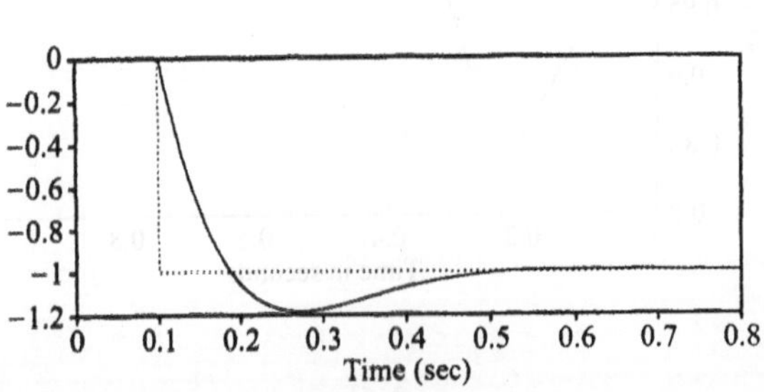

Figure 5.70 Step response of a conventional digital phase-locked loop. (Reproduced from Wang [57].)

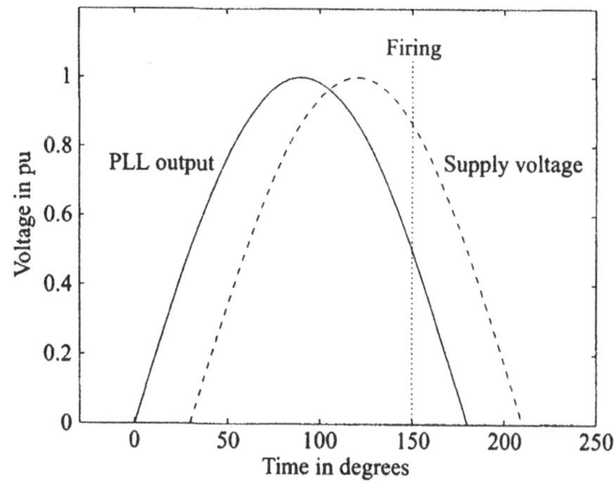

Figure 5.71 Influence of phase-locked loop on firing angle.

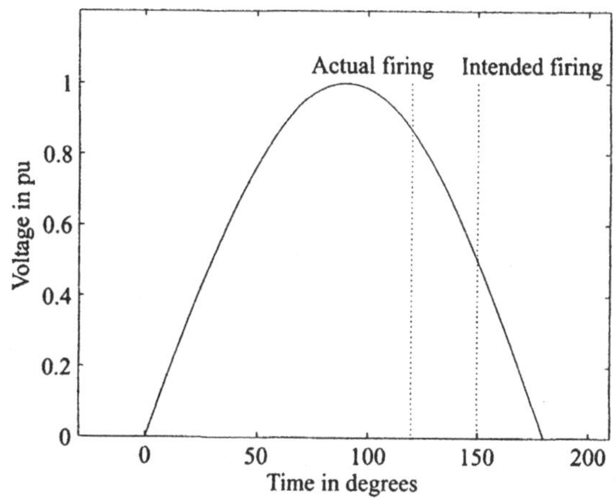

Figure 5.72 Influence of phase-locked loop on firing angle: with actual voltage as a reference.

compared to the reference voltage. Because of this the thyristors are fired at a wrong point of the supply-voltage sine wave. This is shown in Fig. 5.71 for a negative phase-angle jump. The during-sag voltage lags the pre-sag voltage; thus the zero-crossing of the actual supply voltage comes later than the zero crossing of the PLL output. In Fig. 5.72 the sine wave of the actual voltage is used as a reference: due to the negative phase-angle jump $\Delta\phi$, the thyristors are fired at an angle $\Delta\phi$ earlier than intended.

5.4.4.1 Balanced Sags. For balanced sags the phase-angle jump is equal in the three phases; thus the shift in firing angle is the same for all three voltages. If the shift is less than the intended firing-angle delay, the output voltage of the rectifier will be higher than it would be without phase-angle jump. This assumes that the phase-angle jump is negative, which is normally the case. A negative phase-angle jump will thus somewhat compensate the drop in voltage due to the sag. For a positive phase-angle jump the output voltage would be reduced and the phase-angle jump would aggravate the effects of the sag.

For a firing angle equal to α the pre-sag armature voltage equals

$$V_a = \cos(\alpha) \tag{5.52}$$

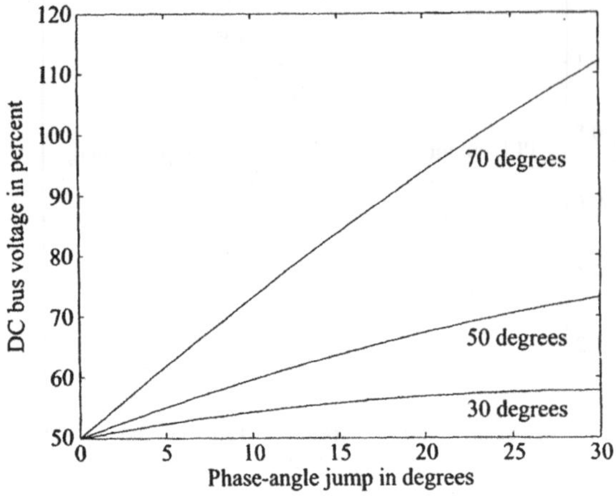

Figure 5.73 Influence of phase-angle jump on the armature voltage, for different firing angles.

The voltage is rated to the armature voltage for zero firing angle. For a sag with magnitude V (in pu) and phase-angle jump $\Delta\phi$, the during-event armature voltage is

$$V_a' = V \times \cos(\alpha - \Delta\phi) \tag{5.53}$$

The phase-angle jump is assumed negative, $\Delta\phi$ is its absolute value. The ratio between V_a' and V_a is the relative magnitude of the sag in the armature voltage. This is plotted in Fig. 5.73 for firing-angle delays of 30°, 50°, and 70°. A during-event magnitude V of 50% has been assumed, and the phase-angle jump is varied between zero and 30°. According to Fig. 4.86 this is the range one can expect for a 50% sag. For large firing-angle delays the armature voltage is low; thus a jump in phase-angle can increase the voltage significantly. For a 70° firing-angle delay and phase-angle jumps of 20° and higher the during-event voltage is even higher than the pre-event voltage. Whether this actually makes the sag less severe depends on the behavior of the field voltage. When a diode rectifier is used to power the field winding, the field voltage will not be influenced by the phase-angle jump. The consequence of the phase-angle jump is that the field voltage drops more than the armature voltage, similar to sag I discussed in the previous section. This can lead to large overcurrents in the armature winding and to overspeed. When a controlled rectifier is used there is a risk of missing pulses which would make the field voltage much lower than the armature voltage.

If the shift is larger than the intended firing-angle delay, the actual firing will take place before the free-firing point. As the forward voltage over the thyristors is still negative it will not commence conducting. How serious this effect is depends on the duration of the firing pulse. The use of a short pulse will make the drive more sensitive. Note that either the armature or the field rectifier is operated at its maximum voltage so that at least one of them always will be prone to missing pulses.

5.4.4.2 Unbalanced Sags. For unbalanced sags the situation becomes rather complicated. In most cases the different phases show positive as well as negative phase-angle jumps. Thus for some phases the phase-angle jump can be an improvement, for others not. Some phases might miss their firing pulses, others not. The armature winding might be influenced differently from the field current as we already saw before.

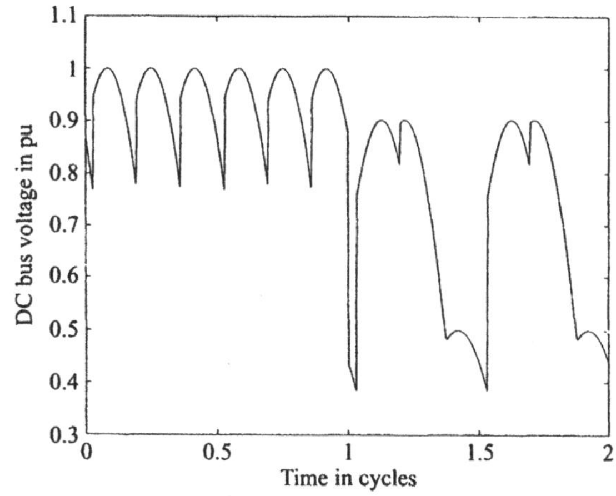

Figure 5.74 DC voltage for sag type D, with rectifier operating at 10° firing angle.

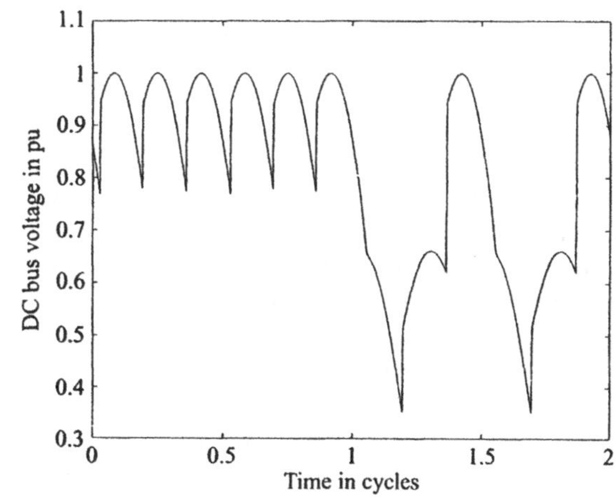

Figure 5.75 DC voltage for sag type C, with rectifier operating at 10° firing angle.

Figures 5.74 and 5.75 show the dc bus voltage before and during a voltage sag, in case the rectifier is operated at a firing angle of 10°. Figure 5.74 shows the effect of a type D sag of 50% magnitude. As all three voltages go down in magnitude the maximum dc voltage also drops. The two voltage pulses belonging to the least-affected phases come very close after each other. In the phasor diagram they move away from each other, so that the voltage maxima of the rectified voltage come closer. The consequence is that the commutation between these two phases takes place at a natural commutation point. The firing of the thyristor has taken place already before that moment in time. There is thus a risk for a missing pulse which would even more distort the dc bus voltage. Figure 5.75 shows the effect of a type C sag of 50% magnitude.

5.4.5 Commutation Failures

The moment a thyristor is fired and forwardly biased, it starts conducting. But the current through the conductor does not immediately reach its full value because of the inductive nature of the source. Consider the situation shown in Fig. 5.76, where the

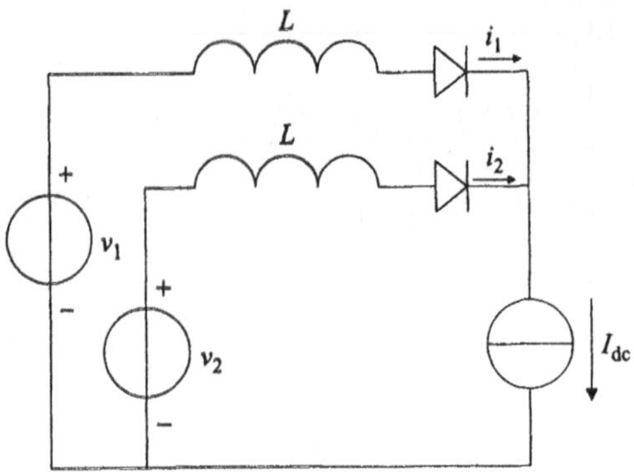

Figure 5.76 Origin of commutation delay.

current commutates from phase 1 to phase 2. The driving voltages in these two phases are shifted by 120°:

$$v_1(t) = \hat{V}\cos\left(\omega_0 t + \frac{\pi}{3}\right) \tag{5.54}$$

$$v_2(t) = \hat{V}\cos\left(\omega_0 t - \frac{\pi}{3}\right) \tag{5.55}$$

At time zero the two driving voltages are the same, thus the line-to-line voltage is zero, which corresponds to the free-firing point. For a firing-delay angle α, thyristor 2 is fired at $\omega_0 t = \alpha$. This is the moment the current through thyristor 1 starts to rise and the current through thyristor 2 starts to decay. The change in current is described through the following differential equation (note that both thyristors conduct, thus the two phases are shorted):

$$v_1(t) - L\frac{di_1}{dt} + L\frac{di_2}{dt} = v_2(t) \tag{5.56}$$

with L the source inductance. We can assume the armature current I_{dc} to be constant; thus the changes in i_1 and i_2 compensate each other:

$$\frac{di_1}{dt} + \frac{di_2}{dt} = 0 \tag{5.57}$$

after which i_2 can be obtained from the differential equation:

$$\frac{di_2}{dt} = \frac{\sqrt{3}\hat{V}\sin(\omega_0 t)}{2L} \tag{5.58}$$

with the following solution:

$$i_2(t) = \frac{\sqrt{3}\hat{V}}{2\omega_0 L}[\cos(\alpha) - \cos(\omega_0 t)], \qquad t > \frac{\alpha}{\omega_0} \tag{5.59}$$

Commutation is complete and thyristor 1 ceases to conduct when $i_2(t) = I_{dc}$. Commutation takes longer for smaller values of $\hat{V}$, thus during voltage sags, and for a firing-delay angle α closer to 180°, thus for the drive being in regenerative mode. The maximum current the supply voltage is able to cummutate is found from (5.59) as

$$I_{max} = \frac{\sqrt{3}\hat{V}}{2\omega_0 L}(1 + \cos\alpha) \tag{5.60}$$

If this is less than the actual armature current, a commutation failure occurs: both thyristors will continue to conduct, leading to a phase-to-phase fault. This will cause blowing of fuses or damage of the thyristors. The risk of commutation failure is further increased by the increased armature current during and after the sag.

A negative phase-angle jump reduces the actual firing angle, thus lowering the risk of commutation failure. A positive phase-angle jump makes a commutation failure more likely. Unbalanced faults cause a combination of positive and negative phase-angle jumps, thus increasing the risk in at least one phase.

5.4.6 Overview of Mitigation Methods for DC Drives

Making dc drives tolerant against voltage sags is more complicated than for ac drives. Three potential solutions, to be discussed below, are adding capacitance to the armature winding, improved control system, and self-commutating rectifiers.

5.4.6.1 Armature Capacitance. Installing capacitance to the armature winding, on dc side of the three-phase rectifier, makes that the armature voltage no longer drops instantaneously upon sag initiation. Instead the armature voltage decays in a similar way to the field voltage. To obtain a large time constant for the decay of the armature voltage requires a large capacitor for the armature winding. Note that the power taken by the armature winding is much larger than the power taken by the field winding. For three-phase unbalanced sags it may be sufficient to keep up the voltage during one half-cycle.

Keeping up the armature voltage will still not solve the problem of missing pulses due to phase-angle jumps and commutation failures. Another disadvantage of any amount of armature capacitance is that it makes the drive react slower to the control system. Changes in motor speed are obtained through changes in firing angle. The armature capacitance slows down the response of the armature current and torque on a change in firing angle. When the drive application requires fast changes in torque and speed, the armature capacitance should be small.

5.4.6.2 Improved Control System. Any control system for a dc drive ultimately controls the firing angle of a controlled rectifier. This may be the armature rectifier, the field rectifier, or both. Due to the nature of a thyristor rectifier it is unlikely that the control system will have an open-loop time constant less than two cycles. We saw before that the drop in armature current and torque takes place much faster than this. It is thus not possible to prevent the transient in armature current and torque.

Two straightforward quantities to be controlled are armature voltage and motor speed. Controlling the armature voltage enables the use of a simple controller with a small open-loop time constant. For the controller to work, sufficient margin must be available in the rectifier to bring the armature voltage back to 100%. If sags down to 50% magnitude have to be mitigated, the normal operating voltage on dc side of the rectifier should not exceed 50% of maximum. The result is that only half of the control range of the rectifier can be used for speed control. The other half is needed for voltage-sag mitigation.

Speed control is the commonly-used method of control for dc drives. The voltage sag will cause a drop in speed. The speed controller detects this and reduces the firing angle to compensate. If the firing angle is zero the controller can no longer increase the speed. Speed control will not mitigate the transients in torque and current but it may reduce the variations in speed.

A disadvantage of both control techniques is that they will lead to a severe transient in armature current and torque upon voltage recovery.

5.4.6.3 Improved Rectifiers. The control of the drive may be significantly improved by using a self-commutating rectifier. These rectifiers enable control of the output voltage on a sub-cycle timescale. This will prevent the drop in armature voltage and thus the severe drop in torque. Using advanced control techniques it may also be possible to install additional enery storage which is only made available during a reduction in the supply voltage.

By using self-commutating rectifiers it may also be possible to use a sophisticated control system that detects and mitigates phase-angle jumps. With such a control system, the reference signal should no longer be obtained from a phase-locked loop but from the measured supply voltage through a suitable digital filter.

5.4.6.4 Other Solutions. Other solutions include a more critical setting of the undervoltage and overcurrent protection; the use of components with higher overcurrent tolerance; and disabling the firing of the thyristors to prevent tripping on overcurrent. All these solutions are only feasible when the load can tolerate rather large variations in speed.

5.5 OTHER SENSITIVE LOAD

5.5.1 Directly Fed Induction Motors

Despite the growth in the number of adjustable-speed drives, the majority of induction motors are still directly fed; i.e., the motor terminals are connected to the constant frequency, constant voltage, supply. It will be clear that speed control of the motor is not possible. Directly fed induction motors are rather insensitive to voltage sags, although problems could occur when too many motors are fed from the same bus.

The drop in terminal voltage will cause a drop in torque for an induction motor. Due to this drop in torque the motor will slow down until it reaches a new operating point. If the terminal voltage drops too much the load torque will be higher than the pull-out torque and the motor will continue to slow down. An induction motor is typically operated at half its pull-out torque. As the pull-out torque is proportional to the square of the voltage, a voltage drop to 70% or less will not lead to a new stable operating point for the induction motor. The drop in speed is seldom a serious concern for directly fed induction motors. These kind of motors are used for processes that are not very sensitive to speed variations; and the variation in speed is seldom more than 10%. The effect of voltage sags on induction motors has already been discussed in Section 5.3 under the assumption that both motor and load torque remain constant. In most practical cases the load torque decreases and the motor torque increases when the motor slows down. The actual drop in speed will thus be less than indicated.

Although the induction motor is normally rather insensitive to voltage sags, there are a few phenomena that could lead to process interruption due to a sag.

- Deep sags lead to severe torque oscillations at sag commencement and when the voltage recovers. These could lead to damage to the motor and to process interruptions. The recovery torque becomes more severe when the internal flux is out of phase with the supply voltage, thus when the sag is associated with a phase-angle jump.

- At sag commencement the magnetic field will be driven out of the airgap. The associated transient causes an additional drop in speed for deep sags. During this period the motor contributes to the short-circuit current and somewhat mitigates the sag. This effect has been discussed in Section 4.8.
- When the voltage recovers, the airgap field has to be built up again. In weaker systems this can last up to 100 ms, during which the motor continues to slow down. This could become a problem in systems where the motor load has grown over the years. Where in the past a voltage sag would not be a problem, now "suddenly" the process can no longer withstand the speed drop due to a sag. As deep sags are rare it can take a long time before such a problem is discovered.
- When the voltage recovers, the motor takes a high inrush current: first to build up the airgap field (the electrical inrush), next to reaccelerate the motor (the mechanical inrush). This inrush can cause a post-fault sag with a duration of one second or more, and lead to tripping of undervoltage and overcurrent relays. Again this problem is more severe for a weak supply, and can thus become a problem when the amount of motor load increases.
- For unbalanced sags the motor is subjected to a positive sequence as well as to a negative-sequence voltage at the terminals. The negative-sequence voltage causes a torque ripple and a large negative-sequence current.

5.5.2 Directly Fed Synchronous Motors

A synchronous motor has similar problems with voltage sags as an induction motor: overcurrents, torque oscillations, and drop in speed. But a synchronous motor can actually lose synchronism with the supply. An induction motor is very likely able to reaccelerate again after the fault: it might take too long for the process, the current might be too high for the motor (or its protection), or the supply might be too weak, but at least it is in theory possible. When a synchronous motor loses synchronism it has to be stopped and the load has to be removed before it can be brought back to nominal speed again.

The loss of synchronism of a synchronous motor is ruled by the equation for the transport of power P from the supply to the motor:

$$P = \frac{V_{sup} E \sin\phi}{X} \tag{5.61}$$

with V_{sup} the supply voltage, E the back-EMF in the motor, ϕ the angle between the back-EMF and the supply voltage, and X the reactance between the supply and the synchronous motor (this includes the leakage reactance of the motor and the source reactance of the supply). This relation is shown in Fig. 5.77. For a given motor load the operating point will be such that the power taken by the load equals the power transported to the motor. This point is indicated in Fig. 5.77 as "normal operating point." When the voltage drops, e.g., during a sag, the power transported to the motor becomes smaller than the power taken by the load. As a result the motor slows down, which means that the angle ϕ increases. The angle will settle down at a new operating point,

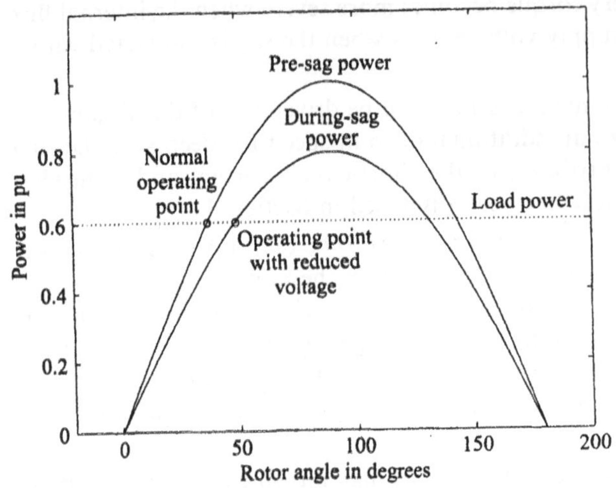

Figure 5.77 Power transfer to a synchronous motor as a function of the rotor angle.

indicated by "operating point with reduced voltage," where again the power to the motor and the power taken by the load are in balance.

It follows from Fig. 5.77 that for deep sags there is no longer a stable operating point. In that case the rotor angle will continue to increase until the supply voltage recovers. If the angle has increased too much the motor loses synchronism. Looking at Fig. 5.78 we see two operating points: the normal operating point, labeled as "stable" and a second point labeled as "instable." In the latter point, both power flows are again equal so the motor would be able to operate at constant speed. But any small deviation will make that the motor drifts away from this operating point: either to the left (when it will end up in the stable operating point) or to the right (when it will lose synchronism). The motor loses synchronism the moment its rotor angle exceeds this instable operating point.

There is a second curve plotted in Fig. 5.78, which indicates the power transfer during the sag. In this case there is no stable operating point during the sag and the motor will continue to slow down until the voltage recovers. At that moment the motor

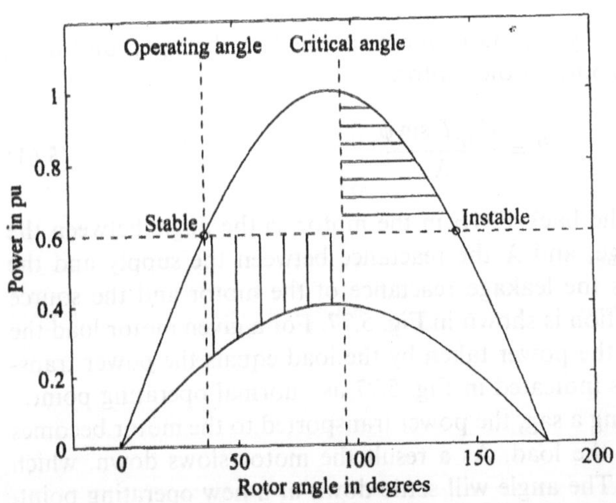

Figure 5.78 Power transfer in normal situation and for a deep sag.

will start to accelerate again but as it still rotates slower than the airgap field (thus slower than the frequency of the supply voltage) its rotor angle will continue to increase. The maximum rotor angle is reached the moment the motor speed comes back to nominal. As long as this angle is smaller than the angle for the instable operating point, the motor does not lose synchronism. The figure shows the maximum angle at the end of the sag which does not lead to an instable situation; this angle is indicated as "critical angle." According to the so-called "equal-area-criterion" the two shaded parts in the figure are equal in area [207].

The highest possible steady-state rotor angle equals 90°—this occurs when the motor load equals the maximum power which can be transported to the motor. If the motor load is only half this maximum value, a drop in voltage to 50% will bring the operating point back to the top of the sine wave again. This 50% is, however, *not* the deepest sag the motor can withstand for a long time. The drop in voltage causes the motor to slow down, thus when the rotor angle reaches 90° it does not stop but will continue to increase until the voltage recovers. The deepest long-duration sag can be found from Fig. 5.79. Again the equal-area criteria tells us that the two shaded parts have the same area.

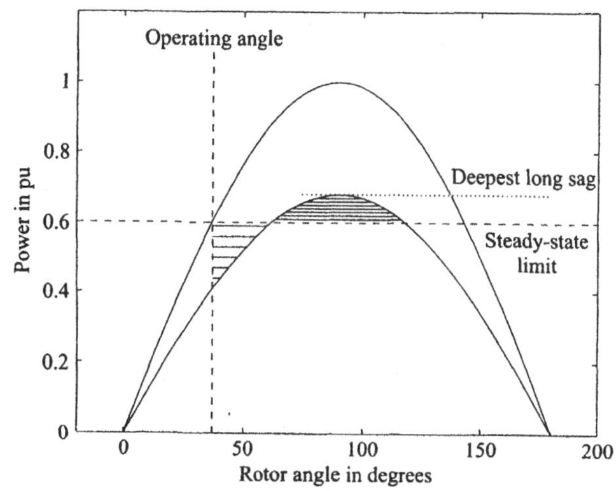

Figure 5.79 Power transfer in normal situation and for the deepest long-duration sag.

5.5.3 Contactors

Contactors are a very common way of connecting motor load to the supply. The supply voltage is used to power an electromagnet which keeps the contact in place. When the supply voltage fails the contact opens, preventing the motor from suddenly restarting when the supply voltage comes back. This works fine for long interruptions where the unexpected starting of motors can be very dangerous. But contactors also drop out for voltage sags and short interruptions where such a behavior is not always acceptable. Test results for contactors are presented in [34]. The measured voltage tolerance curve for a contactor is shown in Fig. 5.80. We see that the contactor tolerates any voltage sag down to about 70%. When the sag magnitude is below 70% for longer than a few cycles, the contactor drops out. We also see the remarkable effect that the voltage tolerance becomes better for deeper sags: a zero voltage can be tolerated for 3.5 cycles but a 50% voltage only for one cycle. This effect is probably due to the experimental setup. Sags were generated by switching between a normal supply and the out-

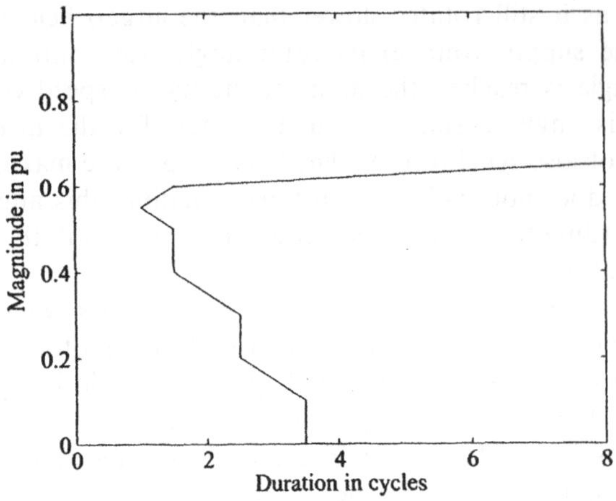

Figure 5.80 Voltage-tolerance curve for a contactor. (Data obtained from [34].)

put of a variable-output transformer. It is not the voltage but the current through the coil that causes the force keeping the contactor closed. The moment the current drops below a certain value the contactor will start to drop out. For lower voltages the current path through the transformer is smaller, thus there is less resistance to damp the current. As the current damps more slowly for smaller voltages, the contactor will not drop out as fast as for medium voltages. This shows that for contactors the supply characteristics can significantly influence the voltage tolerance.

The fact that it is the current and not the voltage that determines the dropping out of the contactor follows also from the dependence of the voltage tolerance on the point-on-wave of sag commencement. The contactor of Fig. 5.80 tolerates a 3.4 cycle sag starting at voltage zero, but only a 0.5 cycle sag starting at voltage maximum. As the contactor coil is mainly inductive the current has a maximum at voltage zero and is zero at voltage maximum.

The influence of the point-on-wave of sag commencement has been further studied by Turner and Collins [38], reporting a voltage tolerance of 30 ms for sag commencements within 30° of the voltage zero crossing, reducing to less than 8 ms for sags commencing at voltage maximum.

Note that all this refers to so-called ac contactors. An alternative is to use dc contactors which are fed from a separate dc system with their own battery backup. These contactors do normally not drop out during voltage sags. However, they require a separate dc system and an alternative protection against unexpected restart of the motor.

5.5.4 Lighting

Most lamps just flicker when a voltage dip occurs. Somebody using the lamp will probably notice it, but it may not be considered as something serious. It is different when the lamp completely extinguishes and takes several minutes to recover. In industrial environments, in places where a large number of people are gathered, or with street lighting, this can lead to dangerous situations.

Dorr et al. [36] have studied the voltage tolerance of high-pressure sodium lamps. Voltage sags can extinguish the lamp, which must cool down for one to several minutes before restarting. The voltage-tolerance curves for three lamps are shown in Fig. 5.81. For voltages below 50% the lamps already extinguish for a sag of less than two cycles.

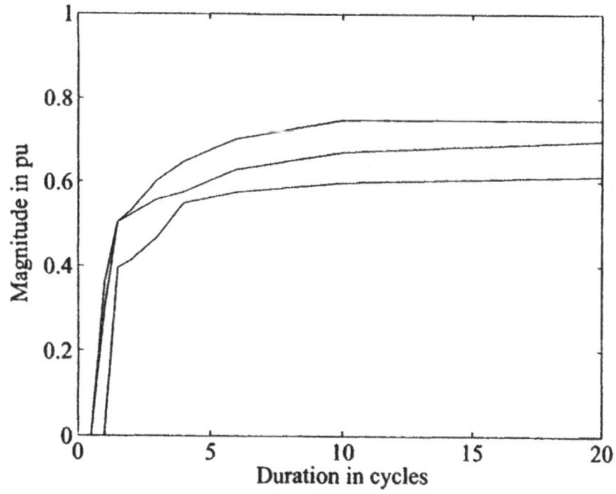

Figure 5.81 Voltage tolerance of high-pressure sodium lamps. (Data obtained from Dorr et al. [36].)

The lamps took about one minute to restrike, and another three minutes before the full light intensity was reached again. The voltage tolerance of the lamp is further dependent on the age. When lamps age they need a larger voltage to operate; they will thus extinguish already for a lower drop in voltage. The minimum voltage for longer sags varied from 45% for new lamps to 85% for lamps at the end of their useful life.

6

Voltage Sags—Stochastic Assessment

In this chapter we discuss methods to describe, measure, and predict the severity of the voltage sag problem: how many times per year will the equipment trip. There are two methods available that quantify the severity of the problem: power quality monitoring and stochastic prediction. Power quality monitoring gives mainly information about common events. For less common events stochastic prediction is more suitable. In this chapter both are discussed in detail.

After explaining the need for stochastic assessment, the various ways of presenting the voltage sag performance of the supply are discussed. The chapter continues with some aspects of voltage sag monitoring, including the results of a number of large surveys. Finally, two methods for stochastic prediction of voltage sags are discussed, together with a few examples. The method of fault positions is suitable for implementation in computer software and is the preferred tool for studies in meshed transmission systems. For radial distribution systems and hand calculations, the method of critical distances is more suitable.

6.1 COMPATIBILITY BETWEEN EQUIPMENT AND SUPPLY

Stochastic assessment of voltage sags is needed to find out whether a piece of equipment is compatible with the supply. A study of the worst-case scenario is not feasible as the worst-case voltage disturbance is a very long interruption. In some cases, a kind of "likely-worst-case-scenario" is chosen, e.g., a fault close to the equipment terminals, cleared by the primary protection, not leading to an interruption. But that will not give any information about the likelihood of an equipment trip. To obtain information like that, a "stochastic compatibility assessment" is required. Such a study typically consists of three steps:

1. **Obtain system performance.** Information must be obtained on the system performance for the specific supply point: the expected number of voltage sags with different characteristics. There are various ways to obtain this

information: contacting the utility, monitoring the supply for several months or years, or doing a stochastic prediction study. Both voltage sag monitoring and stochastic prediction are discussed in detail in this chapter. Note that contacting the utility only shifts the problem, as also the utility needs to perform either monitoring or a stochastic prediction study.

2. **Obtain equipment voltage tolerance.** Information has to be obtained on the behavior of the piece of equipment for various voltage sags. This information can be obtained from the equipment manufacturer, by doing equipment tests, or simply by taking typical values for the voltage tolerance. This part of the compatibility assessment is discussed in detail in Chapter 5.
3. **Determine expected impact.** If the two types of information are available in an appropriate format, it is possible to estimate how often the piece of equipment is expected to trip per year, and what the (e.g., financial) impact of that will be. Based on the outcome of this study one can decide to opt for a better supply, for better equipment or to remain satisfied with the situation. An essential condition for this step is that system performance and equipment voltage tolerance are presented in a suitable format. Some possible formats are discussed in Section 6.2.

An example of a stochastic compatibility assessment is given, based on Fig. 6.1. The aim of the study is to compare two supply alternatives and two equipment tolerances. The two supply alternatives are indicated in Fig. 6.1 through the expected number of sags as a function of the sag severity: supply I is indicated through a solid line; supply II through a dashed line. We further assume the following costs to be associated with the two supply alternatives and the two devices (in arbitrary units):

supply I	200 units/year
supply II	500 units/year
device A	100 units/year
device B	200 units/year

We also assume that the costs of an equipment trip are 10 units.

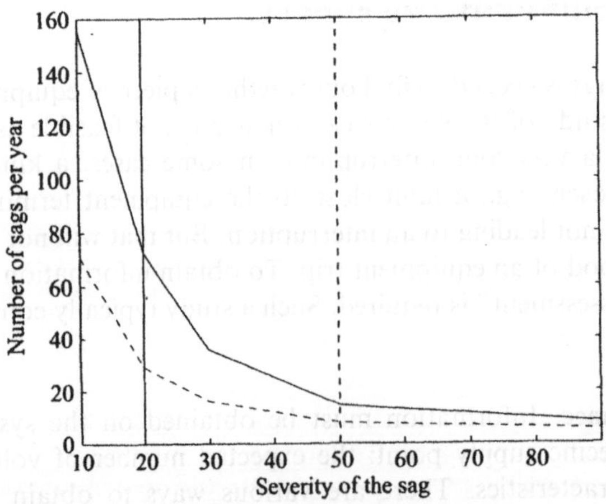

Figure 6.1 Comparison of two supply alternatives (solid curve: supply I, dashed curve: supply II) and two equipment tolerances (solid vertical line: device A, dashed line: device B).

From Fig. 6.1, one can read the number of spurious trips per year, for each of the four design options, at the intersection between the supply curve and the device (vertical) line. For device A and supply I we find 72.6 spurious equipment trips per year, etc. The results are shown in Table 6.1.

TABLE 6.1 Number of Spurious Trips per Year for Four Design Alternatives

	Supply I	Supply II
Device A	72.6	29.1
Device B	14.6	7.9

Knowing the number of trips per year, the annual costs of each of the four design options, and the costs per spurious trip, it is easy to calculate the total annual costs. For the combination of device A and supply I these costs are

$$72.6 \times 10 + 100 + 200 = 1026 \text{ units/year}$$

The results for the four design options are shown in Table 6.2. From this table it follows that the combination of supply I and device B has the lowest annual costs.

TABLE 6.2 Total Costs per Year for Four Design Alternatives

	Supply I	Supply II
Device A	1026	891
Device B	546	779

Note the stochastic character of the assessment. An expected value (the expected number of equipment trips per year multiplied by the cost of one equipment trip) is added to a deterministic value (the annual cost of supply and device). Assume that the voltage tolerance for a device is the same under all circumstances; the voltage tolerance is thus a deterministic quantity. But the number of sags will vary from year to year. We further assume the occurrence of a sag to be independent of the occurrence of other sags. In that case the number of sags in any given year follows a Poisson distribution. Let N be the number of sags in any given year and μ the expected number of sags (as indicated in Table 6.1). The probability that $N = n$ for a Poisson distribution is found from

$$Pr\{N = n\} = e^{-\mu} \frac{\mu^n}{n!} \tag{6.1}$$

For the four design alternatives in Table 6.1 this distribution has been plotted in Fig. 6.2. It follows from the figure, for example, that the number of trips of design BII (supply II in combination with device B) varies between 2 and 18, and for design BI between 7 and 26. It is thus not sure that in a given year, design BII gives less trips than design BI.

From the probability density function for the number of trips (Fig. 6.2) the probability density function for the total costs per year can be calculated, resulting in

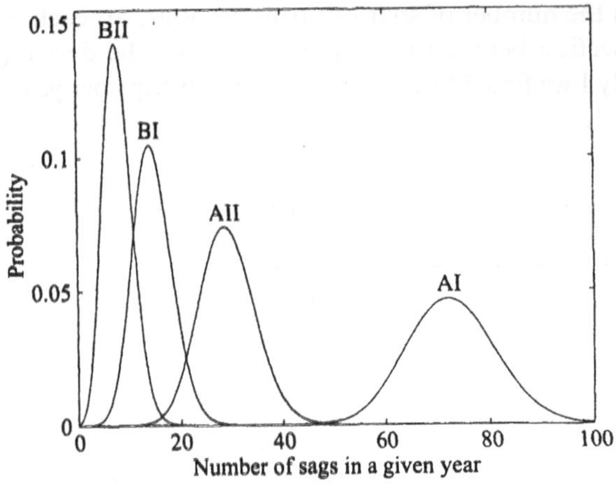

Figure 6.2 Probability density function of the number of sags per year for four design alternatives.

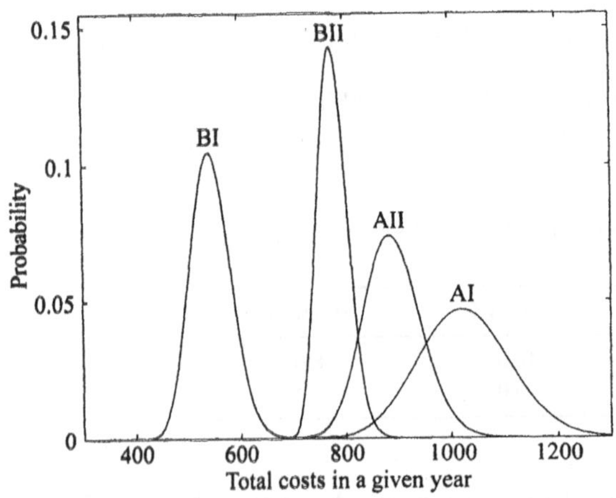

Figure 6.3 Probability density function of the costs per year for four design alternatives.

Fig. 6.3. This figure shows that design BI is clearly better than any of the other design options.

6.2 PRESENTATION OF RESULTS: VOLTAGE SAG COORDINATION CHART

In this section we discuss a number of ways to present the supply performance. The discussion concentrates on the presentation of results obtained from power quality monitoring. The same technique can be applied to the results of a stochastic assessment study.

6.2.1 The Scatter Diagram

Every power quality monitor will at least give magnitude and duration as an output for a sag. When the supply is monitored for a certain period of time, a number of sags will be recorded. Each sag can be characterized by a magnitude and a duration and be plotted as one point in the magnitude-duration plane. An example of the resulting scatter diagram is shown in Fig. 6.4. The scatter diagram is obtained from

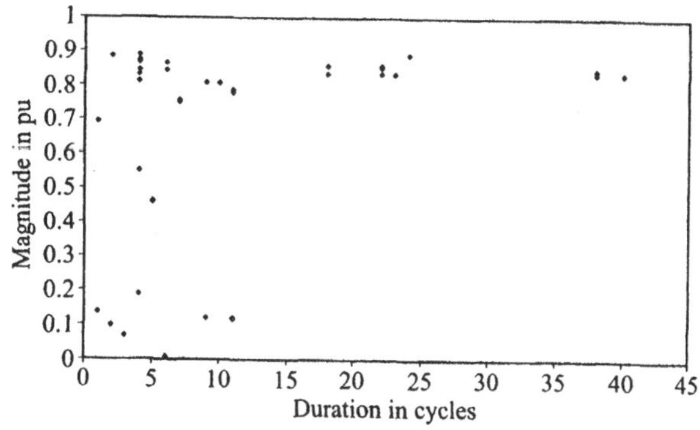

Figure 6.4 Scatter diagram obtained by one year of monitoring at an industrial site.

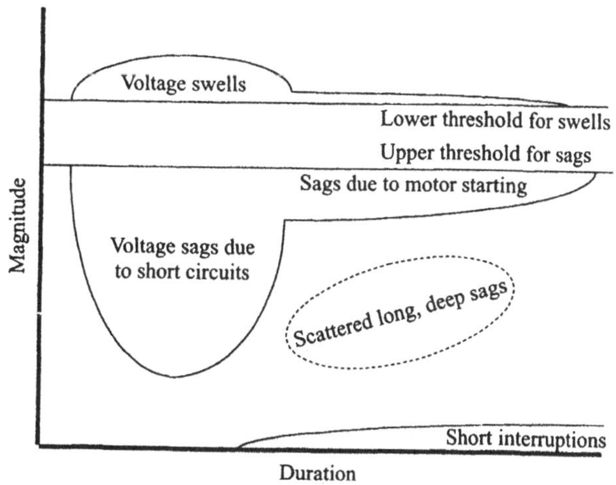

Figure 6.5 Scatter diagram as obtained from a large power quality survey.

one year of monitoring at an industrial site [155]. For a large power quality survey, the scatter diagrams of all the sites can be combined. A stylized version of the resulting scatter diagram is shown in Fig. 6.5. In this figure not only voltage sags, but also interruptions and voltage swells are indicated.

In Fig. 6.5 we see a number of heavily populated regions:

- Voltage sags due to short circuits, with durations up to a few hundred milliseconds and magnitudes from 50% upwards. Deeper and longer sags are present but rare.
- Voltage sags due to motor starting, with durations of a few seconds and longer, and magnitudes from 80% upwards.
- Short interruptions due to fast reclosing, with voltage magnitude zero and durations from about 10 cycles onward.
- Voltage swells with similar durations as sags due to short circuits, but magnitudes up to 120%.

Next to these densely populated areas there are scattered, long, deep sags, likely due to the errors made in recording duration of sags with a long, post-fault sag. These long, deep sags consist of a short, deep sag followed by a long, shallow sag. This points to one of the shortcomings of the commonly used method of sag characterization: the lowest rms value as sag magnitude and the number of cycles below the threshold as the sag duration.

No reliable information has been published about the number of sags with a large non-rectangular part. It is mentioned in [156] that about 10% of sags in the U.S. distribution systems are non-rectangular. Another indication that this effect is not very severe is the fact that the duration of most sags corresponds to typical fault-clearing times in the system.

6.2.2 The Sag Density Table

The scatter diagram is very useful to give a qualitative impression of the supply performance, but for a quantitative assessment other ways of presentation are needed. A straightforward way of quantifying the number of sags is through a table with magnitude and duration ranges. This is done in Table 6.3 for data obtained from a large power quality survey [20]. Each element in the table gives the number of events with magnitude and duration within a certain range; e.g., magnitude between 40 and 50% and duration between 400 and 600 ms. Each element gives the density of sags in that magnitude and duration range; hence the term "sag density table" or "sag density function." A combination of magnitude range and duration range is called a "magnitude-duration bin."

The sag density function is typically presented as a bar chart. This is done in Fig. 6.6 for the data shown in Table 6.1. The length of each bar is proportional to the number of sags in the corresponding range. From the bar chart it is easier to get an impression of the distribution of the sag characteristics, but for numerical values the table is more useful. In this case we see from Fig. 6.6 that the majority of sags has a magnitude above 80% and a duration less than 200 ms. There is also a concentration of short interruptions with durations of 800 ms and over.

In Fig. 6.6 all magnitude ranges are of equal size, so are all duration ranges. In most cases the ranges will be of different size. There are more sags of short duration and high magnitude than sags elsewhere in the magnitude-duration plane. Therefore, the resolution is chosen higher for shorter duration sags and for shallow sags. Several examples of the density function in bar-chart form are shown in Section 6.3.

TABLE 6.3 Example of Sag Density Table: Number of Sags per Year

Magnitude	0–200 ms	200–400 ms	400–600 ms	600–800 ms	> 800 ms
80–90%	18.0	2.8	1.2	0.5	2.1
70–80%	7.7	0.7	0.4	0.2	0.5
60–70%	3.9	0.6	0.2	0.1	0.2
50–60%	2.3	0.4	0.1	0.1	0.1
40–50%	1.4	0.2	0.1	0.1	0.1
30–40%	1.0	0.2	0.1	0.0	0.1
20–30%	0.4	0.1	0.1	0.0	0.0
10–20%	0.4	0.1	0.1	0.0	0.1
0–10%	1.0	0.3	0.1	0.0	2.1

Source: Data obtained from [20].

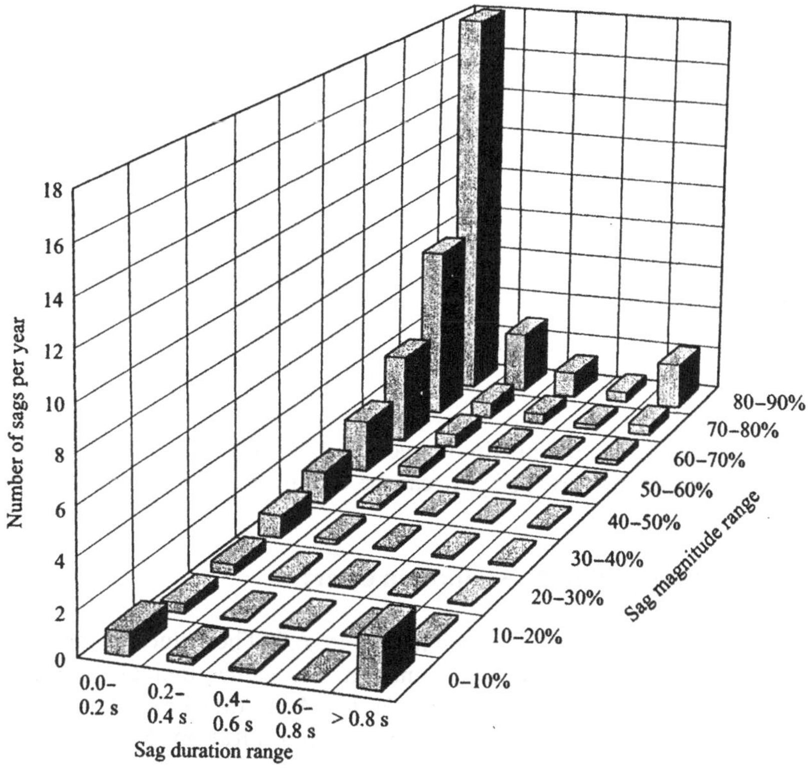

Figure 6.6 Two-dimensional bar chart of the sag density function shown in Table 6.3.

6.2.3 The Cumulative Table

Of interest to the customer is not so much the number of voltage sags in a given magnitude and duration range, but the number of times that a certain piece of equipment will trip due to a sag. It therefore makes sense to show the number of sags worse than a given magnitude and duration. For this a so-called "cumulative sag table" is calculated. Element MD of the cumulative sag table is defined as follows:

$$F_{MD} = \sum_{m=0}^{M} \sum_{d=D}^{d_{max}} f_{md} \tag{6.2}$$

with f_{md} element md of the density table: the number of sags in the duration range d and the magnitude range m; and with F_{MD} element MD of the cumulative table: the number of sags with duration longer than D and magnitude less than M. Durations are summed from the value upward because a longer sag is more severe; magnitudes are summed from the value down to zero because a lower magnitude indicates a more severe sag. This is a direct consequence of the definition of sag magnitude, where a higher magnitude indicates a less severe event.

The cumulative table obtained from the density table in Table 6.3 is shown in Table 6.4. The table shows, e.g., that the rms voltage drops below 60% for longer than 200 ms, on average 4.5 times per year. If the equipment can only tolerate a sag

TABLE 6.4 Example of Cumulative Sag Table, Number of Sags per Year

Magnitude	0	200 ms	400 ms	600 ms	800 ms
90%	49.9	13.9	8.4	6.1	5.2
80%	25.4	7.4	4.7	3.6	3.1
70%	15.8	5.5	3.6	2.9	2.6
60%	10.9	4.5	3.1	2.6	2.4
50%	8.0	3.8	2.9	2.5	2.3
40%	6.2	3.4	2.7	2.3	2.3
30%	4.9	3.1	2.6	2.3	2.2
20%	4.2	2.8	2.4	2.2	2.2
10%	3.5	2.5	2.2	2.1	2.1

Source: Data obtained from Table 6.3.

below 60% for 200 ms, it will trip on average 4.5 times per year. From such a table the number of equipment trips per year can be obtained almost directly.

6.2.4 The Voltage Sag Coordination Chart

Table 6.4 is shown as a bar chart in Fig. 6.7. The values in the cumulative table belong to a continuous monotone function: the values increase toward the left-rear corner in Fig. 6.7. The values shown in Table 6.4 can thus be seen as a two-dimensional function of number of sags versus magnitude and duration. Mathematically speaking,

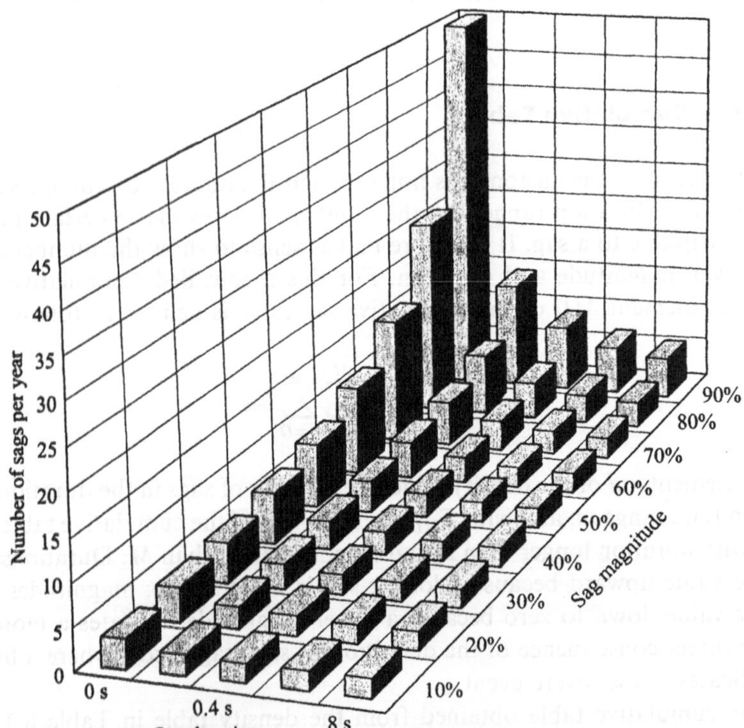

Figure 6.7 Bar chart of the cumulative voltage sag table shown in Table 6.4.

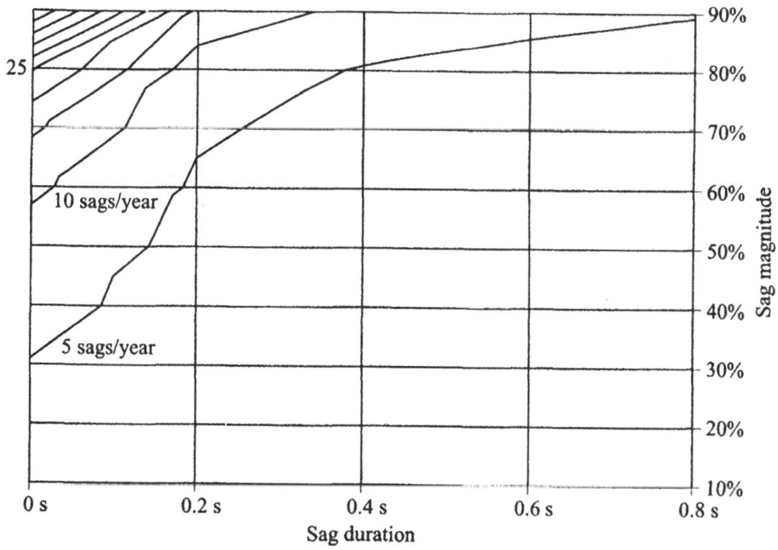

Figure 6.8 Contour chart of the cumulative sag function, based on Table 6.4.

this function is defined for the whole magnitude-duration plane. When obtained from power quality monitoring the function is not continuous. Stochastic prediction techniques will normally also not lead to a continuous function. Whether the function is continuous or not, a common way of presenting a two-dimensional function is through a contour chart. This was done by Conrad for the two-dimensional cumulative sag function, resulting in Fig. 6.8 [20].

The contour chart is recommended as a "voltage sag coordination chart" in IEEE Standard 493 [21] and in IEEE Standard 1346 [22]. In a voltage sag coordination chart the contour chart of the supply is combined with the equipment voltage-tolerance curve to estimate the number of times the equipment will trip. Figure 6.8 has been reproduced in Fig. 6.9 including two equipment voltage-tolerance curves. Both curves are rectangular; i.e., the equipment trips when the voltage drops below a certain voltage for longer than a given duration. Device A trips when the voltage drops below 65% of nominal for longer than 200 ms. According to the definition given before, the number of voltage sags below 65% for longer than 200 ms is equal to the element of the cumulative table for 65%, 200 ms. The values in the cumulative sag table are the underlying function of the contour chart in Figs. 6.8 and 6.9. In short, the number of spurious trips is equal to the function value at the knee of the voltage-tolerance curve, indicated as a circle in Fig. 6.9. For device A this point is located exactly on the five sags per year contour. Thus, device A will trip five times per year. For device B, the knee is located between the 15 and 20 sags per year contours. Now we use the knowledge that the underlying function is continuous and monotone. The number of trips will thus be between 15 and 20 per year; using interpolation gives an estimated value of 16 trips per year.

For a non-rectangular equipment voltage-tolerance curve, as shown in Fig. 6.10, the procedure becomes somewhat more complicated. Consider this device as consisting of two components, each with a rectangular voltage-tolerance curve.

- Component A trips when the voltage drops below 50% for longer than 100 ms; according to the contour chart this happens six times per year.

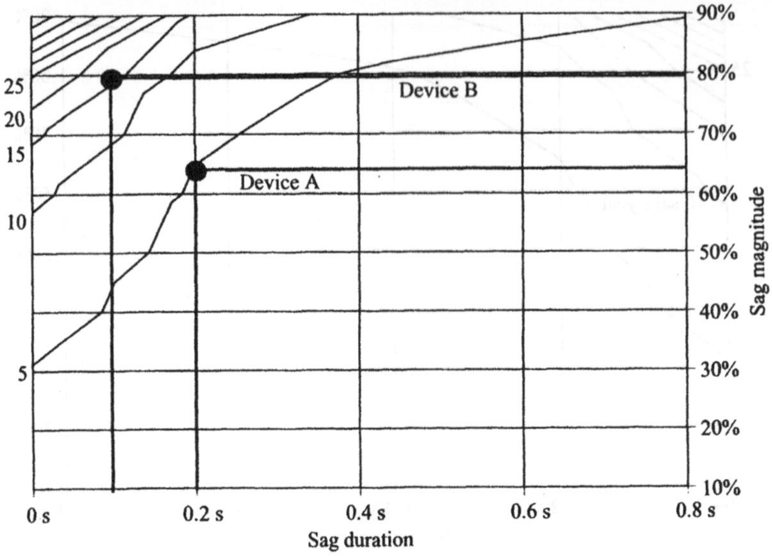

Figure 6.9 Voltage sag coordination chart, reproduced from Fig. 6.8, with two equipment voltage-tolerance curves.

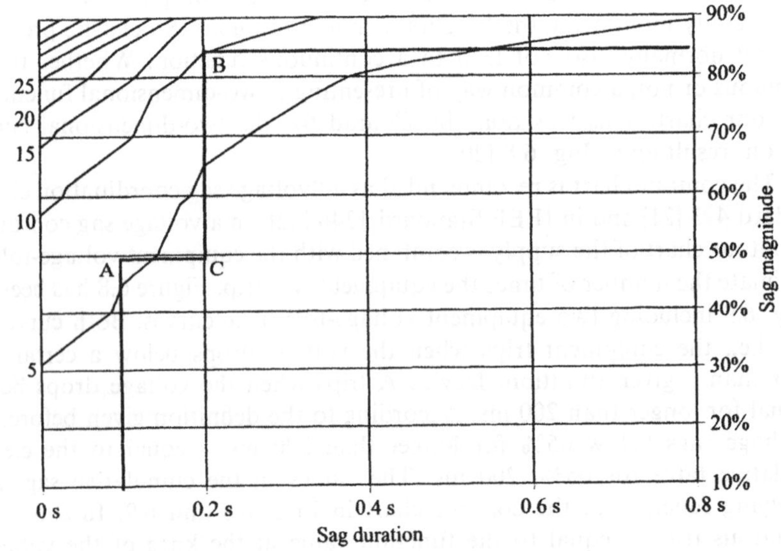

Figure 6.10 Voltage sag coordination chart, reproduced from Fig. 6.8, with non-rectangular equipment voltage-tolerance curve.

- Component B trips when the voltage drops below 85% for longer than 200 ms, which happens 12 times per year.

Adding these two numbers (6 + 12 = 18) would count double those voltage sags for which both components trip. Both components trip when the voltage drops below 50% for longer than 200 ms; about four times per year. This corresponds to point C in the chart. The number of equipment trips is thus equal to

$$F_A + F_B - F_C = 6 + 12 - 4 = 14 \tag{6.3}$$

Note that assuming a rectangular equipment voltage-tolerance curve (100 ms, 85%) would have resulted in the incorrect value of 20 trips per year.

By using this procedure, the voltage sag coordination chart provides for a simple and straightforward method to predict the number of equipment trips.

6.2.5 Example of the Use of the Voltage Sag Coordination Chart

The data obtained from a large survey [68] has been used to plot the sag density bar chart shown in Fig. 6.11. The survey measured the quality of the voltage at the terminals of low-voltage equipment (at the wall outlet) at many sites across the United States and Canada. Figure 6.11 can thus be interpreted as the average voltage quality experienced by low-voltage equipment.

From Fig. 6.11, a voltage sag coordination chart has been obtained, shown in Fig. 6.12. Four equipment voltage tolerances are indicated by the points A, B, C, and D. The meaning of these will be explained next.

Suppose that a computer manufacturer considers different options for the power supply of personal computers. The choice is between two different dc/dc converters, with minimum operating voltages of 100 V and 78 V, and between two capacitor sizes, leading to 5% and 1% dc voltage ripple. Using (5.6) we can calculate the voltage tolerance of the four design options. For a minimum operating voltage of 100 V and a dc voltage ripple of 5% we find a voltage tolerance of 84% (100 V) and 1.5 cycles, etc. The results are shown in column 4 of Table 6.5. The voltage tolerance for the four options (A, B, C, and D) is indicated by the four dots in Fig. 6.12. From this voltage sag

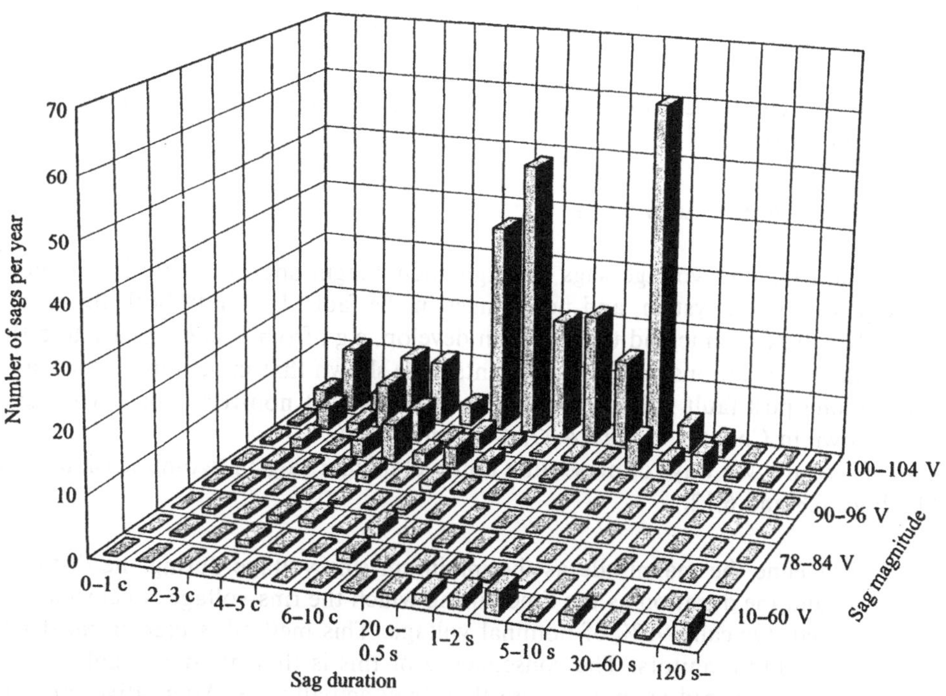

Figure 6.11 Sag density for the average low-voltage supply in the United States and Canada. (Data obtained from Dorr [68].)

TABLE 6.5 Comparison of Four Design Options for the Power Supply of a Personal Computer

Option	Minimum Operating Voltage	dc Ripple	Voltage Tolerance	Estimated Trip Frequency
A	100 V	5%	84%, 1.5 cycles	100/year
B	100 V	1%	84%, 8 cycles	50/year
C	78 V	5%	65%, 3 cycles	25/year
D	78 V	1%	65%, 15 cycles	20/year

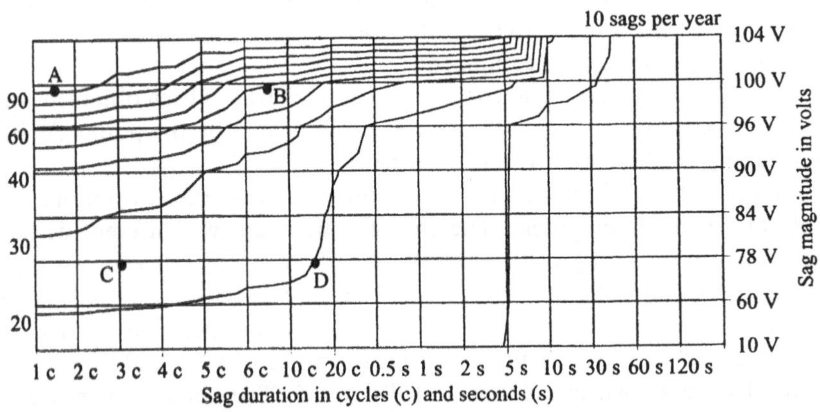

Figure 6.12 Voltage sag coordination chart for the average low-voltage supply in the United States and Canada. (Obtained from the sag density chart in Fig. 6.11.)

coordination chart the trip frequency can easily be estimated, resulting in the last column of Table 6.5.

6.2.6 Non-Rectangular Sags

Characterizing voltage sags through their magnitude and duration assumes a static load, a static system, and no changes in the fault. In reality both the load and the system are dynamic and the fault can develop, e.g., from a single-phase to a three-phase fault. Simulations and measurements have shown that induction motor load can lead to long post-fault voltage sags. A few examples of non-rectangular voltage sags were shown in Chapter 4: Figs. 4.47, 4.48, and 4.130.

There are two ways of presenting non-rectangular sags in two-dimensional charts like Figs. 6.8 and 6.12.

1. Define the magnitude as the minimum rms voltage during the disturbance and the duration as the time during which the rms voltage is below a threshold, typically 90% of nominal voltage. This method is used in most power quality monitors. The consequence of this is that non-rectangular sags are characterized as more severe than they actually are. Alternatives are to use the average or the rms of the one-cycle rms values (the latter is a measure of the energy remaining during the sag).

2. Characterize the voltage quality by the number of times the voltage drops below a given value for longer than a given time. This again results in a graph like Fig. 6.8, but now without the need to characterize sags individually. Such a method was first proposed in [17] and used in [18], and became part of IEEE Std. 493 [21]. A similar method is proposed in [156] for inclusion in contracts between utility and customers. The argument for the latter proposal being that utilities should not be overly punished for non-rectangular sags.

To explain the second method, the cumulative table will be introduced in a different way. We define each element as a counter counting the number of sags worse than the magnitude and duration belonging to this element. Each sag that occurs increases the value of part of the elements by one. The elements whose value is increased are those for which the sag is more severe than the element. In other words, those elements less severe than the sag; in the table, the elements above the sag. This is shown in Fig. 6.13 for a rectangular sag.

Figure 6.14 again shows the grid of points corresponding to the cumulative sag function. But this time a non-rectangular sag is shown. The procedure is exactly the same as before: "The function value should be increased by one for all points above the sag."

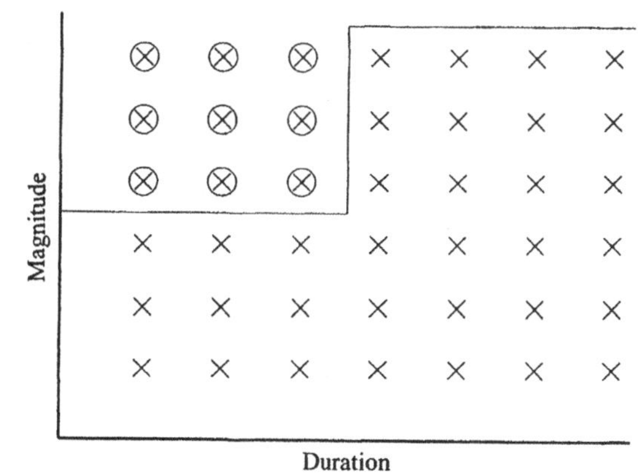

Figure 6.13 Update of cumulative table for rectangular sag.

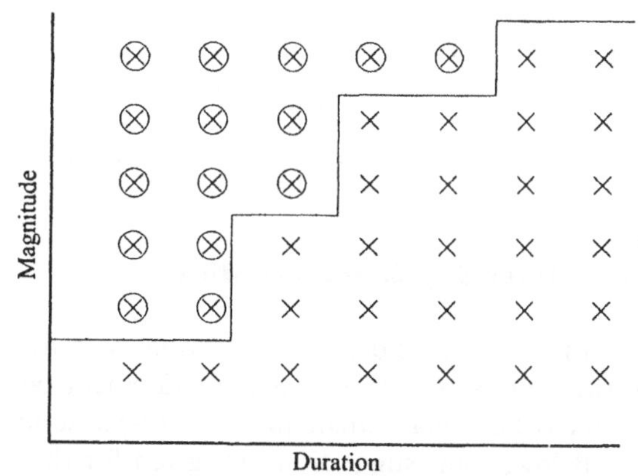

Figure 6.14 Update of cumulative table for non-rectangular sag.

Using this method it is possible to quantify the quality of the supply including non-rectangular sags. But this method cannot be used to characterize individual sags. Note that this is often not a serious concern when one is interested in merely quantifying the supply performance.

Some sags will still escape quantification, as shown in Fig. 6.15. A possible choice here is to measure the time the sag is in each magnitude range in the table, and then increase the points to the left of the table in that magnitude range. This would lead to an equivalent sag as indicated in Fig. 6.15. The method proposed in [156] treats these "very non-rectangular sags" in a similar way. To understand the limitation of the method in Figs. 6.13, 6.14, and 6.15 the term "rectangular voltage-tolerance curve" is introduced. A piece of equipment has a rectangular voltage-tolerance curve if its tripping is determined by one magnitude and one duration. Thus, the equipment trips when the voltage drops below a certain magnitude for longer than a certain duration. The actual shape of the rms voltage versus time has no influence on the equipment behavior. Examples of such equipment are undervoltage relays (e.g., used to protect induction motors) and most non-controlled rectifiers. Also computers and other consumer electronics equipment fit in this category. Many adjustable-speed drives trip due to an undervoltage-time relay at the dc bus or on the ac terminals. Also those can be considered as having a rectangular voltage-tolerance curve.

For equipment with a rectangular voltage-tolerance curve this method directly gives the expected number of spurious trips. For non-rectangular voltage-tolerance curves the method no longer works. That might appear a serious disadvantage until one realizes that a non-rectangular voltage-tolerance curve will normally be obtained for rectangular sags. Applying it directly to non-rectangular sags is prone to uncertainties anyway, no matter which definition of magnitude and duration is used. When assessing the influence of non-rectangular sags on a piece of equipment it is recommended to use a rectangular approximation of the voltage-tolerance curve unless more detailed information on its behavior under non-rectangular sags is available.

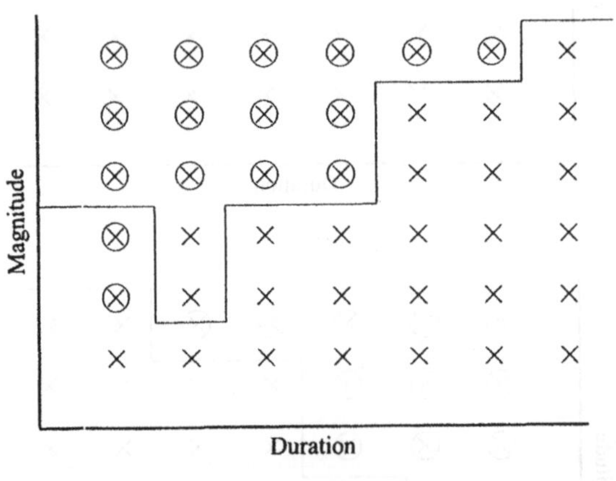

Figure 6.15 Problems in updating the cumulative table for a very non-rectangular sag.

6.2.7 Other Sag Characteristics

In the previous part of this section, we only considered magnitude and duration of the sags. We saw before that the equipment behavior may also be affected by other characteristics: phase-angle jump, three-phase unbalance, point-on-wave of sag initiation. Below, some suggestions are given for the presentation of the results when these additional characteristics need to be incorporated. Note that, unlike magnitude and duration, no monitoring data are available on phase-angle jump, three-phase unbalance, and point-on-wave of sag initiation. This makes that some of the suggestions remain rather theoretical, without the chance to apply them to actual data.

6.2.7.1 Three-Phase Unbalance. We saw in Section 4.4 that three-phase unbalanced sags come in a number of types. The fundamental types were referred to as A, C, and D. The concept of voltage sag coordination chart can be extended to three-phase unbalance by creating one chart for each type, as shown in Fig. 6.16. A contour chart is created for the number of sags more severe than a given magnitude and duration, for each type. Also the equipment voltage-tolerance curve is obtained for each type. In exactly the same way as before, the number of equipment trips can be found for each type; in this example: N_A, N_C, and N_D. The total number of equipment trips N is the sum of these three values:

$$N = N_A + N_C + N_D \tag{6.4}$$

The method can be extended toward other types. The main problem remains to obtain the type of sag from monitoring data. A technique for this has been proposed in [203], [204] which requires the sampled waveforms.

6.2.7.2 Phase-Angle Jumps. Including phase-angle jumps in the compatibility assessment for single-phase equipment creates a three-dimensional problem. The three dimensions are magnitude, duration, and phase-angle jump. Next to this there are two additional complications:

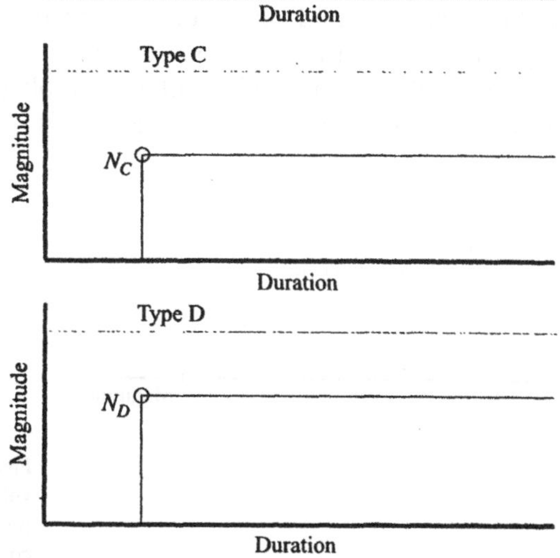

Figure 6.16 Use of the voltage sag coordination chart when three-phase unbalance needs to be considered.

- Phase-angle jumps can be both positive and negative, with the majority of values likely to be found around zero phase-angle jump. Using a cumulative function requires the splitting up of the three-dimensional space in two half-spaces: one for positive phase-angle jump, one for negative phase-angle jump. Note that equipment behavior may be completely different for positive and for negative phase-angle jump.
- An increasing phase-angle jump (in absolute value) not necessarily leads to a more severe event for the equipment. With both magnitude and duration it was possible to indicate a direction in which the event becomes more severe (decreasing magnitude and increasing duration). For phase-angle jumps this is not possible.

Especially the latter complication makes a three-dimensional version of the voltage sag coordination chart not feasible. A possible solution is to split the phase-angle jump axis in a number of ranges, e.g., [−60°, − 30°], [−30°, − 10°], [−10°, + 10°], [+10°, + 30°], [+30°, + 60°]. For each range the number of equipment trips is determined like before. The total number of equipment trips is the sum of the values obtained for each range of phase-angle jump. A plot of magnitude versus phase-angle jump for single-phase equipment was shown in Fig. 4.108. Splitting the phase-angle jump axis in a number of ranges shows that not all charts will contain the whole range of magnitude values. Only in the range around zero phase-angle jump do we expect magnitude values between zero and 100%. The range [+30°, + 60°] may only contain magnitude values around 50% of nominal. An alternative is to split the duration axis in a number of ranges. In a stochastic prediction study this could correspond to the typical fault-clearing time in different parts of the system, e.g., at different voltage levels. For each duration range, a plot of magnitude versus phase-angle jump results, similar to the one plotted in Fig. 4.108. Within this plot, an equipment voltage-tolerance curve can be drawn. A hypothetical example is shown in Fig. 6.17. Note that this curve has a different shape than the voltage-tolerance curve in the magnitude-duration plane. Note further that it is no longer possible to use a cumulative function for the number of events like in the voltage sag coordination chart. Instead a density function must be used, and the number of events outside of the voltage-tolerance curve added.

For three-phase equipment the problem becomes slightly less complicated. Using characteristic magnitude and phase-angle jump results in negative phase-angle jump values only. But a larger (negative) phase-angle jump could still be a less severe event for the equipment. Presenting equipment and supply performance still requires splitting up the phase-angle jump axis or the duration axis.

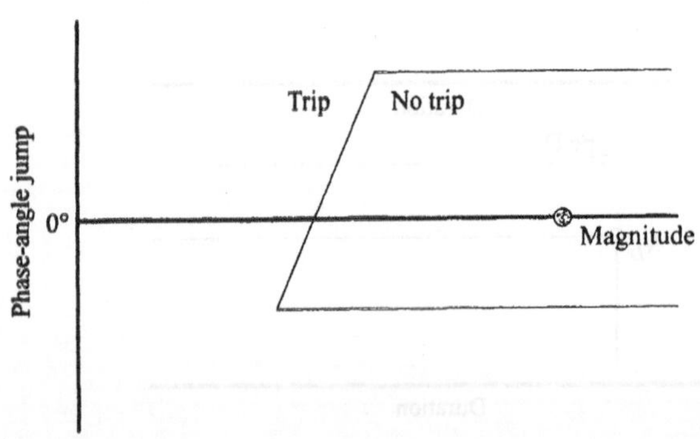

Figure 6.17 Hypothetical example of the voltage-tolerance curve for magnitude against phase-angle jump. The sag duration is considered constant.

6.2.7.3 Point-on-Wave. Point-on-wave characteristics may be easier to include in the compatibility assessment than phase-angle jumps, because the point-on-wave of sag initiation is likely to be independent of the other characteristics. For here we will assume that this is the case. Analysis of monitoring data is needed to check this assumption.

As the point-on-wave of sag initiation is independent of the sag magnitude and duration, there is no need for a three-dimensional treatment. Next to the standard contour chart of magnitude versus duration, a one-dimensional plot is needed for the point-on-wave. A hypothetical example is shown in Fig. 6.18. Note that only values between zero and 90° are shown; other values can be translated into a value in this range.

For a number of values a voltage-tolerance curve needs to be obtained and plotted in the standard voltage sag coordination chart; see Fig. 6.19. The resulting number of equipment trips N_i from each voltage-tolerance curve is weighted by the fraction of sags ζ_i with a point-on-wave value equal to i, and added to get the total number of equipment trips N:

$$N = \sum_i N_i \zeta_i \tag{6.5}$$

In the example shown in Figs. 6.18 and 6.19, this total number of equipment trips is obtained from

$$N = \zeta_0 N_0 + \zeta_{30} N_{30} + \zeta_{60} N_{60} + \zeta_{90} N_{90} \tag{6.6}$$

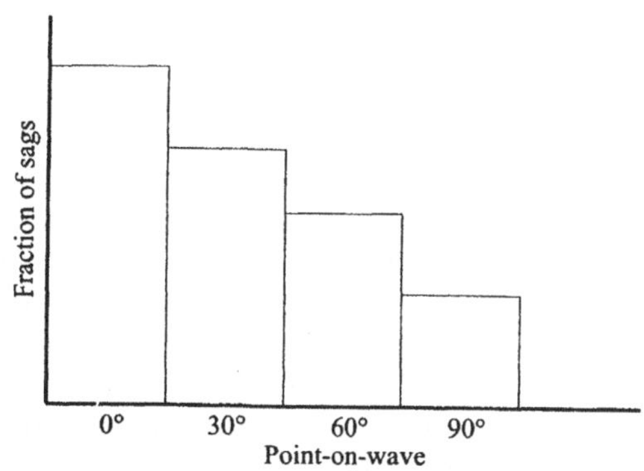

Figure 6.18 Hypothetical example of the fraction of sags with a given point-on-wave value.

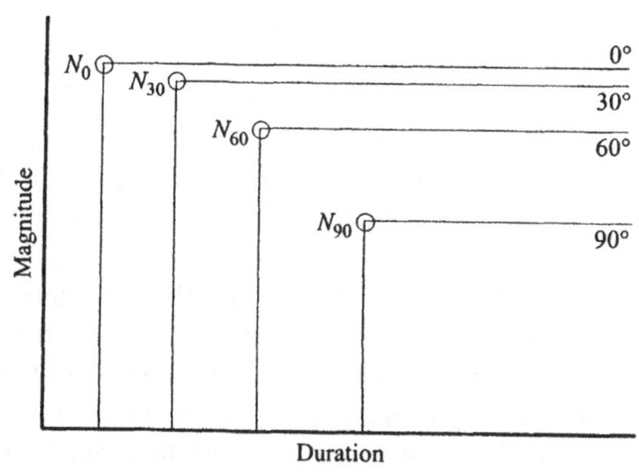

Figure 6.19 Hypothetical example of the voltage-tolerance curves for different point-on-wave of sag initiation.

6.3 POWER QUALITY MONITORING

A common way of obtaining an estimate for the performance of the supply is by recording the disturbance events. For interruptions of the supply this can be done manually as described in Chapter 2. For voltage sags and other short-duration events an automatic recording method is needed. A so-called power quality monitor is an appropriate tool for that, although modern protective relays can perform the same function. Power quality monitors come in various types and for a range of prices. A further discussion about them is beyond the scope of this book.

For each event the monitor records a magnitude and a duration plus possibly a few other characteristics and often also a certain number of samples of raw data: time domain as well as rms values. This could result in an enormous amount of data, but in the end only magnitude and duration of individual events are used for quantifying the performance of the supply.

Two types of power quality monitoring need to be distinguished:

- monitoring the supply at a (large) number of positions at the same time, aimed at estimating an "average power quality": a so-called power quality survey.
- monitoring the supply at one site, aimed at estimating the power quality at that specific site.

Both will be discussed in more detail below.

6.3.1 Power Quality Surveys

Large power quality surveys have been performed in several countries. Typically ten to a hundred monitors are installed at one or two voltage levels spread over a whole country or the service territory of a utility. Because not all substations and feeders can be monitored, a selection has to be made. The selection should be such that the average power quality, as measured, is also representative for the substations and feeders not monitored. Making such a fully representative choice is very difficult if not impossible. Sites come in different types, but it is hard to decide which sites are different from a sag viewpoint without first doing the survey. A further analysis of data from the current generation of surveys will teach us more about the differences between sites. This knowledge can be used for choosing sites in future surveys.

Some aspects of power quality surveys and the way in which the data can be processed, are discussed below by using data from four surveys:

- *The CEA survey.* A three-year survey performed by the Canadian Electrical Association (CEA). A total of 550 sites was monitored for 25 days each. Residential, commercial, and industrial sites were monitored at their 120 V or 347 V service entrance panels. Approximately 10% of the sites had metering on primary side of the service transformer to provide an indication of the power quality characteristics of the utility's distribution system [54], [65], [66].
- *The NPL survey.* A five-year survey performed by National Power Laboratory (NPL). At 130 sites within the continental US and Canada, single-phase line-to-neutral data were connected at the standard wall receptacle. The survey resulted in a total of 1200 monitor months of data [54], [68], [69].
- *The EPRI survey.* A survey performed by the Electric Power Research Institute (EPRI) between June 1993 and September 1995. Monitoring took place in

distribution substations and on distribution feeders at voltages from 4.16 to 34.5 kV. Monitoring at 277 sites resulted in 5691 monitor months of data. In most cases three monitors were installed for each randomly selected feeder: one at the substation and two at randomly selected places along the feeder [54], [70].

- *The EFI survey*. The Norwegian Electric Power Research Institute (EFI, recently renamed "SINTEF Energy Research") has measured voltage sags and other voltage disturbances at over 400 sites in Norway. The majority (379) of the sites were at low-voltage (230 and 400 V), 39 of them were at distribution voltages, and the rest at various voltage levels [67].

The results of these surveys will be presented and discussed in the following paragraphs. For more details about the surveys refer to the various papers cited. These are by far the only surveys, but they were the ones for which detailed results were available. With the exception of the EFI survey all the results presented below were published in the international literature. Especially the paper by Dorr [54] contains very useful information. The amount of results published, even in reports, is still very limited. There must still be gigabytes of very interesting monitoring data stored at utilities all over the world, waiting to be processed. A number of observations can be made from the various surveys, some of which are mentioned below. To explain or check all this, further analysis of the data is needed.

6.3.1.1 Magnitude Versus Duration: CEA Survey. The cumulative number of sags per year, as obtained from the CEA survey is shown in Tables 6.6 and 6.7 for primary as well as secondary side of the service transformer. Bar charts of the sag density function are shown in Figs. 6.20 and 6.22. A voltage sag coordination chart for the secondary side data is shown in Fig. 6.21.

TABLE 6.6 Cumulative Voltage Sag Table for CEA Secondary Side Data: Number of Sags per Year

	Duration						
Magnitude	1 cycle	6 cycles	10 cycles	20 cycles	0.5 sec	1 sec	2 sec
90%	98.0	84.0	84.0	67.3	63.8	35.8	6.6
80%	19.2	9.2	9.2	5.5	5.0	3.2	2.3
70%	14.4	5.7	5.7	4.4	4.2	3.1	2.3
50%	10.5	3.5	3.5	3.2	3.2	2.8	2.2
10%	6.5	2.8	2.8	2.8	2.8	2.6	2.1

Source: Data obtained from Dorr et al. [54].

TABLE 6.7 Cumulative Voltage Sag Table for CEA Primary Side Data: Number of Sags per Year

	Duration						
Magnitude	1 cycle	6 cycles	10 cycles	20 cycles	0.5 sec	1 sec	2 sec
90%	20.3	11.2	10.8	5.5	5.2	1.9	1.3
80%	12.0	5.8	5.4	3.2	3.1	0.9	0.7
70%	9.4	3.6	3.3	2.0	1.9	0.7	0.7
50%	4.8	1.2	1.2	1.1	1.1	0.7	0.7
10%	3.1	1.2	1.2	1.1	1.1	0.7	0.7

Source: Data obtained from Dorr et al. [54].

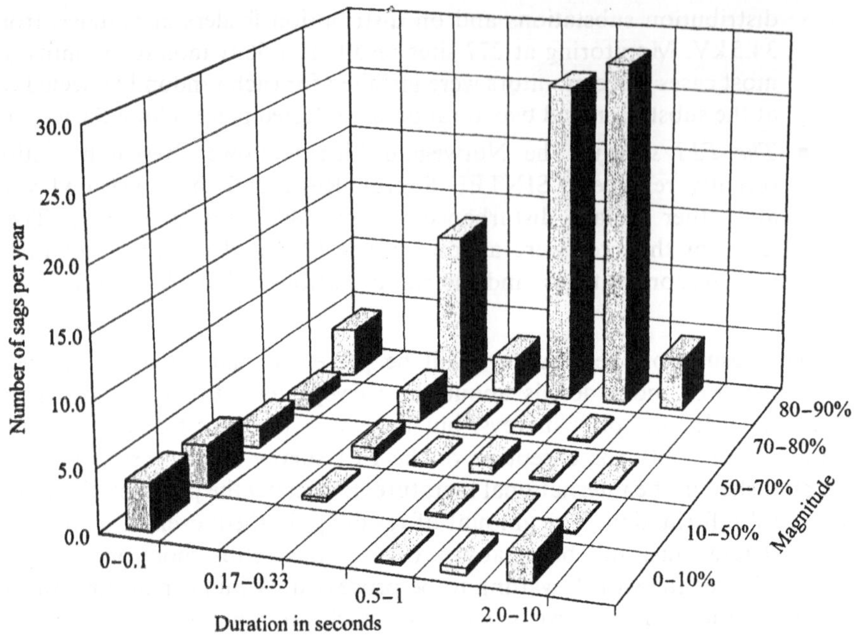

Figure 6.20 Sag density function for CEA secondary side data, corresponding to Table 6.6.

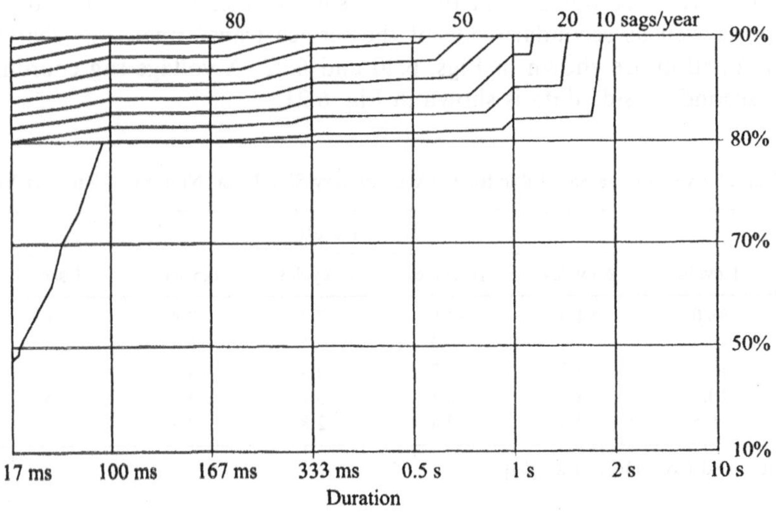

Figure 6.21 Voltage sag coordination chart for CEA secondary side data, corresponding to Table 6.6.

We see that the number of sags on secondary side is significantly higher than the number of sags on primary side. Part of the secondary side sags originates at secondary side, i.e., within the customer premises. The large number of long shallow sags at secondary side can be explained as motor starting on secondary side. As we saw in Section 4.9, these sags are not noticeable (i.e., magnitude above 90%) on primary side of the transformer.

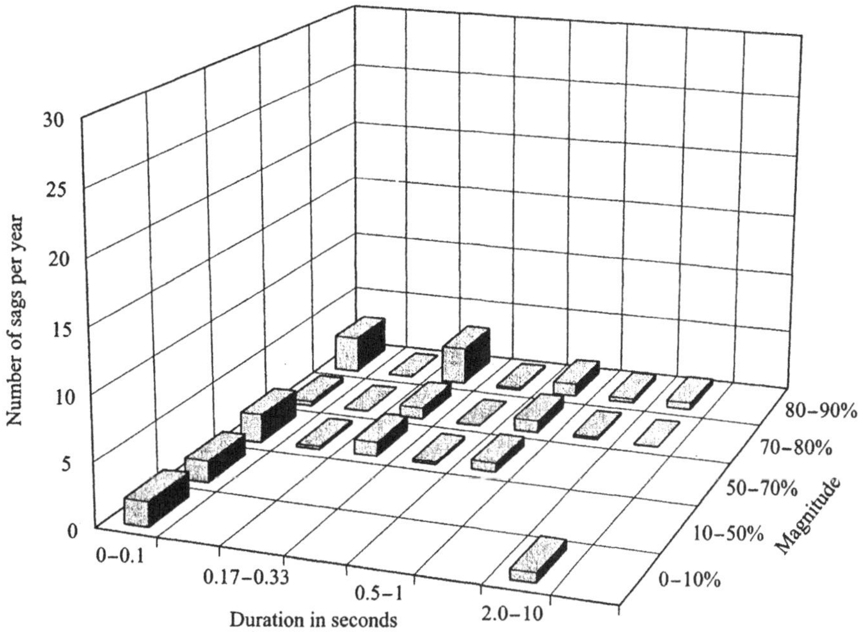

Figure 6.22 Sag density of primary side CEA data, corresponding to Table 6.7.

Another interesting observation is the large number of deep short sags (0–100 ms, 0–50%). The number is less on secondary side, but still significant. A comparison with other surveys shows that this is a typical feature of the CEA survey. Further analysis of the data is needed to explain this.

With any interpretation of the CEA primary side data one should also consider the uncertainty in the results. As mentioned above, about 10% of the 550 sites was located on primary side of a distribution transformer. As each site was monitored for only 25 days, this resulted in only 3.7 monitoring-years of data. The uncertainty in sag frequency is at least a factor of two for each of the bins in the sag density table. In the CEA secondary side data the uncertainty is smaller as the amount of data is equivalent to 38 monitor years.

6.3.1.2 Magnitude Versus Duration: NPL Survey. The number of sags per year, as obtained from the NPL survey, is shown in cumulative form in Tables 6.8 and 6.9. Table 6.8 shows the original data, where each individual event is counted, even if they are due to the same reclosure cycle. In Table 6.9 a 5-minute filter is applied: all events within 5 minutes are counted as one event: the one with the worst magnitude being the one counted. The sag densities are shown in Figs. 6.23 and 6.24 without and with filter, respectively. A voltage sag coordination chart for the filtered data is shown in Fig. 6.25.

Comparing Figs. 6.23 and 6.24, we see that there is some reduction in the number of short interruptions (voltage below 10%) as already discussed in Chapter 3. The most serious reduction is the number of long, shallow sags, the ones attributed to load switching. Apparently load switching sags come in clusters, with on average about 15 events within 5 minutes. This clearly distorts the quality of supply picture as drawn by

TABLE 6.8 Cumulative Voltage Sag Table for NPL Data Without Filter: Number of Sags per Year

	Duration							
Magnitude	1 cycle	6 cycles	10 cycles	20 cycles	0.5 sec	1 sec	2 sec	10 sec
87%	351.0	259.8	211.9	157.9	134.0	108.2	90.3	13.7
80%	59.5	32.3	23.7	19.0	16.2	13.1	10.4	5.8
70%	31.4	23.2	19.4	17.1	15.2	12.7	10.3	5.8
50%	20.9	18.3	16.8	15.4	14.1	12.2	10.2	5.8
10%	15.5	15.2	14.9	14.1	13.2	11.8	9.9	5.7

Source: Data obtained from Dorr et al. [54].

TABLE 6.9 Cumulative Voltage Sag Table for NPL Data with 5-minute Filter: Number of Sags per Year

	Duration							
Magnitude	1 cycle	6 cycles	10 cycles	20 cycles	0.5 sec	1 sec	2 sec	10 sec
87%	126.4	56.8	36.4	27.0	23.0	18.1	14.5	5.2
80%	44.8	23.7	17.0	13.9	12.2	10.0	8.0	4.3
70%	23.1	17.3	14.5	12.8	11.5	9.7	7.9	4.3
50%	15.9	14.1	12.9	11.8	10.6	9.4	7.8	4.3
10%	12.2	12.0	11.7	11.0	10.2	9.0	7.5	4.2

Source: Data obtained from Dorr et al. [54].

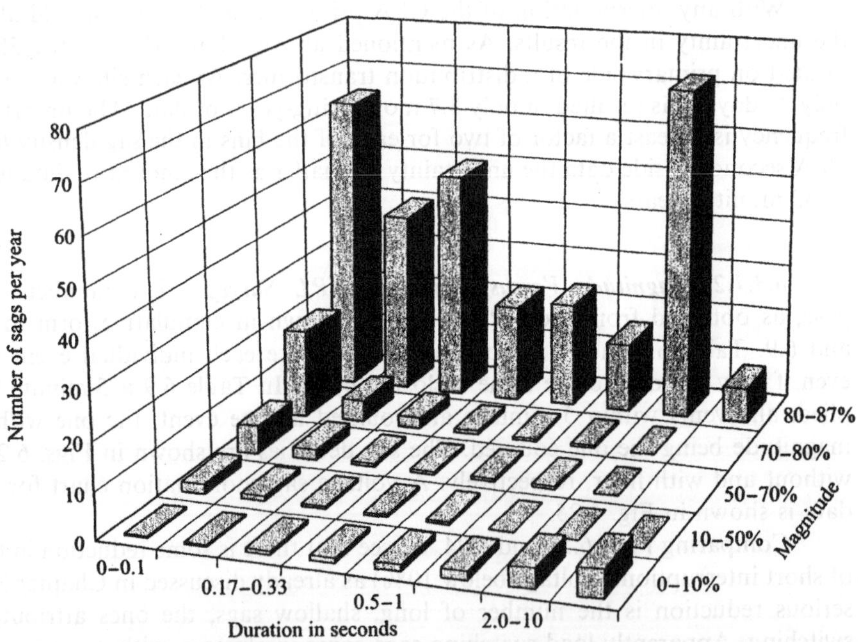

Figure 6.23 Sag density of NPL data, no filter, corresponding to Table 6.8.

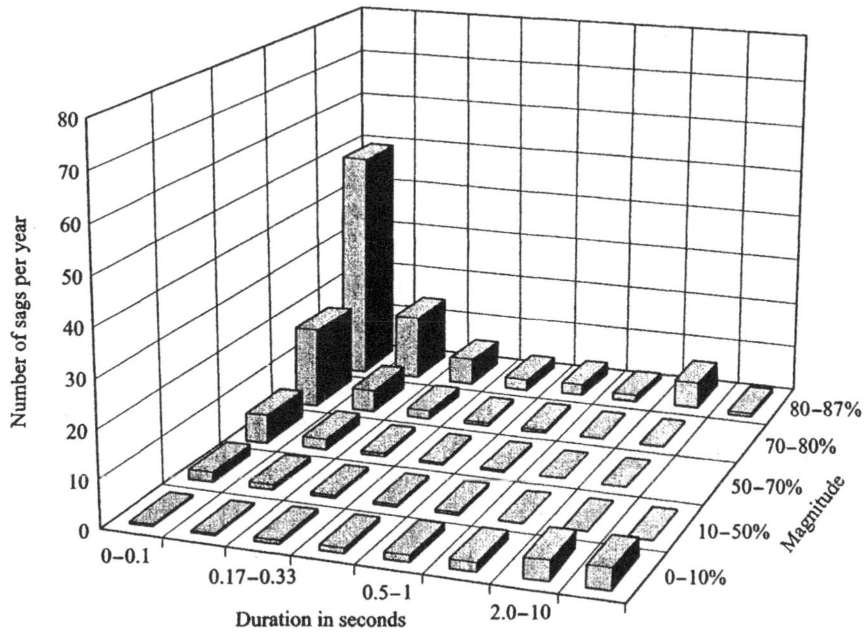

Figure 6.24 Sag density of NPL data, 5-minute filter, corresponding to Table 6.9.

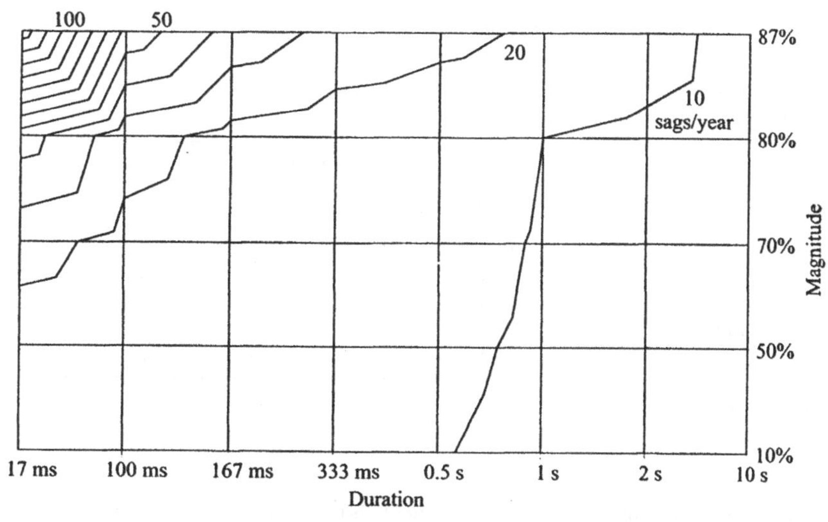

Figure 6.25 NPL data: voltage sag coordination chart, 5-minute filter, corresponding to Table 6.9.

the survey. Further investigation of the data is needed to find out whether most starting events are clustered or whether it is all due to a small number of sites. A comparison between the NPL data and the CEA data shows a much larger number of events for the former. The most likely explanation is the much lower lightning activity in Canada as compared to the United States.

6.3.1.3 *Magnitude Versus Duration: EPRI Survey.* The cumulative number of sags per year, as obtained from the EPRI survey, is shown in Tables 6.10 and 6.11. Table 6.11 gives the results for substations, while Table 6.10 is obtained from measurements along feeders. For both tables a 5-minute filter was applied. The sag density function is shown in Figs. 6.26 and 6.28. Figures 6.27 and 6.29 give the corresponding voltage sag coordination charts.

The differences between the feeder data and the substation data are small: in total only seven events per year, which is about 10% (this is the value in the upper-left corner of the tables). The seven-event difference is found in two areas in the magnitude-duration plane:

- Events up to 10 cycles with magnitudes below 70%. Here we find 13.6 events for the feeders, but only 8.3 for the substation.
- Interruptions of 1 second and longer: 3.4 events for the substation, 5.1 for the feeder.

Where the total number of events is remarkably similar, the relative difference in the number of severe events is significant. Table 6.12 compares the number of events below certain voltage levels, including events recorded at low voltage (NPL survey). Only events with a duration less than 20 cycles (about 300 ms) are included in the comparison: i.e. mainly events due to short circuits. Looking at Table 6.12 we see more interruptions and deep sags on the feeder as compared to the substation. The increased

TABLE 6.10 Cumulative Voltage Sag Table for EPRI Feeder Data with 5-minute Filter: Number of Sags per Year

	Duration							
Magnitude	1 cycle	6 cycles	10 cycles	20 cycles	0.5 sec	1 sec	2 sec	10 sec
90%	77.7	31.2	19.7	13.5	10.7	7.4	5.4	1.8
80%	36.3	17.4	12.4	9.3	7.9	6.4	4.9	1.7
70%	23.9	13.1	10.3	8.3	7.2	6.2	4.8	1.7
50%	14.6	9.5	8.4	7.5	6.6	5.9	4.6	1.7
10%	8.1	6.5	6.4	6.2	5.6	5.1	4.0	1.7

Source: Data obtained from Dorr et al. [54].

TABLE 6.11 Cumulative Voltage Sag Table for EPRI Substation Data with 5-minute Filter: Number of Sags per Year

	Duration							
Magnitude	1 cycle	6 cycles	10 cycles	20 cycles	0.5 sec	1 sec	2 sec	10 sec
90%	70.8	28.1	17.4	11.4	8.6	5.4	3.7	1.5
80%	29.1	14.7	10.1	7.1	5.6	4.3	3.2	1.4
70%	16.1	9.8	7.8	6.0	4.9	4.0	3.0	1.4
50%	7.9	6.6	6.1	5.3	4.4	3.8	2.9	1.4
10%	5.4	5.2	5.1	4.7	3.9	3.4	2.5	1.4

Source: Data obtained from Dorr et al. [54].

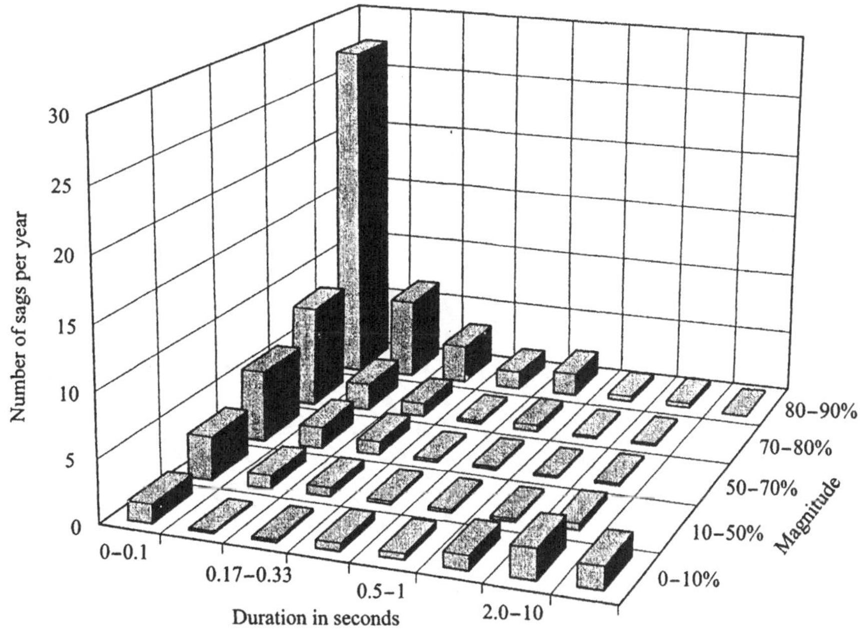

Figure 6.26 EPRI feeder data: sag density function, corresponding to Table 6.10.

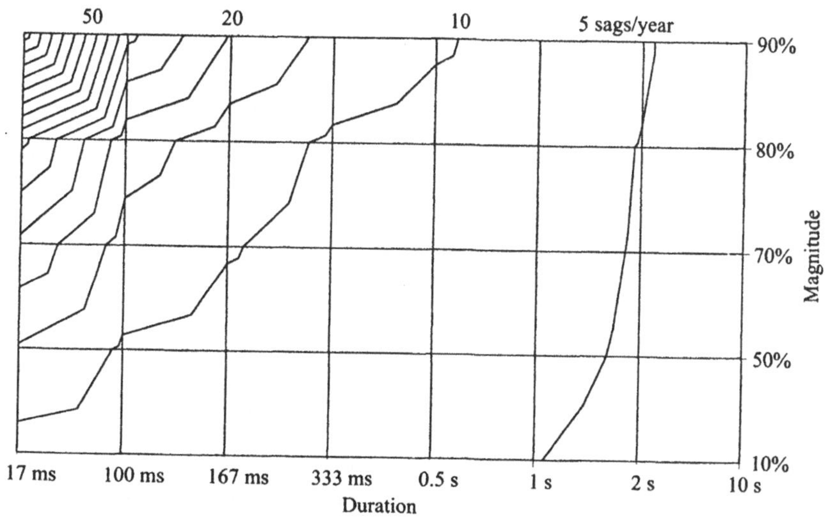

Figure 6.27 EPRI feeder data: voltage sag coordination chart, corresponding to Table 6.10.

number of interruptions is understandable: some interruptions only affect part of the feeder; the closer to the equipment, the higher the number of interruptions simply because the path that can be interrupted is longer. For the increase in the number of deep short sags there is no ready explanation. Three possible explanations, which will probably all somewhat contribute, but for which more investigations are needed to give a definite explanation are:

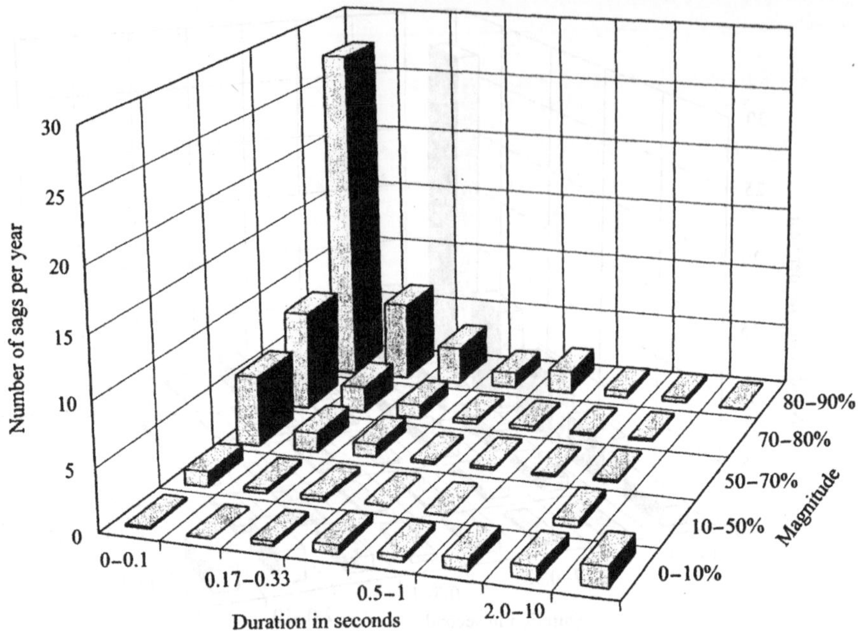

Figure 6.28 EPRI substation data: sag density function, corresponding to Table 6.11.

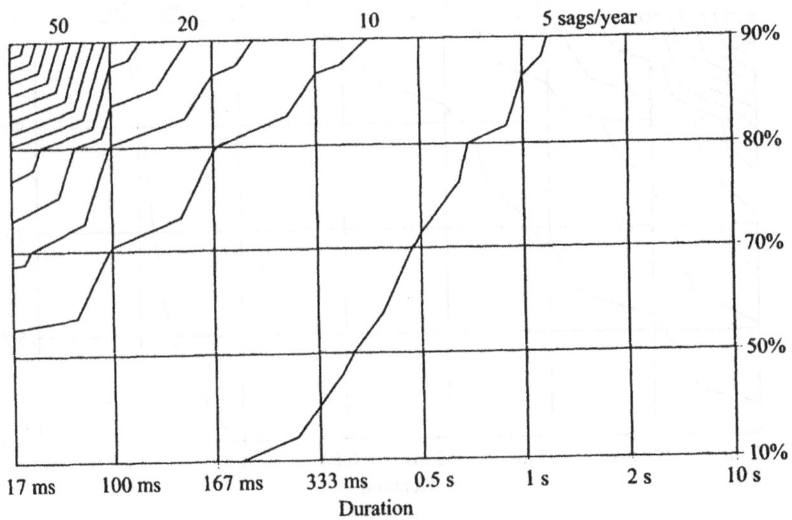

Figure 6.29 EPRI substation data: voltage sag coordination chart, corresponding to Table 6.11.

- Reclosing actions on the feeder beyond the point where the monitor is connected. The monitor on the feeder will record a deeper sag than the one in the substation. This would explain the deep short sags. As the distribution transformer is often Dy-connected, deep sags due to single-phase faults will not transfer fully to low voltage. This explains the smaller number of deep short sags measured at low voltage (NPL survey).

TABLE 6.12 Number of Events with a Duration Less than 20 Cycles: NPL Survey (LV) and EPRI Survey (Feeder, Substation)

	Events per Year		
		Distribution	
Voltage Range	LV	Feeder	Substation
80–90%	68.5	37.2	37.4
70–80%	20.6	11.4	12.0
50–70%	6.2	8.5	7.5
10–50%	2.9	5.8	1.9
0–10%	1.1	1.9	0.7

Source: Data obtained from Dorr et al. [54].

- The normal operating voltage at the feeder is lower. As the sag magnitude is given as a percentage of the nominal voltage, the sag will appear deeper at the feeder than at the substation. Giving the sag magnitude as a percentage of the pre-event voltage would compensate this effect. This may explain the increase in the number of shallow sags along the feeder.
- Induction motor influence. Induction motors slow down more for deeper sags and thus reduce the positive sequence voltage. A reduction in positive sequence voltage would imply a reduction (also) in the lowest phase voltage and thus a reduction in sag magnitude.

Comparing low voltage and medium voltage data we see that the number of shallow sags is much higher at low voltage than at medium voltage, whereas the number of deep sags is smaller at low voltage.

6.3.1.4 Magnitude Versus Duration: EFI Survey. The cumulative voltage sag tables, as obtained by the EFI survey, are shown in Tables 6.13 through 6.16. The sag density functions are presented in Figs. 6.30 through 6.33. Table 6.13 and Fig. 6.30 give the average results for the low-voltage sites, Table 6.14 and Fig. 6.31 refer to the distribution sites.

We see that the average distribution site experiences somewhat less longer-duration events but clearly more short-duration events. The increase in number of interruptions for lower voltage levels is consistent with the findings of U.S. surveys. To

TABLE 6.13 Cumulative Voltage Sag Table for EFI Data, All Low-Voltage Networks: Number of Sags per Year

	Duration (sec)					
Magnitude	0.01	0.1	0.5	1.0	3.0	20.0
90%	74.7	36.5	18.5	12.1	8.6	6.8
70%	26.3	11.9	8.2	7.5	6.8	5.9
40%	16.6	9.8	7.5	7.5	6.8	5.9
1%	9.3	8.2	7.5	7.5	6.8	5.9

Source: Data obtained from Seljeseth [67].

TABLE 6.14 Cumulative Voltage Sag Table for EFI Data, All Distribution Networks: Number of Sags per Year

	Duration (sec)					
Magnitude	0.01	0.1	0.5	1.0	3.0	20.0
90%	112.2	39.2	15.5	7.9	6.0	5.2
70%	40.5	16.9	11.4	6.6	6.0	5.2
40%	15.2	7.6	6.8	6.0	5.7	5.2
1%	7.2	5.7	5.7	5.7	5.7	5.2

Source: Data obtained from Seljeseth [67].

TABLE 6.15 Cumulative Voltage Sag Table for EFI Data, 95% Percentile for Low-Voltage Networks: Number of Sags per Year

	Duration (sec)					
Magnitude	0.01	0.1	0.5	1.0	3.0	20.0
90%	315	128	47	20	11	11
70%	120	39	11	11	11	11
40%	66	25	11	11	11	11
1%	25	11	11	11	11	11

Source: Data obtained from Seljeseth [67].

TABLE 6.16 Cumulative Voltage Sag Table for EFI Data, 95% Percentile for Distribution Networks: Number of Sags per Year

	Duration (sec)					
Magnitude	0.01	0.1	0.5	1.0	3.0	20.0
90%	388	159	57	20	12	12
70%	130	53	22	12	12	12
40%	45	21	12	12	12	12
1%	18	12	12	12	12	12

Source: Data obtained from Seljeseth [67].

understand all effects, one needs to understand the propagation of sags to lower voltage levels, for which the study of more individual events is needed.

Tables 6.15 and 6.16 give the 95% percentile of the sag distribution over the various sites. A stochastic distribution function was created for the total number of sags measured at one single site. The 95% percentile of this distribution was chosen as a reference site. The number of sags at this site is thus exceeded by only 5% of the sites. The 95% value was suggested in Chapter 1 as a way of characterizing the electromagnetic environment (the term used by the IEC for the quality of the supply). Thus, we could say that Table 6.15 characterizes the electromagnetic environment for the Norwegian low-voltage customer.

6.3.1.5 Variation in Time—Lightning Strokes. A large fraction of the voltage sags is due to lightning strokes on overhead lines. Two phenomena play a role here: short circuits due to lightning strokes and triggering of spark gaps due to lightning-

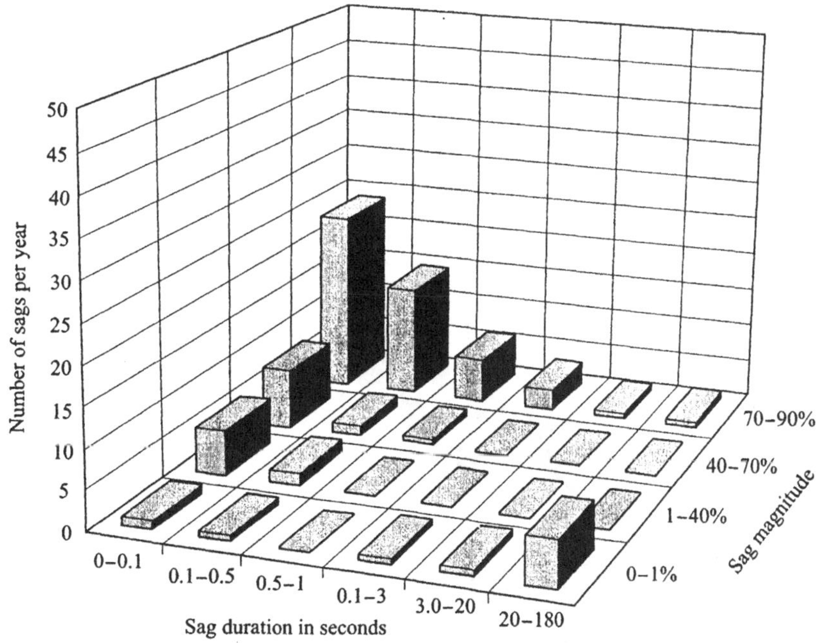

Figure 6.30 Sag density for EFI low-voltage networks, corresponding to Table 6.13.

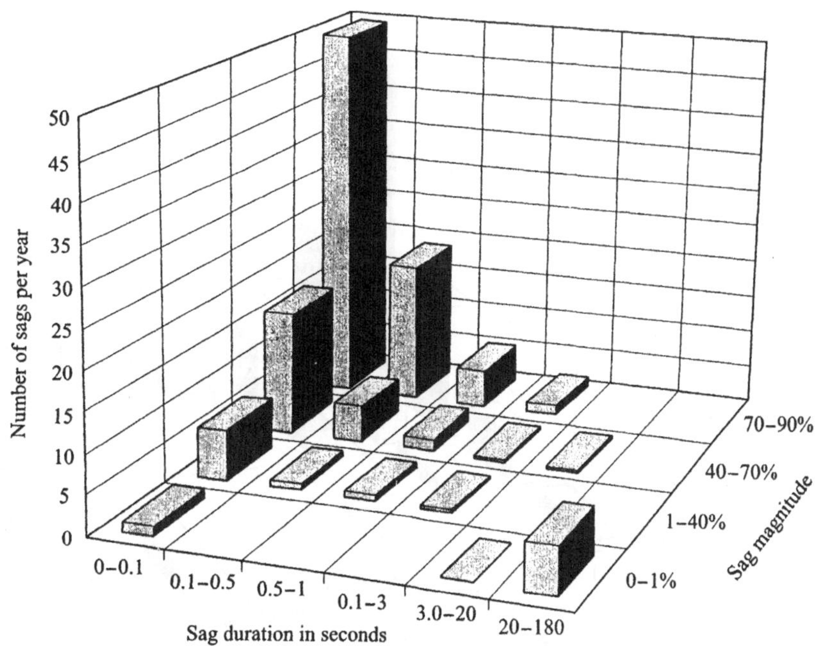

Figure 6.31 Sag density for EFI distribution networks, corresponding to Table 6.14.

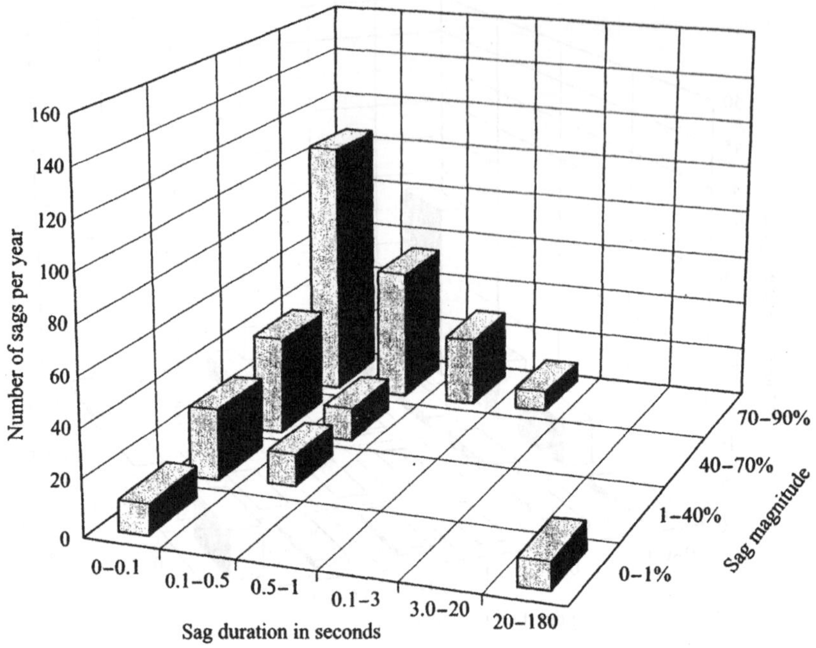

Figure 6.32 Sag density for 95% percentile of EFI low-voltage networks, corresponding to Table 6.15.

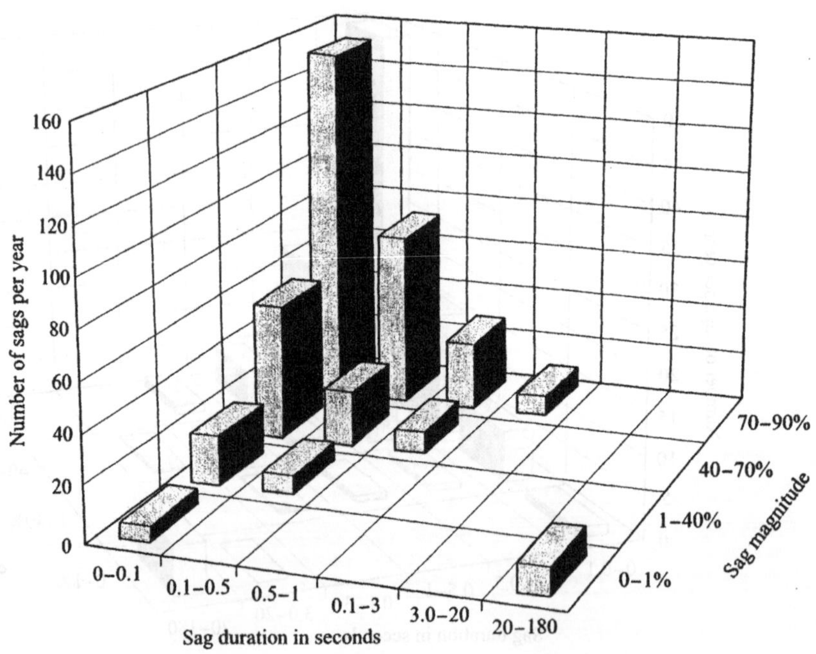

Figure 6.33 Sag density for 95% percentile of EFI distribution networks, corresponding to Table 6.16.

induced overvoltages. The effect of a lightning stroke is to induce a large overvoltage on the line. If this voltage exceeds the insulation withstand level it results in a short circuit, otherwise the voltage peak will start to propagate through the system. If the peak voltage is not high enough to cause a flashover on the line, it might still trigger a spark gap or a (ZnO) varistor. A spark gap mitigates the overvoltage by creating a temporary short circuit, which in its turn causes a sag of one or two cycles. A varistor will only cap the overvoltage. A conclusion from one of the first power quality surveys [72] was that the number of voltage transients did not increase in areas with more lightning; instead the number of voltage sags increased.

For a few sites in the EPRI survey, the sag frequency was compared with the lightning flash density [70]. This comparison showed that the correlation between sags and lightning was much stronger than expected. Plotting the sag frequency against the flash density (number of lightning flashes per km^2 per year) for five sites resulted in almost a straight line. This justifies the conclusion that lightning is the main cause of voltage sags in U.S. distribution systems.

As sags are correlated with lightning and lightning activity varies with time, we expect the number of sags to vary with time. This is shown in Fig. 6.34 for the NPL survey [68]. The sag frequency is at its maximum in summer, when also the lightning activity is highest. This effect has been confirmed in other countries. Also the distribution of sags through the day follows the lightning activity, with its peak in the evening.

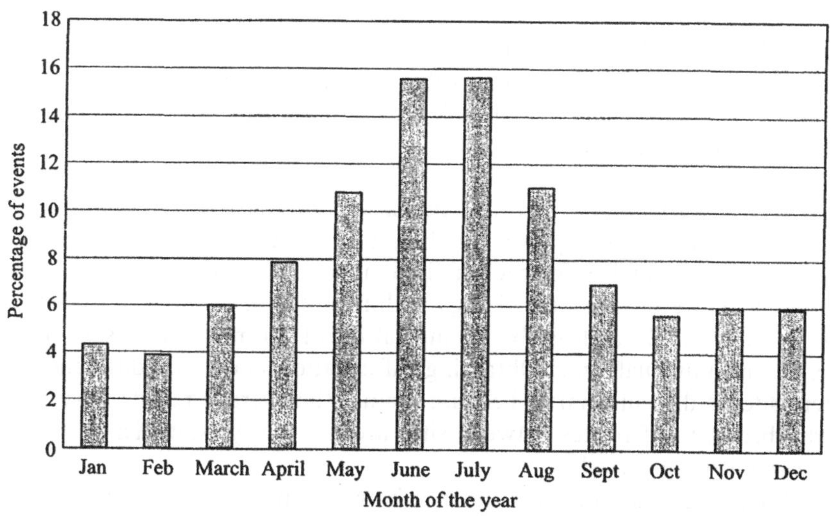

Figure 6.34 Variation of voltage sag frequency through the year. (Data obtained from Dorr [68].)

6.3.1.6 Correcting for Short Monitoring Periods. The variation of the sag frequency through the year indicates that the monitoring period should be at least 1 year to get a good impression of the power quality at a certain site. As weather activity varies from year to year, it is even needed to monitor several years. In case a limited monitoring period is used, it is still possible to get a rough estimate of the average number of sags over a longer period [49]. To do this, fault data are needed over the monitoring period as well as over a longer period of time.

The basic assumption behind the correction method is that voltage sags are due to short circuits: thus that the number of sags is proportional to the number of short-circuit faults. In equation form this reads as

$$\frac{\overline{N}_{sags}}{N_{sags}} = \frac{\overline{N}_{faults}}{N_{faults}} \tag{6.7}$$

where N_{sags} and N_{faults} are the number of sags and faults, respectively, recorded during the monitoring period, and $\overline{N}_{sags}$ and $\overline{N}_{faults}$ the (average) number during a longer period of time. The number of sags over a longer period of time can thus be obtained from

$$\overline{N}_{sags} = N_{sags} \times \frac{\overline{N}_{faults}}{N_{faults}} \tag{6.8}$$

Ideally, one would like to know the number of faults in the area of the system in which the sags originate. Often this information is not available: one is likely to only have fault data over the whole service area of the utility. This method also neglects the above-mentioned short-duration sags due to triggering of overvoltage devices and sags due to transient faults which are not recorded.

The correction method can be improved if the sags can be traced back to the voltage levels at which they originated:

$$\overline{N}_{sags} = \sum_i \left[N_{sags}^{(i)} \times \frac{\overline{N}_{faults}^{(i)}}{N_{faults}^{(i)}} \right] \tag{6.9}$$

with $N_{sags}^{(i)}$ the number of sags during the monitoring period originating at voltage level i, etc. In most cases it will not be possible to trace back all sags. Only for a small number of sites this method might be suitable. It has been used in [49] to quantify the average supply performance in Japan.

6.3.1.7 Variation in Space. The basic assumption of a large power quality survey is that the average power quality, over a number of sites, gives information about the power quality for each individual site. Thus, if the conclusion of the survey is that there are on average 25 sags within a certain magnitude and duration range, this number should at least be an indication of the number of sags at an individual site, in an individual year. Obtaining information about the differences between different sites is difficult; partly because mainly the average results have been published; partly because differences between sites are not always statistically significant after a short monitoring period.

Some indication of the difference between sites is obtained from the EFI survey. The difference between the 95% site and the average of all sites is very large, as can be seen by comparing Tables 6.13 and 6.15. At least 5% of the sites have about four times as many sags as the average of all sites. For those sites the average values do not give much useful information. The problem is that without a prior study it is difficult to know whether the average data applies to a certain site. Further splitting up the data set in different types of sites, e.g., systems with mainly overhead lines and systems with mainly underground cables, can reduce the spread among the sites within one group. But reducing the data set will also increase the statistical error in the estimates.

Information on the spread in power quality among different sites is also given in [72]. Sags and some other voltage disturbances were measured at 24 sites from May 1977 through September 1979, leading to a total of 270 monitor-months of data. The

TABLE 6.17 Distribution Over the Sites of the Number of Sags and Interruptions

	Maximum Number of Sags Longer Than the Indicated Duration				
Number of Sites	1 cycle	100 ms	200 ms	0.5 sec	1 sec
10%	11	6	3	2	0
25%	17	9	5	3	2
50%	25	13	8	5	3
75%	36	19	12	8	5
90%	51	26	17	12	8

Source: Data obtained from [72].

total amount of data of this survey is not very large, but the monitor period at each site is long enough to make some comparison between the different sites. Some of the results are shown in Table 6.17. This table gives, for various minimum durations, the maximum number of sags and interruptions for a certain percentage of sites. As an example: 25% of the sites has fewer than five events per year longer than 200 milliseconds. Also: 80% of the sites has between 11 and 51 events per year longer than one cycle in duration, the remaining 20% of sites are outside of that range. For about half of the sites the median value is a reasonable indicator of the number of sags that can be expected. As already mentioned before, it is hard to know if a site belongs to the 50% average sites or not, without monitoring the supply.

6.3.2 Individual Sites

Monitoring is not only used for large power quality surveys, it is also used for assessing the power quality of individual sites. For harmonics and voltage transients, reliable results can be obtained in a relatively short period of time. Some interesting site surveys in Canadian rural industry have been performed by Koval [58]. One of the conclusions of his studies was that a monitoring period of two weeks gives a good impression of the power quality at a site [59]. Again it needs to be stressed that this holds only for relatively frequent events like voltage transients and motor starting sags and for phenomena like harmonics and voltage fluctuation. Voltage sags and interruptions of interest for compatibility assessment have occurrence frequencies of once a month or less. Much longer monitoring periods are needed for those events.

6.3.2.1 The Required Monitoring Period. To estimate how long the monitoring period needs to be, we assume that the time-between-events is exponentially distributed. This means that the probability of observing an event, in let's say the next minute, is independent of the time elapsed since the last event. Thus, events occur completely independent from each other. Under that condition the number of events captured within a certain period is a stochastic variable with a so-called Poisson distribution.

Let μ be the expected number of events per year, then the observed number of events K, over a monitoring period of n years is a discrete stochastic variable with the following distribution:

$$Pr\left\{K = k\right\} = e^{-n\mu}\frac{(n\mu)^k}{k!} \tag{6.10}$$

This Poisson distribution has an expected value $n\mu$ and a standard deviation $\sqrt{n\mu}$. The result of monitoring is an estimate of the expected number of events per year, obtained as follows:

$$\mu_{est} = \frac{K}{n} \tag{6.11}$$

This estimate has an expected value μ (it is a true estimate) and a standard deviation $\sqrt{\frac{\mu}{n}}$. For a large enough value of $n\mu$ (i.e., for a sufficient number of observed events) the Poisson distribution can be approximated by a normal distribution with expected value μ and standard deviation $\sqrt{\frac{\mu}{n}}$. For a normal distribution with expected value μ and standard deviation σ the so-called 95% confidence interval is between $\mu - 1.96\sigma$ and $\mu + 1.96\sigma$, with σ the standard deviation. The relative error in the estimate of μ after n samples is thus,

$$\frac{1.96\sigma}{\mu} = \frac{1.96}{\sqrt{n\mu}} \approx \frac{2}{\sqrt{N}} \tag{6.12}$$

with $N = n\mu$ the expected number of events in n years, i.e., in the whole observation period. To limit the relative error to ϵ the monitoring period n should fulfill the following inequality:

$$\frac{2}{\sqrt{n\mu}} < \epsilon \tag{6.13}$$

or

$$n > \frac{4}{\mu\epsilon^2} \tag{6.14}$$

For an event with a frequency of μ times per year, the monitoring period should be at least $\frac{4}{\mu\epsilon^2}$ years to obtain an accuracy ϵ.

Table 6.18 gives the minimum monitoring period for various event frequencies and accuracies. Note that sag frequencies are ultimately used to predict equipment trip frequencies. It shows that site monitoring can only give accurate results for very sensitive equipment (high frequency of tripping events). When equipment becomes more compatible with the supply (and thus trips less often) site monitoring can no longer be used to predict the number of trips.

As mentioned before, the approximation of a Poisson distribution by a normal distribution holds for a sample of large size. Nothing was said about what this large size is. A more accurate expression for the uncertainty is obtained by using the so-called Student's t-distribution. Using this distribution gives another factor in (6.12) instead of 1.96. The deviation is small: for 10 events we find a factor of 2.228, which is an increase of 14%; for five events the value is 2.571. For 16 events (50% accuracy according to the

TABLE 6.18 Minimum Monitoring Period Needed to Obtain a Given Accuracy

Event Frequency	50% Accuracy	10% Accuracy	2% Accuracy
1 per day	2 weeks	1 year	25 years
1 per week	4 months	7 years	200 years
1 per month	1 year	30 years	800 years
1 per year	16 years	400 years	10,000 years

approximation) the Student's t-distribution gives an accuracy of 53%. The effect of this on Table 6.18 is small.

6.3.2.2 More Uncertainties. The above reasoning assumes a stationary system with exponentially distributed times between events, thus where events appear completely at random. For a stationary system it is possible to obtain the event frequency with any required accuracy by applying a long-enough monitoring period. In the actual situation there are two more effects which make that monitoring results have a limited predictive value:

- A large fraction of voltage sags is due to bad weather: lightning, heavy wind, snow, etc. The sag frequency is therefore not at all constant but follows the annual weather patterns. But the amount of weather activity also varies significantly from year to year. Due to the relation between voltage sags and adverse weather, the sags come in clusters. To get a certain accuracy in the estimate, one needs to observe more than a minimum number of clusters. It is obvious that this will increase the required monitoring period. To get a long-term average a long monitoring period is needed. A correction made according to (6.8) might increase the accuracy.
- Power systems themselves are not static but change continuously from year to year. This especially holds for distribution networks. The number of feeders connected to a substation can change; or another protective relay is used. Also component failure rates can change, e.g., due to aging; increased loading of components; different maintenance policies; or because the amount of squirrels in the area suddenly decreases.

Despite these disadvantages, site monitoring can be very helpful in finding and solving power quality problems, as some things are simply very hard to predict. In addition, stochastic assessment requires a certain level of understanding of voltage disturbances and their origin. This understanding can only be achieved through monitoring.

6.4 THE METHOD OF FAULT POSITIONS

6.4.1 Stochastic Prediction Methods

The great advantage of stochastic prediction as compared to monitoring is that the required accuracy is obtained right away. With stochastic prediction it is even possible to assess the power quality of a system that does not yet exist; something which is impossible to achieve by power quality monitoring.

Stochastic prediction methods use modeling techniques to determine expected value, standard deviation, etc., of a stochastic variable. With stochastic predictions one should not think of a prediction like *a voltage sag down to 35% will occur at 7:30 in the evening on July 21*. Instead, the kind of predictions are more like *in July one can expect 10 sags below 70%, half of which are expected to occur between 5 and 9 in the evening.*

Stochastic prediction methods have been used for many years to predict frequency and duration of long interruptions as discussed in detail in Chapter 2. For shorter duration events, the use of stochastic prediction techniques is still very uncommon.

Those events tend to have a higher occurrence frequency, making monitoring more feasible. Also the required electrical models have a higher complexity than for long interruptions. A final explanation is that power quality is still very much an industry-driven area, whereas reliability evaluation is much more a university-driven subject.

Stochastic prediction methods are as accurate as the model used and as accurate as the data used. The accuracy of the models can be influenced; the accuracy of the data is often outside our control. Any stochastic prediction study in power systems requires two kinds of data: *power system* data and *component reliability* data. The main data concern is the latter one. Component reliability data can only be obtained through observing the behavior of the component. From a stochastic point of view this is identical to the power quality monitoring of one individual site we discussed earlier. Component reliability data has therefore the same uncertainties as the outcome of power quality monitoring. One could now be tempted to draw the conclusion that we did not gain anything by using stochastic prediction. This conclusion is fortunately not correct. Many utilities have records of component failures over several decades. Components do not need to be considered separately but can be grouped into "stochastically identical" types: like all distribution transformers. This enormously reduces the error in the component failure rate.

Some problems remain of course: maintenance methods change; the failure rate of new components is hard to assess; component loading patterns can change; even weather patterns are prone to change. The same uncertainties are present with power quality monitoring, but with stochastic assessment one is able to somewhat assess the influence of these uncertainties.

6.4.2 Basics of the Method of Fault Positions

The method of fault positions is a straightforward method to determine the expected number of sags. It was proposed independently by a number of authors but probably first used by Conrad [48] whose work has become part of IEEE Std-493 [8], [21]. The method is also used by EdF (Electricité de France) to estimate the number of sags due to faults in their distribution systems [60]. The method of fault positions was combined with Monte Carlo simulation by the author in [61], [63], extended with non-rectangular sags due to motor re-acceleration in [18], [62] and extended with generator outages in [64]. At least one commercial software package is available using the method of fault positions. More packages will almost certainly follow as the method is computationally very simple, although it often requires excessive calculation time. The accuracy of the results can be increased by increasing the number of fault positions. Non-rectangular sags can be taken into account by using dynamic generator and load models; phase-angle jumps by working with complex impedances and voltages; three-phase unbalance by including single-phase and phase-to-phase faults.

6.4.2.1 Outline of the Method. The method of fault positions proceeds, schematically, as follows:

- Determine the area of the system in which short circuits will be considered.
- Split this area into small parts. Short circuits within one part should lead to voltage sags with similar characteristics. Each small part is represented by one fault position in an electric circuit model of the power system.

- For each fault position, the short-circuit frequency is determined. The short-circuit frequency is the number of short-circuit faults per year in the small part of the system represented by a fault position.
- By using the electric circuit model of the power system the sag characteristics are calculated for each fault position. Any power system model and any calculation method can be used. The choice will depend on the availability of tools and on the characteristics which need to be calculated.
- The results from the two previous steps (sag characteristics and frequency of occurrence) are combined to obtain stochastical information about the number of sags with characteristics within certain ranges.

6.4.2.2 Hypothetical Example. Consider a 100 km line as shown in Fig. 6.35. Short circuits in this part of the system are represented through eight fault positions. The choice of the fault positions depends on the sag characteristics which are of interest. In this example we consider magnitude and duration. Fault position 1 (representing busbar faults in the local substation) and fault position 2 (faults close to the local substation) will result in the same sag magnitude. But the fault-clearing time is different, therefore two fault positions have been chosen. The fault positions along the line (2, 3, 4, and 5) have similar fault-clearing time but different sag magnitude. Fault positions 6, 7, and 8 result in the same sag magnitude but different duration.

For each fault position a frequency, a magnitude, and a duration are determined, as shown in Table 6.19. Failure rates of eight faults per 100 km of line per year and 10 faults per 100 substations per year have been used. It should be realized here that not all fault positions along the line represent an equal fraction of the line: e.g., position 5 represents 25 km (between 5/8th and 7/8th of the line) but position 6 only 12.5 km (between 7/8th and 1).

The resulting sags (1 through 8 in Table 6.19) are placed in bins or immediately in a cumulative form. Table 6.20 shows how the various sags fit in the bins. Filling in the frequencies (failure rates) leads to Table 6.21 and its cumulative equivalent shown in Table 6.22. Alternatively it is possible to update the cumulative table after each fault

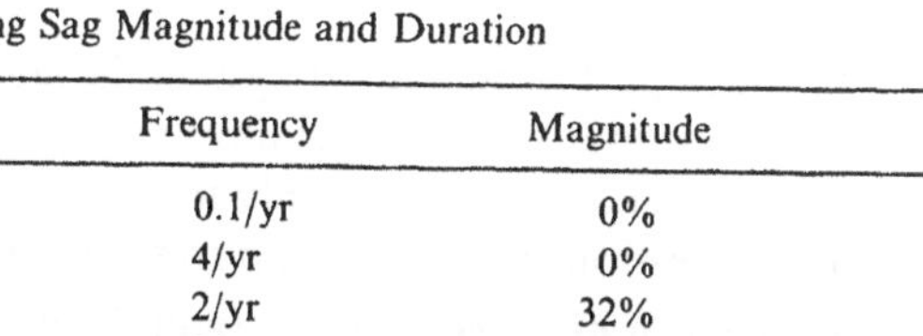

Figure 6.35 Part of power system with fault positions.

TABLE 6.19 Fault Positions with Resulting Sag Magnitude and Duration

Fault Position	Frequency	Magnitude	Duration
1 Busbar fault in local substation	0.1/yr	0%	180 ms
2 Fault on a line close to local substation	4/yr	0%	80 ms
3 Fault at 25% of the line	2/yr	32%	90 ms
4 Fault at 50% of the line	2/yr	49%	105 ms
5 Fault at 75% of the line	2/yr	57%	110 ms
6 Fault at 100% of local line	1/yr	64%	250 ms
7 Fault at 0% of remote line	2/yr	64%	90 ms
8 Busbar fault in remote substation	0.1/yr	64%	180 ms

TABLE 6.20 Fault Positions Sorted for Magnitude and Duration Bins

	0–100 ms	100–200 ms	200–300 ms
60–80%	7	8	6
40–60%		4 and 5	
20–40%	3		
0–20%	2	1	

TABLE 6.21 Table with Event Frequencies for Example of Method of Fault Positions

	0–100 ms	100–200 ms	200–300 ms
60–80%	2.0	0.1	1.0
40–60%		4.0	
20–40%	2.0		
0–20%	4.0	0.1	

TABLE 6.22 Cumulative Table for Example of Method of Fault Positions

	0 ms	100 ms	200 ms
80%	13.2	5.2	1.0
60%	10.1	4.1	0.0
40%	6.1	0.1	0.0
20%	4.1	0.1	0.0

position. As we have seen in Section 6.2 this is needed anyway when non-rectangular sags are considered. Please note that this is a completely fictitious example. No calculation at all has been used to obtain the magnitude and durations in Table 6.19.

6.4.3 Choosing the Fault Positions

The first step in applying the method of fault positions is the choice of the actual fault positions. It will be obvious that to obtain more accurate results, more fault positions are needed. But a random choice of new fault positions will probably not increase the accuracy, only increase the computational effort.

Three decisions have to be made when choosing fault positions:

1. **In which part of the power system do faults need to be applied?** Only applying faults to one feeder is certainly not enough; applying faults to all feeders in the whole country is certainly too much. Some kind of compromise is needed. This question needs to be addressed for each voltage level.
2. **How much distance between fault positions is needed?** Do we only need fault positions in the substations or also each kilometer along the lines ? Again this question needs to be addressed for each voltage level.
3. **Which events need to be considered?** For each fault position, different events can be considered. One can decide to only study three-phase faults, only

single-phase faults, or all types of faults. One can consider different fault impedances, different fault-clearing times, or different scheduling of generators, each with its own frequency of occurrence and resulting sag characteristics.

Below are some suggestions for the choice of the fault positions. A number of those suggestions are borrowed from the method of critical distances to be discussed in Section 6.5. In this section only the results will be used; for more theoretical background one is advised to read Section 6.5 first.

The main criterion in choosing fault positions is: *a fault position should represent short-circuit faults leading to sags with similar characteristics.* This criterion has been applied in choosing the fault positions in Fig. 6.35 and Table 6.19.

6.4.3.1 Distance between Fault Positions. To understand how the distance between fault positions influences the result, consider the sag magnitude as a function of the distance between the fault and the substation from which the load is fed. The sag magnitude is plotted in Fig. 6.36. The shape of the curve can be obtained from the equations in Section 6.5. By choosing one fault position to represent a certain range of possible faults, we make the sag magnitude for the whole range equal to the sag magnitude for that one position. The approximated magnitude versus distance is shown in Fig. 6.37. We see that the error is largest when the exact curve is at its steepest, which is close to the load. Here we would need a higher density of fault positions. For more remote faults, the curve becomes more flat, and the error smaller. Further away from the load, a lower density of fault positions would be acceptable.

To quantify this, consider a radial system as shown in Fig. 6.38. A load is fed from a substation with a nominal (phase-to-phase) voltage V_{nom}. The fault current for a terminal fault on the indicated feeder is I_{fault}, thus the source impedance is

$$Z_S = \frac{V_{nom}}{\sqrt{3} \times I_{fault}} \tag{6.15}$$

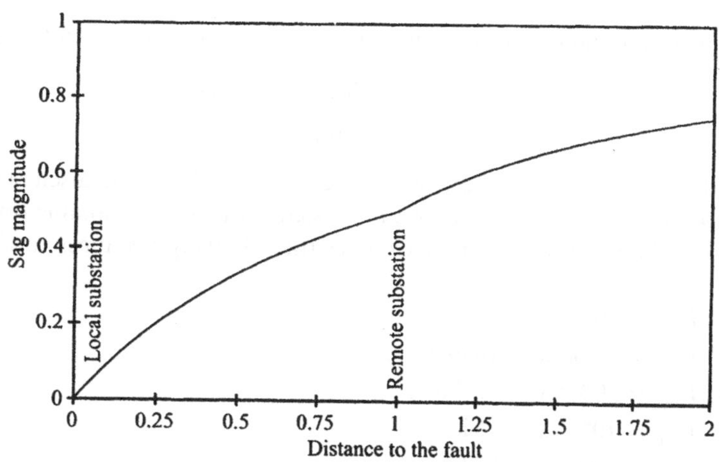

Figure 6.36 Voltage as a function of the distance to the fault.

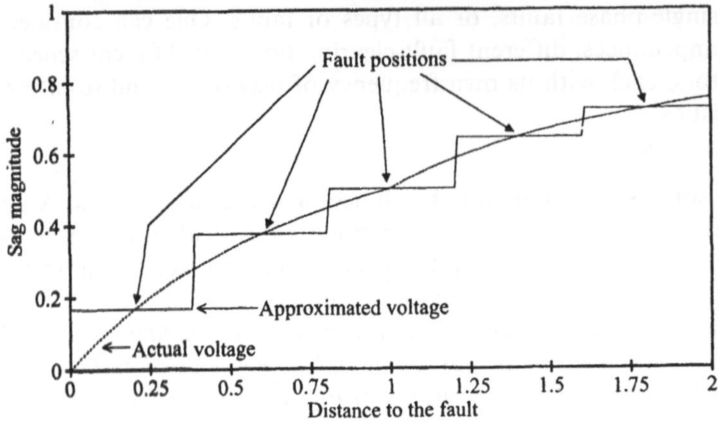

Figure 6.37 Approximated voltage as a function of the distance to the fault.

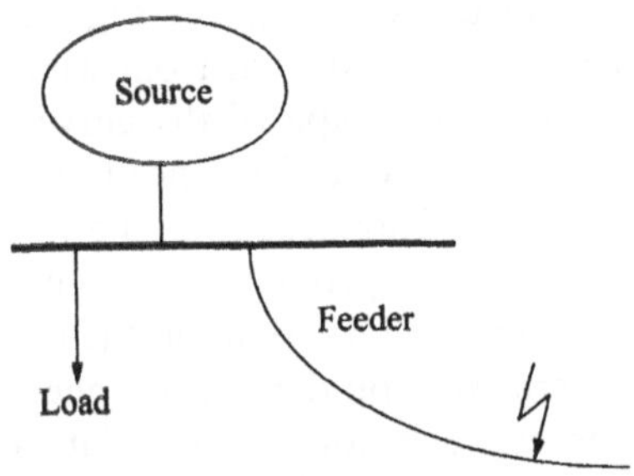

Figure 6.38 Faults in a radial system.

The feeder has an impedance z per unit length and the distance between the substation and the fault is x, leading to a feeder impedance of $Z_F = zx$. The voltage at the substation during the fault (as a fraction of the pre-fault voltage) is found from

$$V_{sag} = \frac{Z_F}{Z_S + Z_F} = \frac{xz}{\frac{V_{nom}}{\sqrt{3}I_{fault}} + xz} \tag{6.16}$$

For a given sag magnitude V_{sag}, we can calculate the distance to the fault:

$$x = \frac{V_{nom}}{\sqrt{3}zI_{fault}} \times \frac{V_{sag}}{1 - V_{sag}} \tag{6.17}$$

Note that some approximations are made here, which will be discussed in Section 6.5.

Consider as an example a 34.5 kV system with 10 kA available fault current and a feeder impedance of 0.3 Ω/km. This gives the following distances to the fault:

- $V_{sag} = 10\%$: $x = 750\,\text{m}$
- $V_{sag} = 20\%$: $x = 1650\,\text{m}$
- $V_{sag} = 50\%$: $x = 6.5\,\text{km}$
- $V_{sag} = 70\%$: $x = 15\,\text{km}$
- $V_{sag} = 80\%$: $x = 27\,\text{km}$
- $V_{sag} = 90\%$: $x = 60\,\text{km}$

If we want to distinguish between a sag down to 10% and one down to 20%, we need fault positions at least every kilometer. But if the borders of the bins in the sag density table are at 50%, 70%, 80%, and 90%, fault positions every 5 km are sufficient. Note also that the required distance between fault positions increases very fast when moving away from the load position. Thus, the required density of fault positions decreases fast for increasing distance to the fault.

Equation (6.17) gives an indication of the distance between fault positions for lines originating in the substation from which the load is fed. For other lines, one or two fault positions per line is normally enough, if the substations are not too close. A possible strategy is to first calculate the resulting sag magnitude for faults in the substation and to insert fault positions in between when the resulting sag magnitude for two neighboring substations differs too much.

Choosing two fault positions per line instead of one could actually speed up the calculations if the fault positions are chosen at the beginning and end of the line. This way, all lines originating from the same substation need only one voltage calculation.

The situation becomes more complicated when networks are meshed across voltage levels, like the transmission voltage levels in the United States and in several other countries. Consider a system like in Fig. 6.39. A safe strategy is to use multiple fault positions on the indicated lines and only one or two fault positions on the other lines, including 138 kV, 230 kV, and 345 kV. Due to the multiple paths for the fault current and the relatively large transformer impedances, faults at 138 kV and higher will not cause very deep sags; and the precise fault position will not have much influence on the sag magnitude. For 230 kV and 345 kV, one fault position per substation is probably still too much. The main problem is that no definite rules can be given for the required number of fault positions. In case computation time is no concern, and the selection of fault positions is automatic, one might simply choose 10 or even more fault positions for each line.

In the above, only the sag magnitude has been used to determine the number of fault positions. Apart from the sag magnitude, the sag duration will also have to be considered. The sag duration depends on the protection used for the various feeders and substation components. It is especially important to consider parts of the system where faults lead to longer fault-clearing time and thus to a longer sag duration. Possible examples are busbars protected by the backup protection of the infeeding lines; faults toward the remote end of a transmission line cleared by the distance protection in its zone 2.

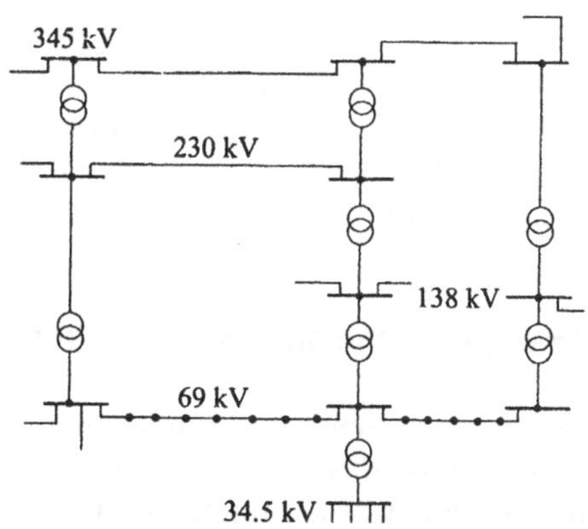

Figure 6.39 Network meshed across voltage levels, with suggested fault positions.

6.4.3.2 Extent of the Fault Positions. In the preceding section, the requirements for the distance between fault positions were discussed. The resulting recommendation was to use one or two fault positions per line for all but those lines which are directly feeding the load. The next question that comes up is: How far do we have to go with this? Is it, e.g., needed to consider a 345 kV substation at 1000 km away? Probably not, but how about one at 200 km? There are two possible ways forward, both of which are not really satisfying:

1. Use (6.17) to estimate at which distance a fault would lead to a sag down to 90%, or any other value for the "most shallow sag of interest." For transmission voltages this will give very large values (600 km for a 345 kV system with 10 kA available fault current), which are probably much higher than actually needed.
2. Start with fault positions in a restricted area, and look at the sag magnitudes for faults at the border of this area. If these magnitudes are below 90%, the area needs to be extended. If the system is available in the right format for a suitable power system analysis package, this might still be the fastest method.

6.4.3.3 Failure of the Protection. Failure of the protection is of concern for voltage sag calculations because it leads to a longer fault-clearing time, and thus a longer sag duration. This longer sag duration, often significantly longer, could be important for the compatibility assessment. The equipment might tolerate the sag when the primary protection clears the fault, but not when the backup protection has to take over.

To include failure of the protection, two events have to be considered for each fault position: one representing clearing by the primary protection, the other fault clearing by the backup. The two events will typically be given different fault frequencies. Alternatively one can use a fixed failure rate of the protection and a fixed fault-clearing time for both the primary and the backup protection. In that case the resulting magnitude distribution only needs to be shifted toward the relevant duration.

6.4.3.4 Multiple Events. The method of fault positions in its basic form only considers short-circuit faults in an otherwise normal system. Multiple events like a fault during the failure of a nearby power station are normally not considered. To include these, fault calculations need to be performed for the system with the power station out of operation. The choice of fault positions becomes even more complicated now. Only those faults need to be considered for which the outage of the power station influences the sag. When an automatic method is used, it is probably simplest to consider all situations. The best strategy appears again to start with generator stations near the load, and move further away from the load until there is no longer any significant influence on the sag magnitude. Significant influence should be defined as likely to affect behavior of equipment.

6.4.4 An Example of the Method of Fault Positions

In this section we discuss an example of the use of the method of fault positions. A small system is used for this: the reason being that the data was readily available and that the data processing was limited so that various options could be studied in a relatively short time. A study in a U.S. transmission system is described in [8], and a study in a large European transmission system in [71], [74].

6.4.4.1 The Reliability Test System. The reliability test system (RTS) was proposed by the IEEE subcommittee on the application of probability methods to compare stochastic assessment techniques for generation and transmission systems [73]. The RTS has been used by Qader [64], [71] to demonstrate the method of fault positions. The reliability test system consists of 24 busses connected by 38 lines and cables, as shown in Fig. 6.40. Ten generators and one synchronous condenser are connected at 138 kV and at 230 kV.

6.4.4.2 Voltages Due to One Fault. Figure 6.41 shows the effect of a fault halfway between busses 2 and 4 on the voltages throughout the system. Only bus 4

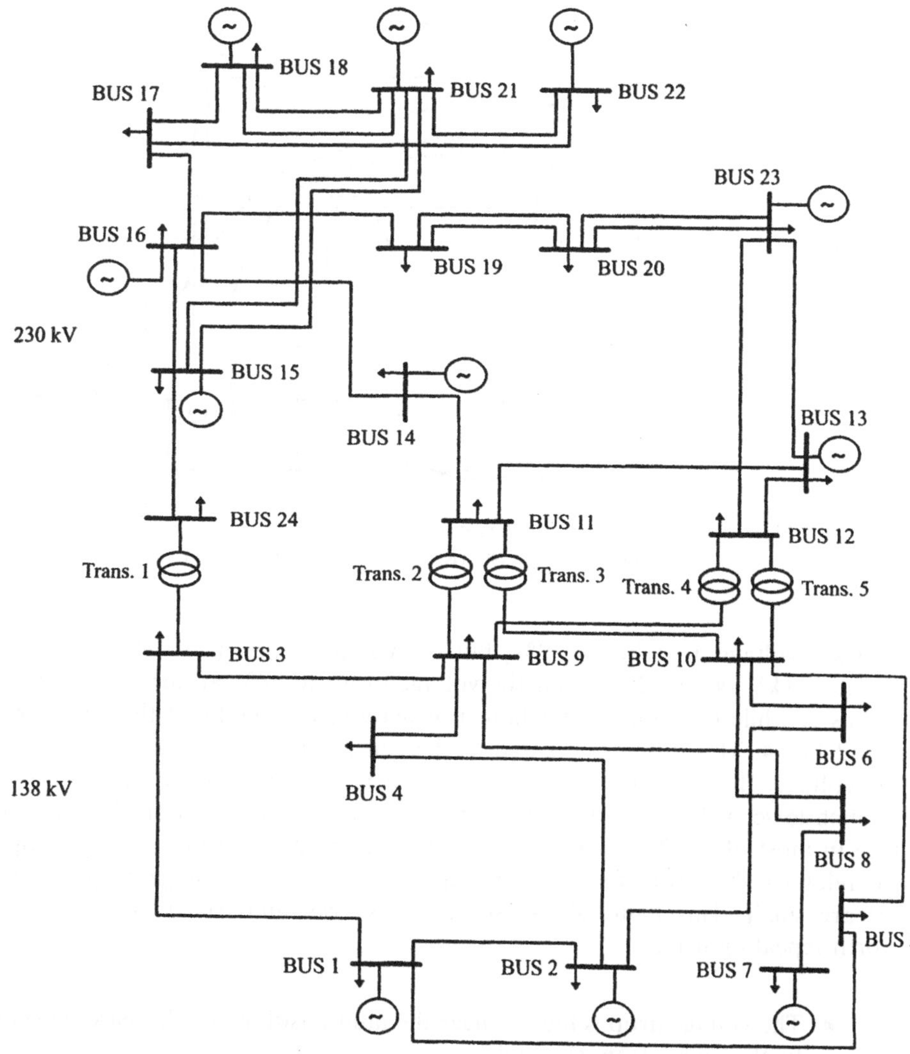

Figure 6.40 Reliability test system. (Reproduced from Qader [71].)

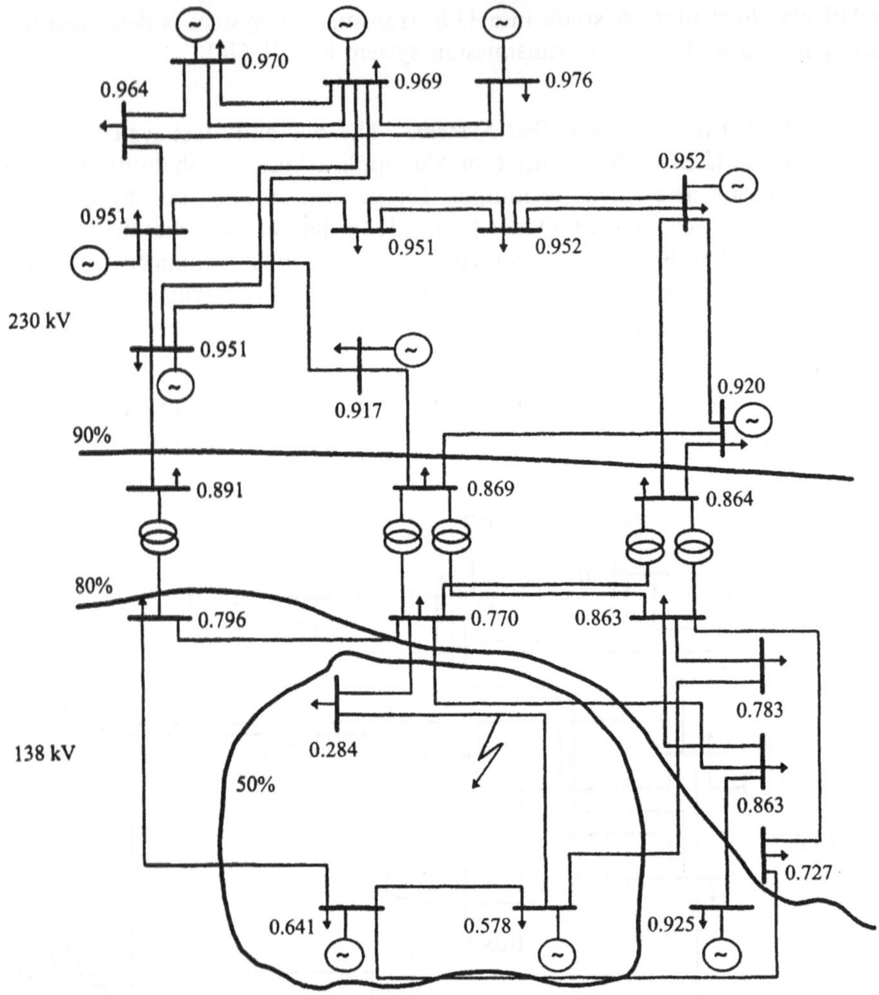

Figure 6.41 Voltage sags at different busses due to a fault halfway between bus 2 and bus 4 in Fig. 6.40. (Reproduced from Qader [71].)

shows a voltage drop below 50%, but the voltage drops below 90% in a large part of the 138 kV system. Note that the voltage drops to 28% at bus 4, but only to 58% at bus 2, while the short-circuit fault is exactly in the middle of the line between bus 2 and bus 4. This difference is due to the generators at bus 1 and bus 2 keeping up the voltage. Bus 4 is far away from any generator station, thus the voltage drops to a much lower value. The dense concentration of generator stations keeps up the voltage in most of the 230 kV system, thus preventing more serious voltage drops. Also, the relatively high transformer impedance makes that the voltage drops at 230 kV level are small. This figure shows some well-known and trivial facts which are still worth repeating here:

- The voltage drop is highest near the fault position and decreases when moving further away from the fault.
- The voltage drop diminishes quickly when moving toward a generator station.

- The voltage drop diminishes when moving across a transformer toward a higher voltage level. This assumes that more generation is connected to higher voltage levels. The high-voltage side of the transformer is closer to the source, so that the voltage drops less in magnitude.

6.4.4.3 Exposed Area. In Fig. 6.41 the fault position was fixed and voltage sags were calculated for all busses. Figure 6.42 gives the reversed situation: the voltage magnitude is calculated for one bus but for many fault positions. In this case, the sag magnitude at bus 4 is calculated. Positions leading to equal sag magnitudes at bus 4 are connected through "contour lines" in Fig. 6.42. Contour lines have been plotted for sag magnitudes of 30%, 50%, 60%, 70%, and 80%. The area in which faults lead to a sag below a certain voltage is called the "exposed area." The term exposed area was originally linked to equipment behavior. Suppose that the equipment

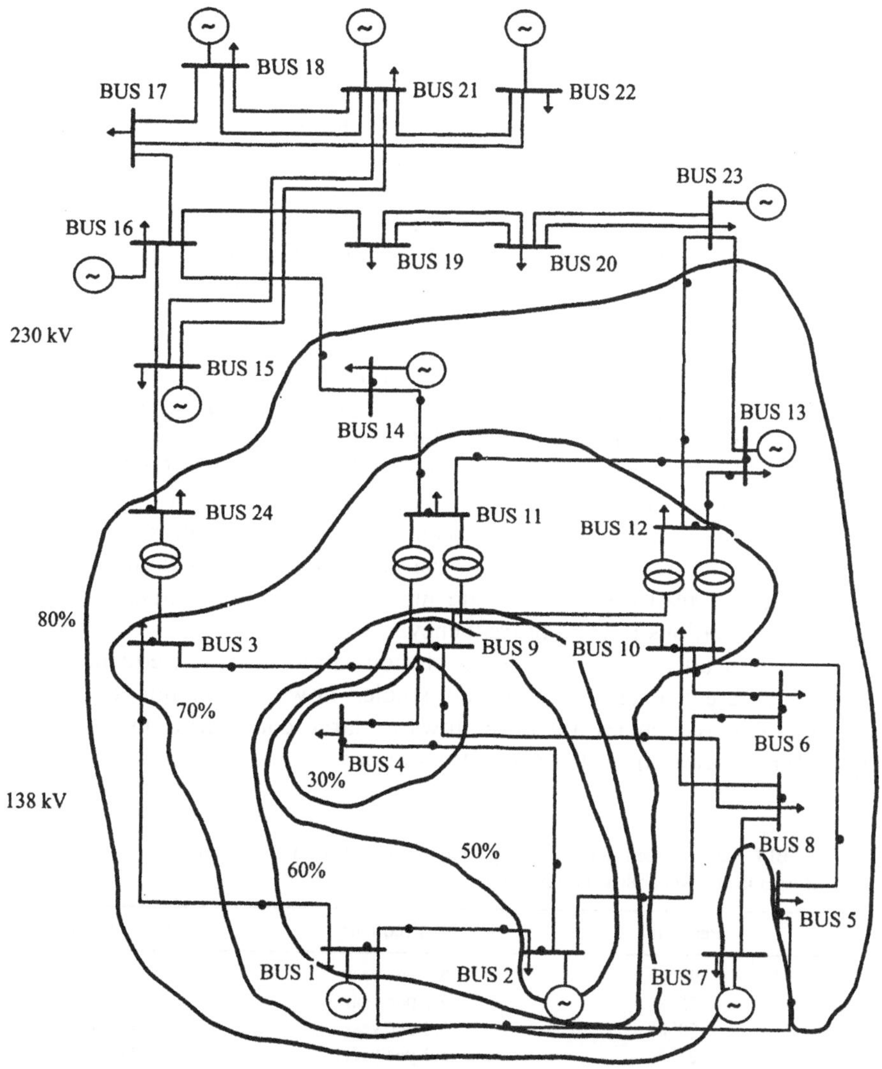

Figure 6.42 Exposed area contours for bus 4. (Reproduced from Qader [71].)

trips when the voltage drops below 60%. In that case the equipment is "exposed" to all faults within the 60% contour in the figure; hence the term exposed area. As faults can only occur on primary components (lines, cables, transformers, busses, etc.), the exposed area is strictly speaking not an area, but a collection of points (the substations) and curves (the lines and cables). But drawing a closed contour helps to visualize the concept. Knowing which primary components are within the exposed area can be more valuable information than the actual number of sags. Suppose there is an overhead line across a mountain prone to adverse weather, within the exposed area. Then it might be worth to consider additional protection measures for this line, or to change the system structure so that this line no longer falls within the exposed area, or to improve equipment immunity so that the exposed area no longer contains this line.

From Fig. 6.42 and other exposed area contours, the following conclusions are drawn:

- The exposed area extends further toward large concentrations of generation, than toward parts of the system without generation.
- The shape of the exposed area contour near transformer stations depends on the amount of generation present on the other side of the transformer. The exposed area typically extends far into higher-voltage networks but rarely into lower-voltage networks. If the fault takes place in a lower-voltage network the voltage drop over the transformer impedance will be large. This assumes that the main generation is at a higher voltage level than the fault. Considering the simple network structures in Chapter 4 explains this behavior.

6.4.4.4 Sag Frequency. These calculations can be performed for all busses, resulting in a set of exposed area contours for each bus. Plotting them in one figure would not result in something easily interpretable. Instead Fig. 6.43 gives the expected number of sags to a voltage below 80% for each bus. The average number of sags per bus is 6.85 per year; the various percentiles are given in Table 6.23. We see that 80% of the busses has a sag frequency within 30% of the average sag frequency for all busses. Note that we assumed the same fault rate (in faults per km per year) for all lines. In reality some lines are more prone to faults than others, which can give larger variations in the sag frequency.

It is difficult to draw general conclusions about the sag frequency, because each system is different. From this and other studies, however, one might, draw the conclusion that sag frequencies are lower towards large concentrations of generation and higher further away from the generator stations.

TABLE 6.23 Percentiles of the Sag Frequency Distribution Over the Busses in the Reliability Test System

Percentile	Sag Frequency	Percent of Average
90%	4.7 per year	70%
75%	5.2 per year	75%
50%	6.8 per year	100%
25%	8.2 per year	120%
10%	9.0 per year	130%

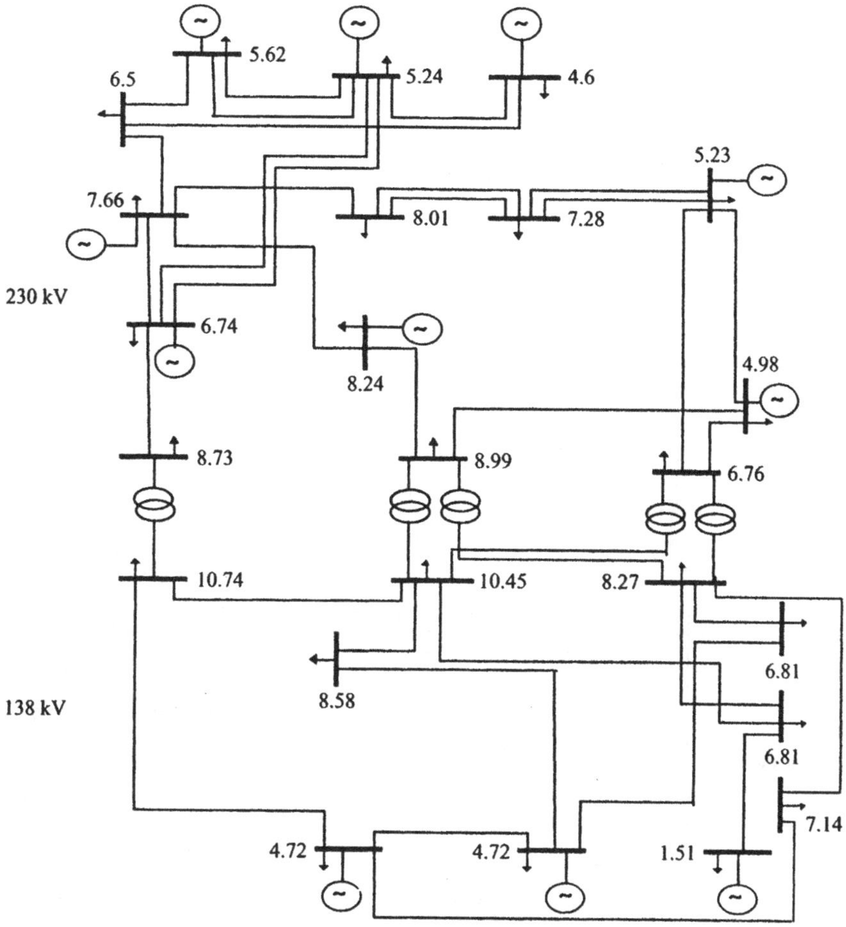

Figure 6.43 Voltage sag frequency for all busses in the RTS: number of sags below 80%. (Reproduced from Qader [71].)

6.4.4.5 Generator Scheduling. In the preceding study it was assumed that all generators were in operation. In reality this is an unlikely situation. We saw that generator stations have a significant influence on the voltages in the system during a fault, and on the sag frequency. To quantify this influence, the calculations in the reliability test system have been repeated for the situation in which all 138 kV substations are out of operation. The resulting sag frequency is shown in Fig. 6.44. Comparing this figure with Fig. 6.43 shows that the sag frequency is increased at all busses but most significantly at the 138 kV busses. The sag frequency is very similar for all 138 kV busses. The reason is that faults in the 138 kV system, and nearby in the 230 kV system, make that the voltage drops below 80% for all 138 kV busses. If the sag frequency is defined as the number of sags below 65% the differences between the 138 kV busses become larger, see Table 6.24.

As a next step it has been assumed that the three 138 kV generators are each out of operation during four months of the year, and that there is no overlap in these periods; thus there are always two 138 kV generators in operation. For each of these periods (i.e., for each combination of one generator out and two in operation) the sag frequency has been calculated in exactly the same way as before. The results for the

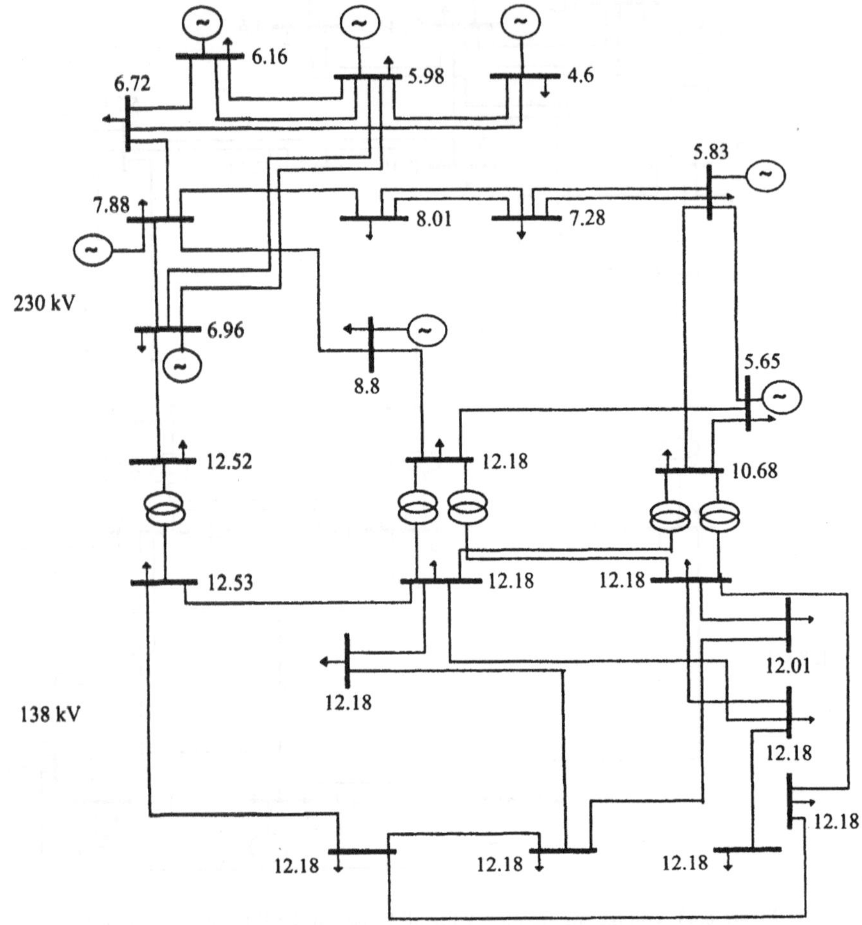

Figure 6.44 Voltage sag frequency (number of sags per year) for all busses in the reliability test system when the 138 kV generators are out of operation. (Reproduced from Qader [71].)

TABLE 6.24 Influence of Generator Scheduling on the Sag Frequency in the Reliability Test System, Number of Sags per Year below 65%

Generator Scheduling	138 kV Bus									
	1	2	3	4	5	6	7	8	9	10
Generator 1 out	2.80	2.77	3.24	3.65	3.42	3.16	0.80	1.47	2.65	3.38
Generator 2 out	2.43	2.79	3.06	3.77	3.44	3.18	0.80	1.49	2.64	3.40
Generator 7 out	1.54	1.40	3.06	2.81	3.20	3.18	4.42	4.42	3.11	3.44
Average	2.26	2.32	3.12	3.41	3.35	3.17	2.01	2.46	2.80	3.41
All generators in	1.34	1.40	2.85	2.19	2.16	2.60	0.80	1.34	2.59	2.81
All generators out	7.37	7.37	6.73	7.43	7.06	5.19	6.66	6.66	5.88	5.96

138 kV busses are shown in Table 6.24. The table shows the number of sags below 65% for all 138 kV substations, for a number of generator scheduling options. The sag frequency for the three 4-month periods mentioned, is given in the rows labeled "generator 1 out," "generator 2 out," and "generator 7 out." The number of sags per year has been calculated as the average of these three sag frequencies, and included in the row labeled "average." For reference the sag frequency is also given for the situation when all generators are in operation ("all generators in") and when all three 138 kV generators are out of operation ("all generators out").

6.5 THE METHOD OF CRITICAL DISTANCES

The method of critical distances does not calculate the voltage at a given fault position, but the fault position for a given voltage. By using some simple expressions, it is possible to find out where in the network a fault would lead to a voltage sag down to a given magnitude value. Each fault closer to the load will cause a deeper sag. The number of sags more severe than this magnitude is the number of short-circuit faults closer to the load than the indicated positions.

We first describe the basic theory and give the outline of the method. A simple example demonstrates how to apply the method. In the derivation of the basic expression, a number of approximations have been made. More exact expressions and expressions for non-radial systems are derived next. Finally the results of the method are compared with the results of the method of fault positions.

6.5.1 Basic Theory

The method of critical distances is based on the voltage divider model for the voltage sag, as introduced in Fig. 4.14. Neglecting load currents and assuming the pre-event voltage to be one, we obtained for the voltage at the point-of-common coupling (pcc) during the fault:

$$V_{sag} = \frac{Z_F}{Z_F + Z_S} \tag{6.18}$$

where Z_F is the impedance between the pcc and the fault, and Z_S the source impedance at the pcc. Let $Z_F = z\mathcal{L}$ with z the feeder impedance per unit length and $\mathcal{L}$ the distance between the pcc and the fault. This results in the following expression for the sag magnitude:

$$V_{sag} = \frac{z\mathcal{L}}{z\mathcal{L} + Z_S} \tag{6.19}$$

The "critical distance" is introduced as follows: the magnitude at the pcc drops below a critical voltage V whenever a fault occurs within the critical distance from the pcc. An expression for the critical distance $\mathcal{L}_{crit}$ is easily be obtained from (6.19):

$$\mathcal{L}_{crit} = \frac{Z_S}{z} \times \frac{V}{1 - V} \tag{6.20}$$

Here it is assumed that both source and feeder impedance are purely reactive (a rather common assumption in power system analysis), or more general: that the angle in the complex plane between these two impedances is zero.

Strictly speaking (6.20) only holds for a single-phase system. For three-phase faults in a three-phase system, the expressions are valid if for Z_S and z the positive-sequence impedances are used. For single-phase faults the sum of positive-, negative-, and zero-sequence impedances should be used; for phase-to-phase faults the sum of positive and negative sequence. The voltage in the expressions above is the phase-to-neutral voltage in the faulted phase in case of a single-phase fault and the voltage between the faulted phases in case of a phase-to-phase fault. We will come back to single-phase faults and phase-to-phase faults below.

Equation (6.20) can be used to estimate the exposed area at every voltage level in the supply to a sensitive load. The exposed area contains all fault positions that lead to a voltage sag causing a spurious equipment trip. The expected number of spurious trips is found by simply adding the failure rates of all equipment within the exposed area.

Transformer impedances are a large part of the source impedance at any point in the system. Therefore, faults on the secondary side do not cause a deep sag on the primary side. To estimate the number of sags below a certain magnitude it is sufficient to add all lengths of lines and cables within the critical distance from the pcc. The total length of lines and cables within the exposed area is called the "exposed length." The resulting exposed length has to be multiplied by the failure rate per unit length to obtain the number of sags per year.

6.5.2 Example—Three-Phase Faults

Consider the 11 kV network in Fig. 6.45. The fault level at the main 11 kV bus is 151 MVA (source impedance 0.663 pu on a 100 MVA base), the feeder impedance is 0.336 Ω/km (0.278 pu/km on the 100 MVA base).

The critical distance for different critical voltages, calculated from (6.20), is given in Table 6.25. The next-to-last column (labeled "exposed length") gives the total feeder length within the exposed area. Figure 6.45 gives the contours of the exposed area for various critical voltages. Each fault between the main 11 kV bus (the pcc) and the 50% contour will lead to a voltage sag at the pcc with a magnitude below 50%. All points on the 50% contour are at a distance of 2.4 km (see Table 6.25) of the main 11 kV bus. The last column in Table 6.25 gives the expected number of equipment trips per year. A value of 0.645 faults per km per year has been used.

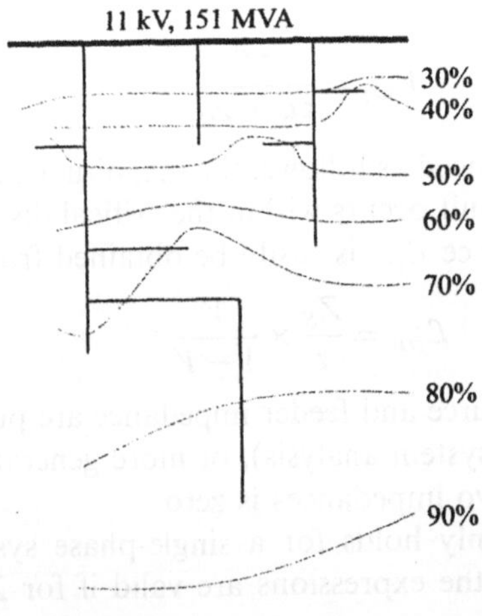

Figure 6.45 An 11 kV network used as an example for the method of critical distances.

TABLE 6.25 Results of Method of Critical Distances, Three-Phase Faults

Critical Voltage	Critical Distance	Exposed Length	Number of Trips per Year
90%	21.4 km	24.0 km	15.5
80%	9.6 km	21.6 km	13.9
70%	5.6 km	16.8 km	10.8
60%	3.6 km	12.2 km	7.9
50%	2.4 km	8.6 km	5.5
40%	1.6 km	5.4 km	3.5
30%	1.0 km	3.0 km	1.9
20%	0.6 km	1.8 km	1.1
10%	0.3 km	0.9 km	0.6

6.5.3 Basic Theory: More Accurate Expressions

To obtain a more accurate expression, we have to consider that both the feeder and the source impedance are complex. The basic expression is again obtained from the voltage divider shown in Fig. 4.14, but with complex voltage and impedances:

$$\overline{V} = \frac{\overline{Z}_F}{\overline{Z}_S + \overline{Z}_F} \tag{6.21}$$

where $\overline{Z}_S = R_S + jX_S$ is the source impedance at the pcc, $\overline{Z}_F = (r + jx)\mathcal{L}$ is the impedance between the fault and the pcc, $\mathcal{L}$ is the distance between the fault and the pcc, $\overline{z} = r + jx$ is the feeder impedance per unit length. The load currents have been neglected; the pre-fault voltage at the pcc equals the source voltage equals 100%.

In Section 4.5 expressions have been derived for the magnitude V and the phase-angle jump as a function of the distance between the pcc and the fault. Equation (4.87) for the magnitude of the voltage reads as follows:

$$V = \frac{\lambda}{1+\lambda} \frac{1}{\sqrt{1 - \frac{2\lambda(1-\cos\alpha)}{(1+\lambda)^2}}} \tag{6.22}$$

with

$$\lambda = \frac{Z_F}{Z_S} = \frac{z \times \mathcal{L}}{Z_S} \tag{6.23}$$

α the angle in the complex plane between source and feeder impedance, the so-called impedance angle:

$$\alpha = \arctan\left(\frac{X_S}{R_S}\right) - \arctan\left(\frac{x}{r}\right) \tag{6.24}$$

and $Z_S = |R_S + jX_S|$, $z = |r + jx|$, $V = |\overline{V}|$, etc.

To obtain an expression for the critical distance, λ needs to be solved from (6.22) for known V. Therefore, this equation is rewritten into the second-order polynomial equation

$$\lambda^2(V^2 - 1) + 2\lambda V^2 \cos\alpha + V^2 = 0 \tag{6.25}$$

The positive solution of this equation can be written as

$$\lambda = \frac{V}{1-V}\left[\frac{V\cos\alpha + \sqrt{1 - V^2\sin^2\alpha}}{V+1}\right] \tag{6.26}$$

Together with (6.23) the desired expression for the critical distance is obtained:

$$\mathcal{L}_{crit} = \frac{Z_S}{z} \times \frac{V}{1-V}\left[\frac{V\cos\alpha + \sqrt{1 - V^2\sin^2\alpha}}{V+1}\right] \tag{6.27}$$

The first part of (6.27)

$$\mathcal{L}_{crit} = \frac{Z_S}{z} \times \frac{V}{1-V} \tag{6.28}$$

is the expression for the critical distance obtained (6.20). For most applications (6.20) is sufficient, especially as the data are not always available to calculate the impedance angle. To assess the error made by using the approximated expression the critical distance has been calculated for different values of α.

Figure 6.46 gives the critical length as a function of the critical voltage for 11 kV overhead lines. A source impedance of 0.663 pu and a feeder impedance of 0.278 pu/km have been used. Note that these are the same values as used in the previous example (Fig. 6.45). We see that the error only becomes significant for large impedance angles (more than 30°). In that case more accurate expressions should be used. In the next section a simple but accurate approximation for the critical distance is derived.

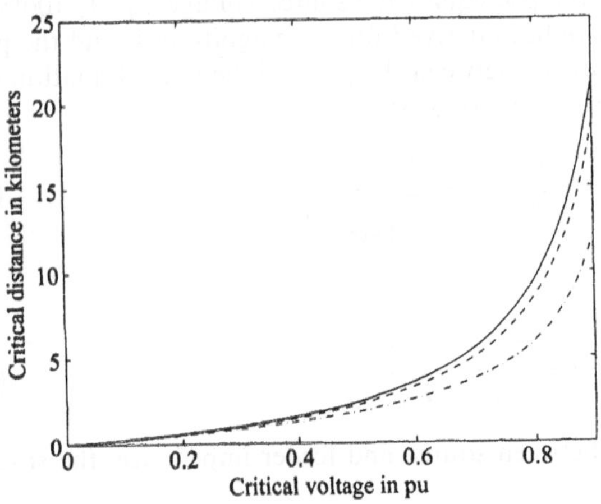

Figure 6.46 Critical distance as a function of the critical voltage for impedance angle 0° (solid line), −30° (dashed line), −60° (dash-dot line).

6.5.4 An Intermediate Expression

In the previous sections an exact and an approximate expression for the critical distance have been derived: (6.27) and (6.20), respectively. The difference between these two expressions is the factor between square brackets in the right-hand side of (6.27):

$$k = \frac{V\cos\alpha + \sqrt{1 - V^2\sin^2\alpha}}{1+V} \tag{6.29}$$

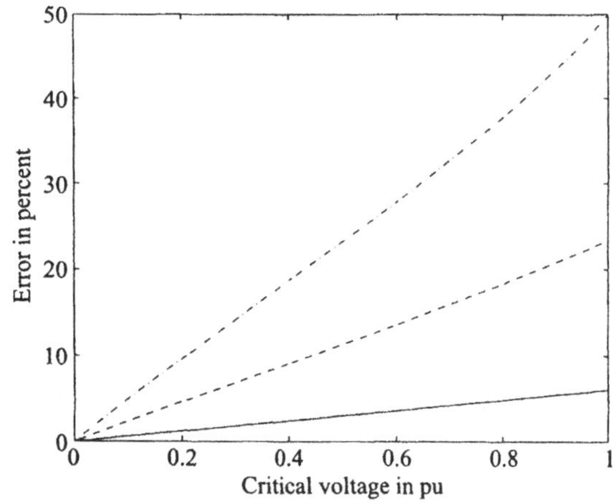

Figure 6.47 Error made in the simplified expression of critical distance; impedance angle: −20° (solid line), −40° (dashed line), and −60° (dash-dot line).

The more this factor deviates from one, the larger the error made by using the simplified expression (6.20). This error has been calculated as $(1-k)*100\%$ and plotted in Fig. 6.47 for three values of the impedance angle. The simplified expression (6.20) overestimates the critical distance (and thus the number of sags) as is also shown in Fig. 6.46. The error is, however, small in most cases, with the exception of systems with large impedance angles like underground cables in distribution systems. A first-order correction to the simplified expression (6.20) can be obtained by approximating (6.29) around $V=0$:

$$k \approx \left\{V\cos\alpha + \left(1 - \frac{1}{2}V^2\sin^2\alpha\right)\right\}(1-V) \tag{6.30}$$

$$k \approx 1 - V(1-\cos\alpha) \tag{6.31}$$

The error made by using approximation (6.31) is shown in Fig. 6.48 for different impedance angles. The error made never exceeds a few percent.

An important conclusion from Fig. 6.48 is that the following expression gives the critical distance in systems with a large impedance angle:

$$\mathcal{L}_{crit} = \frac{Z_S}{z} \times \frac{V}{1-V}\{1 - V(1-\cos\alpha)\} \tag{6.32}$$

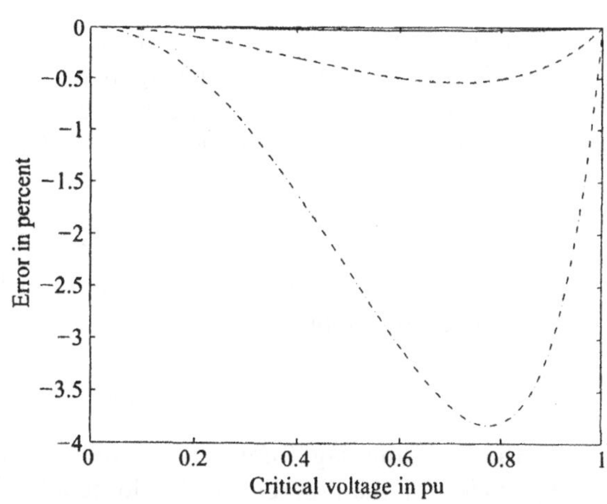

Figure 6.48 Error made by using a first-order approximation for the critical distance; impedance angle: −20° (solid line), −40° (dashed line), and −60° (dash-dot line).

6.5.5 Three-Phase Unbalance

The above reasoning applies to three-phase faults only. For unbalanced faults (single-phase, phase-to-phase) the method needs adjustment. Most of the discussion below follows directly from the treatment of three-phase unbalanced sags in Section 4.4.

6.5.5.1 Phase-to-Phase Faults. Phase-to-phase faults lead to sags of type C or type D, with a characteristic magnitude equal to the initial (phase-to-phase) voltage at the point-of-common coupling. The method of critical distances applies to the voltage at the pcc and can thus be used without modification for phase-to-phase faults. The impedance values to be used are the average of positive- and negative-sequence values. As these are normally about equal, the positive-sequence impedance can be used just like for three-phase faults. In terms of characteristic magnitude: the critical distance for phase-to-phase faults equals the critical distance for three-phase faults.

In case the voltage at the equipment terminals is of interest (e.g., for single-phase equipment), the strategy is to translate this voltage back to characteristic magnitude and apply the equations for the critical distance to the characteristic magnitude. Of importance here is to determine whether a fault at a certain voltage level leads to a type C or type D sag.

Suppose that the fault leads to a type C sag. In that case $\frac{1}{3}$ of the single-phase equipment will not see any sag at all, where $\frac{2}{3}$ will see a sag between 50% and 100%. Let V_{eq} be the sag magnitude at the equipment terminals and V_{char} the characteristic magnitude of the three-phase unbalanced sag. These two magnitudes relate according to

$$V_{eq} = \frac{1}{2}\sqrt{1 + 3V_{char}^2} \tag{6.33}$$

This expression is obtained from Fig. 4.90 when neglecting the characteristic phase-angle jump ($\phi = 0$). Including phase-angle jumps is possible, but would result in rather complicated expressions.

The characteristic magnitude can be obtained from the magnitude at the equipment terminals by using

$$V_{char} = \sqrt{\frac{4}{3}V_{eq}^2 - \frac{1}{3}} \tag{6.34}$$

For $V_{eq} < \frac{1}{2}$, there are no sags. For $\frac{1}{2} < V_{eq} < 1$, (6.20) can be used to calculate the critical distance, with $V = V_{char}$. The resulting sag frequency should be multiplied by $\frac{2}{3}$ to account for the fact that one in three faults does not lead to a sag at the equipment terminals. For a type D sag of magnitude V_{char}, one phase has a magnitude of V_{char} also. The expression for the critical distance can be applied directly, but the resulting sag frequency needs to be multiplied by $\frac{1}{3}$. The two other phases drop to

$$V_{eq} = \frac{1}{2}\sqrt{V_{char}^2 + 3} \tag{6.35}$$

For $V_{eq} < \frac{1}{2}\sqrt{3}$ this gives no contribution. For $\frac{1}{2}\sqrt{3} < V_{eq} < 1$, the critical distance can be calculated by using

$$V_{char} = \sqrt{4V_{eq}^2 - 3} \tag{6.36}$$

and the resulting sag frequency should be multiplied by $\frac{2}{3}$. Note that the two sag frequencies for the type D sag should be added.

6.5.5.2 Example: Phase-to-Phase Faults. Consider the same system as in the example for three-phase faults. We are interested in the number of spurious trips for phase-to-phase (delta) connected single-phase load at 660 V. A Dy-connected 11kV/660V transformer is used. The sag type at the equipment terminals is determined as follows:

- The phase-to-phase fault leads to a three-phase unbalanced sag of type C for star-connected load at 11 kV.
- For delta-connected load at 11 kV the sag is of type D.
- For delta-connected load at 660 V it is of type C.

The calculation of the trip frequency as a function of the equipment voltage tolerance is summarized in Table 6.26. It proceeds as follows:

- For a given critical voltage at the equipment terminals V_{eq}, the critical characteristic magnitude V_{char} is calculated by using

$$V_{char} = \sqrt{\frac{4}{3}V_{eq}^2 - \frac{1}{3}} \tag{6.37}$$

  The result is shown in the second column of Table 6.26. For $V_{eq} < 0.5$ the value under the square root is negative, which means that even for a terminal fault (distance zero), the voltage at the equipment terminals is higher than the critical voltage. The contribution to the exposed length is thus zero, hence the zeros in the first few rows of the table.
- From the critical characteristic magnitude, the critical distance is calculated in the standard way, by using

$$V_{crit} = \frac{Z_S}{z} \times \frac{V_{char}}{1 - V_{char}} \tag{6.38}$$

  with $Z_S = 0.661$ pu and $z = 0.278$ pu/km. The resulting critical distance is given in the third column of Table 6.26.

TABLE 6.26 Method of Critical Distances—Phase-to-Phase Faults, Type C Sags

Sag Magnitude at Equipment Terminals	Characteristic Magnitude	Critical Distance (km)	Exposed Length (km)	Trip Frequency (per year)
0	0	0	0	0
0.1	0	0	0	0
0.2	0	0	0	0
0.3	0	0	0	0
0.4	0	0	0	0
0.5	0	0	0	0
0.6	0.38	1.5	5.0	2.1
0.7	0.57	3.2	11.4	4.9
0.8	0.72	6.1	18.2	7.8
0.9	0.86	14.7	24	10.3

- From the critical distance, the exposed length is calculated for the 11 kV distribution system in Fig. 6.45. The method used for this is the same as shown in Fig. 6.45 for three-phase faults.
- Knowing the exposed length it is possible to calculate the trip frequency. Here it is assumed that the number of phase-to-phase faults is equal to the number of three-phase faults: 0.645 per km per year. This is not a realistic assumption, but it enables an easier comparison of the influence of the different types of fault. Because the voltage is only down on two phases for a type C sag, this fault frequency has to be multiplied by $\frac{2}{3}$ to get the trip frequency. The latter is given in the last row of the table.

Consider, as a second example, that the low-voltage load is connected in star (thus phase-to-neutral single-phase load). The three-phase unbalanced sag will be of type D, with one deep sag and two shallow sags at the equipment terminals. A calculation of the trip frequency using the method of critical distances is summarized in Table 6.27. Only critical voltages between 80% and 96% are shown in the table. The calculation for other voltage values proceeds in a similar way.

- Like for delta-connected load, the calculation starts with the choice of a critical voltage at the equipment terminals. Next, separate calculations are needed for the deep sag and for the shallow sag.
- The calculations for the deep sag (labeled "lowest voltage" in Table 6.27) are almost identical to the calculations for a three-phase fault. The magnitude of the deep sag at the equipment terminals is equal to the characteristic magnitude, so that the standard equation for the critical distance can be used. The only difference is that the fault frequency needs to be divided by three to accommodate for the fact that only one in three voltages shows a deep sag. Thus, from the viewpoint of single-phase equipment: only one in three faults leads to a deep sag. Critical distance, exposed length, and trip frequency for the deep sag are given in columns 2, 3, and 4 of Table 6.27. Note that the exposed length and the trip frequency no longer increase for critical voltages above 84%. This is because the exposed area already includes the whole length of the 11 kV feeders.

TABLE 6.27 Method of Critical Distances—Phase-to-Phase Faults, Type D Sags

	Lowest Voltage			Highest Voltage				
Magnitude Equipment Terminals (pu)	Critical Distance (km)	Exposed Length (km)	Trip Frequency (per year)	Characteristic Magnitude (pu)	Critical Distance (km)	Exposed Length (km)	Trip Frequency (per year)	Total Trip Frequency (per year)
0.80	9.5	21.5	4.6	0	0	0	0	4.6
0.82	10.9	22.9	4.9	0	0	0	0	4.9
0.84	12.5	24	5.2	0	0	0	0	5.2
0.86	14.7	24	5.2	0	0	0	0	5.2
0.88	17.5	24	5.2	0.31	1.1	3.4	1.5	6.7
0.90	21.5	24	5.2	0.49	2.3	8.2	3.5	8.7
0.92	27.4	24	5.2	0.62	3.9	12.8	5.5	10.7
0.94	37.4	24	5.2	0.73	6.4	18.4	7.9	13.1
0.96	57.2	24	5.2	0.83	11.6	23.6	10.1	15.3

- The calculations for the shallow sag proceed fairly similar to the calculations for the delta-connected load. As a first step the critical voltage at the equipment terminals is translated into a critical characteristic magnitude, using the following expression:

$$V_{char} = \sqrt{4V_{eq}^2 - 3} \tag{6.39}$$

resulting in the values in column 5. For $V_{eq} < 0.866$ the characteristic magnitude is set to zero. The shallow sag at the equipment terminals never becomes lower than this value. Calculation of critical distance, exposed length, and trip frequency proceeds like before. For the trip frequency, the fault frequency needs to be multiplied by $\frac{2}{3}$ because only two of the three phases show a shallow sag. The results for the shallow sag are summarized in columns 5 through 8.

- Finally the total trip frequency is the sum of the trip frequency due to deep sags and the trip frequency due to shallow sags. The total trip frequency is given in the last column.

6.5.5.3 Single-Phase Faults—Solidly Grounded Systems. Single-phase faults lead to sags of type B, C, or D at the equipment terminals. The translation from equipment terminal voltages to the voltage to be used in the expressions for the critical distance depends on the type of sag.

A type B sag only occurs in case of equipment connected in star and the single-phase fault at the same voltage level as the equipment (or at a higher level with only YnYn transformers between the fault and the equipment). For a type B sag the terminal voltage can be directly used in the expressions for the critical distance. As only one phase drops in voltage, the resulting sag frequency should be multiplied by $\frac{1}{3}$ for single-phase equipment. For the impedances the sum of positive-, negative-, and zero-sequence values should be used.

Sags of type C or type D occur in all other cases. For these the characteristic magnitude deviates from the initial voltage (the voltage in the faulted phase at the pcc). For solidly grounded distribution systems (where positive- and zero-sequence source impedances are equal), the following relation between characteristic magnitude V_{char} and initial magnitude V_{init} has been derived (4.109):

$$V_{char} = \frac{1}{3} + \frac{2}{3}V_{init} \tag{6.40}$$

Knowing the characteristic magnitude of the three-phase unbalanced sag, and $\frac{1}{3} < V_{char} < 1$, the initial voltage is obtained from

$$V_{init} = \frac{3}{2}V_{char} - \frac{1}{2} \tag{6.41}$$

The characteristic magnitude needs to be translated to an initial magnitude, by using (6.41). In case the magnitude at the equipment terminals is of importance, a second translation has to be made: from magnitude at the equipment terminals to characteristic magnitude. This translation proceeds in exactly the same way as for phase-to-phase faults.

6.5.5.4 Example: Single-Phase Faults in a Solidly Grounded System. When considering single-phase faults, we need to include the zero-sequence impedance of source and feeder. For a solidly grounded distribution system we can assume that positive- and zero-sequence source impedance are equal. But this cannot be assumed for the feeder impedances. From Table 4.4 we get 1.135 pu/km for the zero-sequence feeder impedance, and 0.278 pu/km for the positive-sequence impedance. In the calculations we use the sum of positive-, negative-, and zero-sequence impedance leading to $Z_S = 1.989$ pu for the source and $z = 1.691$ pu/km for the feeder.

The calculation of the critical distance for single-phase faults from a given critical characteristic magnitude is summarized in Table 6.28.

- The first step is the translation from the characteristic voltage to the initial voltage, for which expression (6.41) is used. The characteristic magnitude cannot be less than 0.33 pu, hence the zeros in the tables for lower values than this.
- From the critical initial voltage, the critical distance can be calculated by using the standard expression

$$\mathcal{L}_{crit} = \frac{Z_S}{z} \times \frac{V_{init}}{1 - V_{init}} \tag{6.42}$$

  with $Z_S = 1.989$ pu and $z = 1.691$ pu/km.
- From the critical distance, the exposed length and the trip frequency can be calculated like before. For single-phase faults again a fault frequency of 0.645 faults per km per year has been used.

TABLE 6.28 Method of Critical Distances—Single-Phase Faults, Solidly Grounded System

Characteristic Magnitude (pu)	Initial Magnitude (pu)	Critical Distance (km)	Exposed Length (km)	Trip Frequency (per year)
0	0	0	0	0
0.1	0	0	0	0
0.2	0	0	0	0
0.3	0	0	0	0
0.4	0.10	0.1	0.3	0.2
0.5	0.25	0.4	1.2	0.8
0.6	0.40	0.8	2.4	1.5
0.7	0.55	1.4	4.6	3.0
0.8	0.70	2.7	9.8	6.3
0.9	0.85	6.6	18.6	12.0

6.5.5.5 Single-Phase Faults—General Solutions. In resistance-grounded distribution systems, the assumption that positive- and zero-sequence impedance are equal no longer holds. The assumption is also not valid when line impedances are a large part of the source impedance. This is the case in the 400 kV supply in Fig. 4.21, as was shown in Fig. 4.105. To obtain a more general expression for the critical distance, we can use the phase-to-neutral voltage in the faulted phase according to (4.40):

$$V_{an} = 1 - \frac{3Z_{S1}}{(2Z_{F1} + Z_{F0}) + (2Z_{S1} + Z_{S0})} \tag{6.43}$$

The phase-to-neutral voltages in the non-faulted phases are not affected by single-phase faults. We can thus treat the phase-to-neutral voltages the same as the phase-to-ground voltages in a solidly grounded system. The characteristic magnitude is related to the (initial) phase-to-neutral voltage as follows:

$$V_{char} = \frac{1}{3} + \frac{2}{3} V_{an} \tag{6.44}$$

With this knowledge it is possible to translate sag magnitudes at the equipment terminals to characteristic magnitudes and to phase-to-neutral voltages. It is possible to translate phase-to-neutral voltages to phase-to-ground voltages, but one can alternatively derive an expression for the critical distance for phase-to-neutral voltages. For this we introduce positive- and zero-sequence feeder impedance per unit length, z_1 and z_0, respectively, and the distance to the fault $\mathcal{L}$. Expression (6.43) changes into

$$V_{an} = 1 - \frac{3Z_{S1}}{(2z_1 + z_0)\mathcal{L} + (2Z_{S1} + Z_{S0})} \tag{6.45}$$

The distance to the fault $\mathcal{L}_{crit}$ can be obtained for a given (critical) phase-to-neutral voltage V_{an}:

$$\mathcal{L}_{crit} = \frac{(Z_{S1} - Z_{S0}) + V_{an}(2Z_{S1} + Z_{S0})}{(2z_1 + z_0)(1 - V_{an})} \tag{6.46}$$

For $Z_{S1} = Z_{S0}$ we obtain the expression used for solidly grounded distribution systems. Note that normally $Z_{S1} < Z_{S0}$ so that the critical distance can become negative for small values of V_{an}. Even for a terminal fault the phase-to-neutral voltage is not zero. Any critical voltage less than this minimum value will give a negative critical distance. This has no physical meaning, and for calculating the exposed length (and sag frequency) a critical distance of zero should be used. Alternatively one can calculate the critical distance directly from the characteristic magnitude. For this we use $V_{char} = V_1 + V_2$ together with (4.29) and (4.30) which give the positive- and negative-sequence voltages at the pcc due to a single-phase fault. Using the same notation as before, we get the following expression for the characteristic magnitude as a function of the distance to the fault:

$$V_{char} = \frac{z\mathcal{L} + Z_{S0}}{z\mathcal{L} + Z_S} \tag{6.47}$$

with $Z_S = 2Z_{S1} + Z_{S0}$ and $z = 2z_1 + z_0$. Solving the critical distance gives

$$\mathcal{L}_{crit} = \frac{Z_s}{z} \times \frac{V_{char}}{1 - V_{char}} - \frac{Z_{S0}}{z(1 - V_{char})} \tag{6.48}$$

6.5.5.6 Example: Single-Phase Faults in Resistance-Grounded System. In a resistance-grounded system we can no longer assume that positive- and zero-sequence source impedance are equal. From Table 4.3 we get for the zero-sequence source impedance a value of $Z_{S0} = 8.172$ pu. The calculation results are summarized in Table 6.29. The results are only shown for critical voltages between 86% and 98%. For smaller values of the critical voltage, the trip frequency is zero. Single-phase faults in resistance-grounded systems typically lead to very shallow sags. The critical distance is calculated directly from the critical characteristic magnitude by using (6.48) with $Z_S = 9.494$ pu, $Z_{S0} = 8.172$ pu, and $z = 1.691$ pu/km. Calculation of exposed area and trip frequency proceeds like before.

TABLE 6.29 Method of Critical Distances—Single-Phase Faults, Resistance-Grounded System

Characteristic Magnitude (pu)	Critical Distance (km)	Exposed Length (km)	Trip Frequency (per year)
0.86	0	0	0
0.88	0.9	2.7	1.7
0.90	2.2	7.8	5.0
0.92	4.2	13.3	8.9
0.94	7.4	19.4	12.5
0.96	13.9	24	15.5
0.98	33.5	24	15.5

6.5.6 Generator Stations

In Section 4.2.4 expression (4.16) was derived describing the effect of a generator on the sag magnitude. The equivalent circuit used to obtain this is shown in Fig. 4.24. The expression has the following form:

$$(1 - V_{sag}) = \frac{Z_4}{Z_3 + Z_4}(1 - V_{pcc}) \qquad (6.49)$$

To obtain the voltage at the pcc we have to realize that all load currents have been neglected here. There are no pre-fault power flows, and both generators in Fig. 4.24 have exactly the same output voltage, so that they can be replaced by one source in the equivalent scheme. The following expression for the voltage is obtained from this scheme:

$$V_{pcc} = \frac{Z_2}{Z_3 + Z_1//(Z_3 + Z_4)} \qquad (6.50)$$

where $Z_A//Z_B = \frac{Z_A Z_B}{Z_A + Z_B}$ is the parallel connection of Z_A and Z_B. Combining (6.49) and (6.50) gives the following expression for the during-sag voltage experienced by the load

$$V_{sag} = 1 - \frac{Z_1 Z_4}{Z_2(Z_1 + Z_3 + Z_4) + Z_1(Z_3 + Z_4)} \qquad (6.51)$$

To obtain an expression for the critical distance, we substitute $Z_2 = z \times \mathcal{L}$. The critical distance is obtained by solving $V_{sag} = V_{crit}$ toward $\mathcal{L}$. The resulting expression is

$$\mathcal{L}_{crit} = \frac{Z_1}{z}\left\{\frac{Z_4}{Z_1 + Z_3 + Z_4} \times \frac{V_{crit}}{1 - V_{crit}} - \frac{Z_3}{Z_1 + Z_3 + Z_4}\right\} \qquad (6.52)$$

The critical distance in (6.52) is not the distance between the fault and the load, but the distance between the fault and the main supply point.

6.5.7 Phase-Angle Jumps

As we have seen in Chapter 5, some equipment is sensitive to the jump in phase angle between the pre-sag voltage and the during-sag voltage. In that case it is reasonable to find an expression for the critical distance as a function of the "critical phase-angle jump." In other words, at which distance does a fault lead to a sag with a phase-angle jump equal to a given value? To obtain such an expression we start with the expression for the phase-angle jump as a function of the distance to the fault: (4.84) in Section 4.5.

$$\cos\phi = \frac{\lambda + \cos\alpha}{\sqrt{1 + \lambda^2 + 2\lambda\cos\alpha}} \tag{6.53}$$

where α is the angle in the complex plane between the feeder and the source impedance and λ the ratio between their absolute values:

$$\lambda = \frac{z\mathcal{L}}{Z_S} \tag{6.54}$$

To obtain an expression for the critical distance, we need to solve λ from (6.53) for given phase-angle jump ϕ. Taking the square of both sides of (6.53) and using $\sin^2 = 1 - \cos^2$ gives the following second-order algebraic equation for λ:

$$\lambda^2 + 2\lambda\cos\alpha + 1 = \frac{\sin^2\alpha}{\sin^2\phi} \tag{6.55}$$

This can be solved by using the standard expression for the roots of a second-order polynomial, or by further rewriting the expression. In any way it will lead to the following (positive) root:

$$\lambda = \frac{\sin\alpha}{\tan\phi} - \cos\alpha \tag{6.56}$$

Combining (6.56) with (6.54) gives the following expression for the critical distance for a critical phase-angle jump ϕ:

$$\mathcal{L}_{crit} = \frac{Z_S}{z}\left\{\frac{\sin\alpha}{\tan\phi} - \cos\alpha\right\} \tag{6.57}$$

6.5.8 Parallel Feeders

Voltage sags on parallel feeders and other loops have been discussed in Section 4.2.4. There we saw that most faults on parallel feeders toward the load, lead to deep sags. It is an acceptable approximation to make the sag magnitude zero for all faults on the parallel feeders. In case of long feeders (feeder impedance more than two or three times the source impedance) some additional calculation is needed. It is possible to derive an expression for the critical distance for parallel feeders from (4.18) but that expression would be too complicated to be of any use. Instead a simplified calculation is proposed.

The voltage profile along the feeder can be approximated as a (second-order) parabola:

$$V_{sag} \approx 4V_{max}\,p(1-p) \tag{6.58}$$

with p indicating the position of the fault along the feeder, $0 \le p \le 1$, and V_{max} the maximum sag voltage due to a fault anywhere on the feeder. There is no simple expression for V_{max}; it needs to be obtained graphically from Fig. 4.34 or Fig. 4.35. When the maximum value is known, the "critical fraction" is readily obtained:

$$p_{crit} \approx 1 - \sqrt{1 - \frac{V_{crit}}{V_{max}}}, \qquad V_{crit} \le V_{max} \tag{6.59}$$

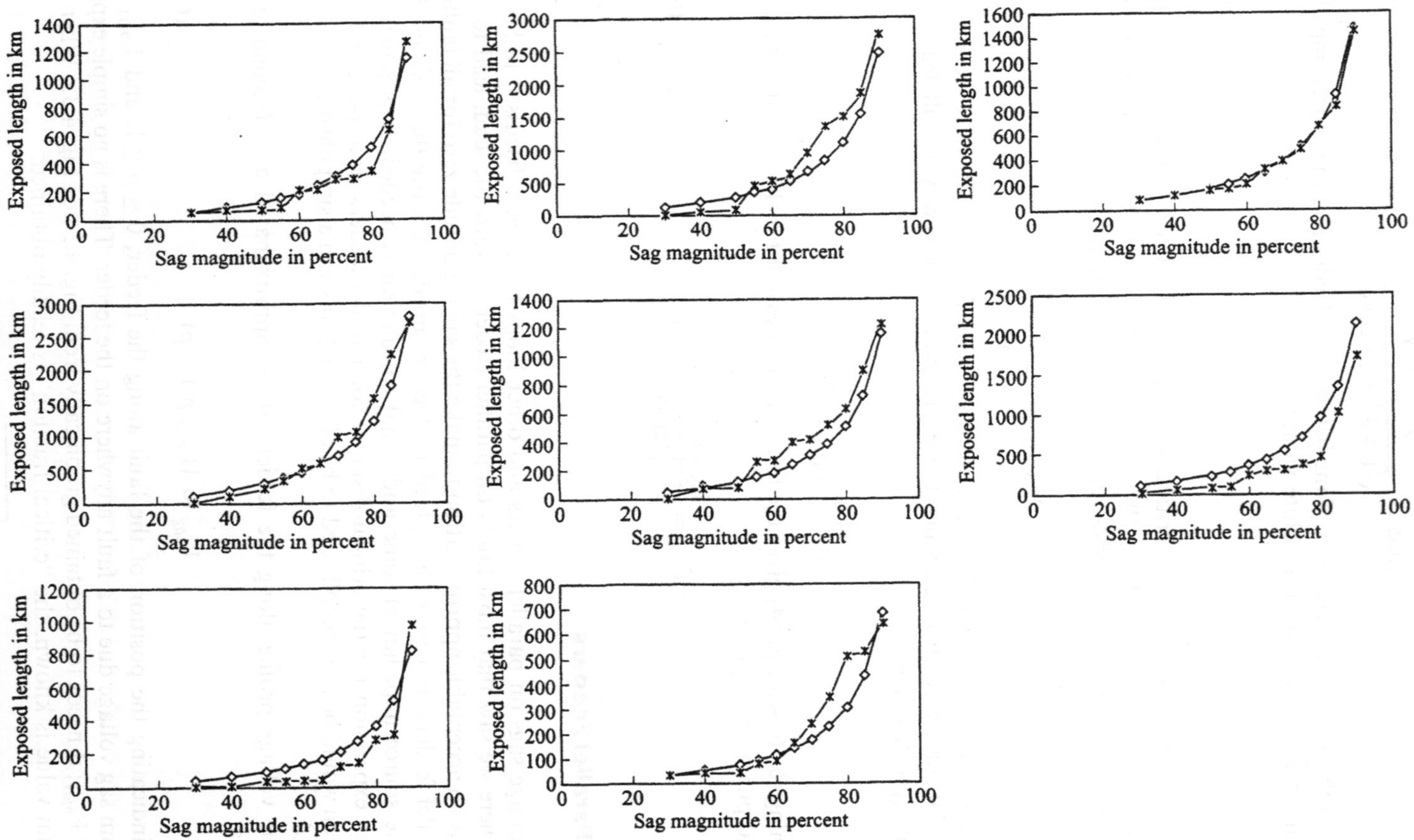

Figure 6.49 Exposed length for nine 400 kV substations: comparison between the method of fault positions (crosses) and the method of critical distances (diamonds).

The contribution of the feeder to the exposed length equals the critical fraction times the feeder length. For $V_{crit} > V_{max}$ the whole feeder contributes to the exposed length.

6.5.9 Comparison with the Method of Fault Positions

The transmission system study performed by Qader [71], [74] resulted in the number of sags as a function of magnitude for all substations in the U.K. 400-kV transmission system. The method of fault positions was used for this study. For a number of substations those results have been compared with the results obtained by using the method of critical distances. The critical distance was calculated as a function of the sag magnitude V by using the approximated expression

$$\mathcal{L}_{crit} = \frac{Z_S}{z} \frac{V}{1 - V} \tag{6.60}$$

where Z_S is the source impedance and z the feeder impedance per unit length. All the lines originating at the substation are assumed infinitely long; the exposed length is simply the critical distance times the number of lines.

The source impedance Z_S is calculated by assuming that all lines contribute equally to the short-circuit current for a busbar fault. During a fault on one of these lines, only $(N - 1)$ out of N lines contribute to the short-circuit current. Thus, the source impedance in p.u. equals

$$Z_S = \frac{N}{N - 1} \frac{S_{base}}{S_{fault}} \tag{6.61}$$

with N the number of lines originating at the substation, S_{base} the base power, and S_{fault} the short-circuit power for a substation fault. The exposed length is found from

$$\mathcal{L}_{exp} = N \times \mathcal{L}_{crit} = \frac{N^2}{N - 1} \frac{\frac{S_{base}}{S_{fault}}}{z} \frac{V}{1 - V} \tag{6.62}$$

The exposed length for the nine substations is shown in Fig. 6.49, where the crosses indicate the results of the method of fault positions. There are obviously differences between the results of the two methods, with the method of fault positions viewed as the most accurate one. But for the method of fault positions a large part of the national grid needs to be modeled. All the data needed for the method of critical distances is, from equation (6.62):

- number of lines originating from the substation;
- fault level of the substation; and
- feeder impedance per unit length.

All this data can be obtained without much difficulty.

Another interesting observation from (6.62) concerns the variation in sag frequency among different substations. The main variation can be brought back to fault level, number of lines originating at the substation, and fault frequency.

7

Mitigation of Interruptions and Voltage Sags

This chapter gives an overview of methods to mitigate voltage sags and interruptions. After a general discussion of the various forms of mitigation, we concentrate on power system design and on mitigation equipment to be installed between the power system and the sensitive equipment. Especially the latter is under fast development since a few years. An attempt is made to give a neutral overview of the various options, knowing that new developments are very hard to predict. Power system design is a more traditional area, although new developments in power electronics are also expected to have an impact here.

7.1 OVERVIEW OF MITIGATION METHODS

7.1.1 From Fault to Trip

In the previous chapters we discussed voltage magnitude events (voltage sags, short interruptions, and long interruptions) in considerable detail: their origin, methods of characterization, monitoring and prediction, and their effects on equipment. In this chapter we look at existing and future ways of mitigating voltage magnitude events. To understand the various ways of mitigation, the mechanism leading to an equipment trip needs to be understood. Figure 7.1 shows how a short circuit leads to an equipment trip. The equipment trip is what makes the event a problem; if there were no equipment trips, there would not be any voltage quality problem. The underlying event of the equipment trip is a short-circuit fault: a low-impedance connection between two or more phases, or between one or more phases and ground. At the fault position the voltage drops to a low value. The effect of the short circuit at other positions in the system is an event of a certain magnitude and duration at the interface between the equipment and the power system. The short-circuit fault will always cause a voltage sag for some customers. If the fault takes place in a radial part of the system, the protection intervention clearing the fault will also lead to an interruption. If there is sufficient redundancy present, the short circuit will only lead to a voltage sag. If the resulting event exceeds a certain severity, it will cause an equipment trip. Admittedly, not only

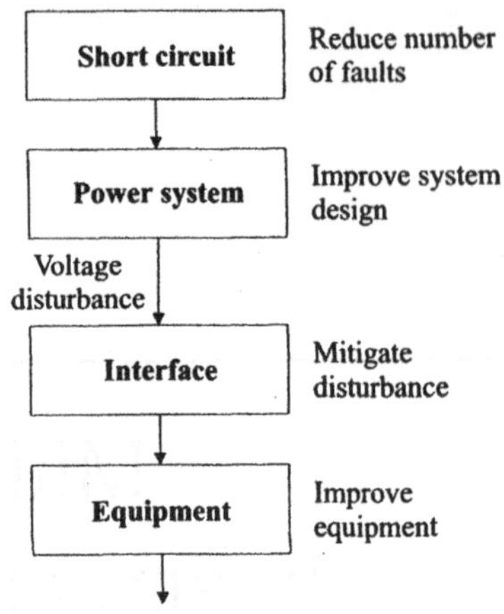

Figure 7.1 The voltage quality problem and ways of mitigation.

short circuits lead to equipment trips, but also events like capacitor switching or voltage sags due to motor starting. But the large majority of equipment trips will be due to short-circuit faults. Most of the reasoning to follow also applies to any other event potentially leading to an equipment trip.

Figure 7.1 enables us to distinguish between the various mitigation methods:

- reducing the number of short-circuit faults.
- reducing the fault-clearing time.
- changing the system such that short-circuit faults result in less severe events at the equipment terminals or at the customer interface.
- connecting mitigation equipment between the sensitive equipment and the supply.
- improving the immunity of the equipment.

These four types of mitigation are discussed briefly next. Power system design and mitigation equipment at the system–equipment interface are discussed in detail in the remainder of this chapter. Power engineers have always used a combination of these mitigation methods to ensure a reliable operation of equipment. Classically the emphasis has been on reducing the number of interruptions, while recently emphasis has shifted toward mitigating voltage sags.

7.1.2 Reducing the Number of Faults

Reducing the number of short-circuit faults in a system not only reduces the sag frequency but also the frequency of sustained interruptions. This is thus a very effective way of improving the quality of supply and many customers suggest this as the obvious solution when a voltage sag or short interruption problem occurs. Unfortunately, the solution is rarely that simple. A short circuit not only leads to a voltage sag or interruption at the customer interface but may also cause damage to utility equipment and plant. Therefore most utilities will already have reduced the fault frequency as far as economically feasible. In individual cases there could still be room for improvement, e.g., when the majority of trips is due to faults on one or two distribution lines. Some examples of fault mitigation are:

- Replace overhead lines by underground cables. A large fraction of short-circuit faults is due to adverse weather or other external influences. Underground cables are much less affected by external phenomena (with the obvious exception of excavation). The fault rate on an underground cable is an order of magnitude less than for an overhead line. The effect is a big reduction in the number of voltage sags and interruptions. A disadvantage of underground cables is that the repair time is much longer.
- Use covered wires for overhead line. A recent development is the construction of overhead lines with insulated wires. Normally the wires of an overhead line are bare conductors. With covered wires, the conductors are covered with a thin layer of insulating material. Even though the layer is not a full insulation, it has proven to be efficient in reducing the fault rate of overhead lines [208], [212]. Also other types of conductors may reduce the fault rate [213].
- Implement a strict policy of tree trimming. Contact between tree branches and wires can be an important cause of short-circuit faults, especially during heavy loading of the line. Due to the heating of the wires their sag increases, making contact with trees more likely. Note that this is also the time during which the consequences of a short circuit are most severe.
- Install additional shielding wires. Installation of one or two shielding wires reduces the risk of a fault due to lightning. The shielding wires are located such that severe lightning strokes are most likely to hit a shielding wire. A lightning stroke to a shielding wire is normally conducted to earth through a tower.
- Increase the insulation level. This generally reduces the risk of short-circuit faults. Note that many short circuits are due to overvoltages or due to a deterioration of the insulation.
- Increase maintenance and inspection frequencies. This again generally reduces the risk of faults. If the majority of faults are due to adverse weather, as is often the case, the effect of increased maintenance and inspection is limited.

One has to keep in mind, however, that these measures may be very expensive and that its costs have to be weighted against the consequences of the equipment trips.

7.1.3 Reducing the Fault-Clearing Time

Reducing the fault-clearing time does not reduce the number of events but only their severity. It does not do anything to reduce the number or duration of interruptions. The duration of an interruption is determined by the speed with which the supply is restored. Faster fault-clearing does also not affect the number of voltage sags but it can significantly limit the sag duration.

The ultimate reduction in fault-clearing time is achieved by using current-limiting fuses [6], [7]. Current-limiting fuses are able to clear a fault within one half-cycle, so that the duration of a voltage sag will rarely exceed one cycle. If we further realize that fuses have an extremely small chance of fail-to-trip, we have what looks like the ultimate solution. The recently introduced static circuit breaker [171], [175] also gives a fault-clearing time within one half-cycle; but it is obviously much more expensive than a current-limiting fuse. No information is available about the probability of fail-to-trip. Additionaliy several types of fault-current limiters have been proposed which not so much clear the fault, but significantly reduce the fault-current magnitude within one or two cycles.

One important restriction of all these devices is that they can only be used for low- and medium-voltage systems. The maximum operating voltage is a few tens of kilovolts. Static circuit breakers show the potential to be able to operate at higher voltage levels in the future.

But the fault-clearing time is not only the time needed to open the breaker but also the time needed for the protection to make a decision. Here we need to consider two significantly different types of distribution networks, both shown in Fig. 7.2.

The top drawing in Fig. 7.2 shows a system with one circuit breaker protecting the whole feeder. The protection relay with the breaker has a certain current setting. This setting is such that it will be exceeded for any fault on the feeder, but not exceeded for any fault elsewhere in the system nor for any loading situation. The moment the current value exceeds the setting (thus for any fault on the feeder) the relay instantaneously gives a trip signal to the breaker. Upon reception of this signal, the breaker opens within a few cycles. Typical fault-clearing times in these systems are around 100 milliseconds. To limit the number of long interruptions for the customers, reclosing is used in combination with (slow) expulsion fuses in the laterals or in combination with interruptors along the feeder. This type of protection is commonly used in overhead systems. Reducing the fault-clearing time mainly requires a faster breaker. The static circuit breaker or several of the other current limiters would be good options for these systems. A current-limiting fuse to protect the whole feeder is not suitable as it makes fast reclosing more complicated. Current-limiting fuses can also not be used for the protection of the laterals because they would start arcing before the main breaker opens. Using a faster clearing with the main breaker enables faster clearing in the laterals as well.

The network in the bottom drawing of Fig. 7.2 consists of a number of distribution substations in cascade. To achieve selectivity, time-grading of the overcurrent relays is used. The relays furthest away from the source trip instantaneously on overcurrent. When moving closer to the source, the tripping delay increases each time with typically 500 ms. In the example in Fig. 7.2 the delay times would be 1000 ms, 500 ms, and zero (from left to right). Close to the source, fault-clearing times can be up to several seconds. These kind of systems are typically used in underground networks and in industrial distribution systems.

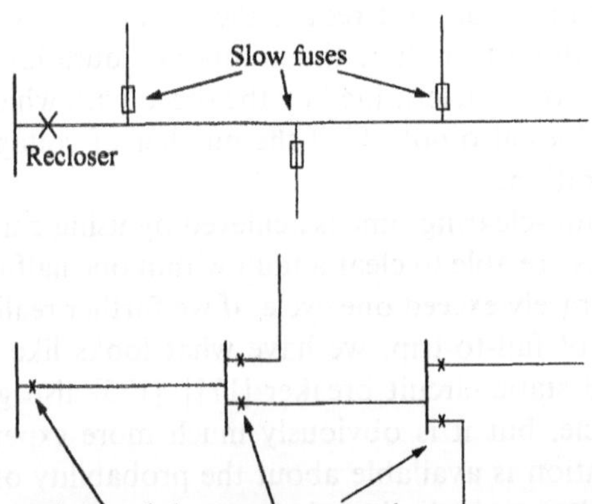

Figure 7.2 Distribution system with one circuit breaker protecting the whole feeder (top) and with a number of substations (bottom).

The fault-clearing time can be reduced by using inverse-time overcurrent relays. For inverse-time overcurrent relays, the delay time decreases for increasing fault current. But even with these schemes, fault-clearing times above one second are possible. The various techniques for reducing the fault-clearing time without loosing selectivity are discussed in various publications on power system protection, e.g., [176] and [10].

To achieve a serious reduction in fault-clearing time one needs to reduce the grading margin, thereby allowing a certain loss of selectivity. The relay setting rules described in most publications are based on preventing incorrect trips. Future protection settings need to be based on a maximum fault-clearing time. A method of translating a voltage-tolerance curve into a time-current curve is described in [167]. The latter curve can be used in combination with relay curves to obtain the various settings. The opening time of the downstream breaker is an important term in the expression for the grading margin. By using faster breakers, or even static circuit breakers, the grading margin can be significantly reduced, thus leading to a significant reduction in fault-clearing time. The impact of static circuit breakers might be bigger in these systems than in the ones with one breaker protecting the whole feeder.

In transmission systems the fault-clearing time is often determined by transient-stability constraints. These constraints are much more strict than the thermal constraints in distribution systems, requiring shorter fault-clearing times, rarely exceeding 200 ms. This also makes that further reduction of the fault-clearing time becomes much more difficult. Some remaining options for the reduction of the fault-clearing time in transmission systems are:

- In some cases faster circuit breakers could be of help. This again not only limits the fault-clearing time directly but it also limits the grading margin for distance protection. One should realize however that faster circuit breakers could be very expensive.
- A certain reduction in grading margin is probably possible. This will not so much reduce the fault-clearing time in normal situations, but in case the protection fails and a backup relay has to intervene. When reducing the grading margin one should realize that loss of selectivity is unacceptable in most transmission systems as it leads to the loss of two or more components at the same time.
- Faster backup protection is one of the few effective means of reducing fault-clearing time in transmission systems. Possible options are to use intertripping for distance protection, and breaker-failure protection.

7.1.4 Changing the Power System

By implementing changes in the supply system, the severity of the event can be reduced. Here again the costs can become very high, especially for transmission and subtransmission voltage levels. The main mitigation method against interruptions is the installation of redundant components.

Some examples of mitigation methods especially directed toward voltage sags are:

- Install a generator near the sensitive load. The generators will keep the voltage up during a sag due to a remote fault. The reduction in voltage drop is equal to the percentage contribution of the generator station to the fault current. In case a combined-heat-and-power station is planned, it is worth to consider the position of its electrical connection to the supply.

- Split busses or substations in the supply path to limit the number of feeders in the exposed area.
- Install current-limiting coils at strategic places in the system to increase the "electrical distance" to the fault. One should realize that this can make the sag worse for other customers.
- Feed the bus with the sensitive equipment from two or more substations. A voltage sag in one substation will be mitigated by the infeed from the other substations. The more independent the substations are the more the mitigation effect. The best mitigation effect is by feeding from two different transmission substations. Introducing the second infeed increases the number of sags, but reduces their severity.

The number of short interruptions can be prevented by connecting less customers to one recloser (thus, by installing more reclosers), or by getting rid of the reclosure scheme altogether. Short as well as long interruptions are considerably reduced in frequency by installing additional redundancy in the system. The costs for this are only justified for large industrial and commercial customers. Intermediate solutions reduce the duration of (long) interruptions by having a level of redundancy available within a certain time. The relations between power system design, interruptions, and voltage sags are discussed in detail in Sections 7.2 and 7.3. The former mainly considers methods of reducing the duration of an interruption, where the latter discusses relations between sag frequency and system design.

7.1.5 Installing Mitigation Equipment

The most commonly applied method of mitigation is the installation of additional equipment at the system–equipment interface. Recent developments point toward a continued interest in this way of mitigation. The popularity of mitigation equipment is explained by it being the only place where the customer has control over the situation. Both changes in the supply as well as improvement of the equipment are often completely outside of the control of the end-user.

Some examples of mitigation equipment are:

- Uninterruptible power supplies (UPSs) are extremely popular for computers: personal computers, central servers, and process-control equipment. For the latter equipment the costs of a UPS are negligible compared to the total costs.
- Motor-generator sets are often depicted as noisy and as needing much maintenance. But in industrial environments noisy equipment and maintenance on rotating machines are rather normal. Large battery blocks also require maintenance, expertise on which is much less available.
- Voltage source converters (VSCs) generate a sinusoidal voltage with the required magnitude and phase, by switching a dc voltage in a particular way over the three phases. This voltage source can be used to mitigate voltage sags and interruptions.

Mitigation equipment is discussed in detail in Section 7.4.

7.1.6 Improving Equipment Immunity

Improvement of equipment immunity is probably the most effective solution against equipment trips due to voltage sags. But it is often not suitable as a short-time solution. A customer often only finds out about equipment immunity after the equipment has been installed. For consumer electronics it is very hard for a customer to find out about immunity of the equipment as he is not in direct contact with the manufacturer. Even most adjustable-speed drives have become off-the-shelf equipment where the customer has no influence on the specifications. Only large industrial equipment is custom-made for a certain application, which enables the incorporation of voltage-tolerance requirements.

Several improvement options have been discussed in detail in Chapter 5. Some specific solutions toward improved equipment are:

- The immunity of consumer electronics, computers, and control equipment (i.e., single-phase low-power equipment) can be significantly improved by connecting more capacitance to the internal dc bus. This will increase the maximum sag duration which can be tolerated.
- Single-phase low-power equipment can also be improved by using a more sophisticated dc/dc converter: one which is able to operate over a wider range of input voltages. This will reduce the minimum voltage for which the equipment is able to operate properly.
- The main source of concern are adjustable-speed drives. We saw that ac drives can be made to tolerate sags due to single-phase and phase-to-phase faults by adding capacitance to the dc bus. To achieve tolerance against sags due to three-phase faults, serious improvements in the inverter or rectifier are needed.
- Improving the immunity of dc adjustable-speed drives is very difficult because the armature current, and thus the torque, drops very fast. The mitigation method will be very much dependent on restrictions imposed by the application of the drive.
- Apart from improving (power) electronic equipment like drives and process-control computers a thorough inspection of the immunity of all contactors, relays, sensors, etc., can also significantly improve the process ridethrough.
- When new equipment is installed, information about its immunity should be obtained from the manufacturer beforehand. Where possible, immunity requirements should be included in the equipment specification.

For short interruptions, equipment immunity is very hard to achieve; for long interruptions it is impossible to achieve. The equipment should in so far be immune to interruptions, that no damage is caused and no dangerous situation arises. This is especially important when considering a complete installation.

7.1.7 Different Events and Mitigation Methods

Figure 7.3 shows the magnitude and duration of voltage sags and interruptions resulting from various system events. For different events different mitigation strategies apply.

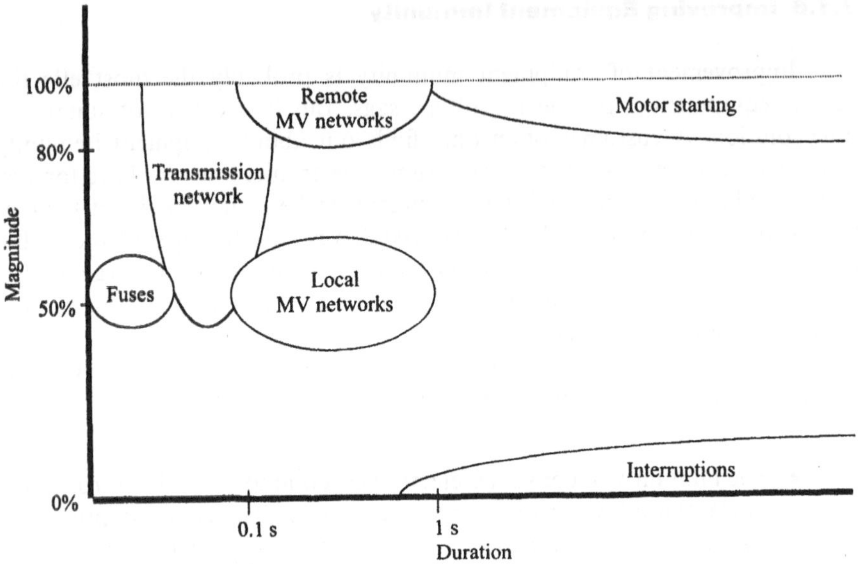

Figure 7.3 Overview of sags and interruptions.

- Sags due to short-circuit faults in the transmission and subtransmission system are characterized by a short duration, typically up to 100 ms. These sags are very hard to mitigate at the source and also improvements in the system are seldom feasible. The only way of mitigating these sags is by improvement of the equipment or, where this turns out to be unfeasible, installing mitigation equipment. For low-power equipment a UPS is a straightforward solution; for high-power equipment and for complete installations several competing tools are emerging.
- As we saw in Section 7.1.3, the duration of sags due to distribution system faults depends on the type of protection used, ranging from less than a cycle for current-limiting fuses up to several seconds for overcurrent relays in underground or industrial distribution systems. The long sag duration makes that equipment can also trip due to faults on distribution feeders fed from another HV/MV substation. For deep long-duration sags, equipment improvement becomes more difficult and system improvement easier. The latter could well become the preferred solution, although a critical assessment of the various options is certainly needed. Reducing the fault-clearing time and alternative design configurations should be considered.
- Sags due to faults in remote distribution systems and sags due to motor starting should not lead to equipment tripping for sags down to 85%. If there are problems the equipment needs to be improved. If equipment trips occur for long-duration sags in the 70%–80% magnitude range, improvements in the system have to be considered as an option.
- For interruptions, especially the longer ones, improving the equipment immunity is no longer feasible. System improvements or a UPS in combination with an emergency generator are possible solutions here. Some alternatives are presented in Sections 7.2 and 7.3.

7.2 POWER SYSTEM DESIGN—REDUNDANCY THROUGH SWITCHING

This and the next section discuss some of the relations between structure and operation of power systems and the number of voltage sags and interruptions. The reduction of interruption frequency is an important part of distribution system design and as such it is treated in detail in a number of books and in many papers. Often cited books on distribution system design are "Electricity Distribution Network Design" by Lakervi and Holmes [114] and "Electric Power Distribution System Engineering" by Gönen [164]. Other publications treating this subject in part are [23], [115], [116], [165], [209], [214]. Many case studies have appeared over the years in conferences and transactions of the IEEE Industry Applications Society and to a lesser degree in the publications of the Power Engineering Society and of the Institute of Electrical Engineers.

7.2.1 Types of Redundancy

The structure of the distribution system has a big influence on the number and duration of the interruptions experienced by the customer. The influence of the transmission system is much smaller because of the high redundancy used. Interruptions originating in the distribution system affect less customers at a time, but any given customer has a much higher chance of experiencing a distribution-originated interruption than a transmission-originated one. The large impact of interruptions originating in the transmission system makes that they should be avoided at almost any cost. Hence the high reliability of transmission systems.

Number and duration of interruptions is determined by the amount of redundancy present and the speed with which the redundancy can be made available. Table 7.1 gives some types of redundancy and the corresponding duration of the interruption. Whether the supply to a certain load is redundant depends on the time scale at which one is looking. In other words, on the maximum interruption duration which the load can tolerate.

When a power system component, e.g., a transformer, fails it needs to be repaired or its function taken over by another component before the supply can be restored. In case there is no redundant transformer available, the faulted transformer needs to be repaired or a spare one has to be brought in. The repair or replacement process can take several hours or, especially with power transformers, even days up to weeks. Repair times of up to one month have been reported.

TABLE 7.1 Various Types of Redundancy in Power System Design

	Duration of Interruption	Typical Applications
No redundancy	Hours through days	Low voltage in rural areas
Redundancy through switching		
– Local manual switching	1 hour and more	Low voltage and distribution
– Remote manual switching	5 to 20 minutes	Industrial systems, future public distribution
– Automatic switching	1 to 60 seconds	Industrial systems
– Solid state switching	1 cycle and less	Future industrial systems
Redundancy through parallel operation	Voltage sag only	Transmission systems, industrial systems

In most cases the supply is not restored through repair or replacement but by switching from the faulted supply to a backup supply. The speed with which this takes place depends on the type of switching used. The various types will be discussed in detail in the remainder of this section.

A smooth transition without any interruption takes place when two components are operated in parallel. This will however not mitigate the voltage sag due to the fault which often precedes the interruption. Various options and their effect on voltage sags are discussed in Section 7.3.

7.2.2 Automatic Reclosing

Automatic reclosing was discussed in detail in Chapter 3. Automatic reclosing after a short-circuit fault reduces the number of long interruptions by changing them into short interruptions. Permanent faults still lead to long interruptions, but on overhead distribution lines this is less than 25% of the total number of interruptions. We saw in Chapter 3 that the disadvantage of the commonly used method of automatic reclosing is that more customers are affected by a fault. A long interruption for part of a feeder is changed into a short interruption for the whole feeder. This is not inherent to automatic reclosing, but to the method of fuse saving used. If all fuses would be replaced by reclosers, the number of short interruptions would be significantly reduced. A customer would only experience a short interruption for what would have been a long interruption without reclosing. This would of course make the supply more expensive, which is not always acceptable for remote rural areas.

7.2.3 Normally Open Points

The simplest radial system possible is shown in Fig. 7.4: a number of feeders originate from a distribution substation. When a fault occurs on one of the feeders, the fuse will clear it, leading to an interruption for all customers fed from this feeder. The supply can only be restored after the faulted component has been repaired or replaced. Such systems can be found in rural low-voltage and distribution systems with overhead feeders. Protection is through fuses in the low-voltage substations. Repair of a faulted feeder (or replacement of a blown fuse) can take several hours, repair or replacement of a transformer several days. As the feeders are overhead they are prone to weather influences; storms are especially notorious for it can take days before all feeders have been repaired.

A commonly used method to reduce the duration of an interruption is to install a normally open switch, often called "tie switch." An example is shown in Fig. 7.5.

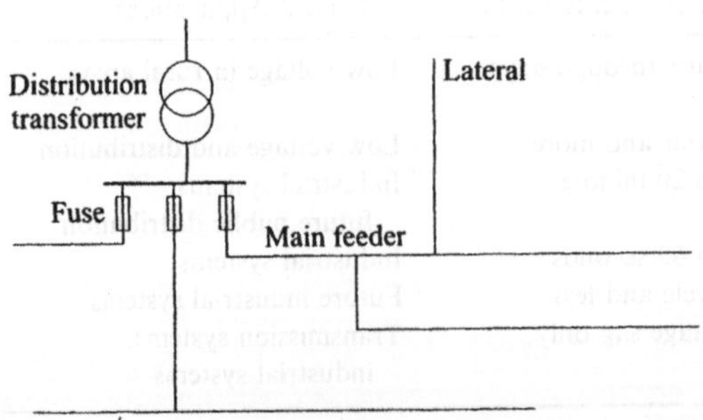

Figure 7.4 Power system without redundancy.

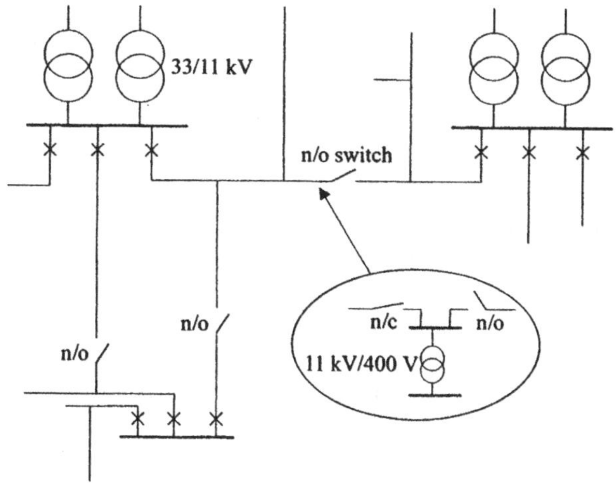

Figure 7.5 Distribution system with redundancy through manual switching.

The system is still operated radially; this prevents the fault level from getting too high and enables the use of (cheap) overcurrent protection. If a fault occurs it is cleared by a circuit breaker in the substation. The faulted section is removed, the normally open switch is closed, and the supply can be restored. The various steps in the restoration of the supply are shown in Fig. 7.6.

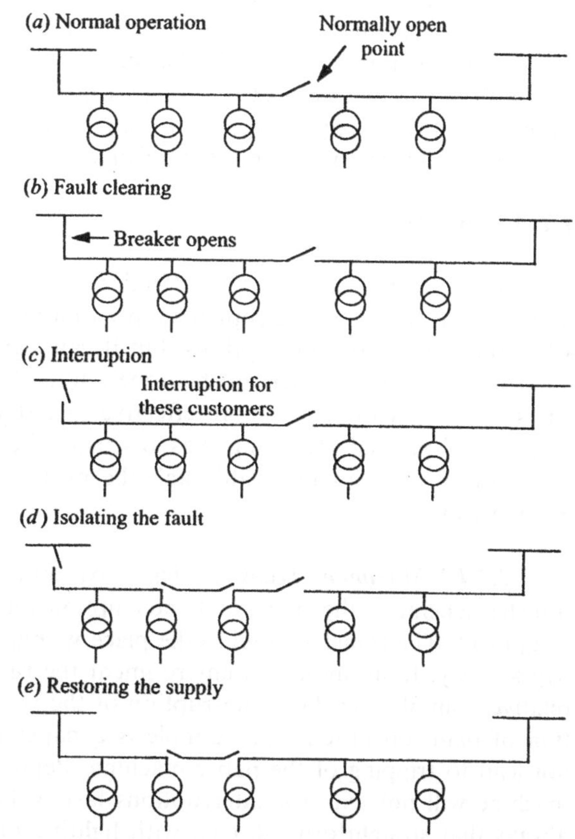

Figure 7.6 Restoration procedure in a distribution system with normally open points. (a) Normal operation, (b) fault clearing, (c) interruption, (d) isolating the fault, (e) restoring the supply.

In normal operation (a) the feeder is operated radially. A normally open switch is located between this feeder and another feeder, preferably fed from another substation. When a fault occurs (b) the breaker protecting the feeder opens leading to an interruption for all customers fed from this feeder (c). After the fault is located, it is isolated from the healthy parts of the feeder (d) and the supply to these healthy parts is restored by closing the circuit breaker and the normally open switch (e). Repair of the feeder only starts after the supply has been restored.

This procedure limits the duration of an interruption to typically one or two hours in case the switching is done locally (i.e., somebody has to go to the switches to open or close them). If fault location and switching is done remotely (e.g., in a regional control center) the supply can be restored in several minutes. Locating the fault may take longer than the actual switching. Especially in case of protection or signaling failure, locating the fault can take a long time. Various techniques are in use for identifying the faulted section of the feeder. More precise fault location, needed for repair, can be done afterwards.

The type of operation shown in Figs. 7.5 and 7.6 is very commonly used in underground low-voltage and medium-voltage distribution systems. The repair of underground cables can take several days so that system operation like in Fig. 7.4 becomes totally unacceptable. Similar restoration techniques are in use for medium-voltage overhead distribution, especially in the more urban parts of the network. The high costs for signaling equipment and communication links make remote switching only suitable for higher voltages and in industrial distribution systems. When customer demands for shorter durations of interruptions continue to increase, remote signaling and switching will find its way into public distribution systems as well.

The additional costs for the system in Fig. 7.5 are not only switching, signaling and communication equipment. The feeder has to be dimensioned such that it can handle the extra load. Also the voltage drop over the, now potentially twice as long, feeder should not exceed the margins. Roughly speaking the feeder can only feed half as much load. This will increase the number of substations and thus increase the costs.

7.2.4 Load Transfer

A commonly used and very effective way of mitigating interruptions is transferring the load from the interrupted supply to a healthy supply. Load transfer does not affect the number of interruptions, but it can significantly reduce the duration of an interruption. Load transfer can be done automatically or manually; automatic transfer is faster and therefore more effective in reducing the interruption duration. An example of manual switching was discussed before. Here we will concentrate on automatic transfer of load, although the proposed schemes are equally suitable for manual transfer.

7.2.4.1 Maximum Transfer Time. An important criterion in the design of any transfer scheme is the maximum interruption duration that can be tolerated by the equipment. The transfer should take place within this time, otherwise the load would trip anyway. In an industrial environment the rule for the maximum transfer time is relatively simple: the short interruption of the voltage should not lead to an interruption of plant operation. An example is a paper mill, where the interruption should not lead to tripping of the paper machine. Below a certain interruption duration the machine will not trip, for interruptions lasting longer it will trip. The choice is not always that straightforward, e.g., with lighting of public buildings. A general rule is that one should in all cases choose a transfer time such that the transfer does not lead to unacceptable consequences. What should be considered as unacceptable is simply part of the decision process. In practice the load of a power system is not constant, and decisions about transfer time may have to be revised several years later because more sensitive equipment is being used, as, e.g., described in [163].

7.2.4.2 Mechanical Load Transfer. Most transfer schemes use a mechanical switch or circuit breaker to transfer from one supply point to another. A typical configuration as used in industrial distribution systems is shown in Fig. 7.7. Two transformers each supply part of the load. If one of them fails, the normally open switch is closed and the total load is fed from one transformer. Each transformer should be able to supply the total load or a load shedding scheme should be in place. When a short circuit occurs close to the transfer switch, it is essential that the load is not transferred before the fault has been cleared: a so-called "break-before-make" scheme. A "make-before-break" scheme would spread the fault to the healthy supply leading to possible intervention by the protection in both feeders. In case one transformer is taken out of operation for maintenance, a (manual) make-before-break scheme can be used. This reduces the risk of a long interruption due to failure of the transfer switch. During the parallel operation, a short circuit could lead to serious switchgear damage.

The advantages of this scheme compared to parallel operation are that the protection is simpler and that the fault current is lower. As long as the load can tolerate the short interruption during load transfer, the reliability of the supply is similar to that of parallel operation. As we saw in Section 2.8, load interruptions for a transfer scheme are mainly due to failure of the transfer switch and due to any kind of common-mode effect in the two supplies. In an industrial environment, maintenance and excavation activities could seriously effect the supply reliability.

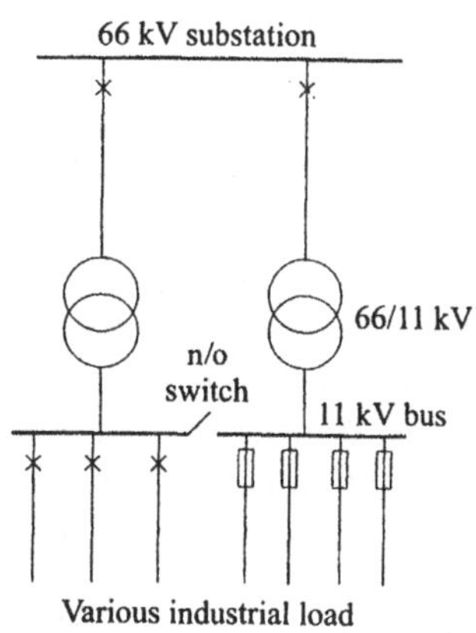

Figure 7.7 Industrial power system with redundancy through automatic switching.

7.2.4.3 Transfer of Motor Load. A problem with automatic switching is the presence of large numbers of induction motors in most industrial systems. When the supply is interrupted, the remaining airgap flux generates a voltage over the motor terminals. This voltage decays in magnitude and in frequency. The switching has to take place either very fast (before the motor voltage has shifted much in phase compared to the system voltage) or very slow (after the motor voltage has become zero). As the first option is expensive, the second one is normally used.

The airgap field in a induction motor decays with a certain time constant which varies from less than one cycle for small motors up to about 100 ms for large motors. The time constant with which the motor slows down is much larger: typically between one and five seconds.

The moment the motor is reconnected, the source voltage will normally not be in phase with the motor voltage. In case they are in opposite phase a large current will flow. This current can be more than twice the starting current of the motor. It can easily damage the motor or lead to tripping of the motor by the overcurrent protection.

The induced voltage has the following form:

$$E = \hat{E} \sin \omega t \tag{7.1}$$

with ω the angular speed of the motor, which decays exponentially:

$$\omega = \omega_0 \left(1 - e^{-\frac{t}{\tau_m}}\right) \tag{7.2}$$

and $\hat{E}$ dependent on the frequency and the exponentially decaying rotor current. Assume for simplicity that the magnitude of the induced voltage remains constant and consider a linear decay in motor speed:

$$\omega \approx \omega_0 \left(1 - \frac{t}{\tau_m}\right) \tag{7.3}$$

This gives for the voltage at the motor terminals:

$$E(t) = \sin\left(\omega_0 \left(1 - \frac{t}{\tau_m}\right) t\right) = \sin\left(\omega_0 t - \frac{\omega_0 t^2}{\tau_m}\right) \tag{7.4}$$

The second term under the sine function is the phase difference between the supply and the induced voltage. As long as this phase difference is less than 60°, the voltage difference between the source and the motor is less than 1 pu. A phase difference of 60° ($\frac{\pi}{3}$) is reached for

$$t = \sqrt{\frac{\tau_m}{6 f_0}} \tag{7.5}$$

For a mechanical time constant $\tau_m = 1$ sec and a frequency of $f_0 = 50$ Hz an angular difference of 60° is reached after 58 ms. In the calculation it is assumed that the motor has not slowed down during the fault. If this is also considered, the value of 60° is reached faster. Only very fast transfer schemes are able to switch within this short time. A second chance at closing the transfer switch is when the angular difference is about 360° (i.e., source and motor are in phase again). This takes place for

$$t = \sqrt{\frac{\tau_m}{f_0}} \tag{7.6}$$

which is 140 ms in the above example. These so-called synchronous transfer schemes are very expensive and may still lead to transfer times above 100 ms. In most cases asynchronous transfer is used where the transfer switch is only closed after the induced voltage has sufficiently decayed, leading to transfer times around one second or longer.

For synchronous machines the airgap field decays with the same time constant as the motor speed, so that the terminal voltage may be present for several seconds. In a system with a large fraction of synchronous motor load, synchronous transfer becomes more attractive. Note that asynchronous transfer will always lead to loss of the synchronous motor load.

7.2.4.4 Primary and Secondary Selective Supplies. Figures 7.8 and 7.9 show two ways of providing a medium-voltage customer with a reliable supply. In a primary selective system (Fig. 7.8) the transfer takes place on the primary side of the transformer. A secondary selective system (Fig. 7.9) is more expensive but there is a much reduced chance of very long interruptions due to transformer failure. A numerical analysis of such a transfer scheme is given in Section 2.8.

The actual transfer is identical to the transfer in the industrial supply shown in Fig. 7.7: the load is transferred from the faulted to the healthy feeder as soon as possible after fault clearing. With a primary selective supply a make-before-break scheme would directly connect two feeders. It is unlikely that the utility allows this. The transfer takes place behind a transformer with the secondary selective supply. The possible consequences of a make-before-break scheme are less severe for the utility.

With the design of primary and secondary selective supplies, it is again very important to determine the tolerance of the load to short interruptions. The choice for a certain type of transfer scheme should depend on this tolerance.

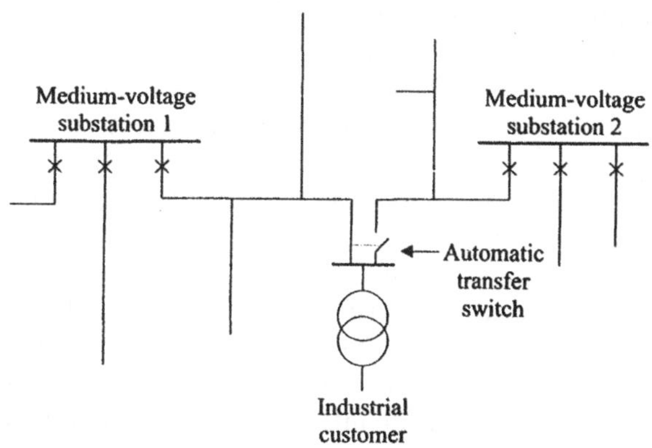

Figure 7.8 Primary selective supply.

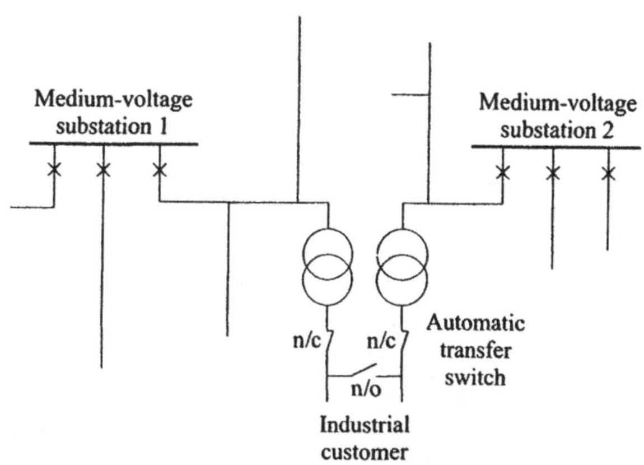

Figure 7.9 Secondary selective supply.

7.2.4.5 Static Transfer Switches. Static transfer switches have been used already for several years in low-voltage applications, e.g., in uninterruptable power supplies to be discussed in Section 7.4. Currently, static transfer switches are also available for medium voltages [166], [171], [173]. A static transfer switch consists of two pairs of anti-parallel thyristors as shown in Fig. 7.10. During normal operation, thyristor pair I is continuously fired, and thus conducting the load current. Thyristor pair II is not fired. In terms of switches, thyristor pair I behaves like a closed switch, pair II like an open switch.

When a disturbance is detected on the normal supply, the firing of thyristor pair I is disabled and the firing of thyristor pair II enabled. The effect of this is that the load current commutates to the backup supply within half a cycle of detecting the disturbance. Actual transfer times are less than 4 ms [166]. The three small figures show the voltages in a stylized way. In reality voltages are sinusoidal, but the principle remains the same. Point A experiences a drop in voltage due to a sag or interruption at time 1. This drop in voltage is also experienced by the load at point C. We assume that the backup supply does not experience this. At time 2, the disturbance is detected, the firing of thyristor pair I is disabled, and the firing of thyristor pair II enabled. At that moment the commutation of the current from the normal supply to the backup supply starts. During commutation the voltage at points A, B, and C is equal as both thyristor pairs are conducting. This voltage is somewhere in between the two supply voltages. At time 3 the commutation is complete (the thyristor current in pair I extinguishes on the first zero crossing after the firing being disabled) and the voltage at B and C comes back to its normal value. Note that the current through the thyristors never exceeds the load current, also not for a fault close to the static switch.

A static transfer switch can be used in any of the transfer schemes discussed before: industrial distribution, primary selective, secondary selective. The speed with which the transfer takes place makes the distinction between synchronized and non-synchronized transfer no longer relevant. Load transfer by a static transfer switch is always synchronized.

To ensure very fast transfer, any voltage sag or interruption in the normal supply should be detected very fast. The commutation of the current from one thyristor pair to the other takes less than half a cycle so that we need a disturbance detection which is equally fast. Static transfer schemes can use the missing voltage or a half-cycle rms value to detect a sag or interruption. For the missing voltage detection scheme, the

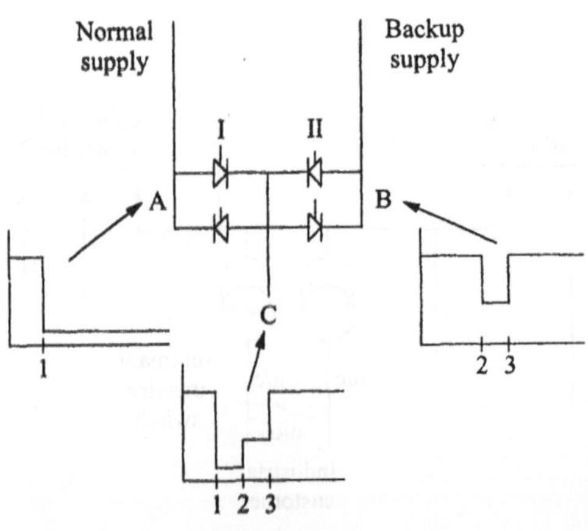

Figure 7.10 Construction and principle of operation of a static transfer switch.

actual voltage is compared on a sample-by-sample basis with the output voltage of a phase-locked-loop (PLL). When the deviation becomes too large for too long, the transfer is initiated. With the rms scheme, transfer is initiated when the rms voltage drops below a certain threshold. The latter scheme is slower as it will lead to an additional half-cycle delay, but it has a smaller chance of incorrect transfer.

A transfer scheme using a static transfer switch enables the duration of a voltage sag to be limited to half a cycle by switching to the backup supply when a sag occurs in the normal supply. For sensitive load, a static transfer switch might be preferable above parallel operation. Voltage sags originating in the transmission system cannot be mitigated by such a transfer scheme as the voltage sag is likely to be present in both supplies; but for sags originating in the distribution system the static transfer scheme is very effective. The main limitations are the unknown reliability of the transfer switch and the degree in which the two sources are independent.

The notch due to load transfer could be a concern, especially for the load on the healthy feeder. When comparing static transfer with parallel operation, a notch of millisecond duration replaces the voltage sags of several cycles duration. When comparing with the mechanical transfer scheme, the notch in the backup supply constitutes a deterioration of the voltage quality, albeit not a severe deterioration. Some utilities do not allow parallel operation of feeders, requiring a so-called "break-before-make" transfer scheme. The static transfer switch as described here is essentially a "make-before-break" scheme. It is impossible to predict how strict utilities will apply this rule on a sub-cycle timescale. As an alternative one could enable firing of thyristor pair II only after the current through pair I has extinguished. Such a break-before-make scheme will obviously make the transfer slower and could actually make the voltage transient in the healthy supply more severe.

A final potential problem with static transfer is that the normal supply and the healthy supply are not exactly in phase. The phase-angle difference could lead to a small phase-angle jump at the load terminals. Values up to 6° have been reported. As long as there are no standards on equipment tolerance to phase-angle jumps, it is hard to assess the impact of this. The successful use of medium-voltage static transfer switches on a number of sites indicates that the equipment is able to tolerate the transient.

7.3 POWER SYSTEM DESIGN—REDUNDANCY THROUGH PARALLEL OPERATION

7.3.1 Parallel and Loop Systems

Figure 7.11 shows a public distribution network with a higher nominal voltage than the one in Fig. 7.5. It serves more customers so it is worth to invest more in reliability. Part of the system is still operated in a radial way with normally open points. These are serving less densely populated areas, and areas with less industrial activity. The majority of the 33 kV system is operated with parallel feeders. Both paths carry part of the load. If one path fails, the other path takes over the supply instantaneously. Also the 33/11 kV transformer and the 33 kV substation bus are operated in parallel. The rating of each component is such that the load can be fully supplied if one component fails.

We see in Fig. 7.11 two types of parallel operation: two feeders in parallel and a loop system. In both cases there is single redundancy. The loop system is significantly cheaper, especially in case of transformer connections. But the voltage control of loop systems is more difficult, and the various loads are more prone to disturbing each

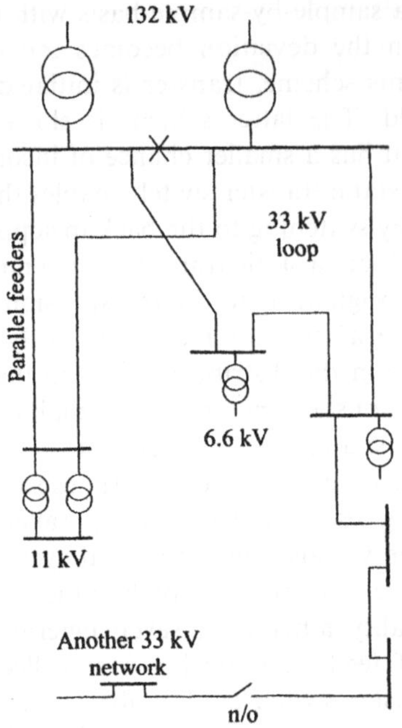

Figure 7.11 Distribution network with redundancy through parallel operation.

other's supply. Loop systems are therefore less popular in industrial systems, although some smaller loops (three or four busses) are used to limit the number of transformers.

7.3.1.1 Design Criteria for Parallel and Loop Systems. The design of parallel and loop systems is based on the so-called $(n-1)$ criterion, which states that the system consisting of n components should be able to operate with only $(n-1)$ components in operation, thus with one component out of operation. This should hold for any one component out of operation. The $(n-1)$ criterion is very commonly used in power system design. It enables a high reliability without the need for stochastic assessment. In some cases (large transmission systems, generator scheduling), $(n-2)$ or $(n-3)$ criteria are used. As we saw in Section 2.8, a thorough assessment of all "common-mode failures" is needed before one can trustfully use such a high-redundancy design criterion.

Here we will concentrate on the $(n-1)$ criterion, also referred to as "single redundancy." This criterion is very commonly used in the design of industrial medium-voltage distribution as well as in public subtransmission systems. The main design rule is that no single event should lead to an interruption of the supply to any of the customers. In an industrial environment the wording is somewhat different: no single event should lead to a production stop for any of the plants. How these basic rules are further developed depends on the kind of system. A list of things that have to be considered is given.

1. The obvious first rule is that no component outage should lead to an interruption. There should thus be an alternate path for the power flow through any component.
2. Not only should there be an alternate path for the power flow, this alternate path should also not lead to an overload situation. In the public supply the

load demand varies significantly during the day. A certain amount of overload can be tolerated for a few hours. In industrial systems the load is typically more constant, so that any overload would be permanent. However in industrial systems it is often easier to reduce the load on a time scale of hours or to start on-site generation.

3. The power system protection should be able to clear any fault without causing an interruption for any of the customers. This requires more complicated protection systems than for radial-operated networks. These protection systems require additional voltage transformers and/or communication links. Also the number of circuit breakers increases: two circuit breakers are needed for each connection between two substations in a looped or parallel system.
4. Voltage fluctuations due to rapid load fluctuations and voltage sags due to motor starting should be within limits for any one component out of operation. This translates into a minimum fault level for any load bus. The switchgear rating dictates a maximum fault level for the system with all components in operation. The optimal use of this margin between maximum and minimum fault levels is one of the main challenges in the design of industrial medium-voltage distribution systems.
5. The electromechanical transient due to a short circuit in the system with all components in operation should not lead to loss of any load. In industrial systems with a large fraction of induction motor load, it must be ensured that these motors are able to re-acellerate after the fault.
6. The voltage sag due to any fault in the system should not lead to tripping of essential load with any of the customers.

From this list it becomes obvious that the design of a parallel or loop system could be a serious challenge. But the reliability demands of large industrial plants are such that no radial system could deliver this. The increased reliability is more than worth the higher installation costs and costs of operation.

7.3.1.2 Voltage Sags in Parallel and Loop Systems. Consider the system shown in Fig. 7.12: three supply alternatives for an industrial plant. In the radial system on the left, the plant is fed through a 25 km overhead line; two more overhead lines originate from the same substation, each with a length of 100 km. In the center figure the plant is fed from a loop by making a connection to the nearest feeder. In the third alternative on the right a separate overhead line has been constructed in parallel with the existing 25 km line. The magnitude of voltage sags due to faults in this system is shown in Fig. 7.13. The calculations needed to obtain this figure are discussed in Section 4.2.4. We will use Fig. 7.13 to assess the number of voltage sags experienced by the plant for the three design alternatives.

For the radial system, the plant will experience interruptions due to faults on 25 km of overhead line, and voltage sags due to faults on 200 km of line. The relation between sag magnitude and distance to the fault is according to the dotted line in Fig. 7.13. Improving the voltage tolerance of the equipment will significantly reduce the exposed length. The exposed length for radial operation is given in Table 7.2 for different equipment voltage tolerances. By simply adding the exposed lengths, it is assumed that the impact of interruptions and voltage sags is the same, which is not always the case. Even if the process trips due to a voltage sag, it might still require power from the supply for a safe shutdown of the plant.

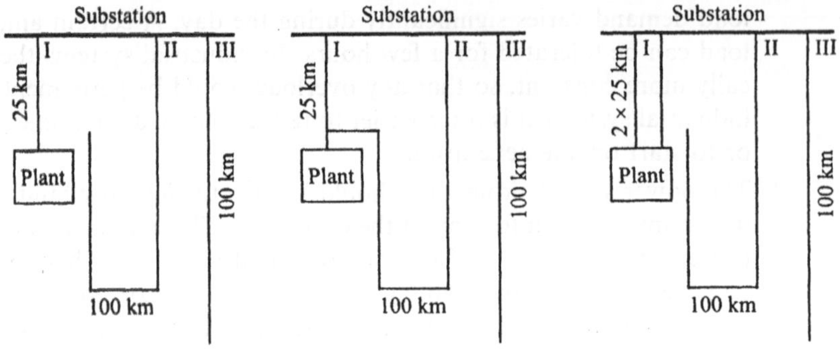

Figure 7.12 Three supply alternatives for an industrial plant: radial (left), looped (center), and parallel (right).

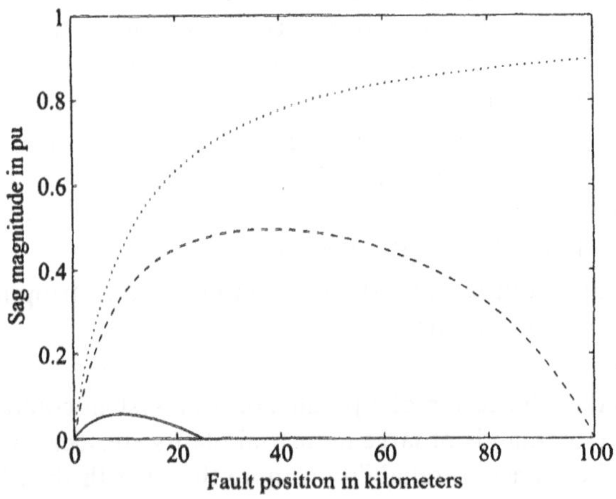

Figure 7.13 Sag magnitude as a function of fault position for faults in the system shown in Fig. 7.12. Solid line: faults on the 25 km branch of a 125 km loop; dashed line: faults on the 100 km branch of a 125 km loop; dotted line: faults on a radial feeder.

TABLE 7.2 Exposed Length for Various Equipment Voltage Tolerances for Radial Operation in Fig. 7.12

	Exposed Length			
Voltage Tolerance	Feeder I	Feeder II	Feeder III	Total
Trips on interruptions only	25 km	–	–	25 km
Trips on sags below 20%	25 km	3 km	3 km	31 km
Trips on sags below 50%	25 km	12 km	12 km	45 km
Trips on sags below 90%	25 km	100 km	100 km	225 km

The calculations have been repeated for looped operation as in the center drawing in Fig. 7.12, resulting in the values shown in Table 7.3. Only for equipment immune to all voltage sags will the number of equipment trips be less than for the radial supply.

The exposed length for the various equipment voltage tolerances is given in Table 7.4 for parallel operation. For a voltage tolerance of 50% this option is preferable above looped operation. Knowledge of the various costs involved is needed to decide if this reduction in trip frequency is worth the investment.

TABLE 7.3 Exposed Length for Various Equipment Voltage Tolerances for Looped Operation in Fig. 7.12

Voltage Tolerance	Exposed Length			
	Feeder I	Feeder II	Feeder III	Total
Trips on interruptions only	–	–	–	–
Trips on sags below 20%	25 km	14 km	3 km	42 km
Trips on sags below 50%	25 km	100 km	12 km	137 km
Trips on sags below 90%	25 km	100 km	100 km	225 km

TABLE 7.4 Exposed Length for Various Equipment Voltage Tolerances for Parallel Operation in Fig. 7.12

Voltage Tolerance	Exposed Length			
	Feeder I	Feeder II	Feeder III	Total
Trips on interruptions only	–	–	–	–
Trips on sags below 20%	50 km	3 km	3 km	56 km
Trips on sags below 50%	50 km	12 km	12 km	74 km
Trips on sags below 90%	50 km	100 km	100 km	250 km

7.3.2 Spot Networks

The basic characteristic of a spot network is that a bus is fed from two or more different busses at a higher voltage level. In the previous section we looked at parallel and loop systems originating at the same bus or at two busses connected by a normally closed breaker. When a bus is fed from two different busses, the same design problems have to be solved as for parallel and loop systems. The $(n-1)$ criterion remains the underlying rule. The magnitude of voltage sags is significantly lower for spot networks, compared to parallel networks. Also the number of interruptions will be somewhat lower, but that difference will not be significant as the number is already low.

7.3.2.1 Magnitude of Voltage Sags. Consider the system in Fig. 7.14: the busbar with the sensitive load is fed from two different busbars at a higher voltage level, Z_{S1} and Z_{S2} are source impedances at the higher voltage level, Z_{t1} and Z_{t2} are transformer impedances, z is the feeder impedance per unit length, $\mathcal{L}$ the distance between bus I and the fault. The two busses can be in the same substation or in two different substations. The reliability in the latter case is likely to be somewhat higher, although it is hard to exactly quantify this difference.

Consider a fault on a feeder originating from bus I at a distance $\mathcal{L}$ from the bus. The magnitude of the voltage at bus I is found from the voltage-divider equation

$$V_I = \frac{z\mathcal{L}}{z\mathcal{L} + Z_{S1}} \tag{7.7}$$

where we neglect the effect of the second source on the voltage at bus I. This is a reasonable assumption as the impedance of the two transformers in series will be much higher than the source impedance at bus I. If we assume the two sources to be

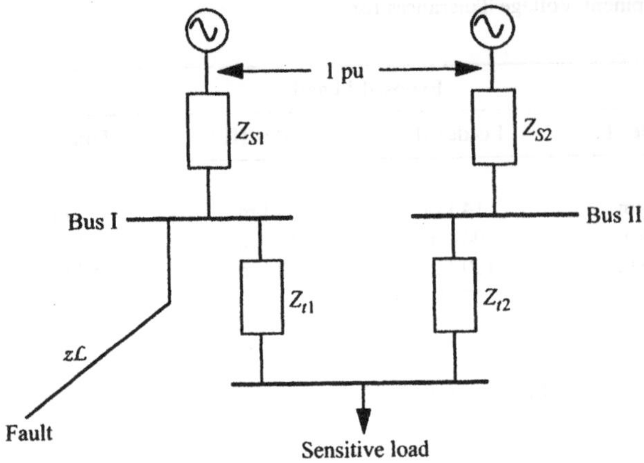

Figure 7.14 Busbar fed from two different busbars at a higher voltage level.

completely independent, so that the source voltage at bus II does not drop due to the fault, the voltage at the load bus is found from

$$V_{sag} = V_I + \frac{Z_{t1}}{Z_{t1} + Z_{t2} + Z_{S1}}(1 - V_I) \tag{7.8}$$

We simplify the expressions somewhat to be better able to assess the effect of the double supply. Assume that $z = Z_{S1}$, which is always possible by choosing the proper distance unit. Assume also that $Z_{t1} = Z_{t2}$ and that $Z_{S1} \ll Z_{t1}$ and $Z_{S2} \ll Z_{t2}$. The voltage at the load bus is, under these assumptions:

$$V_{sag} = \frac{\mathcal{L} + \frac{1}{2}}{\mathcal{L} + 1} \tag{7.9}$$

and at bus I:

$$V_I = \frac{\mathcal{L}}{\mathcal{L} + 1} \tag{7.10}$$

For a radially operated system, without a connection to bus II the voltage at the load bus is equal to the voltage at bus I, given by (7.10). Figure 7.15 compares the voltage magnitude at the load bus for the two design alternatives. It is immediately obvious that the second infeed significantly reduces the voltage drop. The deepest sag will have a magnitude of 50% of nominal. Here it is assumed that the second transformer has the same impedance as the first one. In practice this translates to them having the same rating. If the second transformer has a smaller rating, its impedance will typically be higher and the voltage sag will be deeper.

From the expressions for the voltage versus distance, we can obtain expressions for the critical distance, like in Section 6.5. For the radial system we obtain the same expression as before:

$$\mathcal{L}_{crit} = \frac{V}{1 - V} \tag{7.11}$$

For the system with double infeed, we obtain

$$\mathcal{L}_{crit} = \frac{V - \frac{1}{2}}{1 - V}, \quad V \geq 0.5 \tag{7.12}$$

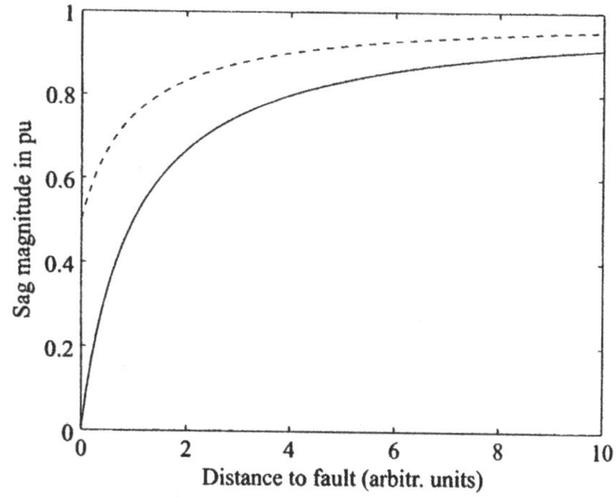

Figure 7.15 Sag magnitude as a distance to the fault, without (solid line) and with (dashed line) a connection to a second substation at a higher voltage level.

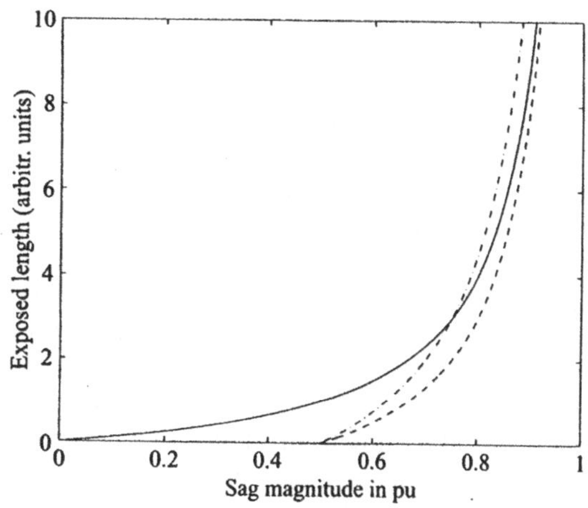

Figure 7.16 Exposed length for radial supply (solid line) and for a connection to a second substation at a higher voltage level: same number of feeders from both substations (dashed line); twice as many feeders from the second substation (dash-dot line).

and $\mathcal{L}_{crit} = 0$ for $V < 0.5$. From the critical distance the exposed length can be calculated, resulting in Fig. 7.16. The main feature is that the exposed length is zero in case the equipment can tolerate a sag down to 50% of nominal. This could be an important piece of information in deciding about the voltage-tolerance requirements for the load. For higher critical voltages (more sensitive equipment) the exposed length depends on the number of feeders originating from the two busses. Let N_1 be the number of feeders fed from bus I and N_2 the number of feeders fed from bus II. The total exposed length for the load fed from both feeders is found from

$$\mathcal{L}_{exp} = (N_1 + N_2)\frac{V - \frac{1}{2}}{1 - V} \tag{7.13}$$

for the spot network and

$$\mathcal{L}_{exp} = N_1 \frac{V}{1 - V} \tag{7.14}$$

for the radial system. In case $N_1 = N_2$, the exposed length for the double infeed is always less than for single infeed. When $N_2 > N_1$ the double-infeed option becomes less attractive when the equipment becomes too sensitive. In the example shown by a dash-dotted line in Fig. 7.16, $N_2 = 2N_1$, the cross-over point is at 75% remaining voltage.

It is important to realize that the second bus does not have to be at another substation. By operating a substation with two busses connected by a normally open breaker, the same effect is achieved. Such a configuration might not be feasible in the public supply as it reduces the reliability for customers fed from a radial feeder. But for industrial distribution systems it is an easy method of reducing the sag magnitude.

7.3.2.2 Public Low-Voltage Systems. An example of a low-voltage spot network is shown in Fig. 7.17. A low-voltage bus is fed by two or more feeders originating from different substations or from busses not operated in parallel. The protection of the feeders takes place by overcurrent protection in the medium-voltage substations and by a sensitive reverse-power relay (the "network protector") at the low-voltage bus. In public systems it is not always possible to supply from different substations. This will still lead to a low number of interruptions, but the number of voltage sags will not be reduced, and will even be somewhat increased due to faults on the parallel feeders.

The system shown in Fig. 7.18 is also referred to as a spot network; others call it a distributed grid network, or simply a secondary network. Such networks are common in the downtown areas of large cities (New York, Chicago, London, Berlin). Distributed low-voltage networks with an operating voltage of 120 V typically use no protection against low-voltage faults. The fault current is so high that every short circuit will burn itself free in a short time. For voltage levels of 200 V and higher, expulsion fuses or current-limiting fuses are used. A network protector is again installed on secondary side of every transformer to prevent backfeed from the low-voltage network into medium-voltage faults. These distributed low-voltage networks offer a high reliability. Outages on any of the distribution feeders will not be noticed by the customers. For the mitigation of sags it is essential that the feeders originate in different substations, otherwise the number of sags will even be increased. Any fault in the low-voltage network will cause a sag for all customers supplied from this network. The use of current-limiting fuses will significantly reduce the sag duration, so that these sags are not of much concern.

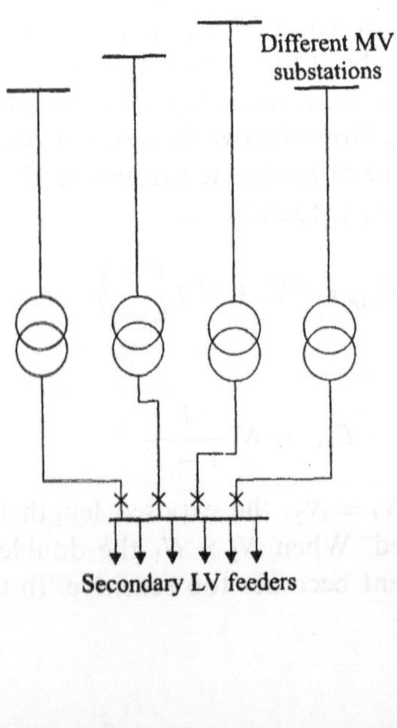

Figure 7.17 Low-voltage spot network.

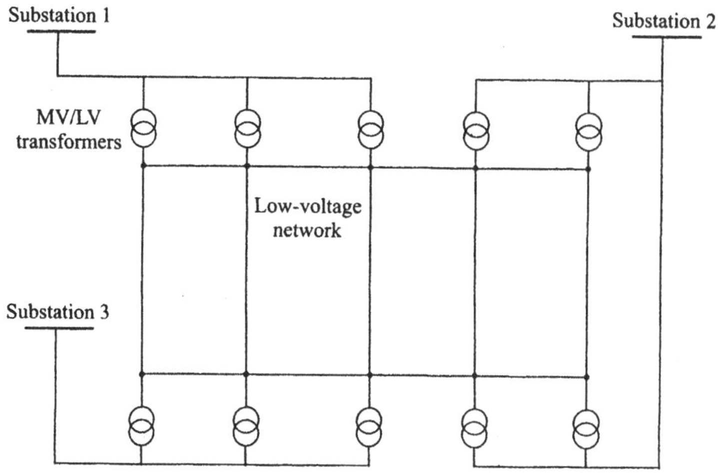

Figure 7.18 Low-voltage distributed grid.

A comparison of different design options for the public supply is given in [165]. Both stochastic prediction techniques and site monitoring were used in the comparison. Spot networks turned out to have much less interruptions than any other network configuration. Looking at the sag frequency, underground networks performed better than overhead networks, experiencing only one third of the number of sags. The supply configuration had only minor effect on the sag frequency.

*7.3.2.3 **Industrial Medium-Voltage Systems.*** In industrial systems spot networks are in use at almost any voltage level; the feeders are typically protected by using differential protection. A configuration with three voltage levels is shown in Fig. 7.19.

At each voltage level, a bus is fed from two different busses at a higher voltage level. These two busses might well be in the same substation, as long as they are not operated in parallel. The effect of this supply configuration has been discussed in Figs. 4.37, 4.38, and 4.39 in Section 4.2.4. By opening the breaker in the substation at an intermediate voltage level, thus changing from parallel operation to a spot supply, the

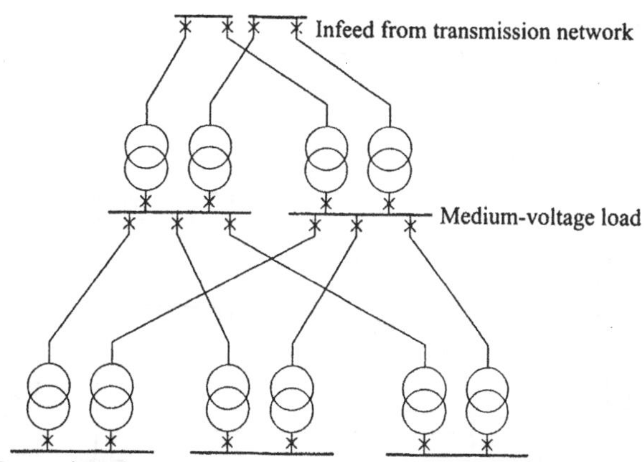

Figure 7.19 Industrial spot network.

magnitude of deep sags is significantly reduced (Fig. 4.39). The effect on shallow sags is more limited.

7.3.2.4 Transmission Systems. Another example of a spot network is the 275 kV system in the UK. These systems form the subtransmission network around the big cities. Each 275 kV system consists of about 10 busses in a loop-like structure, fed at three to five places from the 400 kV national grid. The structure of the grid around Manchester is shown in Fig. 7.20: thick lines indicate 400 kV substations and lines, and thin lines 275 kV.

Similar configurations are used in other European countries, e.g., 150 kV and 400 kV in Italy and Belgium, 150 kV and 380 kV in parts of The Netherlands, 130 kV and 400 kV in Sweden [23]. The number of supply points for the subtransmission systems varies from two through ten. In the United States this type of configuration is in use across all voltage levels, down to 69 kV, as shown in Fig. 6.39.

The effect of supply configurations as shown in Fig. 7.20 is that faults in the 400 kV grid only cause shallow sags at the 275 kV substations. If we neglect the 275 kV line impedances compared to the transformer impedances, the voltage in the 275 kV system is the average of the voltages at the 400 kV sides of the transformers. A fault close to one of the substations will drop the voltage to a low value at this substation, but other substations will be less affected. With nine transformers, the shallow sags will dominate. The effect of this "averaging" is that the customer experiences less deep but more shallow sags. To illustrate this effect, we again consider the transmission system shown earlier in Fig. 4.27. The distance between the substations has been increased to 100 km, all other parameters were kept the same. Figure 7.21 plots the sag magnitude as a function of the fault position; position 0 is a fault in substation 1, position 100 (km) a fault in substation 2. Consider next a subtransmission system fed from substation 1 and substation 2. The voltage in the subtransmission system is approximated by the average voltage in the two transmission substations; this voltage is indicated by the dotted line in Fig. 7.21. Due to the looped operation across the voltage levels, the deepest sags become shallower, and some of the shallow sags deeper.

The disadvantage of the way of operation like in Fig. 7.20 is that faults in the 275 kV networks lead to deep sags. The interconnected operation makes that the exposed area contains more length of lines than in case of radial operation. If these

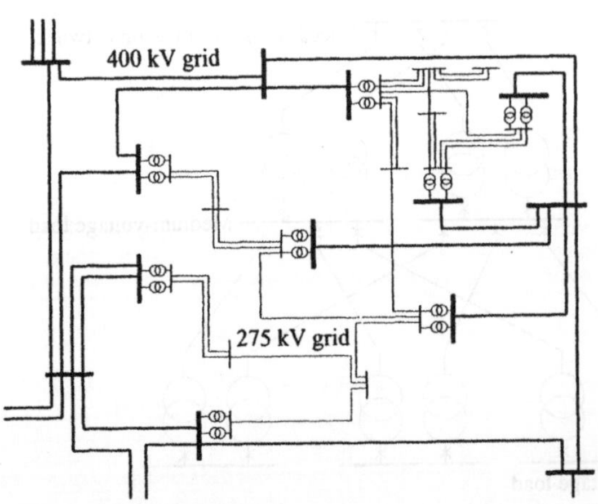

Figure 7.20 Spot network at subtransmission level: 400 kV (thick lines) and 275 kV (thin lines) system in the North of England. (Data obtained from [177].)

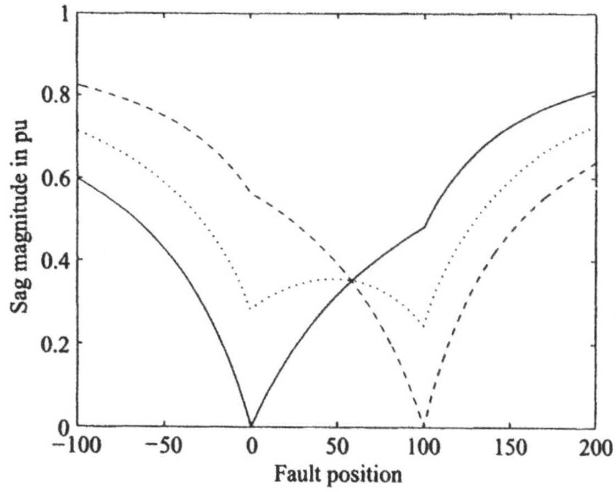

Figure 7.21 Sag magnitude in transmission and subtransmission systems. Solid line: transmission substation 1, dashed line: transmission substation 2, dotted line: subtransmission.

loops cross several voltage levels, like in the United States, the net effect is likely to be a reduction in sag frequency.

7.3.3 Power System Design—On-site Generation

7.3.3.1 Reasons for Installing a Generator. Local generators are used for two distinctly different reasons:

1. Generating electricity locally can be cheaper than buying it from the utility. This holds especially for combined-heat-and-power (CHP) where the waste heat from the electricity generation is used in the industrial process. The total efficiency of the process is typically much higher than in conventional generator stations.
2. Having an on-site generator available increases the reliability of the supply as it can serve as a backup in case the supply is interrupted. Some large industrial plants have the ability to operate completely in island mode. Also hospitals, schools, government offices, etc., often have a standby generator to take over the supply when the public supply is interrupted.

Here we only consider the second situation, which might be an additional advantage next to the economic and environmental benefits of on-site generation. We first assess the effect of the generator on the availability. Suppose that the public supply has an availability of 98%. This might sound high, but an unavailability of 2% implies that there is no supply for 175 hours each year, or on average 29 minutes per day, or 40 4-hour interruptions per year. In other words, 98% availability is for many industrial customers unacceptably low. We assume that an on-site generator is installed which can take over all essential load. Suppose that the on-site generator has an availability of 90%. The supply is guaranteed as long as either the public supply or the generator are available. The methods introduced in Chapter 2 can be used to calculate the reliability of the overall system. The resulting availability is 99.8%, or an unavailability of 18 hours per year, four to five 4-hour interruptions per year. In case a further increase in reliability is needed, one can consider to install two or even three generator units. Each of these is assumed to be able to supply all the essential load. With two generators we obtain an unavailability of 2 hours per year; with three, the unavailability is only 10 minutes per year, neglecting all common-mode effects. As we saw in Chapter 2 the latter assumption is no longer valid for highly reliable systems. Any attempt to further increase the reliability by adding more generator units is unlikely to be successful. Emergency or standby generators are often started when an interruption of the public supply occurs. Instead of calculating unavailabilities it is more suitable to calculate interruption frequencies. Suppose that the public supply is interrupted 40 times per year. The failure to start of an emergency generator is typically somewhere between 1% and 5%. A value of 5% will reduce the number of interruptions from 40 per year to two per year. This assumes that the generator is always available. In reality one has to add another few percent unavailability due to maintenance and repair. The resulting interruption frequency will be around five per year. Again an industrial user is likely to opt for two units, which brings the interruption frequency down to less than one per year.

7.3.3.2 Voltage Sag Mitigating Effects. We saw in Section 4.2.4 and in Section 6.4 that a generator mitigates sags near its terminals. To mitigate sags the generator has to be on-line; an off-line generator will not mitigate any voltage sags. The effect of a generator on the sag magnitude was quantified in Fig. 4.26 and in (4.16). The latter equation is reproduced here:

$$\left(1 - V_{sag}\right) = \frac{Z_4}{Z_3 + Z_4}\left(1 - V_{pcc}\right) \tag{7.15}$$

with Z_3 the impedance between the generator/load bus and the pcc (typically the impedance of a distribution transformer) and Z_4 the (transient) impedance of the generator. If we further assume that $V_{pcc} = \frac{\mathcal{L}}{\mathcal{L}+1}$, with $\mathcal{L}$ the distance to the fault, and introduce $\zeta = \frac{Z_3}{Z_4}$, we get the following expression for the sag magnitude at the load bus as a function of the distance to the fault:

$$V_{sag} = 1 - \frac{1}{(1+\zeta)(1+\mathcal{L})} \tag{7.16}$$

This expression has been used to obtain the curves in Fig. 7.22: the sag magnitude as a function of distance is shown for different values of the impedance ratio ζ. A value $\zeta = 0$ corresponds to no generator; increasing ζ corresponds to increasing generator size or increasing transformer impedance. Consider a typical transformer impedance of 5% of

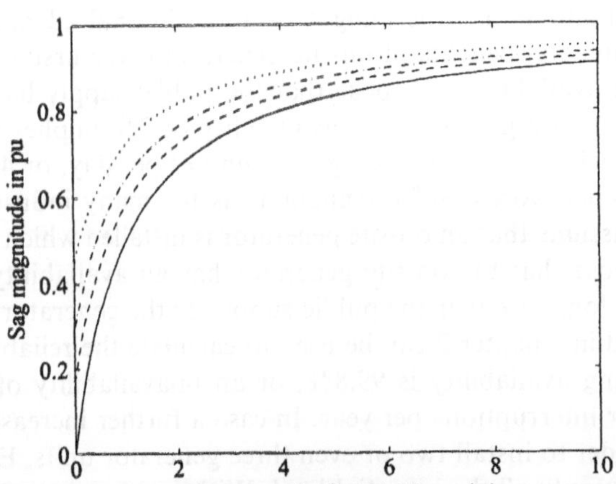

Figure 7.22 Sag magnitude versus distance for different generator sizes. The ratio between transformer and generator impedance used was 0 (solid line), 0.2 (dashed line), 0.4 (dash-dot line), and 0.8 (dotted line).

its rated power, and a typical generator transient impedance of 18%. For equal generator and transformer rating, we find $\zeta = 0.28$; $\zeta = 0.8$ corresponds to a generator size about three times the transformer rating, thus also about three times the size of the load. We saw before that generator capacity of more than three times the load does not have any improving effect on the reliability. It is thus unlikely that the generator capacity is more than three times the load. An exception are some CHP schemes where the industry sells considerable amounts of energy to the utility.

We see in Fig. 7.22 how the generator mitigates the voltage sag. The larger the generator, the more the reduction in voltage drop. From the expression for the sag magnitude as a function of distance, one can again derive an expression for the critical distance:

$$\mathcal{L}_{crit} = \frac{1}{(1+\zeta)(1-V)} - 1 \tag{7.17}$$

This expression has been used to calculate the critical distance for different generator sizes, resulting in Fig. 7.23. The curves are simply the inverse of the curves in Fig. 7.22. We see a reduction in critical distance for each value of the sag magnitude. Note that the installation of an on-site generator does not introduce any additional sags (with the exception of sags due to faults in or near the generator, but those are rare). The sag frequency for the different alternatives can thus be compared by comparing the critical distances.

A better picture of the reduction in sag frequency can be obtained from Fig. 7.24. The various curves show the percentage reduction in sag frequency between the site without generator and the site with a generator. Again three generator sizes have been compared. For small sag magnitudes the reduction in sag frequency is 100%; there are no sags left with these magnitudes. For higher magnitudes the relative reduction becomes less. This mitigation method works best for equipment which already has a certain level of immunity against sags.

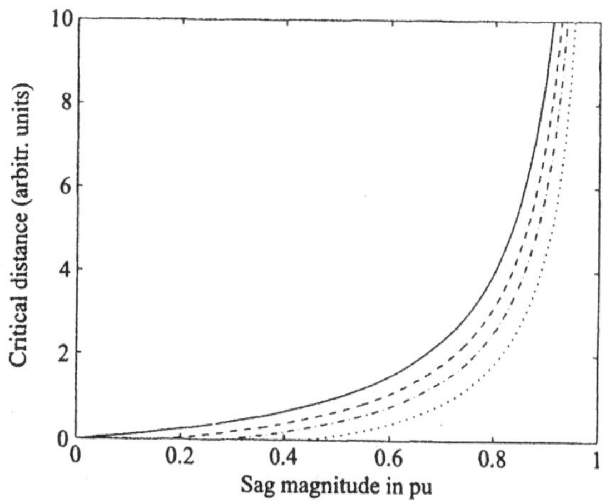

Figure 7.23 Critical distance versus magnitude for different generator sizes. The ratio between transformer and generator impedance used was 0 (solid line), 0.2 (dashed line), 0.4 (dash-dot line), and 0.8 (dotted line).

7.3.3.3 Island Operation. On-site generators are fairly common in large industrial and commercial systems. The on-site generation is operated in parallel with the public supply. When the public supply fails, the on-site generator goes into island operation. This "island" can consist of the whole load or part of the load. The latter situation is shown in Fig. 7.25. The island system should be made more reliable than

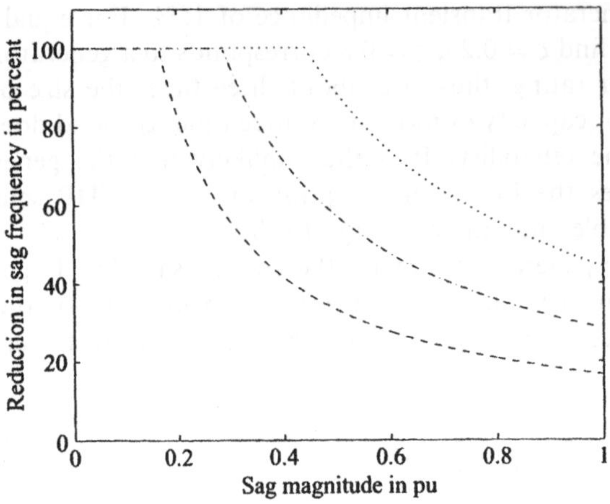

Figure 7.24 Reduction in sag frequency due to the installation of an on-site generator. The ratio between transformer and generator impedance used was 0.2 (dashed line), 0.4 (dash-dot line), and 0.8 (dotted line).

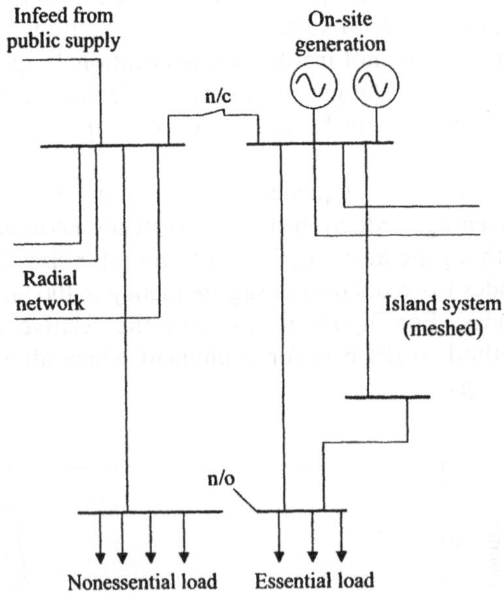

Figure 7.25 Industrial power system with islanding option.

the rest of the industrial distribution system (e.g., by using a meshed network and differential protection). The island system also serves as a backup for the rest of the industrial distribution system. A big problem in large industrial systems is that islanding cannot be tested. One has to wait for an interruption to occur to see if it works.

7.3.3.4 Emergency and Standby Generation. Emergency and standby generators are typically started the moment an interruption is detected. They come online between one second and one minute after the start of the interruption. Note that there is no technical difference between emergency generation and standby generation. The term "emergency generation" is used when there is a legal obligation to have a generator available; in all other cases the term "standby generation" is used [26]. When installing standby generation to improve voltage quality it is important that essential equipment can tolerate the short interruption due to the transfer to the standby generation. Standby generation is often used in combination with a small amount of energy storage supplying the essential load during the first few seconds of an interruption.

7.4 THE SYSTEM–EQUIPMENT INTERFACE

The interface between the system and the equipment is the most common place to mitigate sags and interruptions. Most of the mitigation techniques are based on the injection of active power, thus compensating the loss of active power supplied by the system. All modern techniques are based on power electronic devices, with the voltage-source converter being the main building block. Next we discuss the various existing and emerging technologies, with emphasis on the voltage-source converter. Terminology is still very confusing in this area, terms like "compensators," "conditioners," "controllers," and "active filters" are in use, all referring to similar kind of devices. In the remainder of this section, the term "controller" will be used, with reference to other terms in general use.

7.4.1 Voltage-Source Converter

Most modern voltage-sag mitigation methods at the system–equipment interface contain a so-called voltage-source converter. A voltage-source converter is a power electronic device which can generate a sinusoidal voltage at any required frequency, magnitude, and phase angle. We already saw the voltage-source converter as an important part of ac adjustable-speed drives. In voltage-sag mitigation it is used to temporarily replace the supply voltage or to generate the part of the supply voltage which is missing.

The principle of the voltage-source converter is shown in Fig. 7.26. A three-phase voltage-source converter consists of three single-phase converters with a common dc voltage. By switching the power electronic devices on or off with a certain pattern an ac voltage is obtained. One can use a simple square wave or a pulse-width modulated pattern. The latter gives less harmonics but somewhat higher losses. Details of the

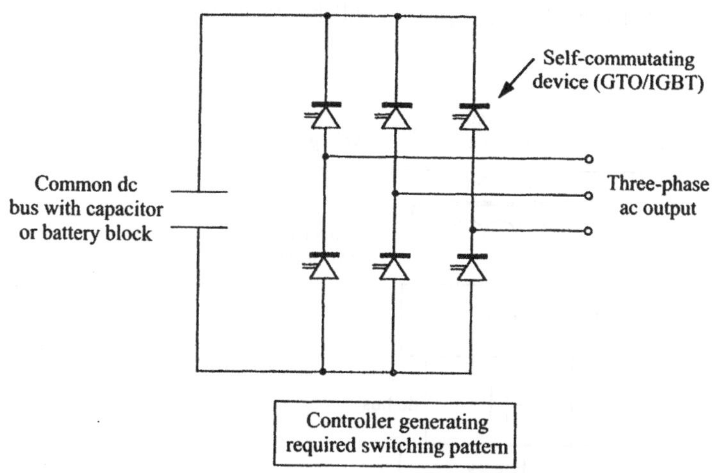

Figure 7.26 Three-phase voltage-source converter.

operation and control of the voltage-source converter can be found in most books on power electronics, e.g., [53], [55].

In circuit-theory models, the voltage-source converter can simply be modeled as an ideal voltage source. To assess the effect of this on voltages and currents, no knowledge is needed about the power electronic devices and the control algorithms. In the forthcoming sections the voltage-source converter is modeled as an ideal voltage source to analyze the mitigation effect of various configurations.

The same voltage-source converter technology is also used for so-called "Flexible AC Transmission Systems" or FACTS [180], [181] and for mitigation of harmonic distortion [179], [182], [183] and voltage fluctuations [170], [178]. In this chapter we will only discuss their use for mitigating voltage sags and interruptions. The whole set of power electronic solutions to power quality problems, including static transfer switches, active harmonic filters, and voltage control, is often referred to as "custom power" [184], [191].

7.4.2 Series Voltage Controllers—DVR

7.4.2.1 Basic Principle. The series voltage controller consists of a voltage-source converter in series with the supply voltage, as shown in Fig. 7.27. The voltage at the load terminals equals the sum of the supply voltage and the output voltage of the controller:

$$\overline{V}_{load} = \overline{V}_{cont} + \overline{V}_{sag} \tag{7.18}$$

A converter transformer is used to connect the output of the voltage-source converter to the system. A relatively small capacitor is present on dc side of the converter. The voltage over this capacitor is kept constant, by exchanging energy with the energy storage reservoir. The required output voltage is obtained by using a pulse-width modulation switching pattern. As the controller will have to supply active as well as reactive power, some kind of energy storage is needed. The term *Dynamic Voltage Restorer (DVR)* is commonly used instead of series voltage controller [184], [185]. In the DVRs that are currently commercially available large capacitors are used as a source of energy. Other potential sources are being considered: battery banks, superconducting coils, flywheels. We will for now assume that there is some kind of energy storage available. The various storage options will be discussed later.

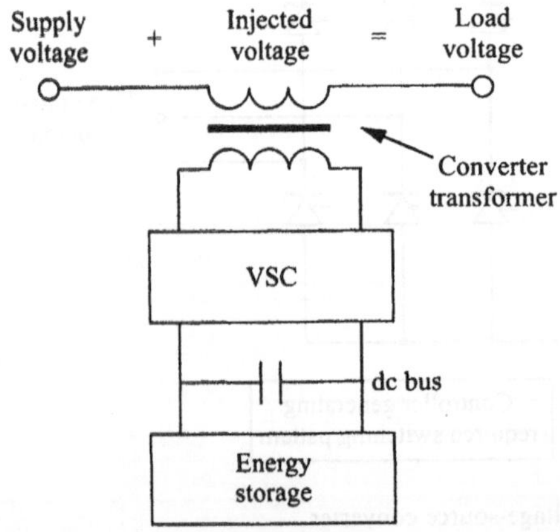

Figure 7.27 Series voltage controller.

The amount of energy storage depends on the power delivered by the converter and on the maximum duration of a sag. The controller is typically designed for a certain maximum sag duration and a certain minimum sag voltage. Some practical aspects of a series voltage controller are discussed in [174].

7.4.2.2 Active Power Injection. To assess the storage requirements we calculate the active power delivered by the controller, using the notation in Fig. 7.28. We assume that the voltage at the load terminals is 1 pu along the positive real axis:

$$\overline{V}_{load} = 1 + 0j \tag{7.19}$$

The load current is 1 pu in magnitude, with a lagging power factor $\cos\phi$:

$$\overline{I}_{load} = \cos\phi - j\sin\phi \tag{7.20}$$

The voltage sag at the system side of the controller has a magnitude V and phase-angle jump ψ:

$$\overline{V}_{sag} = V\cos\psi + jV\sin\psi \tag{7.21}$$

The complex power taken by the load is found from

$$P_{load} + jQ_{load} = \overline{V}_{load}\overline{I}^*_{load} = \cos\phi + j\sin\phi \tag{7.22}$$

The complex power taken from the system is

$$P_{sys} + jQ_{sys} = \overline{V}_{sag}\overline{I}^*_{load} = V\cos(\phi+\psi) + jV\sin(\phi+\psi) \tag{7.23}$$

The active power that needs to be generated by the controller is the difference between the active power taken from the system and the active part of the load:

$$P_{cont} = P_{load} - P_{sys} \tag{7.24}$$

This can be written as

$$P_{cont} = \left[1 - \frac{V\cos(\phi+\psi)}{\cos\phi}\right] \times P_{load} \tag{7.25}$$

For zero phase-angle jump we obtain the following simple expression for the active-power requirement of the controller:

$$P_{cont} = [1 - V]P_{load} \tag{7.26}$$

The active power requirement is linearly proportional to the drop in voltage. When phase-angle jumps are considered the relation is no longer linear and becomes dependent on the power factor also. To assess the effect of phase-angle jump and power factor, we have used the relations between sag magnitude and phase-angle jump as derived in Chapter 4. The active power requirement for different power factor and

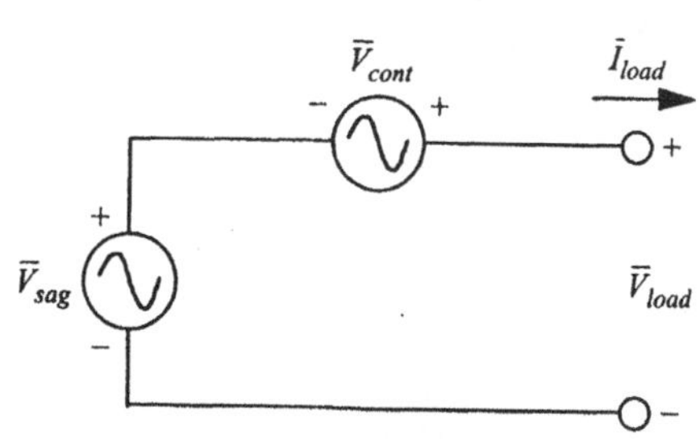

Figure 7.28 Circuit diagram with power system, series controller, and load.

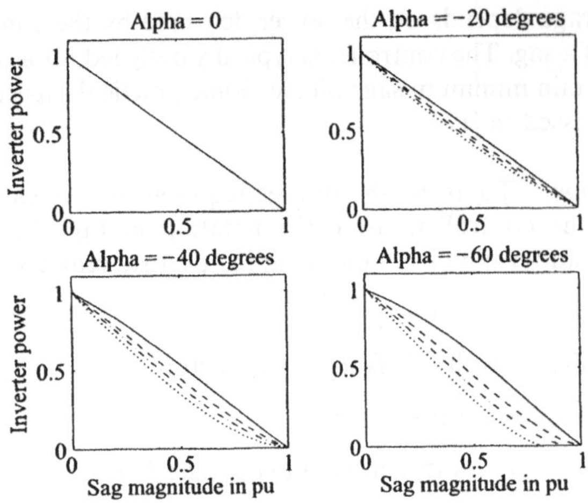

Figure 7.29 Active power requirement for a series voltage controller, for different impedance angles ($\alpha = 0, -20°, -40°, -60°$) and different lagging power factors: 1.0 (solid lines), 0.9 (dashed lines), 0.8 (dash-dot lines), 0.7 (dotted lines).

different phase-angle jump is shown in Fig. 7.29. Sag magnitude and phase-angle jump have been calculated as a function of the distance to the fault by using expressions (4.84) and (4.87). Magnitude and phase-angle jump were calculated for different values of the impedance angle and next filled in in (7.25) to obtain the active power requirement. The latter is plotted in Fig. 7.29 as a function of the sag magnitude V.

As shown in (7.26), the power factor of the load does not influence the active power requirements for sags without phase-angle jumps (upper left). For unity power factor, the phase-angle jump somewhat influences the active power requirement. This is mainly due to the voltage over the controller no longer being equal to $1 - V$. For decreasing power factor and increasing phase-angle jump, the active power requirement becomes less. One should not conclude from this that a low power factor is preferable. The lower the power factor, the larger the load current for the same amount of active power, thus the higher the required rating of the converter.

The reduction in active power requirement with increasing (negative) phase-angle jump is explained in Fig. 7.30. Due to the phase-angle jump the voltage at system side of the controllers becomes more in phase with the load current. The amount of active

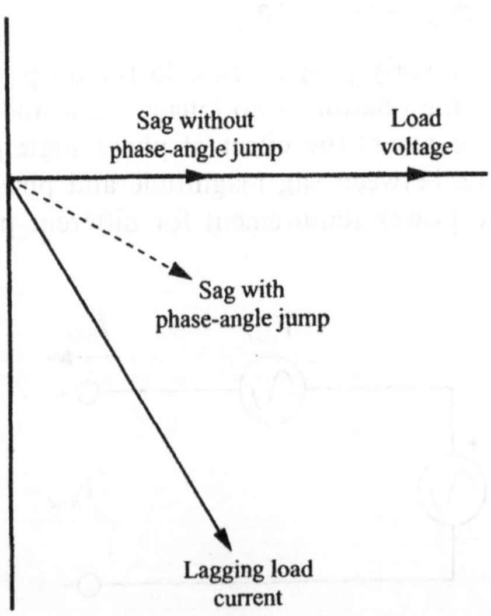

Figure 7.30 Phasor diagram for a series voltage controller. Dashed line: with negative phase-angle jump. Solid line: without phase-angle jump.

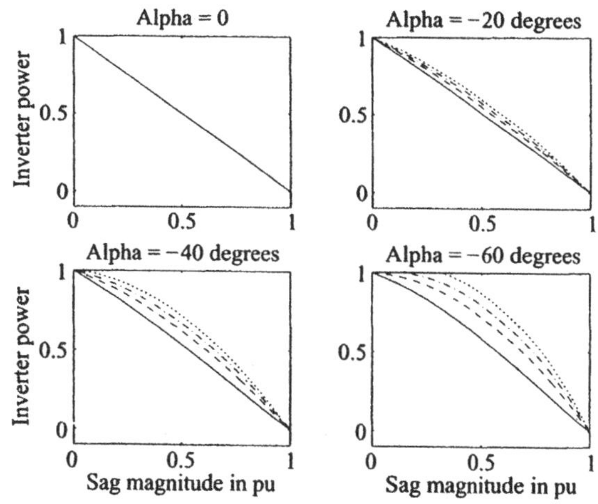

Figure 7.31 Active power requirement for a series voltage controller, for different impedance angles ($\alpha = 0$, $-20°$, $-40°$, $-60°$) and different leading power factors: 1.0 (solid lines), 0.9 (dashed lines), 0.8 (dash-dot lines), 0.7 (dotted lines).

power taken from the supply thus increases and the active power requirement of the controller is reduced. This holds for a negative phase-angle jump and a lagging power factor. For a leading power factor, a negative phase-angle jump increases the active power requirements, as shown in Fig. 7.31.

7.4.2.3 Three-Phase Series Voltage Controllers. The series controllers currently commercially available consist of three single-phase converters with a common dc capacitor and storage reservoir. The power taken from the storage reservoir is the sum of the power in the three phases. For each of the phases, (7.25) can be used to calculate the active power. For a three-phase balanced sag (i.e., a sag due to a three-phase fault) the same amount of power is injected in each phase. The power requirement is multiplied by three. But also the active power taken by the load is three times as large, so that (7.25) still holds, with the difference that P_{load} is the total load in the three phases.

To consider the power requirements for three-phase unbalanced sags, we write (7.25) in a somewhat different form. Let the (complex) remaining voltage (the sag magnitude) be $\overline{V}$, so that the voltage injected by the controller is $1 - \overline{V}$. The load current is $e^{-j\phi}$, which gives for the complex power delivered by the controller:

$$\overline{S} = (1 - \overline{V})e^{j\phi} \tag{7.27}$$

Consider a three-phase unbalanced sag of type C: two phases down in voltage; one phase not affected. To calculate the injected power in phase b, we apply the same line of thought as leading to (7.27). The load voltage in phase b is

$$\overline{V}_{load} = -\frac{1}{2} - \frac{1}{2}j\sqrt{3} \tag{7.28}$$

The complex voltage during the sag is

$$\overline{V}_{sag} = -\frac{1}{2} - \frac{1}{2}j\overline{V}_{char}\sqrt{3} \tag{7.29}$$

with $\overline{V}_{char}$ the complex characteristic voltage of the sag. The voltage injected by the controller is the difference between the load voltage and the sag voltage:

$$\overline{V}_{cont} = \overline{V}_{load} - \overline{V}_{sag} = -\frac{1}{2}j(1 - \overline{V}_{char})\sqrt{3} \tag{7.30}$$

The load current in phase b is shifted over 120° compared to the current in phase a:

$$\overline{I}_{load} = e^{-j\phi}\left(-\frac{1}{2}-\frac{1}{2}j\sqrt{3}\right) \tag{7.31}$$

The complex injected power in phase b is

$$\overline{S}_b = -\frac{1}{2}j(1-\overline{V}_{char})\sqrt{3}e^{j\phi}\left(-\frac{1}{2}+\frac{1}{2}j\sqrt{3}\right) \tag{7.32}$$

For phase c we find

$$\overline{V}_{load} = -\frac{1}{2}+\frac{1}{2}j\sqrt{3} \tag{7.33}$$

$$\overline{V}_{sag} = -\frac{1}{2}+\frac{1}{2}j\overline{V}_{char}\sqrt{3} \tag{7.34}$$

$$\overline{V}_{cont} = +\frac{1}{2}j(1-\overline{V}_{char})\sqrt{3} \tag{7.35}$$

$$\overline{I}_{load} = e^{-j\phi}\left(-\frac{1}{2}+\frac{1}{2}j\sqrt{3}\right) \tag{7.36}$$

$$\overline{S}_c = \frac{1}{2}j(1-\overline{V}_{char})\sqrt{3}e^{j\phi}\left(-\frac{1}{2}-\frac{1}{2}j\sqrt{3}\right) \tag{7.37}$$

Adding the complex powers in phase b and phase c gives the total injected power (the voltage in phase a is not affected by the sag):

$$\overline{S}_{cont} = \frac{3}{2}(1-\overline{V}_{char})e^{j\phi} \tag{7.38}$$

This is identical to (7.27), except for the factor $\frac{3}{2}$. Repeating the calculations for a three-phase unbalanced sag of type D, gives exactly the same injected power as for a type C sag. For the analysis of three-phase unbalanced sags we have neglected the zero-sequence component. This is an acceptable approximation at the terminals of end-user equipment, but not always in medium-voltage distribution, where DVRs are currently being installed. Adding a zero-sequence voltage to all three-phase voltages in the above reasoning will lead to an additional term in the complex power expressions for the three phases. These additional terms add to zero, so that the zero-sequence voltage does not affect the total active power demand of the series controller.

The power injected during a three-phase sag is three times the power injected in one phase. By comparing (7.38) with (7.27) we can conclude that the power injected during a sag of type C or type D is half the power injected during a balanced sag with the same characteristic magnitude, phase-angle jump, and duration.

7.4.2.4 Single-Phase Series Voltage Controllers. For single-phase controllers, the actual voltage in one phase (the voltage at the equipment terminals in the terminology from Chapter 4) determines the amount of active power which needs to be injected. This is not only determined by the characteristic magnitude but also by the type of sag and the phase to which the controller is connected.

What matters to a single-phase controller are the injected powers in each of the three phases, i.e. the real part of $\overline{S}_b$ in (7.32) and of $\overline{S}_c$ in (7.37). These calculations have been performed for three-phase unbalanced sags of type C and type D, resulting in Figs. 7.32 and 7.33, respectively. For each sag type only two phases have been plotted: the two phases with the deep sag for type C, and the two phases with the shallow sag for type D. The third phase for a type C sag does not require any injected power; the active power requirements for the third phase of a type D sag are identical to (7.25). Both in Fig. 7.32 and in Fig. 7.33 the injected power has been plotted for two values of the impedance angle (0 and 30°) and four values of the power factor of the load current (1.0, 0.9, 0.8, 0.7). We can conclude from the figures that the power factor has significant influence on the power injection. The characteristic phase-angle jump makes that the two phases behave slightly differently, but does not change the overall picture.

For a single-phase controller, the characteristic voltage does not have much practical meaning. Therefore the active power requirements have been plotted in a different way in Figs. 7.34 and 7.35. The horizontal axis is the absolute value of the complex voltage during the sag; in other words, the sag magnitude at the equipment terminals. The different curves in each subplot give the relation between sag magnitude and injected power for each of the phases of a type C or type D three-phase unbalanced sag. This leads to a maximum of five curves, two from a type C sag, three from a type D sag. We see that there is no general relation between the injected power and the sag

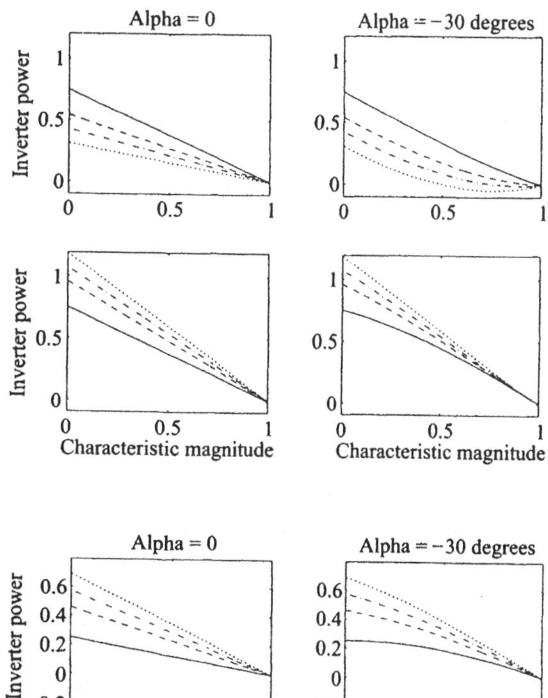

Figure 7.32 Active power requirements for a single-phase series voltage controller, for two phases of a type C unbalanced sag, for impedance angle zero (left) and −30° (right). Power factor 1.0 (solid lines), 0.9 (dashed), 0.8 (dash-dot), 0.7 (dotted).

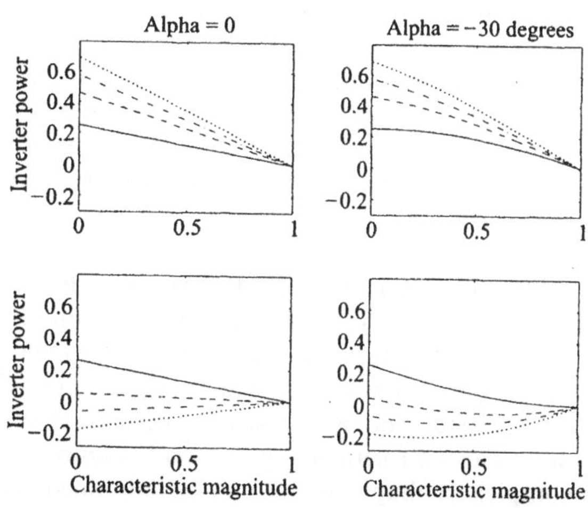

Figure 7.33 Active power requirements for a single-phase series voltage controller, for two phases of a type D unbalanced sag, for impedance angle zero (left) and −30° (right). Power factor 1.0 (solid lines), 0.9 (dashed), 0.8 (dash-dot), 0.7 (dotted).

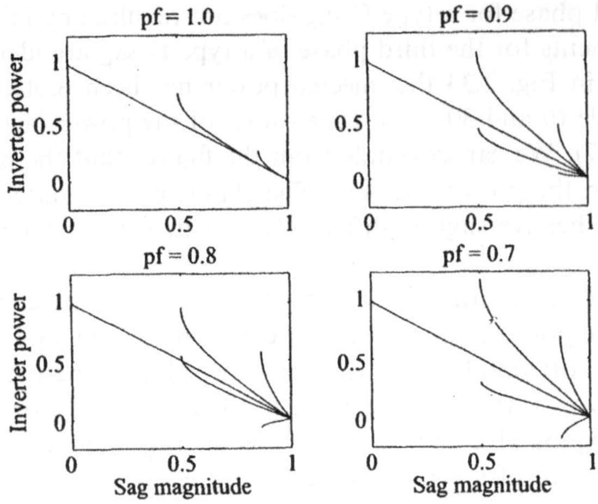

Figure 7.34 Active power requirements for a single-phase series voltage controller as a function of the sag magnitude—for zero impedance angle and four values of the power factor of the load current.

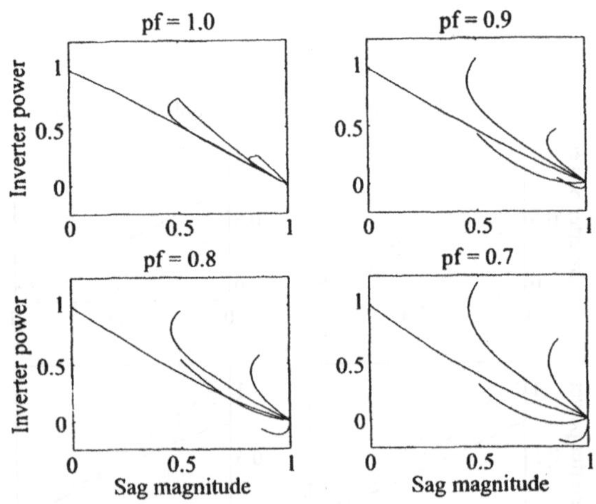

Figure 7.35 Active power requirements for a single-phase series voltage controller as a function of the sag magnitude—for an impedance angle equal to −30° and four values of the power factor of the load current.

magnitude, especially for small values of the power factor. Note also that for low power factor, a zero-magnitude sag is not the one with the highest active power requirements.

Figures 7.34 and 7.35 have been reproduced in Figs. 7.36 and 7.37 with yet another horizontal axis. The active power requirements have been plotted as a function of the absolute value of the complex missing voltage (see Section 4.7.1). We see also that the missing voltage does not uniquely determine the injected power. The load power factor and, to a lesser extent, the characteristic phase-angle jump influence the injected power as well and should thus be considered in dimensioning the energy storage of the controller.

7.4.2.5 Effect of the Voltage Rating. The voltage rating of the voltage-source converter directly determines the maximum voltage (magnitude) which can be injected. This in turn determines against which sags the load is protected. In the above calculations, it was assumed that the load voltage would remain exactly at its pre-event value. This is not strictly necessary: small voltage drop and some phase-angle jump can be tolerated by the load. Figure 7.38 shows how the protected area of the complex (voltage) plane can be obtained for a given voltage rating. The voltage

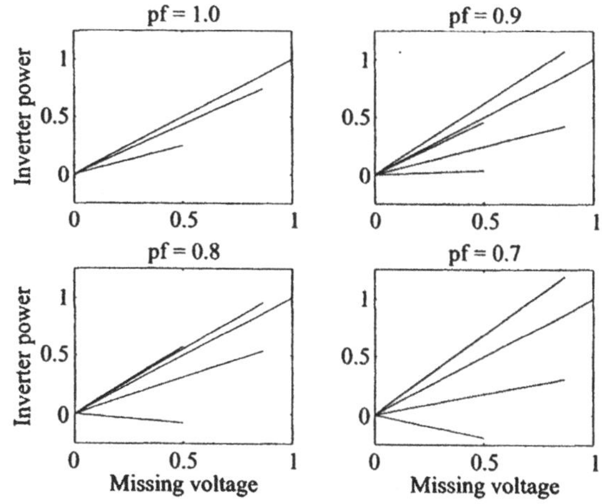

Figure 7.36 Active power requirements for a single-phase series voltage controller as a function of the missing voltage—for zero impedance angle and four values of the power factor of the load current.

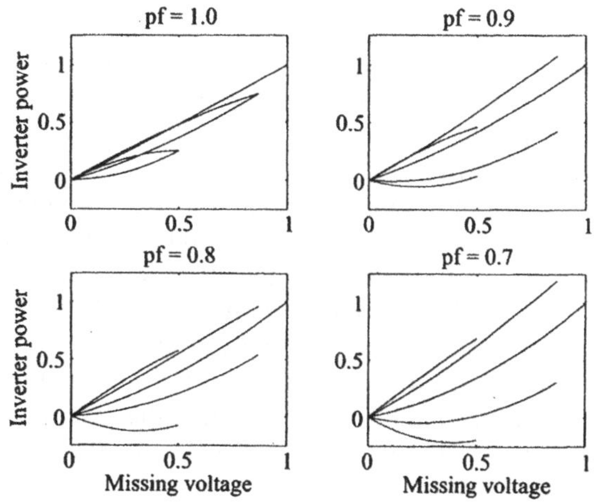

Figure 7.37 Active power requirements for a single-phase series voltage controller as a function of the missing voltage—for an impedance angle equal to −30° and four values of the power factor of the load current.

rating of the voltage-source converter is translated to the same base as the load voltage. The actual rating depends on the turns ratio of the converter transformer.

The voltage tolerance, as indicated in the figure, gives the lowest voltage magnitude and the largest phase-angle jump for which the load can operate normally. The sag voltage should not deviate more than the maximum injectable voltage (i.e., the voltage rating of the converter) from the voltage tolerance. This leads to the dashed curve, which gives magnitude and phase-angle jump of the worst sags that can be mitigated by the controller; i.e., the voltage tolerance of the combination of load and controller. The possible range of sags is indicated by a thick solid line. The range of sags can either be the range for a variety of supplies, like in Fig. 4.96, or for a specific supply, like in Fig. 4.108. It is very well possible to cover the whole range of possible sags by choosing a large enough voltage rating. However, the number of sags decreases for lower magnitudes, and the costs of the controller increase with increasing voltage rating. Therefore the series controllers currently in use have a minimum voltage of typically 50%, so that sags with a magnitude below 50% of nominal are not protected. With reducing costs of power electronics, it is very well possible that future controllers will cover the whole range of possible sags.

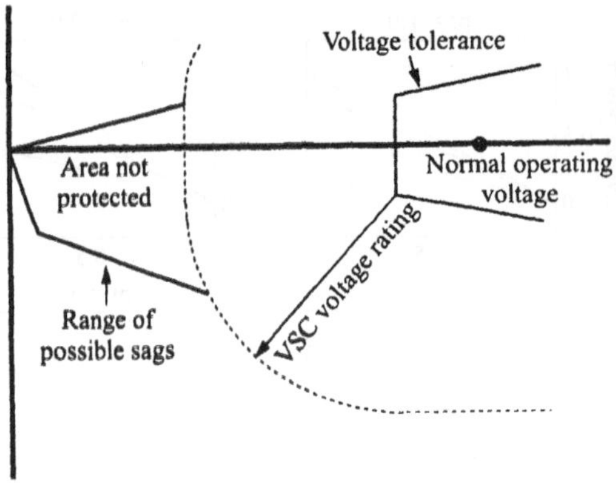

Figure 7.38 Part of the complex (voltage) plane protected by a series voltage controller with the indicated voltage rating.

7.4.2.6 Effect of the Storage Capacity. The voltage rating of the controller determines which range of magnitude and phase-angle jump of sags can be mitigated. For a given magnitude and phase-angle jump the active power requirement is found from (7.25). The active power requirement and the amount of energy storage determine the longest sag duration which can be mitigated.

During the design of a series controller, a sag magnitude and a sag duration are chosen. The sag magnitude gives the voltage rating, the sag duration gives the required storage capacity. Together they determine the "design point" in Fig. 7.39. The voltage tolerance of the load without controller is shown as a dashed line (in this example the voltage tolerance of the load is 200 ms, 90%). The influence of the phase-angle jump is neglected here. (Including the phase-angle jump would give a range of voltage-tolerance curves, both with and without the controller.) Any sag with a magnitude above the design magnitude and with a duration less than the design duration, will be mitigated by the controller: i.e., the resulting load voltage will be above the voltage-tolerance curve of the load. Sags longer than the design duration are only tolerated if they do not deplete the storage capacity. Neglecting the phase-angle jump, we can use (7.26) for the injected power:

$$P_{cont} = (1 - V)P_{load} \tag{7.39}$$

The energy needed to ride through a sag of magnitude V and duration T is

$$\mathcal{E} = (1 - V)TP_{load} \tag{7.40}$$

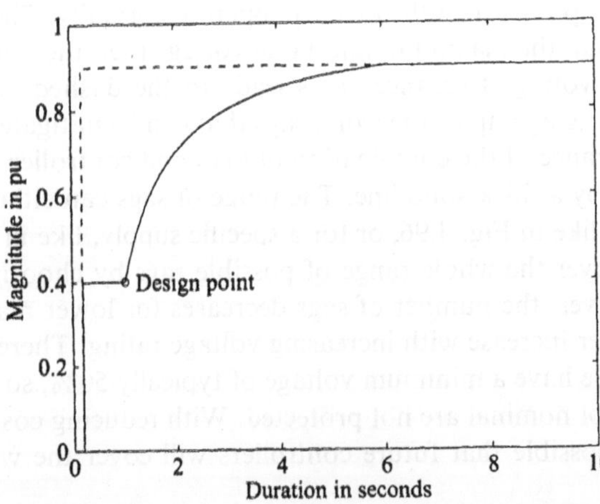

Figure 7.39 Voltage-tolerance curve without (dashed line) and with (solid line) series voltage controller. The design point gives the lowest magnitude and the longest duration which the load-controller combination is able to tolerate.

Let (T_0, V_0) be the design point. The available energy storage is

$$\mathcal{E}_{avail} = (1 - V_0)T_0P_{load} \tag{7.41}$$

The minimum sag magnitude V_{min} for a duration T is found from

$$\mathcal{E}_{avail} = (1 - V_{min})TP_{load} \tag{7.42}$$

This gives the following expression for the voltage-tolerance curve:

$$V_{min} = 1 - (1 - V_0)\frac{T_0}{T} \tag{7.43}$$

This is is shown in Fig. 7.39 as the curve from the design point toward the right and upward. The voltage-tolerance curve of the load with controller gets its final shape by realizing that any sag tolerated without controller can also be tolerated with controller. The area between the curves is the gain in voltage tolerance due to the controller. To assess the reduction in number of trips, a sag density chart is needed.

7.4.2.7 Interruptions. A series voltage controller does not function during an interruption. It needs a closed path for the load current, which is not always present during an interruption. If there is load present upstream of the controller and downstream of the circuit breaker causing the interruption, this load will form a path through which the converter current can close, as shown Fig. 7.40.

The series controller will aim to keep the voltage V_1 and thus the current I_{load} constant. The effect is that the current I_{load} is forced into the upstream load impedance Z_2 leading to a voltage $V_2 = Z_2I_{load}$ on system side of the controller, but in opposite phase compared to V_1. Using $V_1 = Z_1I_{load}$ we get

$$V_2 = \frac{Z_2}{Z_1}V_1 \tag{7.44}$$

with Z_1 the impedance of the load to be protected by the controller. If the upstream load is smaller than the protected load, $Z_2 > Z_1$, this could lead to dangerous overvoltages. With the existing devices this effect is limited in two ways:

- The voltage difference over the controller is $V_1 + V_2$ which is significantly larger than 1 pu if $Z_2 > Z_1$. For a controller with a maximum output voltage of 0.5 pu (a typical value) the resulting voltage over the upstream load can never be more than 0.5 pu.

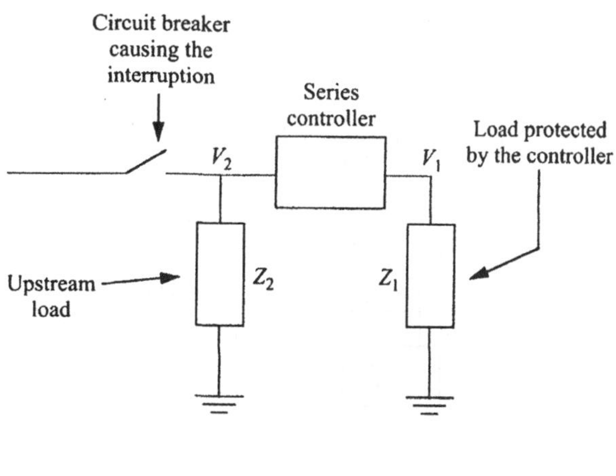

Figure 7.40 Series voltage controller with upstream load during an interruption.

- The energy reservoir is limited, so that this overvoltage will disappear within a few seconds. Note that both the protected load and the upstream load will deplete the energy reservoir.

This could, however, become a problem in the future when the rating of voltage controllers increases, both in injected voltage and in stored energy. The effect of the sudden inversion of the voltage on the upstream load should be studied as well.

7.4.3 Shunt Voltage Controllers—StatCom

A shunt-connected voltage controller is normally not used for voltage sag mitigation but for limiting reactive power fluctuations or harmonic currents taken by the load. Such a controller is commonly referred to as a "Static Compensator" or "StatCom." Alternative terms in use are "Advanced Static Var Compensator" (ASVC) and "Static Condensor" (StatCon). A StatCom does not contain any active power storage and thus only injects or draws reactive power. Limited voltage sag mitigation is possible with the injection of reactive power only [57], [157], [210], but active power is needed if both magnitude and phase angle of the pre-event voltage need to be kept constant.

The principle of a shunt voltage controller is shown in Fig. 7.41. The actual controller has the same configuration as the series controller. But instead of injecting the voltage difference between the load and the system, a current is injected which pushes up the voltage at the load terminals, in a similar way to the sag mitigation by a generator discussed in Section 7.2.

The circuit diagram used to analyze the controller's operation is shown in Fig. 7.42. The load voltage during the sag can be seen as the superposition of the voltage due to the system and the voltage change due to the controller. The former is the voltage as it would have been without a controller present, the latter is the change due to the injected current.

Assume that the voltage without controller is

$$\overline{V}_{sag} = V\cos\psi + jV\sin\psi \tag{7.45}$$

The load voltage is again equal to 1 pu:

$$\overline{V}_{load} = 1 + 0j \tag{7.46}$$

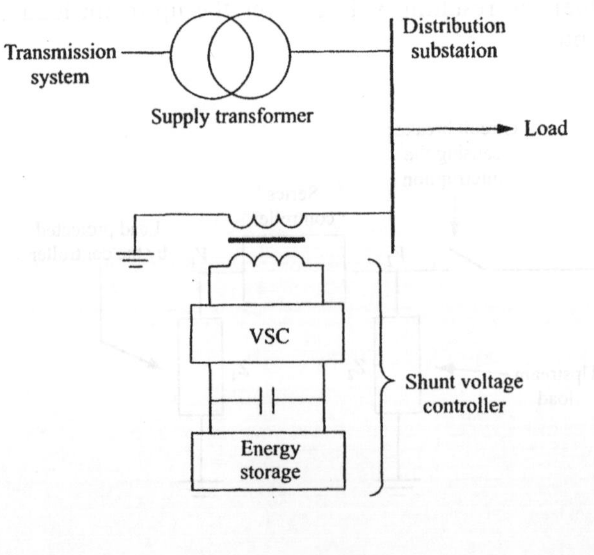

Figure 7.41 Shunt voltage controller.

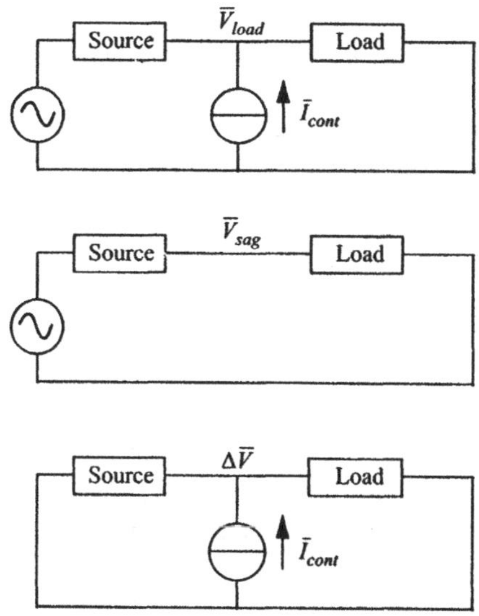

Figure 7.42 Circuit diagram with power system, series controller, and load. Full circuit (top), voltages without controller (center), effect of the controller (bottom).

The required change in voltage due to the injected current is the difference between the load voltage and the sag voltage:

$$\Delta\overline{V} = 1 - V\cos\psi - jV\sin\psi \tag{7.47}$$

This change in voltage must be obtained by injecting a current equal to

$$\overline{I}_{cont} = P - jQ \tag{7.48}$$

with P the active power and Q the reactive power injected by the controller. The active power will determine the requirements for energy storage. Let the impedance seen by the shunt controller (source impedance in parallel with the load impedance) be equal to

$$\overline{Z} = R + jX \tag{7.49}$$

The effect of the injected current is a change in voltage according to

$$\Delta\overline{V} = \overline{I}_{cont}\overline{Z} = (R + jX)(P - jQ) \tag{7.50}$$

The required voltage increase (7.47) and the achieved increase (7.50) have to be equal. This gives the following expression for the injected complex power:

$$P - jQ = \frac{1 - V\cos\psi - jV\sin\psi}{R + jX} \tag{7.51}$$

Splitting the complex power in a real and an imaginary part, gives expressions for active and reactive power:

$$P = \frac{R(1 - V\cos\psi) - VX\sin\psi}{R^2 + X^2} \tag{7.52}$$

$$Q = \frac{RV\sin\psi + X(1 - V\cos\psi)}{R^2 + X^2} \tag{7.53}$$

The main limitation of the shunt controller is that the source impedance becomes very small for faults at the same voltage level close to the load. Mitigating such sags through a shunt controller is impractical as it would require very large currents. We therefore only consider faults upstream of the supply transformer. The minimum value of the source impedance is the transformer impedance. One can think of this configuration as a dedicated supply to a sensitive load (e.g., an automobile plant), where the task of the controller is to mitigate sags originating upstream of the transformer.

The results of some calculations for this configuration are shown in Figs. 7.43 and 7.44. Four different values for the source impedance (transformer impedance) have been used: 0.1, 0.05, 0.033, and 0.025 pu. For the load impedance a value of 1 pu resistive has been chosen. For a 0.05 pu source impedance, the fault level is 20 times the load power. Fault levels of 10 to 40 times the load are typical in distribution systems.

Figure 7.43 shows the amount of active power injected by the controller to maintain the voltage at its pre-event value. We see that for zero impedance angle the active power requirement is independent of the source impedance. This does not hold in general, but only for this specific case with a pure reactance in parallel with a pure resistance. For increasing impedance angle we see an increase in active power, especially for smaller values of the source impedance. The reactive power shown in Fig. 7.44 is rather independent of the impedance angle. The reactive power requirements decrease significantly with increasing source impedance. As the (reactive) source impedance

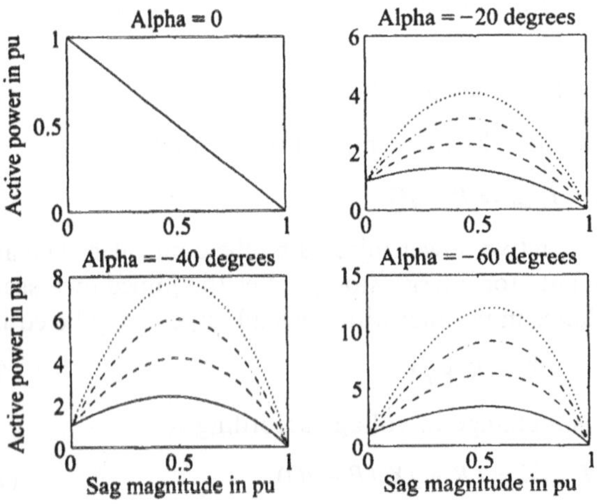

Figure 7.43 Active power injected by a shunt voltage controller, for different impedance angles (0, −20° −40°, −60°) and different source impedances: 0.1 pu (solid line), 0.05 pu (dashed line), 0.033 pu (dash-dot line), 0.025 pu (dotted line).

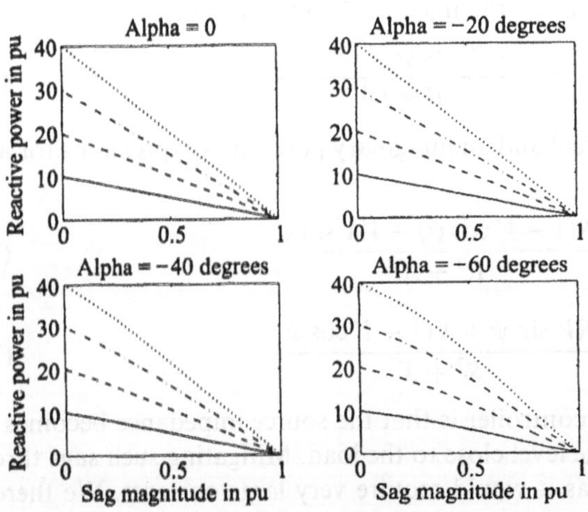

Figure 7.44 Reactive power injected by a shunt voltage controller, for different impedance angles (0, −20°, −40°, −60°) and different source impedances: 0.1 pu (solid line), 0.05 pu (dashed line), 0.033 pu (dash-dot line), 0.025 pu (dotted line).

increases, less injected current is needed to get the same change in voltage. Note the difference in vertical scale between Figs 7.43 and 7.44. The reactive power exceeds the active power injected in all shown situations.

The current rating of the controller is determined by both active and reactive power. From (7.52) and (7.53) we find for the absolute value of the injected current:

$$I_{cont} = \sqrt{\frac{1 - 2V\cos\psi + V^2}{R^2 + X^2}} \tag{7.54}$$

We see that an increasing phase-angle jump (increasing ψ, decreasing $\cos\psi$) increases the current magnitude. The current magnitude is plotted in Fig. 7.45 in the same format as the active power in Fig. 7.43 and the reactive power in Fig. 7.44.

Comparing Fig. 7.45 with Fig. 7.44 shows that the current magnitude is mainly determined by the reactive power. Like the reactive power, the current magnitude is only marginally affected by the phase-angle jump.

The large increase in active power injected with increasing phase-angle jump is explained in Fig. 7.46. The injected voltage is the required voltage rise at the load due to

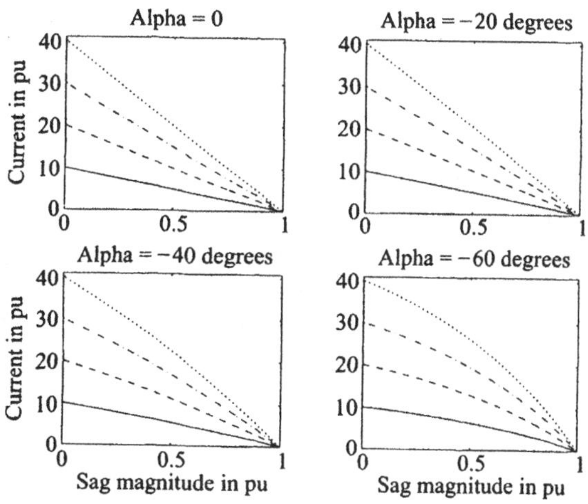

Figure 7.45 Magnitude of the current injected by a shunt voltage controller, for different impedance angles (0, −20°, −40°, −60°) and different source impedances: 0.1 pu (solid line), 0.05 pu (dashed line), 0.033 pu (dash-dot line), 0.025 pu (dotted line).

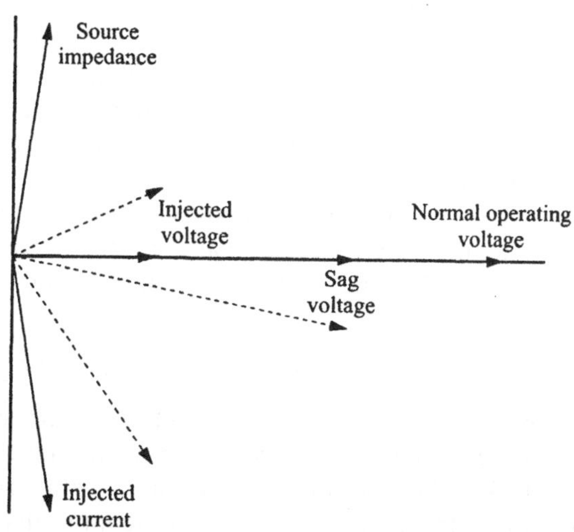

Figure 7.46 Phasor diagram for shunt voltage controller. Solid lines: without phase-angle jump. Dashed lines: with phase-angle jump.

the injection of a current into the source impedance. This injected voltage is the difference between the normal operating voltage and the sag voltage as it would be without controller. The injected current is the injected voltage divided by the source impedance. In phasor terms: the argument (angle, direction) of the injected current is the argument of the injected voltage minus the argument of the source impedance. The source impedance is normally mainly reactive. In case of a sag without phase-angle jump, the injected current is also mainly reactive. A phase-angle jump causes a rotation of the injected voltage as indicated in the figure. This leads to a rotation of the injected current away from the imaginary axis. From the figure it becomes obvious that this will quickly cause a serious increase in the active part of the current (i.e., the projection of the current on the load voltage). The change in the reactive part of the current is small, so is the change in current magnitude.

7.4.3.1 Disadvantages of the Shunt Controller. It is clear from the above reasoning that the main disadvantage of the shunt controller is its high active power demand. In case of a large load with a dedicated supply from a transmission network, a shunt controller might be feasible. Voltage sags in transmission networks show smaller phase-angle jumps, and the transformer losses are very small. The latter have not been taken into consideration in the above calculations, as they are rarely more than a few percent of the load. If the load is supplied through an underground cable network, these losses could dominate the active power requirement of the controller. Another disadvantage of the shunt controller is that it not only increases the voltage for the local load but for all load in the system. Again for a load with a dedicated supply through a large transformer, this effect is small, but for a load fed from a distribution feeder with many other customers it is not feasible to install a shunt controller. In case of a load fed from a distribution feeder, the controller will not be able to mitigate sags originating at distribution level. The source impedance during the sag will simply be too small to enable any serious increase in voltage.

The behavior of the shunt voltage controller during an interruption depends on the amount of load involved in the interruption. When the supply is interrupted, the injected current closes through the load, and the (active and reactive) power demands are formed by the total load involved in the interruption. If this is only the load to be protected, the controller will have no problem providing this power. If a lot more load is interrupted the controller will probably reach its current limits or its energy reservoir will be depleted very fast.

If the controller is able to maintain the load during the interruption, synchronization problems can occur when the voltage comes back. If the supply voltage differs significantly in phase with the voltage generated by the controller, large currents will start to flow leading to relay tripping and/or equipment damage. A phase difference of 60° gives an rms voltage of 1 pu over the terminals of the recloser. A phase difference of 180° gives 2 pu over the terminals. Consider that the nominal system frequency is 60 Hz and that the voltage comes back after 3 seconds. If we want to limit the angular difference to 30°, the relative error in frequency should not be more than:

$$\frac{30^\circ}{3\,\text{s} \times 60\,\text{cycles/s} \times 360\,^\circ/\text{cycle}} = 5 \times 10^{-4} \tag{7.55}$$

From this it follows that the frequency needs to be between 59.97 and 60.03 Hz. To operate the voltage-source converter within this frequency range is not a problem: modern clocks achieve accuracies which are several orders of magnitude better than this. But the system frequency can easily deviate more than 0.03 Hz from its nominal value.

The main advantage of a shunt controller is that it can also be used to improve the current quality of the load. By injecting reactive power, the power factor can be kept at unity or voltage fluctuations due to current fluctuations (the flicker problem) can be kept to a minimum. The shunt controller can also be used to absorb the harmonic currents generated by the load. In case such a controller is present, it is worth considering the installation of some energy storage to mitigate voltage sags. It will be clear from the previous chapters that a stochastic assessment of the various options is needed.

7.4.4 Combined Shunt and Series Controllers

The series controller, as discussed before, uses an energy storage reservoir to power part of the load during a voltage sag. We saw that the series controller cannot mitigate any interruptions, and that it is normally not designed to mitigate very deep sags (much below 50% of remaining voltage). There is thus normally some voltage remaining in the power system. This voltage can be used to extract the required energy from the system. A series-connected converter injects the missing voltage, and a shunt-connected converter takes a current from the supply. The power taken by the shunt controller must be equal to the power injected by the series controller. The principle is shown in Fig. 7.47. Series- and shunt-connected converters have a common dc bus. The change in stored energy in the capacitor is determined by the difference between the power injected by the series converter and the power taken from the supply by the shunt converter. Ensuring that both are equal minimizes the size of the capacitance.

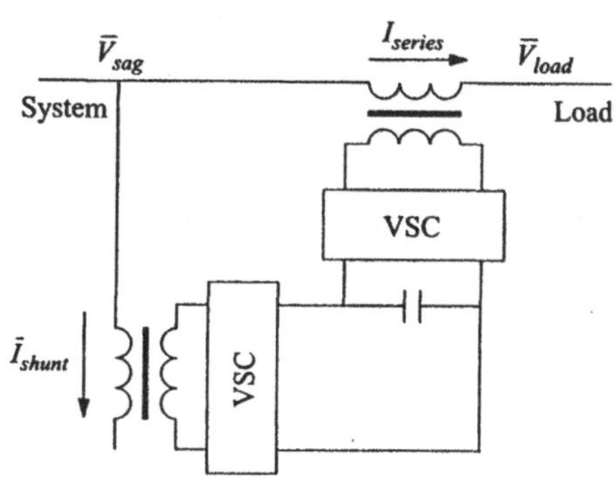

Figure 7.47 Shunt-series-connected voltage controller: the shunt-connected converter is placed on system side of the series controller.

7.4.4.1 Current Rating. The active power taken from the supply by the shunt-connected converter is

$$P_{shunt} = \mathcal{R}\{\overline{V}_{sag}\overline{I}^{*}_{shunt}\} \tag{7.56}$$

We assume that the shunt-connected converter takes a current from the supply with magnitude I_{shunt} and in phase with the system voltage

$$\overline{I}_{shunt} = I_{shunt}\cos\psi + jI_{shunt}\sin\psi \tag{7.57}$$

where ψ is the phase-angle jump of the sag. Taking the current in phase with the system voltage minimizes the current amplitude for the same amount of active power. The active power taken from the supply is

$$P_{shunt} = VI_{shunt} \tag{7.58}$$

with V the sag magnitude. The active power injected by the series controller was calculated before, (7.25):

$$P_{series} = \left[1 - \frac{V\cos(\phi + \psi)}{\cos\phi}\right] P_{load} \tag{7.59}$$

The power taken by the shunt-connected converter P_{shunt} should be equal to the power injected by the series-connected converter P_{series}. This gives the following expression for the magnitude of the shunt current:

$$I_{shunt} = \left[\frac{1}{V} - \frac{\cos(\phi + \psi)}{\cos\phi}\right] P_{load} \tag{7.60}$$

The results of this equation are shown in Fig. 7.48 in the same format and with the same parameter values as before (e.g., Fig. 7.29). The magnitude of the shunt current has been plotted for values up to 4 pu, i.e. four times the active part of the load current. The influence of phase-angle jump and power factor is similar to their influence on the active power as shown in Fig. 7.29. But the overriding influence on the shunt current is the sag magnitude. The less voltage remains in the system, the more current is needed to get the same amount of power. As the power requirement increases with decreasing system voltage, the fast increase in current for decreasing voltage is understandable.

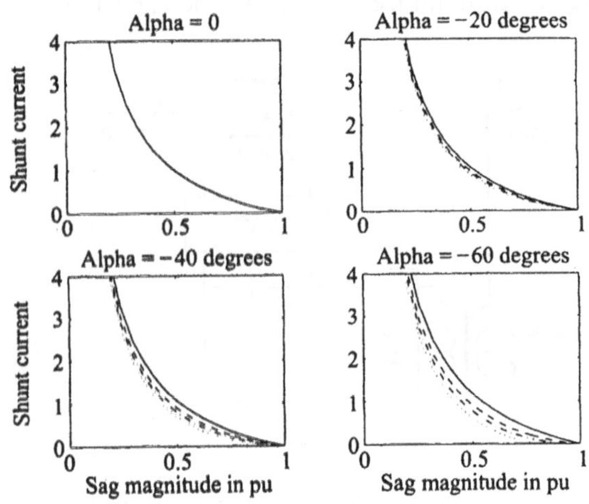

Figure 7.48 Shunt current for a shunt-series voltage controller, for different impedance angles (0, −20°, −40°, −60°) and different leading power factors: 1.0 (solid lines), 0.9 (dashed lines), 0.8 (dash-dot lines), 0.7 (dotted lines).

7.4.4.2 Shunt Converter on Load Side. Figure 7.49 again shows a shunt-series controller. The difference with Fig. 7.47 is that the shunt current is taken off the load voltage.

To assess the effect of this, we again calculate the requirements for the shunt and series currents. We use the same notation as before:

$$\overline{V}_{load} = 1 + 0j \tag{7.61}$$

$$\overline{I}_{load} = \cos\phi - j\sin\phi \tag{7.62}$$

$$\overline{V}_{sag} = V\cos\psi + jV\sin\psi \tag{7.63}$$

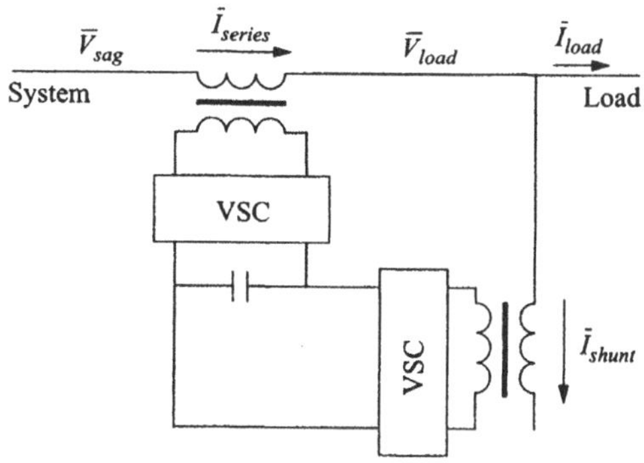

Figure 7.49 Shunt-series connected voltage controller; the shunt-connected converter is placed on load side of the series controller.

We assume that the shunt current is taken at a lagging power factor $\cos\xi$:

$$\overline{I}_{shunt} = I\cos\xi - jI\sin\xi \tag{7.64}$$

The total current taken off the supply, through the series-connected converter, is

$$\overline{I}_{series} = \overline{I}_{shunt} + \overline{I}_{load} = \cos\phi + I\cos\xi - j\sin\phi - jI\sin\xi \tag{7.65}$$

The active power taken off the supply should be equal to the power taken by the load. The power injected by the series converter is taken off again by the shunt converter. As there is no active power storage, the total active power still has to come off the supply. This gives the following expression:

$$\mathcal{R}\{\overline{V}_{sag}\overline{I}^{*}_{series}\} = \cos\phi \tag{7.66}$$

From this the following expression for the shunt current can be obtained:

$$I = \frac{\cos\phi - V\cos(\phi + \psi)}{V\cos(\psi + \xi)} \tag{7.67}$$

To minimize the shunt current, the angle ξ is taken such that $\psi + \xi = 0$; thus the shunt current is in phase with the supply voltage. If we further rate the shunt current to the active part of the load current, we obtain

$$I = \frac{1}{V} - \frac{\cos(\psi + \phi)}{\cos\phi} \tag{7.68}$$

which is exactly the same current as for a system-side shunt.

7.4.4.3 Single-Phase Controller. For a single-phase controller, we have again calculated the inverter current as a function of the sag magnitude in a similar way as for Figs. 7.34 and 7.35. The results are shown in Figs. 7.50 and 7.51 for different power factor of the load current. Fig. 7.50 is for sags without phase-angle jumps (zero impedance angle), Fig. 7.51 for sags with a serious phase-angle jump (an impedance angle equal to −30°). The overall behavior is dominated by the fast increase in current for deep sags. But for small power factor, especially, the phase-angle jump also plays an important role.

7.4.4.4 Advantages and Disadvantages. The main advantage of the shunt-series controller is that it does not require any energy storage. It can be designed to mitigate any sag above a certain magnitude, independent of its duration. This could result in a relatively cheap device, able to compete with the UPS (see below) for the

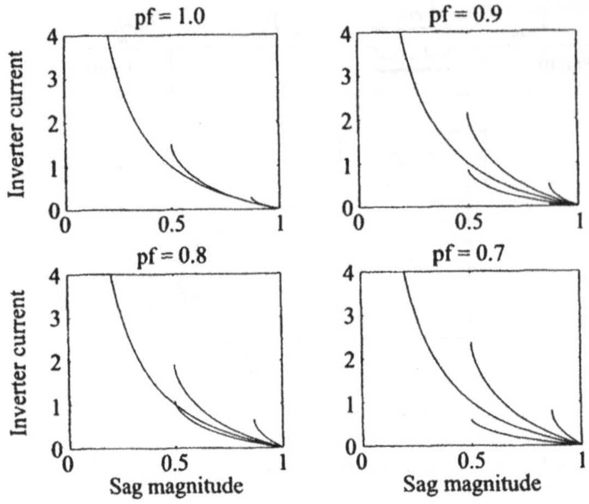

Figure 7.50 Shunt current for a single-phase shunt-series voltage controller as a function of the sag magnitude, for zero impedance angle and four values of the power factor of the load current.

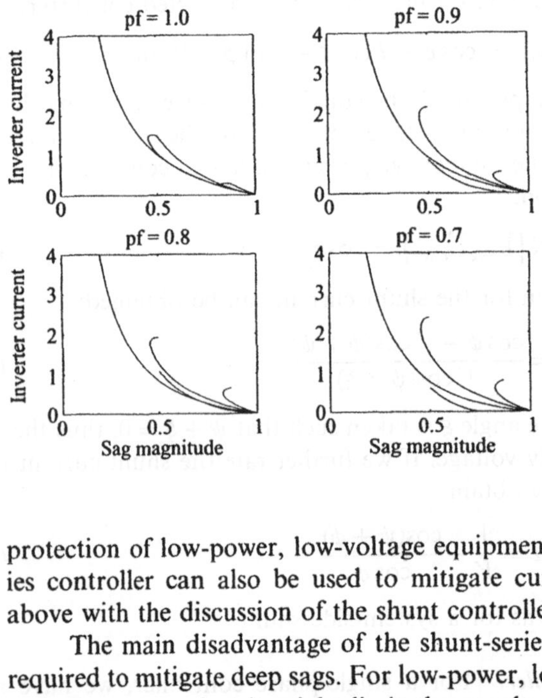

Figure 7.51 Shunt current for a single-phase shunt-series voltage controller as a function of the sag magnitude, for impedance angle −30° and four values of the power factor of the load current.

protection of low-power, low-voltage equipment. The shunt converter of a shunt-series controller can also be used to mitigate current quality problems, as mentioned above with the discussion of the shunt controller.

The main disadvantage of the shunt-series controller is the large current rating required to mitigate deep sags. For low-power, low-voltage equipment this will not be a serious concern, but it might limit the number of large power and medium-voltage applications.

7.4.5 Backup Power Source—SMES, BESS

One of the main disadvantages of a series controller is that it cannot operate during an interruption. A shunt controller operates during an interruption, but its storage requirements are much higher. We saw that the shunt-connected controller operates perfectly when only the controller and the protected load are interrupted. The controller is in that case only feeding the protected load. This principle can be used by creating the right interruption. This results in the shunt-connected backup power source as shown in Fig. 7.52. The configuration is very similar to the shunt

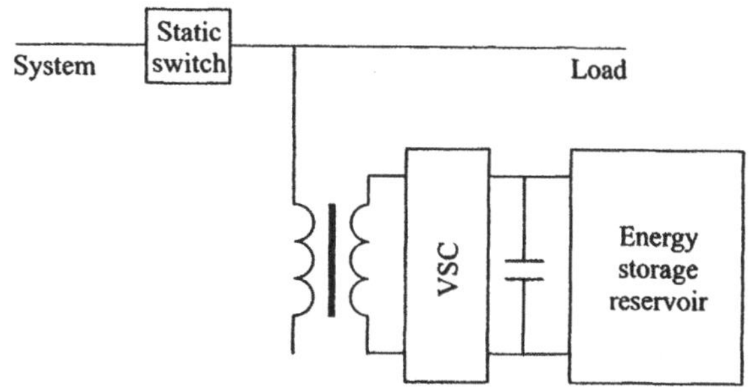

Figure 7.52 Shunt-connected backup power source.

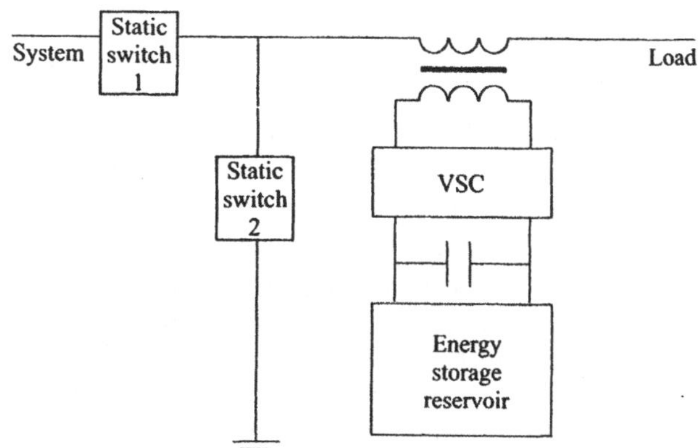

Figure 7.53 Series-connected backup power source.

controller. The difference is the static switch which is present between the system and the load bus. The moment the system voltage drops below a pre-set rms value, the static switch opens and the load is supplied from the energy storage reservoir through the voltage-source converter. Various forms of energy storage have been proposed. A so-called *superconducting magnetic energy storage* (*SMES*) stores electrical energy in a superconducting coil [57], [158], [159], [160], [161], [162]. A *BESS* or *battery energy storage system* uses a large battery bank to store the energy [186], [187], [188]. For small devices the energy storage is not a problem, but using a SMES, BESS, or any other way of storage at medium voltage will put severe strains on the storage. A backup power source is only feasible if it can ride through a considerable fraction of short interruptions. Looking at some statistics for short interruptions, Figs. 3.5, 3.6, and 3.7, shows that the amount of storage should be able to supply the load for 10 to 60 seconds. Less storage would not give any serious improvement in the voltage tolerance compared to the series controller.

All backup power sources suggested in the literature use a shunt connection, but it is also feasible to use a series connection as in Fig. 7.53. This device could operate as a series controller for sags and as a backup power source for interruptions. The moment a deep sag is detected, static switch 1 opens and static switch 2 closes.

7.4.6 Cascade Connected Voltage Controllers—UPS

The main device used to mitigate voltage sags and interruptions at the interface is the so-called *uninterruptable power supply* (*UPS*). The popularity of the UPS is based on its low costs and easy use. For an office worker the UPS is just another piece of equipment between the wall outlet and a computer. All that is needed is to replace the batteries every few years, and as long as one does not power the kettle and the microwave from the same UPS, a virtually problem-free supply is created.

7.4.6.1 Operation of a UPS. The UPS is neither a shunt nor a series device, but what could be described as a cascade connected controller. The basic configuration of a typical UPS is shown in Fig. 7.54. Its operation is somewhat similar to the converter part of an ac adjustable-speed drive (compare Fig. 5.12): a diode rectifier followed by an inverter. The main difference is the energy storage connected to the dc bus of a UPS. In all currently commercially available UPSs the energy storage is in the form of a battery block. Other forms of energy storage might become more suitable in the future.

During normal operation, the UPS takes its power from the supply, rectifies the ac voltage to dc and inverts it again to ac with the same frequency and rms value. The design of the UPS is such that the dc voltage during normal operation is slightly above the battery voltage so that the battery block remains in standby mode. All power comes from the source. The only purpose of the battery block in normal operation is to keep the dc bus voltage constant. The load is powered through the inverter which generates a sinusoidal voltage typically by using a PWM switching pattern. To prevent load interruptions due to inverter failure, a static transfer switch is used. In case the inverter output drops below a certain threshold the load is switched back to the supply.

During a voltage sag or interruption the battery block maintains the voltage at the dc bus for several minutes or even hours, depending on the battery size. The load will thus tolerate any voltage sag or short interruption without problem. For long interruptions, the UPS enables a controlled shutdown, or the start of a backup generator.

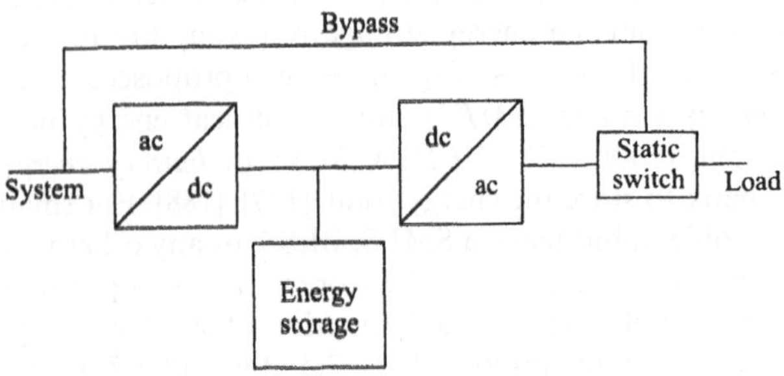

Figure 7.54 Typical configuration of an uninterruptable power supply (UPS).

7.4.6.2 Advantages and Disadvantages. The advantage of the UPS is its simple operation and control. The power electronic components for low-voltage UPSs are readily available and the costs of a UPS are currently not more than the costs of a personal computer. It is probably not worth installing a UPS for each personal computer in an office (making regular backups would be more suitable), but when a computer (or any other low-power device) is an essential part of a production process the costs of the UPS are negligible. As the UPS will mitigate all voltage sags and short interruptions a stochastic assessment is not even needed.

The main disadvantage of the UPS is the normal-operating loss because of the two additional conversions, and the use of batteries. Contrary to general belief, batteries do need maintenance. They should be regularly tested to ensure that they will operate in case of an interruption; also they should not be exposed to high or low temperatures and sufficient cooling should be installed to prevent overheating. All this is not so much a concern for the small UPSs used in an office environment, but for large installations the maintenance costs of a UPS installation could become rather high.

7.4.6.3 Alternatives. As a long-term solution to mitigate voltage sags and interruptions, the UPS is not the most appropriate one. The two additional conversions are not really needed, as can be seen in Fig. 7.55. The top drawing shows the normal configuration: the ac voltage is converted into dc and back to ac by the UPS. In the computer the ac voltage is again converted into dc and next converted to the utilization voltage for the digital electronics. This scheme represents almost any modern consumer electronics device.

Alternatively, one can directly connect the battery block to the dc bus inside the computer. In fact a laptop computer gets its power in such a way. Some mitigation methods for ac adjustable-speed drives also use a direct infeed into the dc bus. From an engineering viewpoint this is a more elegant solution than using a UPS, but the user does not always have the technical knowledge to do this. A solution like this can only be initiated by the equipment manufacturers.

One can extend this idea further, ending up with a dc network for an office building providing backup power to all sensitive equipment. By connecting an array of solar cells to this dc network the situation could arise where the utility supply becomes the backup for the internal dc network.

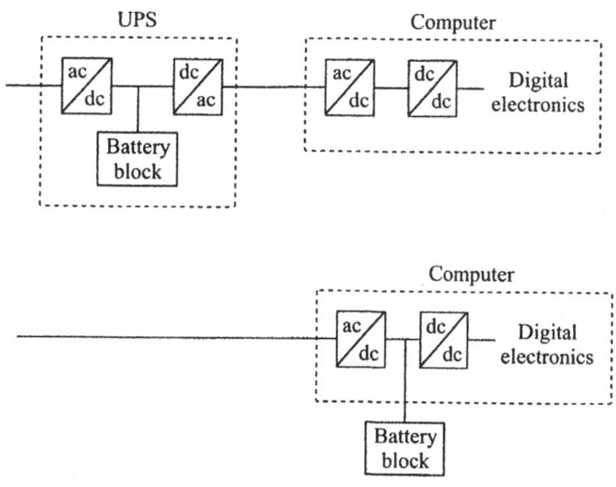

Figure 7.55 Power conversions for a UPS powering a computer, and for an alternative solution.

7.4.6.4 UPS and Backup Generators. Figure 7.56 shows a power system where both UPSs and backup generation are used to mitigate voltage sags and interruptions. The UPS is used to protect sensitive essential load against voltage sags and short interruptions. But especially for large loads, it is not feasible to have more than a few minutes energy supply stored in the batteries. In case of an interruption, the so-called "islanding switch" opens, disconnecting the sensitive load from the utility system. During the interruption the sensitive load is completely powered from a backup generator. This generator can be either running in parallel with the utility

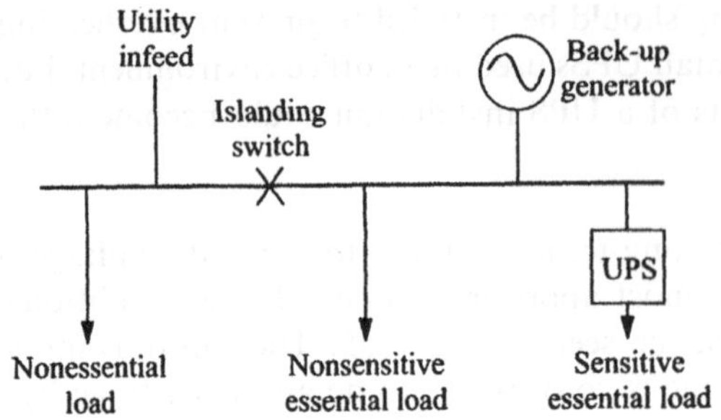

Figure 7.56 UPS combined with backup generation to mitigate voltage sags, short and long interruptions.

supply, or be started the moment an interruption is detected. All essential load is fed from the backup generator, where only the essential load which is sensitive to sags and short interruptions needs to be powered from the UPS. Decreasing the time to switch over to island operation decreases the energy storage requirements in the UPS. The energy storage requirement is proportional to the switch-over time. The UPS only needs to supply the load which cannot tolerate the interruption due to the switch-over to islanding operation. The faster the switch-over, the less load needs to be powered from the UPS.

An interesting example of the use of UPSs in combination with on-site generators to achieve a high reliability is discussed in [172].

7.4.7 Other Solutions

Some mitigation equipment is not based on the voltage-source converter; a few examples are discussed below. Motor-generator sets and ferroresonant transformers have been around for many years to mitigate voltage sags; electronic tap changers form an interesting new technique.

7.4.7.1 Motor-Generator Sets. A motor-generator set is an old solution against voltage sags, making use of the energy stored in a flywheel. The basic principle is shown in Fig. 7.57: a (synchronous or induction) motor and a synchronous generator are connected to a common axis together with a large flywheel. When the power supply to the motor is interrupted, the flywheel makes that the system continues to rotate and thus continues to supply the load. These kind of systems are still in use (and new ones are still being installed) in industrial installations. The ridethrough time of several seconds enables transfer schemes with mechanical switches. The noise of a motor-generator set and the maintenance requirements of the rotating machines are not a concern in most industrial environments. They do however make motor-generator sets unsuitable for an office environment.

In the configuration shown in Fig. 7.57, the normal operation losses are very high which makes this an expensive solution. A number of alternatives have been proposed to limit the losses. One option is to have the motor-generator set operating in no-load when the supply voltage is within its normal range. The moment a sag or interruption is detected, a (static) switch is opened and the generator takes over the supply. A possible configuration is shown in Fig. 7.58.

In normal operation the synchronous machine operates as a synchronous condensor which can, e.g., be used for reactive power compensation or for voltage control. When the supply is interrupted the static switch opens and the synchronous machine

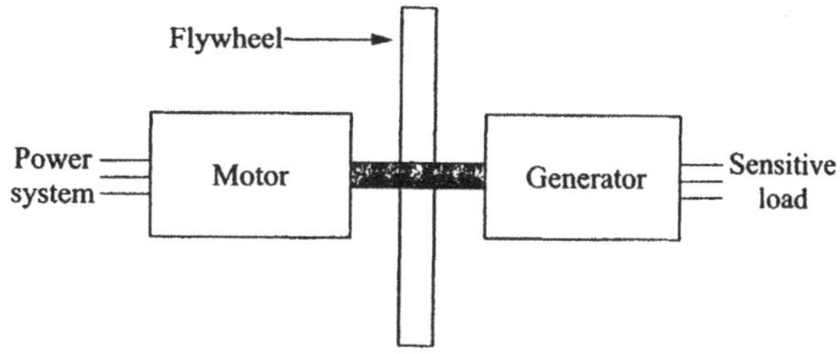

Figure 7.57 Principle of motor-generator set.

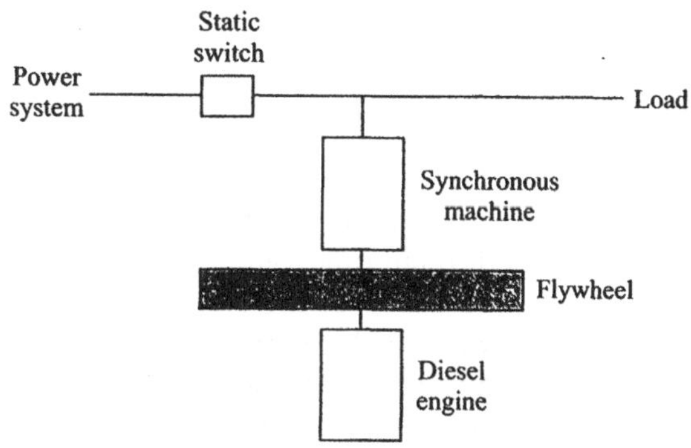

Figure 7.58 Configuration of off-line UPS with diesel engine backup.

starts to operate as a synchronous generator, injecting both active and reactive power. This will provide power for one or two seconds. By using a large reactance between the load and the power system, a certain level of voltage-sag mitigation is achieved. The effect is the same as for an on-site generator. By opening the static switch on an undervoltage it is even possible to operate the synchronous machine as a backup power source during sags as well. While the flywheel provides backup power, the diesel engine is started.

More recent improvements are the use of written-pole motors and the combination of a motor-generator set with power electronics. A written-pole motor is an ac motor in which the magnetic pole pairs are not obtained from windings but instead are magnetically written on the rotor [193]. This enables a constant output frequency of the generator, independent of the rotational speed. The main advantage for use in a motor-generator set is that the generator can be used over a much larger range of speed, so that more energy can be extracted from the flywheel.

A combination of the motor-generator set with power electronic converters is shown in Fig. 7.59. The motor is no longer directly connected to the power system, but through an adjustable-speed drive. This enables starting of the flywheel without causing voltage sags in the system, overspeed of the flywheel increasing the ridethrough time, and loss reduction while the set is in standby. The output of the generator is rectified to a constant dc voltage which can be utilized through a series- or shunt-connected voltage-source converter or directly fed into the dc bus of an adjustable-speed drive. The ac/dc converter enables the extraction of power from the flywheel over a much larger range of speed.

Suppose that a normal motor-generator set gives an acceptable output voltage for a frequency down to 45 Hz (in a 50 Hz system). A frequency of 45 Hz is reached when the speed has dropped to 90%. The amount of energy in the flywheel is still 81% of the energy at maximum speed. This implies that only 19% of the stored energy is used.

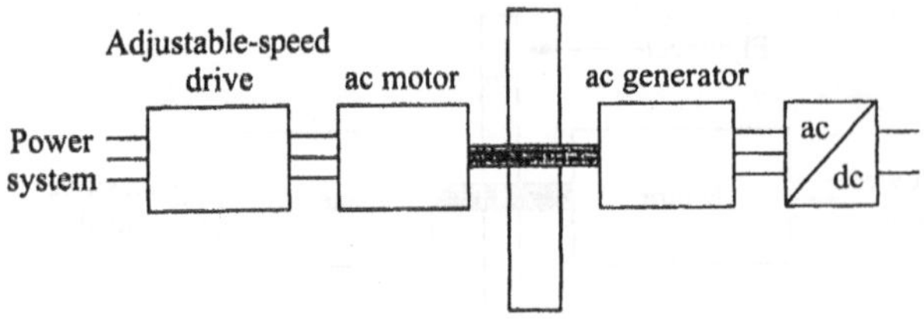

Figure 7.59 Power electronic converters in combination with a motor-generator set.

Suppose that we can generate a constant dc voltage for a speed down to 50%, by using an ac/dc converter. The energy that can be extracted is 75% of the total energy, an increase by a factor of four. The ridethrough time is thus also increased by a factor of four—for example, from 5 to 20 seconds. The ridethrough can be further increased by running the ac motor above nominal speed. By accelerating the flywheel slowly, the mechanical load on the motor can be kept small. As the kinetic energy is proportional to the square of the speed, a rather small increase in speed can already give a serious increase in ridethrough time. Suppose an overspeed of 20%, which increases the energy in the flywheel to 144% of the original maximum. The extraction of energy from the flywheel stops when 25% of the original maximum remains, so that the amount of energy extracted from the flywheel is 119%: a factor of six more than with the original setup. The resulting ridethrough time is 30 seconds.

7.4.7.2 Electronic Tap Changers. Electronic tap changers use fast static switches to change the transformation ratio of a transformer. This can either be a distribution transformer or a dedicated transformer for a sensitive load. The principle of its operation is shown in Fig. 7.60, in this case with three static switches. The number of turns of the four parts of the secondary winding are (top to bottom): 100%, 40%, 20%, and 10% of the nominal turns ratio. By opening or closing these three switches transformation ratios between 100% and 170% can be achieved, with 10% steps. If all three switches are closed, the turns ratio is 100%; with switch 1 closed and 2 and 3 open it is 130%, etc. By using this electronic tap changer, the output voltage is between 95% and 105% of nominal for input voltages down to 56% of nominal. Transformers with electronic tap changers are currently available as

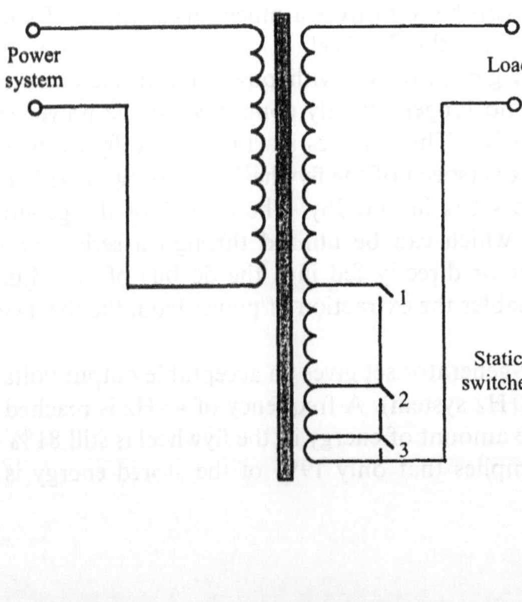

Figure 7.60 Basic principle of the construction of an electronic tap changer.

an additional series component between the source and the load. In future it may be feasible to install electronic tap changers on distribution transformers and save the additional component.

7.4.7.3 Ferroresonant Transformers. A ferroresonant transformer, also known as a constant-voltage transformer, is mainly designed to maintain a constant voltage on its output over a range of input voltage. The basic construction of a ferroresonant transformer is shown in Fig. 7.61. The third winding of a three-winding transformer is connected to a large capacitor. Without this capacitor, the device operates as a normal transformer. The effect of the capacitor is explained through Fig. 7.62. The solid line is the relation between voltage and current for the nonlinear inductance. The dashed line holds for the capacitor. The place where the curves cross is the operating point. Note that these curves give the voltage and current magnitude for one frequency, in this case the power system frequency as that is the frequency exciting the system. This operating point is independent of the supply voltage, thus the flux through the iron core is independent of the supply voltage (assuming that the ferroresonant winding has a smaller leakage than the input winding). The output voltage is related to this flux, thus also independent of the input voltage.

The energy stored in the ferroresonant winding is able to provide some ride-through during voltage dips. A disadvantage of a ferroresonant transformer is its dependence on load changes. The inrush current of the load can lead to a collapse of the flux and a long undervoltage. A modern version of the ferroresonant transformer uses power electronic converters to keep the load current at unity power factor, thus optimizing the operation of the transformer.

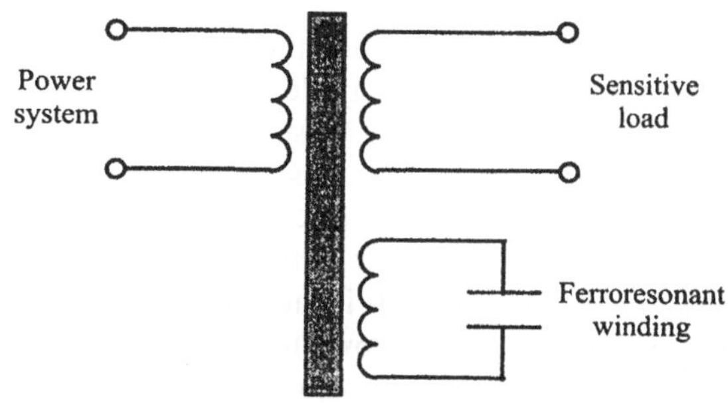

Figure 7.61 Basic principle of the construction of a ferroresonant transformer.

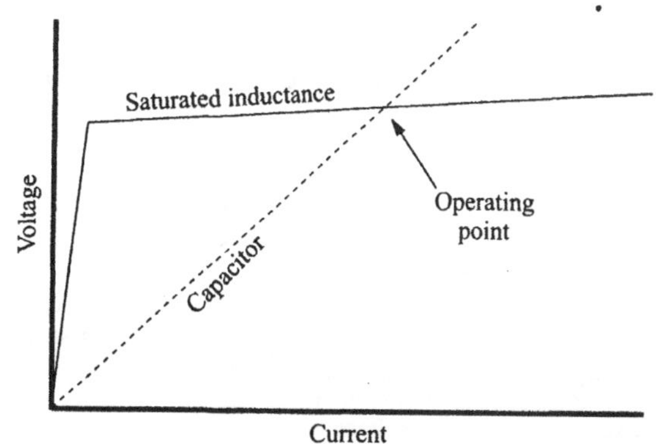

Figure 7.62 Voltage versus current diagram for a saturable inductor (solid line) and for a capacitor (dashed line).

7.4.8 Energy Storage

Several of the controllers discussed above, need energy storage to mitigate a sag. All of them need energy storage to mitigate an interruption. Here we compare different types of energy storage which are currently being used and considered. The comparison is based on three different time scales, related to three different controllers.

- A series voltage controller is only able to mitigate voltage sags. A typical design value is 50%, 1 second; i.e., the controller is able to deliver 50% of nominal voltage for 1 second. In terms of energy-storage requirements this corresponds to full load for 500 ms.
- A (shunt-connected) backup power source is also able to mitigate interruptions. To be able to improve the voltage tolerance significantly a ridethrough between 10 and 60 seconds is needed. We consider the requirement: full load for 30 seconds.
- To achieve very high reliability, sensitive load is typically powered via a UPS which can supply the load for 10 to 60 minutes. During this period, backup generators come on line to take over the supply. The third energy-storage requirement will be full load for 30 minutes.

7.4.8.1 DC Storage Capacitors. Capacitors are mainly used to generate reactive power on an ac system. But in a dc system they can be used to generate active power. The amount of energy stored in a capacitance C with a voltage V is

$$\mathcal{E} = \frac{1}{2}CV^2 \tag{7.69}$$

The voltage decreases when the energy is extracted from the capacitor. Capacitors can thus not be used to supply electric power to a constant-voltage dc bus, as needed for a voltage-source converter. A second (dc/dc) converter is needed between the capacitors and the constant-voltage bus, as shown in Fig. 7.63. Alternatively, the control algorithm of the voltage-source converter can be adjusted to variable dc voltage.

In either case, there will be a minimum voltage below which the converter is no longer able to operate. It is thus not possible to extract all energy from the capacitors. If the converter operates down to 50% of the maximum voltage, 75% of the energy can be extracted. A converter operating down to 25% can extract 94% of the energy.

Consider a medium-voltage controller using 4200 V, 1500 μF storage capacitors. The amount of energy stored in one capacitor is

$$\mathcal{E} = \frac{1}{2}CV^2 = \frac{1}{2} \times 1500\,\mu\text{F} \times (4200\,\text{V})^2 = 13\,\text{kJ} \tag{7.70}$$

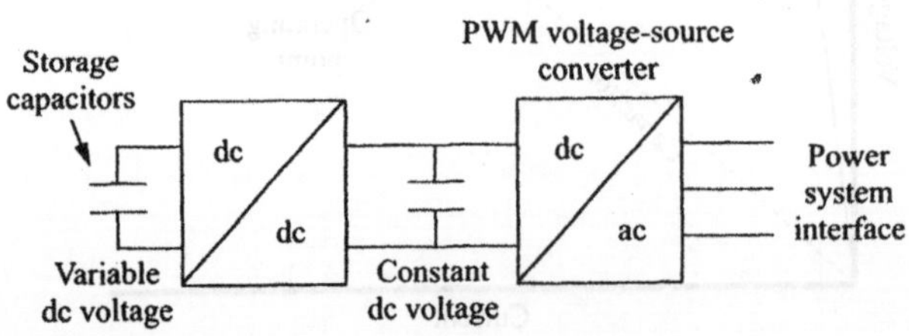

Figure 7.63 Energy extraction from dc storage capacitors.

Suppose that the converter is able to operate down to 50% of voltage. Each capacitor unit is able to supply: 0.75 × 13 kJ = 9.75 kJ.

For a 500 ms ridethrough, each unit can supply 19.5 kW of load. A small medium-voltage load of 500 kW requires 26 capacitor units; a large medium-voltage load of 10 MW over 1000 units. For a 30 second ridethrough each unit can only power 325 W of load, already requiring 1500 units for a small medium-voltage load. Thus dc capacitors are feasible for series controllers with ridethrough up to about 1 second, but not for backup voltage sources requiring ridethrough of 30 seconds and more.

Various energy storage options for adjustable-speed drives are compared in [42]. A price of $35 is given for a 4700 μF, 325 V capacitor. The amount of energy stored in one such capacitor is 250 J, of which 188 J (75%) can be used, enough to power a 375 W load for 500 ms or a 6.25 W load for 30 seconds.

To power a small low-voltage load of 1000 W during 500 ms requires three capacitors costing $105; to power it for half a minute requires 160 capacitors, costing $5600. For a complete low-voltage installation of 200 kW we need 534 capacitors ($18,700) for 500 ms ridethrough and 32,000 capacitors ($1,120,000) for 30 seconds. The conclusion is the same as before: capacitor storage is suitable for 1 second ridethrough but not for 1 minute ridethrough.

7.4.8.2 Batteries. Batteries are a very commonly used method of storing electric energy. They are used in the vast majority of UPSs sold, not only in the small one used to power a single PC but also in larger ones which can power a complete installation. Batteries provide a constant voltage so that they can be directly connected to the voltage-source converter. A 5 MVA, 2.5 MWh battery energy storage system (BESS) has been installed to power critical equipment in a large chemical facility [188]. The amount of stored energy in this system is 9 GJ, much more than in any of the above examples. An even larger installation has been installed in California in 1988 for load-leveling purposes [186]. This BESS is able to supply 10 MW during 4 hours, corresponding to 144 GJ of stored energy. This installation covers an area of 4200 m^2 for the batteries only.

Looking at smaller sizes, consider a car battery with a storage capacity of 1 MJ (12 V, 23 Ah) costing about $50. This simple battery contains enough energy to power a 2 MW load during 500 ms, a 33 kW load during 30 seconds, or a 550 W load during 30 minutes. One car battery contains the same amount of energy as 77 medium-voltage storage capacitors.

The limitation with a battery is not so much the amount of energy stored in it, but the speed with which this energy can be made available. Emptying our car battery in 30 seconds requires a current of 2760 A. The battery will never be able to supply this. If we consider a maximum current of 200 A, the maximum load which can be supplied from one battery is 2400 W. The battery can power this load for 7 minutes, which can be considered as the optimum ridethrough time for this battery. This fits well in equipment to mitigate interruptions for the time until on-site generation becomes available.

The number of batteries needed and the costs of these, are given in Table 7.5 for the load sizes and ridethrough times given before. Only for short ridethrough times will capacitors be able to compete with batteries.

Batteries have a number of disadvantages compared to capacitors, which may compensate the higher costs of the latter. The commonly used lead-acid battery (on which this calculation is based), contains environmentally unfriendly materials, has a limited lifetime (in number of recharging cycles), and requires regular maintenance to ensure a high reliability. The newer types of batteries, which are being developed for use

TABLE 7.5 Number of Batteries (in brackets) and Costs Needed to Power Several Load Sizes for Several Ridethrough Times

	1 kW	200 kW	500 kW	10 MW
500 ms	(1) \$50	(84) \$4200	(209) \$10,000	(4167) \$210,000
30 sec	(1) \$50	(84) \$4200	(209) \$10,000	(4167) \$210,000
30 min	(2) \$100	(364) \$18,000	(910) \$46,000	(18182) \$910,000

in electrical vehicles, do not have these disadvantages but they obviously have higher costs.

7.4.8.3 Supercapacitors. Supercapacitors (or double-layer capacitors) are propagated as a future solution for energy storage to improve equipment voltage tolerance. They have energy densities comparable to batteries, but much longer lifetime and much less maintenance requirements. Their disadvantage is that they are only available for voltages of a few volts. A value of 3.3 F, 5.5 V is mentioned in [189]. The amount of stored energy is 50 J, only 1/5th of the 4700 μF, 325 V capacitor. Like with a battery, there is a limit to the speed with which energy can be extracted from a supercapacitor. For the supercapacitors currently in operation, the discharge time cannot be less than about 1 minute. This makes them somewhat faster than batteries but still much slower than capacitors. The development of supercapacitors is mainly driven by the requirements of electric vehicles, where the amount of stored energy is of more importance than the speed with which it can be extracted.

7.4.8.4 Flywheels. An alternative which is currently being investigated is the storage of energy in fast-spinning flywheels. The classical motor-generator set, discussed before, already uses this principle, but the modern equivalent rotates at a much higher speed. By using magnetic bearings and vacuum sealing of the rotating parts, very high rotational speeds can be achieved [192], values up to 90,000 rpm have been reported [190]. A possible configuration is shown in Fig. 7.64. The flywheel is brought up to speed by an ac adjustable-speed drive. This drive also ensures that rotational speed of the flywheel remains within a certain range during standby operation. During a voltage sag or an interruption the brushless dc generator extracts

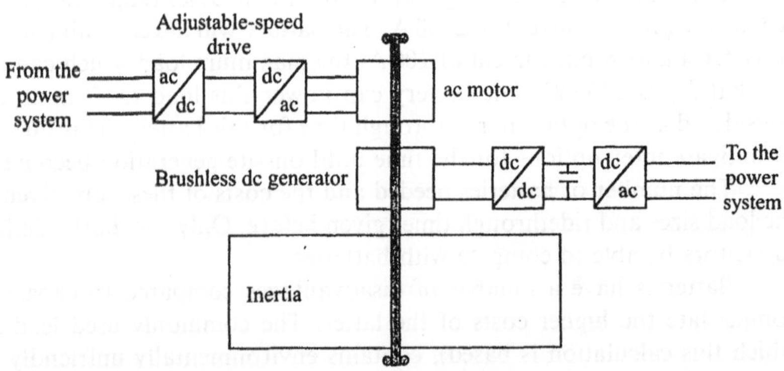

Figure 7.64 Configuration of a flywheel energy storage system and its interface to the power system.

energy from the flywheel and supplies this to the power system via a dc/dc converter and a voltage-source (dc/ac) converter.

Consider a solid cylindrical piece of material with a length of 50 cm and a radius of 25 cm. The inertia of this piece of material, for rotation along the axis of the cylinder, is

$$J = \frac{1}{2} m R^2 \tag{7.71}$$

with m the mass and R the radius of the cylinder. With a specific mass of 2500 kg/m^3 we find for the mass:

$$m = \pi \times 0.25^2 \times 0.50 \times 2500 = 245\,\text{kg}$$

and for the inertia:

$$J = \frac{1}{2} \times 245 \times 0.25^2 = 7.7\,\text{kg}\,\text{m}^2$$

The kinetic energy of an intertia J rotating with an angular velocity ω is

$$\mathcal{E} = \frac{1}{2} J \omega^2 \tag{7.72}$$

If we rotate our cylinder at the "moderate" speed of 3000 rpm ($\omega = 2\pi \times \frac{3000}{60} = 314\,\text{rad/s}$, the amount of kinetic energy stored in the rotating cylinder is

$$\mathcal{E} = \frac{1}{2} \times 7.7 \times 314^2 = 380\,\text{kJ}$$

This energy cannot be extracted completely, as the energy conversion becomes inefficient below a certain speed. Suppose this to be 50% of the maximum speed. The amount of useful energy is again 75% of total energy, in this case $0.75 \times 380\,\text{kJ} = 285\,\text{kJ}$. This flywheel is thus able to power a 570 kW load for 500 ms, a 9.5 kW load for 30 seconds, or a 160 W load for 30 minutes.

Increasing the rotational speed to 25,000 rpm by using the newest technologies, increases the amount of stored energy to

$$\mathcal{E} = \frac{1}{2} \times 7.7 \times 2618^2 = 26\,\text{MJ}$$

The useful energy of $0.75 \times 26\,\text{MJ}$ is enough to power a 40 MW load for 500 ms, a 650 kW load for 30 seconds, or an 11 kW load for 30 minutes.

7.4.8.5 Superconducting Coils. It is well known that an inductor L, carrying a current i, contains an amount of energy in its magnetic field equal to

$$\mathcal{E} = \frac{1}{2} L i^2 \tag{7.73}$$

This would make an inductor an alternative form of energy storage, next to the capacitor. The reason that inductor storage is not commonly used is that the current causes high losses in the wire making up the inductor. The losses due to a current i are equal to

$$P_{loss} = \frac{1}{2} R i^2 \tag{7.74}$$

with R the total series resistance. Suppose that we can achieve an X/R ratio of 100 for the inductor. In that case we find for the losses:

$$\frac{P_{loss}}{\mathcal{E}} = \frac{\frac{1}{2}Li^2}{\frac{1}{2}Ri^2} = \frac{R}{L} = \frac{\omega}{\frac{X}{R}} \approx 3\,\mathrm{s}^{-1} \tag{7.75}$$

To compensate for the resistive losses, the energy contents in the coil has to be supplied three times a second.

A solution suggested several years ago is to store the energy in a superconducting coil. The resistance of a superconductor is (exactly) zero so that the current will flow forever without any reduction in magnitude. A possible configuration for such a *superconducting magnetic energy storage* (*SMES*) is shown in Fig. 7.65. The variable current through the superconducting coil is converted to a constant voltage. The constant-voltage dc bus is connected to the (ac) power system by means of a voltage-source converter. The coil current closes through the dc/dc converter which causes a small loss. The configuration of SMES devices is discussed in more detail in [57], [158], [160], [162], [169].

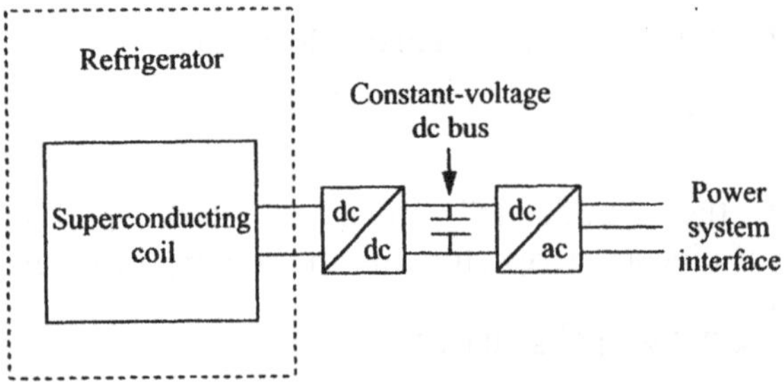

Figure 7.65 Energy storage in a superconducting coil and interface with the power system.

One application [158] uses a 1000 A current through a 1.8 H inductor. The energy stored in the magnetic field is

$$\mathcal{E} = \frac{1}{2} \times 1.8 \times 1000^2 = 900\,\mathrm{kJ} \tag{7.76}$$

Assume that the dc/dc converter operates for currents down to 50% of the maximum current. The usable energy is in this case 0.75 × 900 kJ = 675 kJ. This is enough to power a 1.35 MW load for 500 ms, a 22.5 kW load for 30 seconds, or a 375 W load for 30 minutes. The device described in [158] operates as a shunt-connected backup power source; it is used to mitigate voltage sags and short interruptions with durations up to a few seconds.

Commercial applications of SMES devices are reported for stored energy up to 2.4 MJ and power ratings up to 4 MVA. The devices currently in operation use low-temperature superconductors with liquid helium as a cooling medium. A demonstration SMES using high-temperature superconductors has been built which is able to store 8 kJ of energy. This is still two orders or magnitude away from the devices using low-temperature superconductors, but the manufacturer expects to build 100 kJ devices in the near future. A study after the costs of SMES devices now and in 10 years' time, is described by Schoenung et al. [168]. For example, a 3 MW, 3 MJ unit would cost $2,200,000 now, but "only" $465,000 in 10 years' time. The main cost reduction is based on the so-called learning curve due to the production of about 300 units in 10 years. By using the data in [168] the costs have been plotted as a function of the stored energy, resulting in Fig. 7.66.

In Table 7.6 the costs of energy storage in a SMES are compared with the costs of batteries and capacitors. The costs of the power electronic converters have not been

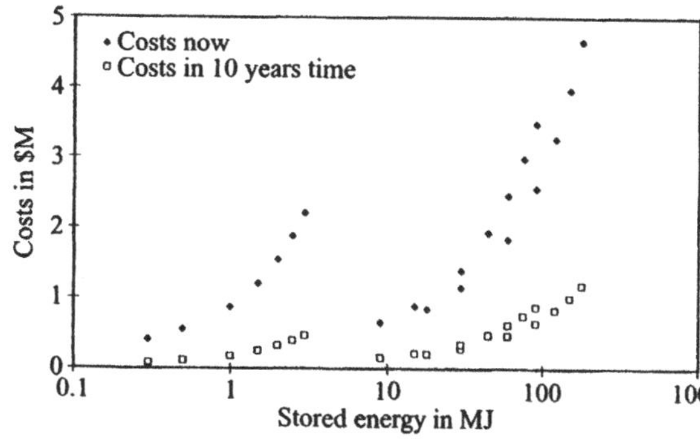

Figure 7.66 Costs of superconducting magnetic energy storage (SMES) including the power system interface, as a function of the amount of stored energy. (Data obtained from [168].)

TABLE 7.6 Costs Comparison of SMES, BESS and Capacitors

Power	Ridethrough Time	Costs of Energy Storage		
		SMES	BESS	Capacitors
300 kW	1 sec	$183,000	$6300	$56,000
	60 sec	$389,000	$6300	$3,350,000
3 MW	1 sec	$411,000	$63,000	$558,000
	60 sec	$1,064,000	$63,000	$33,500,000

included, as these are similar for all energy storage methods. The costs of a battery energy storage system (BESS) is based on the same batteries as used before: 1 MJ of storage, 2400 W of power for $50. The costs of capacitor storage is based on 188 J of storage for $35 as used before. Additional costs of construction, wiring, protection, cooling, etc., have not been included for the capacitors or for the batteries.

We see that, with current prices, battery storage remains by far the cheapest solution, even if we consider a factor of two to three for additional costs. But the lifetime of a battery is limited in number of discharge cycles, and batteries contain environmentally unfriendly products. When the costs of SMES devices go down and the costs of batteries go up in the future, the former will become a more attractive option for high-power short-time ridethrough. For short-time ridethrough capacitor storage is still more attractive, especially if one realizes that we used low-voltage capacitors where medium-voltage capacitors are likely to form a cheaper option.

Note that the amount of energy stored in an SMES is similar to the amount of energy stored in a battery. The main difference is that the energy in a superconducting coil can be made available much faster. The units currently in operation are able to extract 1 MJ of energy from the coil in 1 second. The limitation in energy extraction is the voltage over an inductor when the current changes:

$$v_{ind} = L\frac{di_{dc}}{dt} \tag{7.77}$$

The energy extraction P_{load} is related to the change in current according to

$$\frac{d}{dt}\left\{\frac{1}{2}Li_{dc}^2\right\} = P_{load} \tag{7.78}$$

which gives for the voltage over the inductor:

$$v_{ind} = \frac{P_{load}}{i_{dc}} \tag{7.79}$$

With constant energy extraction (constant P_{load}), the induced voltage increases with decreasing current. For a 500 kW load and a minimum current of 500 A, the voltage over the coil is

$$v_{ind} = \frac{500\,\text{kW}}{500\,\text{A}} = 1000\,\text{V} \tag{7.80}$$

For a 3 MW unit we get $v_{ind} = 6\,\text{kV}$. The dc/dc converter should be able to operate with this voltage over its input terminals.

8

Summary and Conclusions

This chapter summarizes the conclusions from the previous chapters. Next to that some thoughts are given concerning the future of this area of power engineering. Just like in the rest of the book, the emphasis is on voltage sags and interruptions.

8.1 POWER QUALITY

In Chapter 1 the term "power quality" and several related terms are defined. Power quality is shown to consist of two parts: "voltage quality" and "current quality." The voltage quality describes the way in which the power supply affects equipment; as such it is part of the quality of supply. Current quality describes the way in which the equipment affects the power system and is part of the so-called "quality of consumption." The term electromagnetic compatibility (EMC) has a large overlap with "power quality" and the terms can often be used as synonyms.

An overview is given of the various types of power quality disturbances. An important distinction is made between "variations" and "events." Variations are a continuous phenomenon, e.g., the variation of the power system frequency. Measuring voltage and current variations requires continuous recording of their values. Events only occur occasionally: voltage sags and interruptions are typical examples. Measuring voltage and current events requires a triggering process: e.g., the rms voltage becoming less than a pre-defined threshold. These two types of power quality disturbances also require different analysis methods: average and standard deviation for variations; frequency of occurrence for events.

The main subject of this book is formed by voltage sags and interruptions: the two most important examples from a family of voltage events known as "voltage magnitude events." Voltage magnitude events are deviations from the normal magnitude (rms value) of the voltage with a rather well-defined starting and end time. The majority of these events can be characterized by one magnitude and one duration. Different initiating events and different restoration processes lead to different ranges of magnitude and duration. Based on these ranges, a classification of voltage magnitude events is proposed.

8.1.1 The Future of Power Quality

There is one question that always comes up when thinking about the future of power quality: "Will the power quality problem still be among us in 10 years time?" It may well be that equipment will be improved in such a way that it no longer is sensitive to the majority of voltage disturbances and that it no longer produces serious current disturbances. In other words, equipment will have become fully compatible with the power supply. At the moment, however, there is no indication that this will happen soon. Equipment appears to be as sensitive and polluting as ever. A browse through the advertisements in power-quality oriented journals shows that the emphasis is on mitigation equipment (surge suppressors, UPSs, custom power) and on power-quality measurement equipment. Advertisements in which equipment with improved voltage tolerance is offered are extremely rare.

The main drive for improved equipment is likely to come from standards, in particular the IEC standards on electromagnetic compatibility. When the standards on harmonic currents produced by end-user equipment (IEC 61000-3-2 and -3-4) become widely accepted, the harmonic distortion problem may be the first one to move to the background.

Voltage quality events like voltage sags will take even longer to become part of equipment standards. At least voltage sags are reasonably understood nowadays (read Chapters 4, 5, and 6). Higher frequency phenomena like switching transients are less well understood, more difficult to model, and their statistics probably show more variations among different customers. Still they cause equipment problems. High-frequency disturbances may well become the next big power-quality issue.

8.1.2 Education

An important aspect of power quality is education: education of those who come in touch with power quality problems as well as new generations of engineers. Power quality may bring power engineering education closer to the actual aim of power engineering: generating electrical energy and delivering it to electrical end-user equipment. Educating a new generation of engineers is obviously a task for universities. And with engineers I am not only referring to power engineers. Every student in electrical, electronic, and mechanical engineering should know about potential problems due to the connection of equipment to the power supply. Note that these are the persons to use electrical equipment and to design future equipment. When they are aware of potential compatibility problems, they are more likely to come up with equipment that is compatible with the supply.

Postgraduate education is important and not necessarily a task for a university. Several companies offer good power-quality courses that enable people in industry to solve the problems they encounter. However, universities are better suited to give theoretical backgrounds needed to solve future problems, next to providing an understanding of existing problems.

8.1.3 Measurement Data

From the beginning, power quality has been an area very much based on measurements and observations. The standard tools in use at universities, simulations and theoretical analysis, are much less used in the power quality work. In fact, the amount of university research on power quality is still very limited. This will certainly change in the near future; power quality will not only find its way into education but also into university research. There is a serious risk here that a gap will develop between the heavily measurement-based power-quality practice and the very much theory- and simulation-based university research. Such a situation may be prevented if utilities make much more of their data available for university research and education. A very good example is set by IEEE Project group 1159.2. At their Website (accessible through www.standards.ieee.org) a number of voltage recordings are available for downloading. I would like to see much more utilities making data available in this way: not only the actual voltage and current recordings but also some basic data about the kind of event and the kind of power system involved.

8.2 STANDARDIZATION

In the second part of Chapter 1, power quality standards are discussed. The IEC set of standards on electromagnetic compatibility offers the opportunity to seriously solve several power quality problems. The standards describe various power-quality disturbances, define testing techniques and give requirements for equipment and system performance. A large number of standards is still under development and even more are required to fully standardize equipment as far as power quality and EMC is concerned. In Chapter 1 some suggestions are given for the extension of the concept "compatibility level" from variations to events.

The European voltage characteristics standard, EN 50160, is described in detail. The standard gives a good description of the voltage quality for voltage variations, but is rather weak for voltage events.

8.2.1 Future Developments

Developments in this area will unfortunately take a long time, so that power quality problems will be around for at least several more years. This is simply inherent to the standard-setting process. During my work on some IEEE standards, it became clear that one can only take one step at a time. The first step, making people aware of power quality problems, has been taken both within the IEC and within the IEEE. The recently published IEEE standard on compatibility between electronic process equipment and the power system (IEEE Std.1346-1998) may be the first of a long series of IEEE standards on this subject. Also Chapter 9 of the 1997 edition of the IEEE Gold Book (IEEE Std.493-1997) will contribute to the power-quality awareness. It is interesting to notice that both documents were already being used and referred to several years before they actually became accepted as standard documents. The same has happened with several IEC standards, noticeably the one limiting the harmonic current distortion by low-power equipment (IEC 61000-3-2). Both IEEE and IEC should make their draft documents available to a much wider audience. This will not only widen the discussion but also speed up the acceptance process of the standard.

The European voltage characteristics standard EN 50160 is one of the first documents quantifying the voltage quality experienced by customers. Despite all its shortcomings, the publication of this standard has triggered more coordinated measurement campaigns than before. The future will bring the publication of local equivalents of EN 50160.

8.2.2 Bilateral Contracts

An area related to power quality standards, but likely showing much faster development, is formed by the bilateral contracts between utilities and customers. Several examples are already in place where the utility pays compensation to its customers when the quality of supply drops below a certain level. The typical contract defines a maximum-acceptable number per year for each event type, e.g., two long interruptions, five short interruptions. When this number is exceeded within a certain year, the utility pays a predefined amount of compensation for each additional event. The initial contracts only contained interruptions, but voltage sags have been implemented in a number of contracts as well. When setting up these contracts, a precise definition of the various events is essential. Next to these bilateral contracts, utilities are likely to come up with general compensation schemes for customers with a bad voltage quality. When utilities refuse to take these steps they may be forced into worse constructions by political and legal developments outside of their control.

The concept of bilateral contracts is likely to be extended to the interface between transmission and distribution systems. At this interface voltage quality becomes even more two-directional than at the utility–customer interface. Voltage disturbances may originate in either system.

8.3 INTERRUPTIONS

A long interruption is an interruption of the power supply followed by a manual restoration. When the supply is restored automatically the result is a short interruption. Long interruptions are discussed in Chapter 2, short interruptions in Chapter 3. Long interruptions are by far the most serious voltage quality disturbance. Most utilities keep a record of frequency and duration of long interruptions. Unfortunately much of this very useful data is not generally accessible. A positive exception to this is the United Kingdom where utilities are obliged to publish data on the supply performance. Currently this only includes interruption data but it is likely to be extended to other types of events.

Short interruptions are shown to be due to a combination of automatic reclosing and a system design aimed at limiting the number of reclosers. Automatic reclosing makes that a long interruption becomes a short interruption and is as such a mitigation method. But limiting the number of reclosers makes that customers experience a short interruption that otherwise would have experienced a voltage sag. Removing the whole reclosure scheme is a deterioration of the supply for some customers but an improvement for others.

A detailed analysis is presented of voltages and currents associated with single-phase tripping. It is shown that single-phase tripping leads to less severe voltage events at the equipment terminals, but it may also lead to a higher percentage of second trips. A number of pilot schemes should be set up where single-phase tripping is used for the first attempt and three-phase tripping for the second attempt.

8.3.1 Publication of Interruption Data

In the future more utilities will publish interruption frequency and supply availability. For customers to be able to assess the compatibility between equipment and supply, it is essential that utilities publish the supply performance. As interruption data are already available, this will be the first to be published. A likely development is that utilities publish more than just frequency and availability over the whole country. Details like "worst-served customers," regional variations, and distribution of the interruption duration will give more insight into the quality of supply experienced by individual customers. Publication of more statistics will inevitably lead to a comparison between different utilities and regions. To obtain a fair comparison, many years of observation may be needed. Alternatively a standardized reliability evaluation tool can be used to predict the supply performance. As most interruptions originate in the distribution system, relatively simple techniques may be sufficient.

The increase in observation data will probably not include data on short interruptions, at least not initially. Getting data on short interruptions for all customers requires an extensive monitoring effort. For short interruptions, prediction methods may be the only suitable way of getting data for all customers. These prediction methods may be "calibrated" through monitoring at a limited number of sites.

8.4 RELIABILITY

The second part of Chapter 2 summarizes the various aspects of power system reliability and the stochastic analysis techniques currently in use: network modeling, Markov models, and Monte Carlo simulation. Various examples are given for each of these techniques. Different aspects are given for the reliability analysis of generation, transmission, and distribution systems (the three so-called "hierarchical levels"). For the industrial power supply a systematic methodology is given that can be used to obtain the reliability of the supply. This methodology consists of six layers, partly corresponding to the hierarchical levels but also including power quality and equipment failure.

8.4.1 Verification

Power system reliability has two distinctly different faces: the observed reliability and the predicted reliability. Observed reliability, i.e., keeping records of number and duration of interruptions, is the domain of the utilities; predicted reliability, i.e., reliability evaluation, is the domain of universities; without much overlap between these two sides. A comparison between observed and predicted reliability is needed to move forward in reliability evaluation. For this, utilities should provide the data and universities the analysis and prediction techniques. Only such a comparison will give a clear answer about the accuracy of the various stochastic prediction techniques. Such a comparison will also lead to a wider acceptance of stochastic prediction techniques and to a wider use of them within the utilities.

8.4.2 Theoretical Developments

Potential developments on the theoretical side are the inclusion of nonexponential repair-time distributions and of common-mode effects. In both cases the data requirements are high. This again calls for a closer cooperation between utilities and universities. Much of the theoretical work on power system reliability has been directed toward transmission systems. In the near future, distribution networks will become much more a focus of the research. The main theoretical bottleneck is again the distribution of the interruption duration. By using the exponential distribution erroneous results are obtained, especially for the number of very long interruptions.

8.5 CHARACTERISTICS OF VOLTAGE SAGS

In Chapter 4 the various characteristics of voltage sags are discussed. After the more "classical" characteristics, magnitude and duration, two newer characteristics, phase-angle jump and three-phase unbalance, are treated in considerable detail. Techniques are presented to calculate these sag characteristics for a given fault and load position and fault type. The techniques are applied to an example supply consisting of several voltage levels.

Phase-angle jump and three-phase unbalance are discussed in detail in Chapter 4. Especially three-phase unbalance is an important characteristic. The currently used definition of sag magnitude is not suitable for three-phase equipment. The definition of sag magnitude is generalized for three-phase unbalanced sags leading to a classification of three-phase unbalanced sags into seven types, of which two types (C and D) cover the majority of sags. A three-phase unbalanced sag is quantified through a characteristic complex voltage which is independent of voltage level or load connection. Magnitude and phase-angle jump are absolute value and argument, respectively, of the characteristic complex voltage. The possible range in magnitude and phase-angle jump is calculated, for single-phase as well as for three-phase equipment, for the example supply as well as in general.

Chapter 4 concludes with a treatment of two additional sag characteristics, point-on-wave and missing voltage, a discussion about load influence on voltage sags, and a brief treatment of voltage sags due to induction motor starting.

8.5.1 Definition and Implementation of Sag Characteristics

The various characteristics discussed here and others recently introduced, need to be applied to measured voltage sags. This will give information about their statistics and about the range of values that can be expected. The next step will be that these additional characteristics are implemented in commercially available power quality monitors. Before that stage is reached, it is essential that all sag characteristics are uniquely defined. This will prevent confusion due to different manufacturers using different definitions. Missing voltage may become a compromise between the different magnitude definitions used on both sides of the Atlantic (voltage drop versus remaining voltage). The disadvantage of using missing voltage is that the majority of single-phase equipment is affected by the remaining voltage, not by the missing voltage. The application of point-on-wave characteristics may be limited to a small group of equipment. But in any case, all these characteristics describe part of the quality of supply and statistical information about them should be part of the outcome of voltage sag surveys.

8.5.2 Load Influence

An area that has been somewhat forgotten in the various voltage sag studies is the effect of load on the voltage sag characteristics. A qualitative study of the effect of large induction motors is described in Chapter 4. For a quantitative study of all types of load, a detailed analysis of measured voltage sags is needed. Such a study should include large and small motor and electronic load as well as embedded generation. The effect of the load determines how the sag characteristics change when a voltage sag propagates from high voltage to low voltage. Observations have shown that a sag with a magnitude (remaining voltage) of 40% at 132 kV is seen as a sag with a magnitude of 60% at 400 V.

8.6 EQUIPMENT BEHAVIOR DUE TO VOLTAGE SAGS

In Chapter 5 the effect of voltage sags on equipment is discussed. The emphasis is on single-phase rectifiers (computers, consumer electronics, process controllers), ac adjustable-speed drives, and dc adjustable-speed drives. Single-phase rectifiers are affected by magnitude and duration of the voltage sag. They trip when the voltage drops below a certain magnitude for longer than a certain duration (resulting in a so-called "rectangular voltage-tolerance curve"). The voltage tolerance of the equipment can easily be improved by adding additional capacitance to the internal dc bus. Using a voltage regulator that can operate down to a lower voltage is a more elegant but also more difficult solution.

For three-phase rectifiers, as used in ac adjustable-speed drives, it is mainly the characteristic magnitude and the sag type that affect the dc bus voltage and thus the drive behavior. The amount of capacitance currently in use in ac drives is too small for the sag duration to have any influence. Making the drive tolerant against balanced sags requires serious improvements in the design of the PWM inverter. For balanced sags the dc-bus voltage drops to a lower value (equal to the sag magnitude, in pu) within one or two cycles. For three-phase unbalanced sags the size of the dc-bus capacitance is very important. If the capacitance is large enough (in the upper range of the amount of capacitance currently in use) the dc-bus voltage will not drop below 80% for any three-phase unbalanced sag. If the drive is able to stay on-line, the effect of the sag on the load will be very small.

DC adjustable-speed drives are shown to be very sensitive to voltage sags. The armature current and the torque drop to zero almost immediately, even for a rather shallow sag. As dc drives are typically used for speed-sensitive processes, the drop in speed associated with the zero torque will easily lead to a disruption of the process.

8.6.1 Equipment Testing

An important future step is the development of a testing protocol for equipment. This will enable the customer to compare the voltage tolerance of different devices. For single-phase equipment it is probably sufficient to test for different magnitude and duration. Possible exceptions are contactors (affected by point-on-wave) and equipment with controlled rectifiers (affected by phase-angle jump).

Testing of three-phase equipment will be much more complicated: even for non-controlled rectifiers, the characteristic phase-angle jump affects the dc-bus voltage. For dc drives the three phases are no longer equivalent so that the number of tests required increases by a factor of three. Three-phase equipment needs to be tested for several types of three-phase unbalanced sags and for a range of magnitude, duration, and phase-angle jump. Further analysis of monitoring results is needed to obtain realistic values for the range of characteristics to be included in the tests.

Another problem that needs to be solved is the definition of the test criterion. Whether a certain reaction is acceptable depends to a large extend on the process driven by the drive. A possible solution is to give the variation in speed and torque as a function of the sag characteristics. This will enable an assessment of the effect of the sag on the process when using a certain drive.

8.6.2 Improvement of Equipment

Improvement of equipment offers the only long-term solution to the power quality problem. As shown in Chapter 5, the effect of the sag can be mitigated for many devices by installing additional capacitance. There are some drawbacks with this, the first being the additional costs. A risk of additional capacitance is that the inrush current on voltage recovery becomes more severe. This may lead to blowing of fuses or to damage on power electronic components.

Installing additional capacitance has its limits. It is not feasible for making drives tolerate balanced sags and it is in most cases not feasible at all for dc drives. More advanced rectifiers, inverters, and control algorithms are needed to achieve this. There is not yet a drive toward improved equipment but somewhere in the (hopefully not too remote) future this will happen. Possible driving forces are a standardized testing protocol; equipment immunity requirements as part of the EMC standards; and, of course, a demand for improved equipment from the side of the customer.

8.7 STOCHASTIC ASSESSMENT OF VOLTAGE SAGS

Chapter 6 discusses the stochastic and statistical treatment of the compatibility between equipment and supply. Data about the performance of the supply can be obtained from power quality monitoring and from stochastic prediction studies. Monitoring may give a more accurate picture of the kind of disturbances to be expected, but stochastic prediction will give results in a much shorter time.

Different methods are discussed to present the results of a stochastic assessment study (either power quality monitoring or stochastic prediction). The so-called "voltage-sag coordination chart" is shown to be a useful instrument for the compatibility assessment. The results of a number of large power-quality surveys are presented and compared. One of the conclusions is that a further treatment is needed of the propagation of voltage sags from the fault position to lower voltage levels. The above-mentioned effect of load on the sag characteristics will play an important role in such studies.

Two methods are presented for the stochastic prediction of voltage sags: the method of fault positions and the method of critical distances. The method of fault positions is suited for computerized calculations in large meshed (transmission) systems. The method of critical distances is suitable for simple hand calculations and for calculations in radial (distribution) systems.

8.7.1 Other Sag Characteristics

All the techniques discussed in Chapter 6 concentrate on magnitude and duration of voltage sags. To cover a wider range of equipment, new techniques have to be developed for the other sag characteristics: phase-angle jump; three-phase unbalance; point-on-wave. Some suggestions are given in the text. A problem with these additional sag characteristics is that the equipment's reaction to them is not known, not even in a qualitative way.

8.7.2 Stochastic Prediction Techniques

Stochastic prediction techniques will continue to be further developed: both detailed computerized techniques using the method of fault positions as well as simplified methods like the method of critical distances. These developments will reduce the gap between power quality and reliability evaluation. In fact, stochastic prediction of voltage sags may be considered as part of the reliability evaluation of the power supply. Stochastic prediction of voltage sags based on the method of fault positions is likely to become a standard part of power-system analysis software, next to load flow, short-circuit current calculations, transient stability, etc. Calculating the expected number of voltage sags may become as common as calculating the short-circuit current or the normal operating voltage.

It is likely that the first commercially available programs will only give results for magnitude and duration. But soon more characteristics may become part of the calculation results: three-phase unbalance being the most essential one.

The method of critical distances will continue to play a role. It may become part of the stochastic prediction software, e.g., to estimate the extent of and distance between the fault positions. The method of critical distances remains much more powerful than the method of fault positions for fast "back-of-the-envelope" calculations. An example of the latter is the simple expression derived in the last section of Chapter 6. This expression estimates the number of sags due to faults in a meshed transmission system. The drawback with this expression is that there is (not yet) any theoretical basis for it. Further studies and comparisons may teach us about this expression's accuracy level.

8.7.3 Power Quality Surveys

Power quality surveys will also continue to be performed. In fact quite a large number of them is going on at the moment, even though the statistics are not actually being collected in all cases. The number of publications of survey results will however become less, as they are likely to show "more of the same." This is an unfortunate but understandable development. There is a small hope however that the data will be made available for further research, e.g., resulting in statistics for three-phase unbalance, phase-angle jump, point-on-wave, and any other possible sag characteristic. Such data provide very useful results needed to assess the voltage-tolerance requirements of equipment.

The amount of survey results published, even in internal reports, is still very limited. There must be gigabytes of very interesting monitoring data stored at utilities all over the world, waiting to be processed. Only ten years ago it was very difficult to get power system measurements for research purposes. Soon the situation may be that there is a surplus of data for which there are no direct applications. This should of course not stop any utility from installing monitors. The only way of getting an accurate picture of the quality of supply at any given location (i.e., not only sags and interruptions but the whole spectrum of disturbances) is still by means of measuring.

8.7.4 Monitoring or Prediction?

Both monitoring and stochastic prediction are mentioned as a way of obtaining information about the supply performance. Monitoring is still the method most commonly used: it gives not only information on voltage sags but also on other voltage events and variations. Much of this information is still very hard to obtain by stochastic prediction. For voltage sags, however, powerful prediction techniques exist and obtaining accurate results through monitoring may take many years. For individual sites stochastic prediction is most suitable; to obtain the average power quality over a large area (e.g., a whole country) monitoring is more suitable. By comparing monitoring and prediction results the trust in prediction techniques is likely to grow, and the comparison can be used to further develop the prediction techniques.

8.8 MITIGATION METHODS

In Chapter 7 various methods for the mitigation of voltage sags and interruptions are discussed. This is the ultimate aim of any power quality investigation: to solve the problem. The chapter starts with an overview of mitigation methods. Each method is briefly discussed: reducing the number of faults; reducing the fault-clearing time; changing the power system; installing mitigation equipment; and improving equipment immunity. For different types of events, different mitigation methods are most suitable: improving the equipment for short-duration events, improving the system for long-duration events.

Power system design and mitigation equipment are discussed in more detail. The two improvement methods in power system design are parallel operation of components and switching to an alternative supply. Until a few years ago, the latter would only be suitable as a mitigation method against long interruptions. For sag mitigation only certain types of parallel operation were suitable. The introduction of the medium-voltage static switch makes it possible to mitigate voltage sags by very quickly switching to a healthy supply. This may make radial operation a more reliable supply alternative than parallel operation.

Several types of mitigation equipment are discussed in Chapter 7. The emphasis is on shunt and series controllers based on power-electronic voltage-source converters. Through these converters it is possible to compensate for the drop in system voltage or even to temporarily take over the supply completely. For not too deep voltage sags it is possible to compensate the drop in voltage magnitude by injecting reactive power only, but for a full compensation both reactive and active power are needed. The latter calls for a certain amount of energy storage. A number of energy storage options are discussed in the last section of Chapter 7: both classical ones (batteries, capacitors) as well as some of the more recently introduced ones (superconducting coils, high-speed flywheels, supercapacitors). A comparison of the various options shows that batteries and capacitors remain the most-suitable options: capacitors for ridethrough times around one second; batteries for ridethrough times of 10 minutes and longer.

The most commonly used method remains the installation of mitigating equipment at the utility–customer interface or at the equipment terminals. The uninterruptible power supply has become a standard piece of equipment in many installations. This simply takes away lots of worries about the quality of the supply. It is also in many cases the only possible solution: many customers do not have the possibility to opt for improved equipment or for an improved power supply. A recent development is the installation of large mitigation equipment at the utility–customer interface protecting a whole plant against supply disturbances. This may be the cheapest short-term solution, but it should not be used as an excuse to stop the installation and development of less-sensitive equipment.

8.9 FINAL REMARKS

Power quality is an area of power engineering that did not exist only 10 years ago. Power quality and reliability have for many years been part of power system design and operation, but they were rarely considered as a separate area. Being a new area, the developments in power quality are fast and difficult to predict. A new device may be invented tomorrow solving all voltage sag problems.

A more likely development is that sensitive equipment will stay among us for a long time to come. Certainly short and long interruptions will remain a problem. The power quality area will further expand and likely develop into two new areas: a non-technical area covering "customer–utility interactions" and a technical one that will merge with electromagnetic compatibility ("equipment–system interactions"). An additional spin-off of the developments in power quality will be that power system education and research will be much more measurement based than in the past.

Regardless of what the future will bring, power quality in all its varieties will offer utilities, equipment manufacturers, customers, and universities a very interesting field of study, on which lots of cooperation is needed and possible.

Bibliography

[1] T. S. Key, Diagnosing power-quality related computer problems, *IEEE Transactions on Industry Applications*, vol. 15, no. 4, July 1979, pp. 381–393.

[2] M. F. McGranaghan, D. R. Mueller, and M. J. Samotej, Voltage sags in industrial power systems, *IEEE Transactions on Industry Applications*, vol. 29, no. 2, March 1993, pp. 397–403.

[3] IEEE recommended practice for monitoring electric power quality, *IEEE Std. 1159–1995*, New York: IEEE, 1995.

[4] Electromagnetic compatibility (EMC), Part 2: Environment, Section 2: Compatibility levels for low-frequency conducted disturbances and signalling in public low-voltage power supply systems, *IEC Std. 61000-2-2*.

[5] Measurement guide for voltage characteristics, UNIPEDE report 23002 Ren 9531. UNIPEDE documents can be obtained from UNIPEDE 28, rue Jacques Ibert, 75858 Paris Cedex 17, France.

[6] R. Wilkins and M. H. J. Bollen, The role of current limiting fuses in power quality improvement, *3rd Int. Conf. on Power Quality: End-use applications and perspectives*, October 1994, Amsterdam.

[7] Lj. Kojovic and S. Hassler, Application of current limiting fuses in distribution systems for improved power quality and protection, *IEEE Transactions on Power Delivery*, vol. 12, no. 2, April 1997, pp. 791–800.

[8] L. E. Conrad (chair), Proposed Chapter 9 for predicting voltage sags (dips) in revision to IEEE Std 493, the Gold Book, *IEEE Transactions on Industry Applications*, vol. 30, no. 3, May 1994, pp. 805–821.

[9] J. A. Demcko and S. Sullivan, Power quality problems and solutions at Arizona public service company, *7th IEEE Int. Conf. on Harmonics and Quality of Power (ICHPQ)*, Las Vegas, NV, October 1996, pp. 348–353.

[10] *Protective Relays Application Guide*. GEC Alsthom Measurements Ltd, Stafford, U.K., 1987.

[11] R. C. Dugan, L. A. Ray, D. D. Sabin, G. Baker, C. Gilker, and A. Sundaram, Impact of fast tripping of utility breakers on industrial load interruptions, *IEEE Industry Applications Society Annual Meeting*, Denver, CO, October 1994, pp. 2326–2333. A revised version of this paper appeared as "Fast tripping of utility breakers and industrial load interruptions," in *IEEE Industry Applications Magazine*, vol. 2, no. 3, May–June 1996, pp. 55–64.

[12] E. W. Gunther and H. Mehta, A survey of distribution system power quality—Preliminary results, *IEEE Transactions on Power Delivery*, vol. 10, no. 1, January 1995, pp. 322–329.

[13] L. Evans, A. Levy, D. Start, B. H. Turner, and M. J. Williams, *System utilization consultancy group: Project team su1 1—voltage sags*, CEGB Draft Interim report, December 1987, not published.

[14] H. Seljeseth, A. Pleym, K. Sand, and H. Seljeseth, The Norwegian power quality programme, *3rd Int. Conf. on Power Quality: End-use applications and perspectives*, October 1994, Amsterdam. KEMA, Arnhem, The Netherlands, 1994.

[15] M. H. J. Bollen, T. Tayjasajant, and G. Yalcinkaya, Assessment of the number of voltage sags experienced by a large industrial customer, *IEEE Transactions on Industry Applications*, vol. 33, no. 6, November 1997, pp. 1465–1471.

[16] The Excel file containing these measurements was obtained from a web-site with test data set up by R. L. Morgan for IEEE project group P1159.2, with the aim of testing methods of sag characterization. http://grouper.ieee.org/groups/1159/2/index.html.

[17] M. H. J. Bollen, The influence of motor reacceleration on voltage sags, *IEEE Transactions on Industry Applications*, vol. 31, no. 4, July 1995, pp. 667–674.

[18] G. Yalçinkaya and M. H. J. Bollen, Stochastic assessment of frequency, magnitude and duration of non-rectangular sags in a large industrial distribution system, *Power Systems Computation Conference*, August 1996, Dresden, Germany, pp. 1028–1034.

[19] This figure was obtained from the Power Quality monitoring demonstration at the Electrotek Concepts Website. http://www.electrotek.com.

[20] L. E. Conrad and M. H. J. Bollen, Voltage sag coordination for reliable plant operation, *IEEE Transactions on Industry Applications*, vol. 33, no. 6, November 1997, pp. 1459–1464.

[21] IEEE recommended practice for the design of reliable industrial and commercial power systems (The Gold Book), *IEEE Std. 493–1997*.

[22] IEEE recommended practice for evaluating electric power system compatibility with electronic process equipment, *IEEE Std. 1346–1998*.

[23] M. N. Eggleton, E. van Geert, W. L. Kling, M. Mazzoni, and M. A. M. M. van der Meijden, Network structure in sub-transmission systems—features and practices in different countries, *12th Int. Conf. on Electricity Distribution* (*CIRED*), June 1993, paper 6.9, pp. 1–8.

[24] P. M. Anderson, *Analysis of Faulted Power Systems*, New York: IEEE Press, 1995.

[25] Voltage dips, short interruptions and voltage variations immunity tests, *IEC Std. 61000-4-11*.

[26] IEEE recommended practice for emergency and standby power systems for industrial and commercial applications (IEEE Orange Book), New York, *IEEE Std. 446-1995*.

[27] A. Mansoor, E. R. Collins, M. H. J. Bollen, and S. Lahaie, Behaviour of adjustable-speed drives during phase-angle jumps and unbalanced sags, *PQA-97 Europe*, June 1997, Stockholm, Sweden.

[28] Brief 11: Low-voltage ride-through performance of a personal computer power supply, *EPRI Power Quality Database*, Elforsk, Stockholm, Sweden, 1995.

[29] Brief 7: Undervoltage ride-through performance of off-the-shelf personal computers, *EPRI Power Quality Database*, Elforsk, Stockholm, Sweden, 1995.

[30] A. Mansoor, E. R. Collins, and R. L. Morgan, Effects of unsymmetrical voltage sags on adjustable-speed drives, *7th IEEE Int. Conf. on Harmonics and Quality of Power (ICHPQ)*, Las Vegas, NV, October 1996, pp. 467–472.

[31] M. Couvreur, Improving the immunity of industrial power electronics against voltage dips, *11th Int. Conf. on Electricity Distribution (CIRED)*, 22–26 April 1991, Liege, Belgium.

[32] A. Mansoor and R. J. Ferraro, Characterizing ASD power quality application issues, *PQA-97 North America*, March 3–6, 1997, Columbus, OH.

[33] J. Holtz, W. Lotzhat, and S. Stadfeld, Controlled AC drives with ride-through capacity at power interruption, *IEEE Transactions on Industry Applications*, vol. 30, no. 5, September 1994, pp. 1275–1283.

[34] Brief 10: Low-voltage ride-through performance of AC contactor motor starters, *EPRI Power Quality Database*, Elforsk, Stockholm, Sweden, 1995.

[35] C. Pumar, J. Amantegui, J. R. Torrealday, and C. Ugarte, A comparison between DC and AC drives as regards their behaviour in the presence of voltage dips: New techniques for reducing the susceptibility of AC drives, *Int. Conf. on Electricity Distribution (CIRED)*, June 2–5, 1997, Birmingham, U.K., pp. 9/1–5.

[36] D. S. Dorr, A. Mansoor, A. G. Morinec, and J. C. Worley, Effects of power line voltage variations on different types of 400-W high-pressure sodium ballasts, *IEEE Transactions on Industry Applications*, vol. 33, no. 2, March 1997, pp. 472–476.

[37] J. Lamoree, D. Mueller, P. Vinett, W. Jones, and M. Samotyj, Voltage sag analysis case studies, *IEEE Transactions on Industry Applications*, vol. 30, no. 4, July 1994, pp. 1083–1089.

[38] A. E. Turner and E. R. Collins, The performance of AC contactors during voltage sags, *7th Int. Conf. on Harmonics and Quality of Power (ICHPQ)*, Las Vegas, NV, October 1996, pp. 589–595.

[39] PQTN Brief 39: Ride-through performance of programmable logic controllers, *EPRI Power Electronics Applications Center*, Knoxville, TN, November 1996.

[40] H.G. Sarmiento and E. Estrada, A voltage sag study in an industry with adjustable-speed drives, *IEEE Industry Applications Magazine*, vol. 2, no. 1, January 1996, pp. 16–19.

[41] J.C. Smith, J. Lamoree, P. Vinett, T. Duffy, and M. Klein, The impact of voltage sags on industrial plant loads, *Int. Conf. Power Quality: End-use applications and perspectives (PQA-91)*, pp. 171–178.

[42] R. A. Epperly, F. L. Hoadley, and R. W. Piefer, Considerations when applying ASD's in continuous processes, *IEEE Transactions on Industry Applications*, vol. 33, no. 2, March 1997, pp. 389–396.

[43] W. Sheperd, L. N. Hulley, and D. T. W. Liang, *Power Electronics and Motor Control, 2nd ed.*, Chapter 12, Cambridge University Press, Cambridge, U.K., 1995.

[44] L. Moran, P. D. Ziogas, and G. Joos, Design aspects of synchronous PWM rectifier-inverter systems under unbalanced input voltage conditions, *IEEE Transactions on Industry Applications*, vol. 28, no. 6, November 1992, pp. 1286–1293.

[45] E. P. Wiechmann, J. R. Espinoza, and J. L. Rodriguez, Compensated carrier PWM synchronization: A novel method to achieve self-regulation and AC unbalance compensation in AC fed converters, *IEEE Transactions on Power Electronics*, vol. 7, no. 2, April 1992, pp. 342–348.

[46] P. Rioual, H. Pouliquen, and J.-P. Louis, Regulation of a PWM rectifier in the unbalanced network state using a generalized model, *IEEE Transactions on Power Electronics*, vol. 11, no. 3, May 1996, pp. 495–502.

[47] E. G. Strangas, V. E. Wagner, and T. D. Unruh, Variable speed drives evaluation test, *IEEE Industry Applications Society Annual Meeting*, October 1996, San Diego, CA, pp. 2239–2243. A revised version of this paper appeared in *IEEE Industry Applications Magazine*, vol. 4, no. 1, January 1998, pp. 53–57.

[48] L. Conrad, K. Little, and C. Grigg, Predicting and preventing problems associated with remote fault-clearing voltage dips, *IEEE Transactions on Industry Applications*, vol. 27, no. 1, January 1991, pp. 167–172.

[49] Y. Sekine, T. Yamamoto, S. Mori, N. Saito, and H. Kurokawa, Present state of momentary voltage dip interferences and the countermeasures in Japan, *Int. Conf. on Large Electric Networks (CIGRE)*, 34th Session, September 1992, Paris, France.

[50] D. Dorr, T. Key, and G. Sitzlar, User expectations and manufacturer tendencies for system-compatible design of end-use appliances, *3rd Int. Conf. on Power Quality: End-use applications and perspectives*, October 1994, Amsterdam, The Netherlands.

[51] A. David, J. Maire, and M. Dessoude, Influence of voltage dips and sag characteristics on electrical machines and drives: evaluation and perspective, *3rd Int. Conf. on Power Quality: End-use applications and perspectives*, October 1994, Amsterdam, The Netherlands.

[52] *IEC Standard 1800-3*, Adjustable speed electric drive systems, Part 3: EMC product standard including specific test methods.

[53] N. Mohan, T. M. Undeland, and W. P Robbins, *Power Electronics—Converters, Applications and Design*, John Wiley and Sons, New York, 1995.

[54] D. S. Dorr, M. B. Hughes, T. M. Gruzs, R. E. Jurewicz, and J. L. McClaine, Interpreting recent power quality surveys to define the electrical environment, *IEEE Transactions on Industry Applications*, vol. 33, no. 6, November 1997, pp. 1480–1487.

[55] Kj. Torborg, *Power Electronics—in Theory and Practice*, Lund, Sweden: Studentliteratur, 1993. Published in the United States by Chartwell Bratt, Ltd.

[56] E. R. Collins and R. L. Morgan, A three-phase sag generator for testing industrial equipment, *IEEE Transactions on Power Delivery*, vol. 11, no. 1, January 1996, pp. 526–532.

[57] P. Wang, The use of FACTS devices to mitigate voltage sags, PhD thesis, Department of Electrical Engineering and Electronics, University of Manchester Institute of Science and Technology, Manchester, U.K., July 1997.

[58] D. O. Koval and J. J. Leonard, Rural power profiles, *IEEE Transactions on Industry Applications*, vol. 30, no. 2, March–April 1994, pp. 469–475.

[59] D.O. Koval, How long should power system disturbance site monitoring be significant? *IEEE Transactions on Industry Applications*, vol. 26, no. 4, July–August 1990, pp. 705–710.

[60] J. Marquet, CREUTENSI: Software for determination of depth, duration and number of voltage dips (sags) on medium voltage networks, *EDF (Electricité de France)*, Clamart, France, September 1992.

[61] M. H. J. Bollen, Method for reliability analysis of industrial distribution systems, *IEE Proceedings-C*, vol. 140, no. 6, November 1993, pp. 497–502.

[62] M. H. J. Bollen and P. E. Dirix, Simple motor models for reliability/power quality assessment of industrial power systems, *IEE Proceedings—Generation, Transmission, Distribution*, vol. 143, no.1, January 1996, pp. 56–60.

[63] M. H. J. Bollen, Reliability analysis of industrial power systems taking into account voltage sags, *IEEE Industry Applications Society Annual Meeting*, Toronto, Canada, October 1993, pp. 1461–1468.

[64] M. R. Qader, M. H. J. Bollen, and R. N. Allan, Stochastic prediction of voltage sags in the reliability test system, *PQA-97 Europe*, June 1997, Elforsk, Stockholm, Sweden.

[65] M. B. Hughes and J. S. Chan, Early experiences with the Canadian national power quality survey, *Transmission and Distribution International*, vol. 4, no. 3, September 1993, pp. 18–27.

[66] D. O. Koval, R. A. Bocancea, and M. B. Hughes, Canadian national power quality survey: Frequency of industrial and commercial voltage sags, *IEEE Transactions on Industry Applications*, vol. 35, no. 5, September 1998, pp. 904–910.

[67] H. Seljeseth and A. Pleym, Spenningskvalitetsmålinger 1992 until 1996 (voltage quality measurements, 1992 to 1996, in Norwegian), Report EFI TR A4460 published by EFI, 7034 Trondheim, Norway.

[68] D. S. Dorr, Point of utilization power quality study results, *IEEE Transactions on Industry Applications*, vol. 31, no. 4, July 1995, pp. 658–666.

[69] R. E. Jurewicz, Power quality study—1990 to 1995, *Int. Telecommunications Energy Conf. (INTELEC)*, October 1990, Orlando, FL, pp. 443–450.

[70] E. W. Gunther and H. Mehta, A survey of distribution system power quality—preliminary results, *IEEE Transactions on Power Delivery*, vol. 10, no. 1, January 1995, pp. 322–329.

[71] M. R. Qader, Stochastic assessment of voltage sags due to short circuits in electrical networks, PhD thesis, UMIST, Manchester, U.K., 1997.

[72] M. Goldstein and P. D. Speranza, The quality of U.S. commercial AC power, *Int. Telecommunications Energy Conf. (INTELEC)*, October 1982; Washington, DC, pp. 28–33.

[73] IEEE reliability test system, *IEEE Transactions on Power Apparatus and Systems*, vol. 98, no. 6, November 1979, pp. 2047–2054.

[74] M. R. Qader, M. H. J. Bollen, and R. N. Allan, Stochastic prediction of voltage sags in a large transmission system, *IEEE Transactions on Industry Applications*, vol. 35, no. 1, January 1999, pp. 152–162.

[75] R. C. Dugan, M. F. McGranaghan, and H. W. Beaty, *Electric Power Systems Quality*, McGraw Hill, New York, 1996.

[76] W. E. Kazibwe and M. H. Sendaula, *Electric Power Quality Control Techniques*, Van Nostrad Reinhold, New York.

[77] G. T. Heydt, Electric power quality, West LaFayette, In: *Stars in a Circle*, 1991. Only obtainable from Stars in a circle publications, 2932 SR 26W, West LaFayette, IN, 47906.

[78] IEEE recommended practice for powering and grounding sensitive electronic equipment, *IEEE Std. 1100-1992.*

[79] Electromagnetic compatibility (EMC), Part 1: General, Section 1: Application and interpretation of fundamental definitions and terms, IEC 61000-1-1.

[80] *European standard EN-50160*, Voltage characteristics of electricity supplied by public distribution systems, *CENELEC*, Brussels, Belgium, 1994.

[81] *IEC Std. 61000-4-15*, Flickermeter—functional and design characteristics, *IEC*, Geneva, Switzerland, 1997.

[82] IEEE Standard 519, Recommended practices and requirements for harmonic control in electrical power systems, *ANSI/IEEE Std. 519-1992*, IEEE, New York, 1993.

[83] CIGRE Working Group 26-05, Harmonics, characteristic parameters, method of study, estimates of existing values in the network, *Electra*, no. 77, July 1981, pp. 35–54.

[84] R. Billinton and R. N. Allan, *Reliability Evaluation of Power Systems*, 2nd ed., New York: Plenum, 1996.

[85] R. Billinton and R. N. Allan, *Reliability Assessment of Large Electric Power Systems*, Boston: Kluwer, 1988.

[86] R. Billinton and W. Li, *Reliability Assessment of Electric Power Systems Using Monte Carlo Simulation*, New York: Plenum, 1994.

[87] J. Endreyni, *Reliability Modeling in Electric Power Systems*, New York: John Wiley and Sons, Ltd., 1978.

[88] F. W. Kloeppel, *Zuverlässigkeit von Elektroenergiesystemen*, Leipzig, Germany: Verlag für Grondstoffenindustrie, 1990.

[89] H. D. Kochs, *Zuverlässigkeit Elektrotechnische Anlagen*, Berlin: Springer, 1984.

[90] R. N. Allan, R. Billinton, and S. H. Lee, Bibliography on the application of probability methods in power system reliability evaluation 1977–1982, *IEEE Transactions on Power Apparatus and Systems*, vol. 103, no. 2, February 1984, pp. 275–282.

[91] R. N. Allan, R. Billinton, S. M. Shahidehpour, and C. Singh, Bibliography on the application of probability methods in power system reliability evaluation 1982–1987, *IEEE Transactions on Power Systems*, vol. 3, no. 4, February 1994, pp. 1555–1564.

[92] R. N. Allan, R. Billinton, A. M. Briepohl, and C. H. Grigg, Bibliography on the application of probability methods in power system reliability evaluation 1987–1991, *IEEE Transactions on Power Systems*, vol. 9, no. 1, February 1994, pp. 41–49.

[93] Composite power system reliability: Phase 1—scoping study. Final report, EPRI EL-5290, Palo Alto, CA: Electric Power Research Institute, December 1987.

[94] CIGRE Working Group 38.03, *Power System Reliability Analysis—Application Guide.* Paris: CIGRE publications, 1988.

[95] H. H. Kajihara, Quality power for electronics, *Electro-Technology*, vol. 82, no. 5, November 1968, p. 46.

[96] D. J. Hucker, Aircraft a.c. electric system power quality, *Proceedings of the IEEE National Aerospace Electronics Conference*, May 1970, Dayton, OH, pp. 426–430.

[97] R. K. Walter and H. Heinzmann, A customer discusses airborne static power conversion, *Conference Record of the 4th Annual Meeting of the IEEE Industry and General Applications Group*, October 1969, Detroit, MI, pp. 611–616.

[98] D. L. Plette, The effects of improved power quality on utilization equipment, *Proceedings of the IEEE National Aerospace Electronics Conference*, May 1969, Dayton, OH, pp. 243–250.

[99] R. H. McFadden, Power system analysis—what it can do to industrial plants, *Conference Record of the 5th Annual Meeting of the IEEE Industry and General Applications Group*, October 1970, Chicago, IL, pp. 189–199. This paper also appeared in *IEEE Transactions on Industry and General Applications*, vol. 7, no. 2, March 1971, pp. 181–188.

[100] P. M. Knoller and L. Lonnstam, Voltage quality and voltage tendency recorders, *Siemens-Review*, vol. 36, no. 8, August 1969, pp. 302–303.

[101] A. Lidholm, Mättekniska hjälpmedel för bestämning av spänningsgodheten i lägspänningsnät (Measuring techniques applicable to the determination of voltage quality in low-voltage networks, in Swedish), *ERA*, vol. 42, no. 5, 1969, pp. 99–101.

[102] B. A. Konstantinov and G. L. Bagiev, Financial losses due to deterioration of voltage quality, *Electric-Technology-USSR*, vol. 1, 1970, pp. 119–123.

[103] R. D. Hof, The "dirty power" clogging industry's pipeline, *Business Week*, April 8, 1991.

[104] J. Douglas, Quality of power in the electronics age, *EPRI Journal*, vol. 10, no. 9, November 1985, pp. 6–13.

[105] W. E. Kazibwe, R. J. Ringlee, G. W. Woodzell, and H. M. Sendaula, Power quality: A review, *IEEE Computer Applications in Power*, vol. 3, no. 1, January 1990, pp. 39–42.

[106] F. Martzloff, Power quality work at the International Electrotechnical Commission, *PQA-97 Europe*, June 1997, Stockholm, Sweden, Elforsk: Stockholm, Sweden.

[107] M. H. J. Bollen, Literature search for reliability data of components in electric distribution networks, Eindhoven University of Technology Research Report 93-E-276, August 1993.

[108] Basnivo för elkvalitet (Basic level for quality of electricity, in Swedish), Göteborg Energi nät AB, Gothenburg, Sweden, January 1997.

[109] The Office of Electricity Regulation (OFFER), Birmingham, U.K., annually publishes two reports on the quality of the supply: Report on distribution and transmission system performance; and Report on customer services.

[110] P. C. M. van Kruining, J. H. P. Lommert, H. H. Overbeek, and R. J. R. Waumans, eds., *Elektriciteitsdistributienetten* (Electricity distribution networks, in Dutch), Deventer, The Netherlands: Kluwer Techniek, 1996.

[111] R. J. R. Waumans, *Openbare Netten voor Electriciteitsdistributie*, Deventer, The Netherlands: Kluwer, 1986.

[112] M. H. A. J. Hendriks Boers and R. M. L. Frenken, Help! De klant kan kiezen (Help, the customer is allowed to choose, in Dutch), *Energietechniek*, vol. 75, no. 12, December 1997, pp. 682–685.

[113] M. H. J. Bollen and A. Boyd, Instability problems in small power systems with more than one centre of generation, *Universities Power Engineering Conf.*, September 1995, Greenwich, U.K., pp. 191–194.

[114] E. Lakervi and E. J. Holmes, *Electricity Distribution Network Design*, 2nd ed., Institution of Electrical Engineers, London, U.K., 1995, Chapters 4 and 12.

[115] J. A. Burke, *Power Distribution Engineering, Fundamentals and Applications*, New York: Marcel Dekker, 1994, Chapters 6 and 7.

[116] F. S. Prabhakara, R. L. Smith, and R. P. Stratford, *Industrial and Commercial Power Systems Handbook*, New York: McGraw-Hill, 1996, Chapter 11.

[117] The INSPEC database has been accessed via the WebSPIRS server.

[118] The IEEE standard dictionary of electrical and electronics terms, 6th ed., *IEEE Std. 100-1996*, IEEE, New York, 1997.

[119] Engineering Recommendation P2/5: Power system design.

[120] Electricity Plan 1997-2006, and Notes to the Electricity Plan 1997-2006, N.V. September/ Dutch Electricity Generating Board, Arnhem, The Netherlands, 1996.

[121] R. L. Capra, M. W. Sangel, and S. V. Lyon, Underground distribution system design for reliability, *IEEE Transactions on Power Apparatus and Systems*, vol. 88, no. 6, June 1969, pp. 834–842.

[122] N.E. Chang, Evaluate distribution system design by cost reliability indices, *IEEE Transactions on Power Apparatus and Systems*, vol. 96, no. 5, September 1977, pp. 1480–1490.

[123] B. H. T. Smeets and M. H. J. Bollen, Stochastic modelling of protection systems—comparison of four mathematical techniques, Eindhoven University of Technology Research Report 95-E-291, June 1995.

[124] J. W. H. Bär, H. L. Doppen, H. Juckers, H. M. A. Konings, P. Wiersma, H. Zuijderduin, and M. J. Voeten, *Onderhoudsanalyse* (Maintenance analysis, in Dutch), VEEN, Commissie Onderhoudsvraagstukken, Arnhem, The Netherlands, 1990.

[125] E. T. Parascos and J. A. Arceri, Reliability engineering and underground equipment failure, costs and manufacturer's analysis, *3rd Annual Reliability Engineering Conf.*, September 1976, Montreal, Canada, pp. 25–30.

[126] D. A. Ducket and C. M. McDonough, A guide for transformer replacement based on reliability and economics, *Rural Electric Power Conference*, April 1990, Orlando, FL.

[127] N. E. Wiseman, Reliability testing, *3rd Annual Reliability Engineering Conf.*, September 1976, Montreal, Canada, pp. 21–24.

[128] T. Kawamura, M. Horikoshi, S. Kabayashi, and K. Hamamoto, Progress of substation maintenance based on records of operation and maintenance, *Int. Conf. on Large High Voltage Electric Systems (CIGRE)*, August 1990, paper 23-102.

[129] G. Wacker and R. Billinton, Customer costs of electric service interruptions, *Proceedings of the IEEE*, vol. 77, no. 6, June 1989, pp. 919–930.

[130] A. Mäkinen, J. Partanen, and E. Lakervi, A practical approach to estimating future outage costs in power distribution networks, *IEEE Transactions on Power Delivery*, vol. 5, no. 1, January 1990, pp. 313–316.

[131] A. M. Shaalan, Electric service interruptions: Impacts and costs estimation, *Electra (CIGRE)*, no. 127, 1989, pp. 89–109.

[132] G. Wacker, E. Wojczynski, and R. Billinton, Interruption cost methodology and results—A Canadian residential survey, *IEEE Transactions on Power Apparatus and Systems*, vol. 102, no. 10, October 1983, pp. 3385–3392.

[133] S. Burns and G. Gross, Value of service reliability, *IEEE Transactions on Power Systems*, vol. 5, no. 3, August 1990, pp. 825–834.

[134] IEEE Project Group 1159.2, *Recommended practices for the characterization of a power quality event*, draft 1, March 1998.

[135] P. C. Krause, O. Wasynczuk, and S. D. Sudhoff, *Analysis of electric machinery*, New York: IEEE Press, 1995.

[136] G. Yalçinkaya, The influence of induction motors on voltage sags due to short circuits, PhD thesis, University of Manchester Institute of Science and Technology, Manchester, U.K., October 1997.

[137] G. Yalçinkaya, M. H. J. Bollen, and P. A. Crossley, Characterisation of voltage sags in industrial distribution systems, *IEEE Transactions on Industry Applications*, vol. 34, no. 4, July 1998, pp. 682–688.

[138] E. Camm, Preventing nuisance tripping during overvoltages caused by capacitor switching, in: P. Pilay, Ed., "Motor drive/power systems interactions", *IEEE Industry Applications Society Tutorial Course*, October 1997.

[139] C. M. Warren, The effect of reducing momentary outages on distribution reliability indices, *IEEE Transactions on Power Delivery*, vol. 7, no. 3, July 1992, pp. 1610–1617.

[140] Standards coordinating committee SCC-22: Charter, www.standards.ieee.org.

[141] T. Larsson, Voltage source converters for mitigation of flicker caused by arc furnaces, PhD thesis, Royal Institute of Technology, Department of Electric Power Engineering, Stockholm, Sweden, 1998.

[142] P. M. Anderson and R. G. Farmer, Series compensation of power systems, Section 7.3, *Lamp Flicker*, *PBLSH!*, Encinitas, CA, 1996.

[143] J. G. Kappenman and V. D. Albertson, Bracing for the geomagnetic storms, *IEEE Spectrum*, vol. 27, no. 3, March 1990, pp. 27–33.

[144] A. Lui, Ferroresonant overvoltages in electrical distribution systems, MSc thesis, University of Manchester Institute of Science and Technology, Manchester, U.K., March 1997.

[145] H. J. Koglin, E. Roos, H. J. Richter, U. Scherer, and W. Wellssow, Experience in the reliability evaluation of high voltage networks, *Int. Conf. on Large High Voltage Electric Networks (CIGRE)*, September 1986, Paris, France.

[146] M. Mitra and S. K. Basu, On some properties of the bathtub failure rate family of lifetime distributions, *Microelectronics and Reliability*, vol. 36, no. 5, May 1996, pp. 679–684.

[147] L. V. Skof, Customer interruption costs vary widely, *Electrical World*, vol. 188, no. 2, July 15, 1977, pp. 64–65.

[148] E. C. Ifeachor and B. W. Jervis, *Digital Signal Processing: A Practical Approach*, Wokingham, England: Addison-Wesley, 1993.

[149] Cigre Working Group 34.01, Reliable fault clearance and back-up protection, Final report, August 1997.

[150] K. Benson and J. R. Chapman, Boost converters provide power dip ride-through for ac drives, *Power Quality Assurance Magazine*, July 1997.

[151] PQTN Brief No. 34, Performance of an ASD ride-through device during voltage sags, EPRI PEAC, Knoxville, TN, May 1996.

[152] J. Svensson, Synchronisation methods for grid-connected voltage source converters, *IEE Proceedings—Electric Power Applications*, in print.

[153] J. Svensson, Grid-connected voltage source converter–control principles and wind energy applications, PhD thesis, Chalmers University of Technology, School of Electrical and Computer Engineering, Gothenburg, Sweden, 1998. Technical Report No. 331.

[154] R. de Graaff, Simulating the behavior of AC and DC adjustable-speed drives during voltage sags, project report, Chalmers University of Technology, Dept. of Electric Power Engineering, Gothenburg, Sweden, September 1997.

[155] L. E. Conrad, personal communication.

[156] D. L. Brooks, R. C. Dugan, M. Waclawiak, and A. Sundaram, Indices for assessing utility distribution system RMS variation performance, *IEEE Transactions on Power Delivery*, vol. 13, no. 1, January 1998, pp. 254–259.

[157] P. Wang, N. Jenkins, and M. H. J. Bollen, Experimental investigation of voltage sag mitigation by an advanced Static VAr Compensator, *IEEE Transactions on Power Delivery*, vol. 13, no. 4, October 1998, pp. 1461–1467.

[158] J. Lamoree, L. Tang, C. DeWinkel, and P. Vinett, Description of micro-SMES system for protection of critical customer facilities, *IEEE Transactions on Power Delivery*, vol. 9, no. 2, April 1994, pp. 984–991.

[159] C. S. Hsu and W. J. Lee, Superconducting magnetic energy storage for power system applications, *IEEE Transactions on Industry Applications*, vol. 29, no. 5, September 1993, pp. 990–996.

[160] R. H. Lasseter and S. G. Jalali, Power conditioning systems for superconductive magnetic energy storage, *IEEE Transactions on Energy Conversion*, vol. 6, no. 3, September 1991, pp. 381–387.

[161] R. H. Lasseter and S. G. Jalali, Dynamic response of power conditioning system for superconductive magnetic energy storage, *IEEE Transactions on Energy Conversion*, vol. 6, no.3, September 1991, pp. 388–391.

[162] I. D. Hassan, R. M. Bucci, and K. T. Swe, 400 MW SMES power conditioning system development and simulation, *IEEE Transactions on Power Electronics*, vol. 8, no. 3, July 1993, pp. 237–249.

[163] T. W. Diliberti, V. E. Wagner, J. P. Staniak, S. L. Sheppard, and T. L. Orfloff, Power quality requirements of a large industrial user: a case study, *IEEE Industrial and Commercial Power Systems Technical Conference*, Detroit, MI, May 1990, pp. 1–4.

[164] T. Gönen, *Electric Power Distribution System Engineering*, New York: McGraw-Hill, 1986.

[165] R. C. Settembrini, J. R. Fisher, and N. E. Hudak, Reliability and quality comparisons of electric power distribution systems, *IEEE Transmission and Distribution Conf.*, September 1991, Dallas, TX, pp. 704–712.

[166] R. W. DeDoncker, W. T. Eudy, J. A. Maranto, H. Mehta, and J. W. Schwartzenberg, Medium voltage subcycle transfer switch, *Power Quality Assurance*, July 1995, pp. 46–51.

[167] T. H. Ortmeyer and T. Hiyama, Coordination of time overcurrent devices with voltage sag capability curves, *IEEE Int. Conf. on Harmonics and Quality of Power (ICHPQ)*, Las Vegas, NV, October 1996, pp. 280–285.

[168] S. M. Schoenung, W. R. Meier, and R. L. Bieri, Small SMES technology and cost reduction estimates, *IEEE Transactions on Energy Conversion*, vol. 9, no. 2, June 1994, pp. 231–237.

[169] J. J. Skiles, R. L. Kustom, K.-P. Ko, V. Wong, K.-S. Ko, F. Vong, and K. Klontz, Performance of a power conversion system for superconducting magnetic energy storage (SMES), *IEEE Transactions on Power Systems*, vol. 11, no. 4, November 1996, pp. 1718–1723.

[170] A. van Zyl, J. H. R. Enslin, and R. Spée, Converter-based solution to power quality problems on radial distribution lines, *IEEE Transactions on Industry Applications*, vol. 32, no. 6, November 1996, pp. 1323–1330.

[171] N. Woodley, M. Sarkozi, F. Lopez, V. Tahiliani, and P. Malkin, Solid-state 13-kV distribution class circuit breaker: planning, development and demonstration, *IEE Conf. on Trends in Distribution Switchgear*, November 1994, London, U.K., pp. 163–167.

[172] R. J. Lawrie, Power system design for ultra-reliability, *Electric Construction and Maintenance*, vol. 89, no. 5, May 1990, pp. 55-64.

[173] J. W. Schwartzenberg and R. W. DeDoncker, 15 kV medium voltage transfer switch, *IEEE Industry Applications Society Annual Meeting*, October 1995, Orlando, FL, pp. 2515–2520.

[174] S. W. Middlekauff and E. R. Collins, System and customer impact: considerations for series custom power devices, *IEEE Transactions on Power Delivery*, vol. 13, no. 1, January 1998, pp. 278–282.

[175] R. K. Smith, P. G. Slade, M. Sarkozi, E. J. Stacey, J. J. Bonk, and H. Mehta, Solid state distribution current limiter and circuit breaker: Application requirements and control strategies, *IEEE Transactions on Power Delivery*, vol. 8, no. 3, July 1993, pp. 1155–1164.

[176] J. A. Aspinall *et al.*, eds., *Power System Protection*, Institution of Electrical Engineers, London, U.K., 1995.

[177] The National Grid Company Seven-Year Statement for the Years 1994/5 to 2000/2001, Coventry, U.K.: National Grid Company, July 1994.

[178] H. Akagi, Y. Kanazawa, and A. Nabae, Instantaneous reactive power compensators comprising switching devices without energy storage components, *IEEE Transactions on Industry Applications*, vol. 20, no. 3, May 1984, pp. 625–630.

[179] W. M. Grady, M. J. Samotyj, and A. H. Noyola, Survey of active power line conditioning methodologies, *IEEE Transactions on Power Delivery*, vol. 5, no. 3, July 1990, pp. 1536–1542.

[180] C. Schauder, M. Gerhardt, E. Stacey, T. W. Cease, A. Edris, T. Lemak, and L. Guygyi, Development of a 100 MVar static condenser for voltage control of transmission systems, *IEEE Transactions on Power Delivery*, vol. 10, no. 3, July 1995, pp. 1486–1496.

[181] L. Guygyi, C. D. Schauder, S. L. Williams, T. R. Rietman, D. R. Torgerson, and A. Edris, The unified power flow controller: A new approach to power transmission control, *IEEE Transactions on Power Delivery*, vol. 10, no. 2, April 1995, pp. 1085–1097.

[182] H. Akagi, Trends in active power line conditioners, *IEEE Transactions on Power Electronics*, vol. 9, no. 3, May 1994, pp. 263–268.

[183] H. Akagi, New trends in active filters for power conditioning, *IEEE Transactions on Industry Applications*, vol. 32, no. 6, November/December 1996, pp. 1312–1322.

[184] N. H. Woodley, M. Sarkozi, A. Sandaram, and G. A. Taylor, Custom power: The utility solution, *Int. Conf. on Electricity Distribution (CIRED)*, May 1995, Brussels, Belgium, vol. 5, pp. 9/1–6.

[185] IEE Colloquium of Dynamic Voltage Restorers—Replacing those missing cycles, Institution of Electrical Engineers, London, U.K., IEE Digest No. 1998/189.

[186] B. Bhargava and G. Dishaw, Application of an energy source power system stabilizer on the 10 MW battery energy storage system at Chino substation, *IEEE Transactions on Power Systems*, vol. 13, no. 1, February 1998, pp. 145–151.

[187] S. J. Chiang, C. M. Liaw, W. C. Chang, and W. Y. Chang, Multi-module parallel small battery energy storage system, *IEEE Transactions on Energy Conversion*, vol. 11, no. 1, March 1996, pp. 146–154.

[188] N. W. Miller, R. S. Zrebiec, R. W. Delmerico, and G. Hunt, Design and commissioning of a 2.5 MWh battery energy storage system, *Int. Conf. on Electricity Distribution (CIRED)*, June 2–5, 1997, Birmingham, U.K., vol. 2, pp. 10/1–5.

[189] P. Pillay, A switched reluctance drive with improved ride-through capability, in: "Motor drive/power system interactions", *IEEE Tutorial Course 97 TP 123*, October 1997.

[190] B. G. Johnson, K. P. Adler, G. V. Anastas, J. R. Dowver, D. B. Eisenhaure, J. H. Goldie, and R. L. Hockney, Design of a torpedo inertial power storage unit (TIPSU), *Intersociety Energy Conversion Engineering Conference*, August 1990, Reno, NV, vol. 4, pp. 199–204.

[191] N. G. Hingorani, Introducing custom power, *IEEE Spectrum*, vol. 32, no. 6, June 1995, pp. 41–47.

[192] J. R. Hull, Flywheels on a roll, *IEEE Spectrum*, vol. 34, no. 7, July 1997, pp. 20–25.

[193] R. T. Morash, J. F. Roesel, and R. J. Barber, Improved power quality with written-pole motor-generators and written-pole motors, *3rd Int. Conf. on Power Quality: End-use applications and perspectives (PQA 94)*, October 1994, Amsterdam, The Netherlands. Paper D-2.10.

[194] J. Arrillaga, D. Bradley, and P. S. Bodger, *Power System Harmonics*, London: John Wiley and Sons, Ltd., 1985.

[195] J. Arrillaga, B. C. Smith, N. R. Watson, and A. R. Wood, *Power System Harmonic Analysis*, Chichester: John Wiley and Sons, Ltd., 1997.

[196] A. Larsson, M. Lundmark, and J. Hagelberg, Increased pollution in the protective earth, *European Conf. on Power Electronics and Applications*, Trondheim, Norway, September 8–10, 1997.

[197] C. R. Heising, Worldwide reliability survey of high-voltage circuit breakers, *IEEE Industry Applications Magazine*, vol. 2, no. 3, May 1996, pp. 65–66.

[198] IEEE standard terms for reporting and analyzing outage occurences and outage states of electrical transmission facilities, *IEEE Std. 859-1993*.

[199] G. A. Taylor, J. E. Hill, A. B. Burden, and K. Mattern, Responding to the changing demands of lower voltage networks—the utilisation of custom power systems, *Int. Conf. on Large Electric Networks (CIGRE)*, Session 1998, Paris, France, August 1998, paper 14-303.

[200] Avbrottskostnader för elkunder (Interruption costs for electricity customers, in Swedish), Svenska Elverksföreningen, Stockholm, Sweden, 1994.

[201] N. S. Tunaboyla and E. R. Collins, The wavelet transform approach to detect and quantify voltage sags, *7th Int. Conf. on Harmonics and Quality of Power*, October 1996, Las Vegas, NV, pp. 619–624.

[202] O. Poisson, P. Rioual, and M. Meunier, Detection and measurement of power quality disturbances using wavelet transform, *8th. Int. Conf. on Harmonics and Quality of Power*, Athens, Greece, October 1998, pp. 1125–1130.

[203] L. D. Zhang and M. H. J. Bollen, A method for characterizing unbalanced voltage dips (sags) with symmetrical components, *IEEE Power Engineering Review*, vol. 18, no. 7, July 1998, pp. 50–52.

[204] L. D. Zhang and M. H. J. Bollen, Characteristics of voltage dips (sags) in power systems, *8th. Int. Conf. on Harmonics and Quality of Power*, Athens, Greece, October 1998, pp. 555–560.

[205] G. Desuilbet, C. Foucher, and P. Fauquembergue, Statistical analysis of voltage dips, *3rd. Int. Conf. on Power Quality: End-use applications and perspectives*, October 1994, Amsterdam, The Netherlands.

[206] B. K. Bose, Ed., *Power Electronics and Variable Frequency Drives—Technology and Applications*, New York: IEEE Press, 1997.

[207] E. W. Kimbark, *Power System Stability*, New York: IEEE Press, 1995.

[208] N. Carter, Improvements in network performance on urban and rural 11 kV networks, In: *Improving Power Quality in Transmission and Distribution*, January 1998, Amsterdam, IBC Technical Services, London, U.K., 1998.

[209] H. L. Willis, *Power Distribution Planning Reference Book*, New York: Marcel Dekker, 1997.

[210] G. T. Heydt, W. Tan, T. LaRose, and M. Negley, Simulation and analysis of series voltage boost technology for power quality enhancement, *IEEE Transactions on Power Delivery*, vol. 13, no. 4, October 1998, pp. 1335–1341.

[211] J. Lundquist, Omriktare och olinjära laster i elektriska nät (Converters and nonlinear load in electric power systems, in Swedish), MSc Report, Högskolan i Skövde, Skövde, Sweden, December 1996.

[212] R. Hart, HV overhead line—the Scandinavian experience, *Power Engineering Journal*, June 1994, pp. 119–123.

[213] M. W. Vogt, T2 ACSR conductors: lessons learned, *IEEE Industry Applications Magazine*, vol. 4, no. 3, May 1998, pp. 37–39.

[214] Recommended practice for electric power distribution for industrial plants (IEEE Red Book), *IEEE Std. 141-1993.*

[215] H. C. Tijms, *Stochastic Models: An Algorithmic Approach.* Chichester, U.K.: Wiley, 1994.

[216] K. K. Kariuki and R. N. Allan, Factors affecting customer outage costs due to electric service interruptions, *IEE Proceeding—Generation, Transmission, Distribution*, vol. 143, no. 6, November 1996, pp. 521–528.

Appendix A: Overview of EMC Standards

IEC 61000: Electromagnetic Compatibility (EMC) consists of 6 parts, each consisting of several sections. Below a list is given of the sections related to power quality, as well as some documents currently (February 1999) under development.

- **Part 1: General**

  – Section 1: Application and interpretation of fundamental definitions and terms.

  – Section 2: Methodology for the achievement of functional safety of electrical and electronic equipment (in preparation).

- **Part 2: Environment**

  – Section 1: Description of the environment—Electromagnetic environment for low-frequency conducted disturbances and signalling in power supply systems.

  – Section 2: Compatibility levels for low-frequency conducted disturbances and signalling in public supply systems.

  – Section 3: Description of the environment—Radiated and non-network-frequency-related conducted disturbances.

  – Section 4: Compatibility levels in industrial plants for low-frequency conducted disturbances.

  – Section 5: Classification of electromagnetic environments.

  – Section 6: Assessment of the emission levels in the power supply of industrial plants as regards low-frequency conducted disturbances.

  – Section 7: Low-frequency magnetic fields in various environments.

  – Section 8: Voltage dips, short interruptions and statistical measurement results (in preparation).

  – Section 12: Compatibility levels for low-frequency conducted disturbances and signalling in public medium-voltage power supply systems (in preparation).

- **Part 3: Limits**
  - Section 1: Overview of emission standards and guides (in preparation).
  - Section 2: Limits for harmonic current emissions (equipment input current ≤ 16A per phase).
  - Section 3: Limitation of voltage fluctuations and flicker in low-voltage supply systems for equipment with rated current ≤ 16A.
  - Section 4: Limitation of emission of harmonic currents in low-voltage power supply systems for equipment with rated current greater than 16A.
  - Section 5: Limitation of voltage fluctuations and flicker in low-voltage power supply systems for equipment with rated current greater than 16 A.
  - Section 6: Assessment of emission limits for distorting loads in MV and HV power systems.
  - Section 7: Assessment of emission limits for fluctuating loads in MV and HV power systems.
  - Section 8: Signalling on low-voltage electrical installations—Emission levels, frequency bands and electromagnetic disturbance levels.
  - Section 9: Limits for interharmonic current emissions (equipment with input power ≤ 16 A per phase and prone to produce interharmonics by design) (in preparation).
  - Section 10: Emission limits in the frequency range 2 ... 9 kHz (in preparation).
  - Section 11: Limitation of voltage changes, voltage fluctuations and flicker in low voltage supply systems for equipment with rated current ≤ 75 A and subject to conditional connection (in preparation).
- **Part 4: Testing and measurement techniques**
  - Section 1: Overview of immunity tests.
  - Section 2: Electrostatic discharge immunity test.
  - Section 3: Radiated, radio-frequency, electromagnetic field immunity test.
  - Section 4: Electrical fast transient/burst immunity test.
  - Section 5: Surge immunity test.
  - Section 6: Immunity to conducted disturbances, induced by radio-frequency fields.
  - Section 7: General guide on harmonic distortion and interharmonics measurement and instrumentation, for power supply systems and equipment connected thereto.
  - Section 8: Power frequency magnetic field immunity test.
  - Section 9: Pulse magnetic field immunity test.
  - Section 10: Damped oscillatory magnetic field immunity test.
  - Section 11: Voltage dips, short interruptions and voltage variations immunity tests.
  - Section 12: Oscillatory waves immunity test.
  - Section 13: Test for immunity to harmonics and interharmonics including mains signalling at a.c. power port (in preparation).
  - Section 14: Voltage fluctuations—Immunity test.
  - Section 15: Flickermeter—Functional and design specifications.

– Section 16: Test for immunity to conducted common mode disturbances in the frequency range 0 Hz to 150 kHz.

– Section 17: Ripple on d.c. input power port, immunity test (in preparation).

– Section 20: TEM cells (in preparation).

– Section 21: Reverberation chambers (in preparation).

– Section 22: Guide on measurement methods for electromagnetic phenomena (in preparation).

– Section 26: Calibration of probes and associated instruments for measuring electromagnetic fields (in preparation).

– Section 27: Unbalance, immunity test (in preparation).

– Section 28: Variation of power frequency, immunity test (in preparation).

– Section 29: Voltage dips, short interruptions and voltage variations on d.c. input power ports, immunity tests (in preparation).

– Section 30: Measurements of power quality parameters (in preparation).

– Section 31: Measurements in the frequency range 2 kHz to 9 kHz (in preparation).

- **Part 5: Installation and mitigation guidelines**

  – Section 1: General considerations.

  – Section 2: Earthing and cabling.

  – Section 6: Mitigation of external EM influences (in preparation).

  – Section 7: Degrees of protection against electromagnetic disturbances provided by enclosures (in preparation).

- **Part 6: Generic standards**

  – Section 1: Immunity for residential, commercial and light-industrial environments.

  – Section 2: Immunity for industrial environments.

  – Section 4: Emission standard for industrial environments.

  – Section 5: Immunity of apparatus for generating stations and high-voltage substations (in preparation).

Appendix B: IEEE Standards on Power Quality

The American standard setting organizations, ANSI and IEEE, do not have such a comprehensive and structured set of power quality standards as the IEC. On the other hand, the IEEE standards give much more practical and some theoretical background on the phenomena. This makes many of the IEEE standard documents very useful reference documents, even outside of the United States. Below follows a list of existing IEEE standards on power quality, and some standard documents currently under development.

- **Std 4-1995** Standard techniques for high-voltage testing.
- **Std 120-1989** Master Test Guide for Electrical Measurements in Power Circuits.
- **Std 141-1993** Recommended practice for electric power distribution for industrial plants.
- **Std 142-1991** Recommended practice for grounding of industrial and commercial power systems, also known as the Green Book.
- **Std 213-1993** Standard procedure for measuring conducted emissions in the range of 300 kHz to 25 MHz from television and FM broadcast receivers to power lines.
- **Std 241-1990** Recommended practice for electric power systems in commercial buildings, also known as the Gray Book.
- **Std 281-1994** Standard service conditions for power system communication equipment.
- **Std 299-1991** Standard method of measuring the effectiveness of electromagnetic shielding enclosures.
- **Std 352-1993** Guide for general principles of reliability analysis of nuclear power generating station safety systems.
- **Std 367-1996** Recommended practice for determining the electric power station ground potential rise and induced voltage from a power fault.

- **Std 376-1993** Standard for the measurement of impulse strength and impulse bandwidth.
- **Std 430-1991** Standard procedures for the measurement of radio noise from overhead power lines and substations.
- **Std 446-1987** Recommended practice for emergency and standby power systems for industrial and commercial applications, also knows as the Orange Book.
- **Std 449-1990** Standard for ferroresonance voltage regulators.
- **Std 473-1991** Recommended practice for an electromagnetic site survey (10 kHz to 10 GHz).
- **Std 493-1997** Recommended practice for the design of reliable industrial and commercial power systems, also known as the Gold Book.
- **Std 519-1992** Recommended practice and requirements for harmonic control in electric power systems.
- **Std 539-1990** Standard definitions of terms relating to corona and field effects of overhead power lines.
- **Std 762-1987** Standard definitions for use in reporting electric generating unit reliability, availability, and productivity.
- **Std 859-1987** Standard terms for reporting and analyzing outage occurrences and outage states of electrical transmission facilities.
- **Std 944-1986** Application and testing of uninterruptible power supplies for power generating stations.
- **Std 998-1996** Guide for direct lightning stroke shielding of substations.
- **Std 1048-1990** Guide for protective grounding of power lines.
- **Std 1057-1994** Standard for digitizing waveform recorders.
- **Std 1100-1992** Recommended practice for powering and grounding sensitive electronic equipment, also known as the Emerald Book.
- **Std 1159-1995** Recommended practice for monitoring electric power quality.
- **Std 1184-1995** Guide for the selection and sizing of batteries for uninterruptible power systems.
- **Std 1250-1995** Guide for service to equipment sensitive to momentary voltage disturbances.
- **Std 1325-1996** Recommended practice for reporting field failure data for power circuit breakers.
- **Std 1313.1-1996** Standard for insulation coordination—definitions, principles, and rules.
- **Std 1346-1998** Recommended practice for evaluating electric power system compatibility with electronics process equipment.
- **Project 1409** Custom power task force.
- **Project 1433** A standard glossary of power quality terminology.
- **Project 1453** Voltage flicker.
- **Std C37.10-1995** Guide for diagnostics and failure investigation of power circuit breakers.
- **Std C37.95-1994** Guide for protective relaying of utility–consumer interconnections.

- **Std C37.100-1992** Standard definitions for power switchgear.
- **Std C.57.110-1986** Recommended practice for establishing transformer capability when supplying nonsinusoidal load currents.
- **Std C57.117-1986** Guide for reporting failure data for power transformers and shunt reactors on electric utility power systems.
- **Std C62.41-1991** Recommended practice on surge voltages in low-voltage ac power circuits.
- **Std C62.45-1992** Guide on surge testing for equipment connected to low-voltage ac power circuits.
- **Std C62.48-1995** Guide on interactions between power system disturbances and surge-protective devices.

Appendix C: Power Quality Definitions and Terminology

This appendix gives an overview of power quality terminology as used in this book and as defined in standard documents. The main source for the latter is the "IEEE Standard dictionary on electrical and electronics terms" (IEEE Std 100-1996). Other sources used are IEC standard 61000-1-1 (Electromagnetic Compatibility: application and interpretation of fundamental definitions and terms); CENELEC standard EN 50160 (Voltage characteristics in public distribution systems), the UIE "Guide to quality of electrical supply for industrial installations" and the book "Reliability evaluation of power systems" (R. Billinton, R.N. Allan, Plenum Press, 1996). The references with the various definitions below are to IEEE standards, unless otherwise noted.

The list below is certainly not consistent, neither is it complete. It does, however, give an overview of the terminology in use as well as the potential pitfalls in defining and using power quality terminology. Currently a number of IEEE standards are under development and are aimed at providing a complete and comprehensive set of definitions for power quality terminology (among others 1159 and 1433). As only early drafts were available these are not included in the list below.

C.1 GENERAL POWER QUALITY TERMINOLOGY

- **Compatibility level** The specified disturbance level at which an acceptable, high probability of electromagnetic compatibility should exist [IEC 61000-1-1].
- **Compatibility margin** The ratio of the immunity limit to the emission limit [IEC 61000-1-1].
- **Conducted disturbance**

  – An electromagnetic disturbance propagated along the line conductors of a distribution system or across transformer windings [EN 50160].

  – An electromagnetic disturbance that reaches a device through a conducting medium.

- **Conducted interference**

  – Interference resulting from conducted radio noise or unwanted radio signals entering a device by direct coupling [539].

  – Electromagnetic interference due to an electromagnetic disturbance reaching the affected device through a conducting medium.
- **Current disturbance** A variation or event during which the current in the system or at the equipment terminals deviates from the ideal sine wave (this book).
- **Current event** One of two classes of current disturbances. A large deviation from the ideal current sine wave, which only occurs occasionally (this book).
- **Current magnitude variation** A current variation in which the magnitude of the load current is not constant (this book).
- **Current phase variation** A current variation in which the load current is not in phase with the system voltage (this book).
- **Current quality** The study or description of deviations of the load or equipment current from the ideal sine wave. The ideal current sine wave is of constant magnitude, constant frequency equal to the voltage frequency, and in phase with the voltage. The term "current quality" is rarely used, but it has been introduced as a complement to "voltage quality" (this book).
- **Current variation** One of two classes of current disturbances. A small slowly varying deviation from the ideal sinusoidal current which is always present but nominally or ideally zero (this book).
- **Electromagnetic compatibility** The ability of an equipment or system to function satisfactorily in its electromagnetic environment without introducing intolerable electromagnetic disturbances to anything in that environment [IEC 61000-1-1].
- **Electromagnetic compatibility level** = compatibility level [IEC].
- **Electromagnetic disturbance** Any electromagnetic phenomenon which may degrade the performance of a device, equipment, or system, or adversely affect living or inert matter [IEC 61000-1-1].
- **Electromagnetic emission** = emission.
- **Electromagnetic environment**

  – The electromagnetic field(s) and or signals existing in a transmission medium [IEEE Electromagnetic Compatibility Society].

  – The totality of electromagnetic phenomena existing at a given location [IEC 61000-1-1].
- **Electromagnetic interference** Degradation of the performance of a device, equipment, or system caused by an electromagnetic disturbance [IEC 61000-1-1].
- **Electromagnetic noise** = electromagnetic disturbance [539].
- **Emission** The phenomenon by which electromagnetic energy emanates from a source [IEC 61000-1-1].
- **Emission level** The level of a given electromagnetic disturbance emitted from a particular device, equipment, or system, measured in a specified way [IEC 61000-1-1].
- **Emission limit** The maximum permissible emission level [IEC 61000-1-1].

- **Emission margin** The ratio of the compatibility level to the emission limit [IEC 61000-1-1].
- **Immunity level** The maximum level of a given electromagnetic disturbance, incident in a specified way on a particular device, equipment, or system, at which no degradation of operation occurs [IEC 61000-1-1].
- **Immunity limit** The minimum required immunity level [IEC 61000-1-1].
- **Immunity margin** The ratio of the immunity limit to the compatibility level [IEC 61000-1-1].
- **Interference** = electromagnetic interference.
- **Interference voltage** Voltage produced by electromagnetic interference [IEEE Electromagnetic Compatibility Society; CISPR–International Special Committee on Radio Interference].
- **Power disturbance** Any deviation from the nominal value (or from some selected thresholds based on load tolerance) of the input ac power characteristics [1100], [1159].
- **Power quality**

  – The study or description of both voltage and current disturbances. Power quality can be seen as the combination of voltage quality and current quality (this book).

  – The concept of powering and grounding sensitive equipment in a manner that is suitable to the operation of that equipment [1100], [1159].
- **Quality of consumption** The complementary term of "quality of service," referring to the customer's responsibilities in the interaction between customer and utility (this book).
- **Quality of service** The non-technical part of the "quality of supply" (this book). The term is also used as a synonym for "quality of supply."
- **Quality of supply** Referring to the utility's responsibilities in the interaction between the utility and the customer. The term "quality of supply" includes a technical part which more or less coincides with the term "voltage quality" and a non-technical part sometimes referred to as "quality of service" (this book).
- **Radiated interference**

  – Radio interference resulting from radiated noise or unwanted signals [IEEE Electromagnetic Compatibility Society].

  – Electromagnetic interference due to an electromagnetic disturbance reaching the affected device in the form of radiation.
- **Unwanted signal** A signal that may impair the measurement or reception of a wanted signal [539].
- **Voltage characteristics** A description of the voltage quality experienced by customers or equipment in a certain area [EN 50160].
- **Voltage disturbance** A variation of event during which the voltage in the system or at the equipment terminals deviates from the ideal sine wave (this book).
- **Voltage event** One of two classes of voltage disturbances. A large deviation from the ideal voltage sine wave, which only occurs occasionally (this book).
- **Voltage magnitude event** A voltage event in which the rms voltage is outside of its normal operating range for a limited period of time (this book).

- **Voltage magnitude variation** A voltage variation in which the voltage magnitude deviates from its ideal or nominal value (this book).
- **Voltage quality** The study or description of deviations of the voltage from the ideal sine wave (this book).
- **Voltage variation**
  – One of two classes of voltage disturbances. A small slowly varying deviation from the ideal sinusoidal voltage which is always present but nominally zero (this book).
  – An increase or decrease of voltage normally due to variation of the total load of a distribution system or a part of it [EN 50160].
- **Wanted signal** A signal that constitutes the object of the particular measurement or reception [539].

C.2 VOLTAGE MAGNITUDE EVENTS

- **Brownout** Used to describe a scheduled long or very-long undervoltage. The use of this term should be avoided.
- **Expected interruption duration** The expected or average duration of a single load interruption event [493].
- **Instantaneous interruption** An interruption with a duration between 0.5 cycles and one-half second [1159], [1250].
- **Interruption**
  – A voltage event in which the voltage is zero during a certain time. The time during which the voltage is zero is referred to as the "duration" of the interruption [493], [1100], [1250].
  – A voltage magnitude event with a magnitude less than 10% of the nominal voltage [1159].
  – The loss of electric power supply to one or more loads, consumers, or other facilities [493], [ANSI 51.1].
- **Long (interruption, undervoltage, overvoltage)** A voltage magnitude event with a duration between a few minutes and a few hours, corresponding to events in the power system followed by a manual restoration of the pre-event situation (this book).
- **Long-duration voltage variation** A voltage magnitude event with a duration longer than 1 minute [1159].
- **Long Interruption** An interruption with a duration longer than three minutes [EN 50160], [UIE].
- **Momentary disturbance** A variation in the level of the steady-state supply voltage that results from surges, sags, circuit and equipment switching, or from the operation of circuit breakers or reclosers resulting from their response to abnormal circuit conditions [1250].
- **Momentary interruption**
  – An interruption with a duration between 0.5 and 2 seconds. [1250]
  – An interruption with a duration between 0.5 and 3 seconds [1159].
  – An interruption with a duration limited to the period required to restore service by automatic switching operations or by manual switching at locations

where an operator is immediately available. Such switching operations must be completed within 5 minutes [346].

- **Overvoltage**

  – An abnormal voltage higher than the normal service voltage, such as might be caused from switching or lightning surges [432].

  – A voltage above the normal rated voltage or the maximum operating voltage of a device or circuit [95].

  – Abnormal voltage between two points of a system that is greater than the highest value appearing between the same two points under normal service conditions [C62.22], [1313.1].

  – An increase of the rms voltage with a duration longer than a few seconds [1250].

  – A voltage magnitude event with a magnitude greater than the nominal voltage and a duration longer than 1 minute [1159].

  – A voltage magnitude event in which the rms voltage is higher than the normal operating range (this book).

- **Permanent forced outage** A forced outage where the component or unit is damaged and cannot be restored to service until repair or replacement is completed [859].
- **Permanent outage** An outage of a power system component which is restored through repair or replacement [859].
- **Recovery time** Time interval needed for the voltage or current to return to its normal operating value, after a voltage or current event [1100].
- **RMS variation** = voltage magnitude event. The term rms variation is confusing as it is not a voltage variation but a voltage event.
- **Short (interruption, undervoltage, overvoltage)** A voltage magnitude event with a duration between a few cycles and a few minutes, corresponding to events in the power system for which the pre-event situation is restored automatically (this book).
- **Short-duration voltage variation** A voltage magnitude event with a duration less than 1 minute [1159].
- **Short Interruption**

  – An interruption with a duration up to three minutes [EN 50160].

  – The disappearance of one or more phases of the supply voltage for a period of time typically not exceeding three minutes [UIE].

- **Supply interruption** A condition in which the voltage at the supply terminals is lower than 1% of the declared voltage [EN 50160].
- **Sustained interruption**

  – An interruption with a duration longer than 2 minutes [1250].

  – An interruption with a duration longer than 1 minute [1159].

  – Any interruption not classified as a momentary interruption [346].

- **Swell**

  – A momentary increase in the power frequency voltage delivered by the mains, outside of the normal tolerances, with a duration of more than one cycle and less than a few seconds [C62.41], [C62.48].

– An rms increase in the ac voltage, at the power frequency, for durations from one half-cycle to a few seconds [1100], [1250].

– A voltage magnitude event with a magnitude above 110% of the nominal voltage, and a duration between 0.5 cycles and 1 minute [1159].

- **Temporary fault** A short-circuit fault that is self-clearing or is cleared by fault-clearing followed by fast reclosing [1250].
- **Temporary forced outage** A forced outage where the unit or component is undamaged and is restored to service by manual switching operations without repair but possibly with on-site inspection [859]. Note the contradiction with "temporary interruption." The use of these terms should be avoided.
- **Temporary interruption**

  – An interruption with a duration between 3 seconds and 1 minute [1159].

  – An interruption with a duration between 2 seconds and 2 minutes [1250].
- **Temporary power frequency overvoltage** = swell [EN 50160].
- **Transient fault** A fault that disappears of its own accord [IEC], [IEEE Power Engineering Society]. The term is also used for a fault that disappears after reclosure and the use of the term should be avoided.
- **Transient forced outage** A forced outage where the unit or component is undamaged and is restored to service automatically [859].
- **Transient outage** An outage of a power system component which is restored automatically [859].
- **Undervoltage**

  – A voltage event in which the rms voltage is outside its normal operating margin for a certain period of time (this book).

  – A voltage magnitude event with a magnitude less than the nominal rms voltage, and a duration exceeding 1 minute [1159].

  – A voltage magnitude event with a duration less than the nominal rms voltage, and a duration longer than a few seconds [1100], [1250].
- **Very long (interruption, undervoltage, overvoltage)** A voltage magnitude event with a duration more than a few hours, corresponding to events in the power system requiring repair or replacement of faulted components before the pre-event situation can be restored (this book).
- **Very short (interruption, undervoltage, overvoltage)** A voltage magnitude event with a duration less than a few cycles, corresponding to transient and self-restoring events in the power system (this book).
- **Voltage interruption** = interruption [1159].
- **Voltage swell** = swell.

C.3 POWER SYSTEM RELIABILITY

- **Active failure** The outage of a primary component associated with a short-circuit fault [Billinton–Allan].
- **Adequacy** The existence of sufficient facilities within a power system to satisfy the customer demand [Billinton–Allan].
- **Adverse weather** Weather conditions that cause an abnormally high failure rate for exposed components during the periods such conditions persist [859].

- **Aging** The change of the failure rate of a stochastic component with time.
- **ASAI** Average service availability index [Billinton–Allan].

$$\text{ASAI} = \frac{\text{customer hours of available service}}{\text{customer hours demanded}} \tag{C.1}$$

- **Automatic outage** An outage occurrence that results from automatic operation of switching devices [859].
- **Availability**

  – The fraction of time during which a system is capable of performing its mission [446], [493], [859], [896.9], [C37.1], [C37.100].

  – The probability that an item will be operational at a randomly selected future instant in time [352], [380], [577], [896.3].
- **CAIDI** Customer average interruption duration index [Billinton–Allan].

$$\text{CAIDI} = \frac{\text{sum of customer interruption durations}}{\text{total number of customer interruptions}} \tag{C.2}$$

  This definition corresponds to the "average duration of an interruption" as defined in (2.3).
- **CAIFI** Customer average interruption frequency index [Billinton–Allan].

$$\text{CAIFI} = \frac{\text{total number of customer interruptions}}{\text{total number of customers affected}} \tag{C.3}$$

  Note that the CAIFI is at least one for any given period.
- **Class 0 unplanned outage** An outage of a generator unit that results from the unsuccessful attempt to place the unit in service [762].
- **Class 1 unplanned outage** An outage of a generator unit that requires immediate removal from the existing state [762].
- **Class 2 unplanned outage** An outage of a generator unit that does not require immediate removal from the in-service state but requires removal within 6 hours [762].
- **Class 3 unplanned outage** An outage of a generator unit that can be postponed beyond 6 hours but requires that the unit be removed from the in-service state before the end of the next weekend [762].
- **Class 4 unplanned outage** An outage of a generator unit that will allow a unit outage to be deferred beyond the end of the next weekend but which requires that a unit be removed from the available state before the next planned outage [762].
- **Common-mode failure** Multiple failures attributable to a common cause [308], [627], [649], [650], [C37.100].
- **Common-mode outage event** A component failure due to a common-mode failure [859].
- **Complete outage stage** The component or unit is completely deenergized or is connected so that it is not serving any of its functions within the power system [859].
- **Constant failure rate period** The possible period during the life of a component during which failures occur at an approximately uniform rate [IEEE Vehicular Technology Society].

- **Dead time** = reclosing interval [C37.100].
- **Degradation failure** A failure that is both gradual and partial. In time, such a failure may develop into a complete failure [1100].
- **Downtime** The time during which a device or system is not capable of meeting performance characteristics [C37.100].
- **Early-failure period** The early period in the life of a component during which the failure rate decreases rapidly [IEEE Reliability Society].
- **Electrical failure** Failure of a circuit breaker, attributable to the application of electrical stresses to the main circuit of the circuit breaker [C37.10].
- **Emergency maintenance** Unscheduled corrective maintenance performed to keep a system operational [1219].
- **Extended planned outage** The scheduled outage state that is the extension of the basic scheduled outage beyond its predetermined duration [762].
- **Failure** The termination of the ability of a power system component to fully perform its required function.
- **Failure of continuously required function** The inability of a component to perform a function that is continuously required [859].
- **Failure of response function** The inability of a component to perform a function that is required as a response to system conditions or to a manually or automatically initiated command [859].
- **Failure rate** A quantity related to a stochastic power system component, giving the number of failures per component per year. The observed failure rate is defined as the number of failures divided by the number of component-years. Mathematically the failure rate is defined from the probability distribution function of the component lifetime.
- **Failure to trip** In the performance of a relay or relay system, the lack of tripping that should have occurred considering the objectives of the relay system design [C37.90], [C37.100].
- **Failure with forced outage** Failure of a transformer that requires its immediate removal from service [C57.115].
- **Failure with scheduled outage** Failure of a transformer for which it must be taken out of service at a selected time [C57.115].
- **False operation probability** The ratio of the number of unintended operations to the number of exposure operations for which the component should not respond [859].
- **False tripping** The tripping of a protection relay that should not have occurred considering the objectives of the relay design [C37.90], [C37.100].
- **Forced interruption** An interruption of the power supply due to a forced outage [IEEE Power Engineering Society].
- **Forced outage** The outage of a power system component resulting from a failure or from incorrect operator intervention [446], [859].
- **Forced outage duration** = repair time.
- **Forced unavailability** The long-term average fraction of time that a component is out of service due to failures [493], [859].

- **Interruption criterion** A criterion used in reliability to assess if a certain system state or event constitutes a failure of the power supply to a given load or group of loads (this book).
- **Interruption frequency** The expected or average number of interruptions per year [493].
- **Interruption to service** The isolation of an electrical load from the system supplying that load, resulting from an abnormality in the system [C37.100].
- **Major storm disaster** Designates weather that exceeds design limits of facilities, and that satisfies all of the following: extensive mechanical damage to facilities; more than a specified percentage of customers out of service (typically 10%); service restoration longer than a specified time (typically 24 hours) [859].
- **Malfunction** The loss of capability to initiate or sustain a required function, often a protective action, or the initiation of undesired spurious action [C37.100].
- **Manual outage** An outage occurrence that results from intentional or inadvertent operator controlled opening of switching devices [859].
- **Mean outage duration** The mean duration of outage occurrences of a specified type [859].
- **Mean time between failures (MTBF)**

  – The expected time interval between failures of an operating device or component [C37.1], [C37.100], [C62.1], [C610.10].

  – The observed average time between failures of a continuously operating device, circuit, or system [599], [352], [859], [C610.10], [IEEE Reliability Society].
- **Mean time to outage** The mean time to outage occurrence of a specified type [859].
- **Mean time to repair**

  – The expected time interval between the failure of a device or component and its return to proper operation [C37.1], [C37.100], [610.12].

  – The observed average time interval between the failure of a device or component and its return to proper operation [352], [380], [610.10], [610.12].
- **Mean time to restoration** = mean outage duration [859].
- **Multiple independent outages** Outage occurrences, each having distinct and separate initiating incidents, where no outage occurrence is the consequence of any other, but the outage states overlap [859].
- **Multiple outage event** An outage event involving two or more components [859].
- **Nonexponential distribution** A component lifetime or repair time distribution which does not result in a constant failure or repair rate.
- **Normal weather** All weather not designated as adverse weather or major storm disaster [859].
- **Operations related outage** A scheduled outage in which the unit or component is removed from service to improve system operating conditions [859].
- **Outage** The state of a power system component when it is not available to perform its intended function due to some event directly associated with that component [346], [493], [IEEE Power Engineering Society].

- **Outage duration** The period from the initiation of an outage until the affected component once again becomes available to perform its intended function [346], [859].
- **Outage event** An event involving the outage occurrence of one or more units or components [859].
- **Outage occurrence** The change in the state of one component or one unit from the in-service state to the outage state [859].
- **Outage rate** = failure rate [346], [859].
- **Outage state** The component or unit is not in the in-service state; that is, it is partially or fully isolated from the system [859].
- **Partial outage state** The component or unit is at least partially energized, or is not fully connected to all of its terminals, or both, so that it is not serving some of its functions within the power system [859].
- **Passive failure** The outage of a primary component not associated with a short-circuit fault [Billinton–Allan].
- **Permanent forced outage** A forced outage where the component or unit is damaged and cannot be restored to service until repair or replacement is completed [859].
- **Permanent outage** An outage of a power system component which is restored through repair or replacement [859].
- **Planned (interruption, outage, etc.)** = scheduled (interruption, outage, etc.).
- **Power system reliability** The area of power engineering covering the stochastic prediction of frequency and duration of supply interruptions. The term is normally used to cover only interruptions, but the same techniques can be applied to other power quality events.
- **Primary outage** An outage occurrence within a related multiple outage event that occurs as a direct consequence of the initiating incident and is not dependent on any other outage occurrence [859].
- **Probability of failure to close on command** The ratio of the number of failures to close and the number of commands to close for a circuit breaker, switch, or recloser [859].
- **Probability of failure to open on command** The ratio of the number of failures to open and the number of commands to open for a circuit breaker, switch, or recloser [859].
- **Reclosing interval** The time between the opening of a circuit breaker and its automatic reclosure [C37.100].
- **Reclosure** The automatic closure of a circuit-interrupting device following automatic tripping [C37.95].
- **Redundant** Referring to a (power system) component whose failure or outage does not lead to an interruption of the supply for any load or customer.
- **Related multiple outage event** A multiple outage event in which one outage occurrence is the consequence of another outage occurrence, or in which multiple outage occurrences were initiated by a single incident, or both. Each outage occurrence in a related multiple outage event is classified as either a primary outage or a secondary outage depending on the relationship between that outage occurrence and its initiating incident [859].
- **Reliability assessment** = reliability evaluation [729].

- **Reliability evaluation** A stochastic study of a system in order to obtain its failure characteristics. In power engineering the system under study is the power supply and the failure characteristics are normally frequency and duration of interruptions.
- **Repair rate** The expected number of repair actions of a given type completed on a given item per unit of time [352].
- **Repair time** The clock time from the occurrence of the failure to the time when the component is restored to service, either by repair of the failed component or by substitution of a spare component for the failed component [493].
- **Reserve shutdown** The state in which a unit is available but not in service [762].
- **SAIDI** System average interruption duration index [Billinton–Allan].

$$\text{SAIDI} = \frac{\text{sum of customer interruption durations}}{\text{total number of customers served}} \tag{C.4}$$

  This definition corresponds to the "average unavailability per customer" as defined in (2.2).
- **SAIFI** System average interruption frequency index [Billinton–Allan].

$$\text{SAIFI} = \frac{\text{total number of customer interruptions}}{\text{total number of customers served}} \tag{C.5}$$

  This definition corresponds to the "average number of interruptions per customer" as defined in (2.1).
- **Scheduled interruption** An interruption of the supply due to a scheduled outage [IEEE Power Engineering Society].
- **Scheduled outage** The outage of a power system component due to intentional operator intervention at a previously selected time [446], [493], [859].
- **Scheduled outage duration** The period from the initiation of a scheduled outage until construction, preventative maintenance, or repair work is completed and the affected component is made available to perform its intended function [493].
- **Scheduled unavailability** The long-term average fraction of time that a component or system is out of service due to scheduled outages [859].
- **Secondary outage** An outage occurrence that is the result of another outage occurrence [859].
- **Security** The ability of a power system to respond to disturbances arising within that system [Billinton–Allan].
- **Single outage event** An outage event involving only one component [859].
- **Starting failure** The inability to bring a unit from some unavailable state or reserve shutdown state to the in-service state within a specified period [762].
- **Switching time** The period from the time a switching operation is required due to a component failure until that switching operation is completed [493], [IEEE Power Engineering Society].
- **System-related outage** A forced outage that results from system effects or conditions and is not caused by an event directly associated with the component or unit being reported [859].

- **Temporary forced outage** A forced outage where the unit or component is undamaged and is restored to service by manual switching operations without repair but possibly with on-site inspection [859].
- **Transient forced outage** A forced outage where the unit or component is undamaged and is restored to service automatically [859].
- **Transient outage** An outage of a power system component which is restored automatically [859].
- **Unavailability**

  – The (observed) fraction of time during which a component or a system does not perform its intended operation. The unavailability of the power supply is normally expressed in minutes per year [493], [859].

  – The probability that a component is in an outage stage at a given moment in time [352], [493].
- **Wearout-failure period** = wearout period [IEEE Reliability Society].
- **Wearout period** The final period in the life of a component during which the failure rate increases rapidly [352].

C.4 VOLTAGE SAGS

- **Balanced sag** An equal drop in the rms value of voltage in the three phases of a three-phase system or at the terminals of three-phase equipment for a duration up to a few minutes. Note that a balanced sag is a special case of the three-phase unbalanced sag (this book).
- **Characteristic complex voltage** A characteristic of a three-phase unbalanced sag, indicating the severity of the sag. For the various types of three-phase unbalanced sags, definitions for the characteristic complex voltage are given. The characteristic complex voltage may generally be defined as the complex voltage in the phase most affected or the complex voltage of the voltage difference most affected, whichever one is more affected. In either case the relevant pre-event voltage or voltage difference is along the positive real axis (this book).
- **Characteristic magnitude** The absolute value of the characteristic complex voltage (this book).
- **Characteristic phase-angle jump** The argument of the characteristic complex voltage (this book).
- **Complex voltage at the equipment terminals** For three-phase equipment, the three complex voltages as experienced at the terminals of a device or as measured at a certain location due to a three-phase unbalanced sag. For each of the three voltages, the pre-event value is along the positive real axis (this book).
- **Critical distance** The distance at which a short-circuit fault will lead to a voltage sag of a given magnitude for a given load position (this book).
- **Duration (of a voltage sag)** The time during which the voltage deviates significantly from the ideal voltage. A further definition of "significant deviation" remains a point of discussion; a typical definition is an rms voltage less than 90% of its nominal value in at least one phase (this book).

- **Envelope voltage** The magnitude of the complex representation of the observed instantaneous voltage [473]. Note that this definition is equivalent to the definition of magnitude (of a voltage sag).
- **Evolving fault** A change in the current during interruption whereby the magnitude of current increases in one or more phases [C37.100].
- **Initial complex voltage** The complex voltage in the faulted phase or between the faulted phases at the point-of-common coupling between a short-circuit fault leading to a voltage sag and the sensitive equipment experiencing this sag. The pre-event voltage is along the positive real axis. The point-of-common coupling is strictly speaking only defined in radial systems, but also in many non-radial systems it is possible to indicate a point-of-common coupling (this book).
- **Initial magnitude** The magnitude of the initial complex voltage (this book).
- **Initial phase-angle jump** The argument of the initial complex voltage (this book).
- **Magnitude (of a voltage sag)** For single-phase equipment, the rms value of the voltage during a voltage sag. The magnitude is rarely exactly constant during the sag, in which case one can give the magnitude as a function of time, or characterize the magnitude through one value, typically the lowest value.
- **Magnitude of the voltage sag at the equipment terminals** Absolute value of the complex voltage at the equipment terminals (this book).
- **Missing voltage** For single-phase equipment, the difference between the actual time-domain voltage during a voltage event and the voltage as it would have been had the event not occurred.
- **Phase-angle jump** For single-phase equipment, the difference between the phase angle of the voltage during an event and the phase angle of the voltage before the event. The phase-angle jump is positive when the during-event voltage leads the pre-event voltage (this book).
- **Phase-angle jump at the equipment terminals** Argument of the complex voltage at the equipment terminals (this book).
- **Point-on-wave of sag initiation** For single-phase equipment, the angle of the voltage wave at the instant of sag initiation. The last upward zero crossing of the pre-event voltage is used as a reference point (this book).
- **Point-on-wave of voltage recovery** For single-phase equipment, the angle of the voltage wave at the instant of voltage recovery. The last upward zero crossing of the pre-event voltage is used as a reference point (this book).
- **Post-fault** A qualifying term that refers to an interval beginning with the clearing of a fault [C37.100].
- **Post-fault voltage sag** The part of a voltage sag during which the rms voltage remains outside its normal operating range, after the instant of voltage recovery. The short-circuit fault which led to the voltage sag is no longer present during the post-fault voltage sag (this book).
- **Pre-fault** A qualifying term that refers to an interval ending with the inception of a fault [C37.100].
- **Sag**

  – = voltage sag.

  – A voltage magnitude event with a magnitude between 10% and 90% of nominal and a duration between 0.5 cycles and one minute [1159].

– A voltage magnitude event with a magnitude less than the nominal voltage and a duration between 0.5 cycles and a few seconds [1100], [1250].

- **Sag initiation** The sudden change in voltage somewhere in the power system or at the equipment terminals, directly attributed to the initiation of a short-circuit fault. The instant of sag initiation can be viewed as the actual start of a voltage sag (this book).
- **Supply voltage dip** A sudden reduction of the supply voltage to a value between 90% and 1% of the declared voltage, followed by a recovery after a short period of time [EN 50160].
- **Three-phase balanced sag** = balanced sag.
- **Three-phase unbalanced sag** A drop in the rms value of the voltage for a duration up to one minute, in at least one phase of a three-phase system or at the terminals of three-phase equipment (this book).
- **Unbalanced fault** A short-circuit or open-circuit fault in which not all three phases are equally involved. Examples are single-phase-to-ground, and phase-to-phase short circuits.
- **Voltage dip** Sudden reduction in the supply voltage by a value of more than 10% of the reference value, followed by a voltage recovery after a short period of time [UIE].
- **Voltage sag** For single-phase equipment, a drop in the rms value of the voltage for up to a few minutes.
- **Voltage sag duration** See duration.
- **Voltage sag magnitude** See magnitude.
- **Voltage recovery** The sudden change in voltage somewhere in the power system or at the equipment terminals, directly attributed to the removal of a short-circuit fault from the healthy part of the power system. The instant of voltage recovery can be viewed as the end of the actual voltage sag. Note that the voltage does not necessarily recover completely to its pre-event value.
- **Voltage tolerance** The immunity of a piece of equipment against voltage magnitude variations (voltage sags, voltage swells, and interruptions) and short-duration overvoltages.
- **Voltage-tolerance curve** The relation between the maximum sag duration and the minimum sag magnitude for which a sensitive component will trip.

C.5 WAVEFORM DISTORTION

- **Characteristic harmonic** Harmonic current component produced by a three-phase power electronic converter during balanced operation [519].
- **Crest factor** The ratio of the peak value of a periodic signal to its rms value [C57.12.80], [120], [145], [194], [1100].
- **Deviation factor** The ratio of the maximum difference between corresponding ordinates of the wave and of the equivalent sine wave when the waves are superposed in such a way as to make this maximum difference as small as possible. Note: The equivalent sine wave is defined as having the same frequency and the same root-mean-square value as the wave being tested [120], [IEEE Power Engineering Society].

- **Deviation from a sine wave** The ratio of the absolute value of the maximum difference between the distorted wave and the crest value of the fundamental [519], [937].
- **Distortion** Non-power frequency components of voltage or current. The term is often used as a synonym for "harmonic distortion."
- **Distortion factor**

  – The ratio of the rms of the harmonic contents of voltage or current to the rms value of the fundamental quantity [120], [519], [1100], [1250].

  – The ratio of the rms of the harmonic contents of voltage or current to the rms value of the full wave [281], [IEEE Power Engineering Society]. Note the difference between these two definitions.
- **Distortion power** A third power term next to active and reactive power, mathematically defined as

$$D = \sqrt{S^2 - P^2 - Q^2} \tag{8.6}$$

  where S is the apparent power, P the active power, and Q the sum of the reactive powers in all harmonic components [270].
- **Form factor** The ratio of the root-mean-square value of a periodic waveform to the absolute value averaged over a full period of the waveform [IEEE Industry Applications Society], [1100], [270], [59], [120].
- **Harmonic** A sinusoidal component of a periodic wave or quantity having a frequency that is an integer multiple of the fundamental frequency. Note: For example, a component, the frequency of which is twice the fundamental frequency, is called a second harmonic [519], [599], [936], [1250], [C62.48], [EN 50160].
- **Harmonic component** = harmonic.
- **Harmonic content** The distortion of a voltage or current wave, expressed in the absolute value of the various harmonic components [446], [539], [644], [IEEE Industry Applications Society].
- **Harmonic distortion** Frequency components of voltage or current that are integer multiples of the power-system frequency [1057], [1100], [1143], [1250], [C62.48].
- **Harmonic factor** = distortion factor [519].
- **Harmonic (voltage or current) distortion** A voltage or current variation in which the steady-state waveshape contains components with frequencies that are an integer multiple of the fundamental frequency.
- **Interharmonic (voltage or current) distortion** A voltage or current variation in which the steady-state waveshape contains a component with a frequency which is not an integer multiple of the fundamental frequency [1159], [EN 50160].
- **Line voltage notch** = periodic voltage notching [519].
- **Maximum theoretical deviation from a sine wave** For a nonsinusoidal wave, the ratio of the arithmetic sum of the amplitudes (rms) of all harmonics in the wave to the amplitude (rms) of the fundamental [519], [936].

- **Noncharacteristic harmonic** Harmonic current component, not being a characteristic harmonic, produced by a three-phase power electronic converter [519], [936].
- **Notch area** The area of the line voltage notch. It is the product of the notch depth, in volts, times the width of the notch measured in microseconds [519].
- **Notch depth** The average depth of the line voltage notch, measured as the deviation from the sine wave of the voltage [519].
- **Notching** A periodic voltage disturbance caused by the normal operation of power electronics devices when current is commutated from one phase to another [1159].
- **Periodic voltage notching** A repetitive voltage disturbance where the voltage one or more times a cycle becomes (significantly) closer to zero than the ideal sine wave.
- **Relative harmonic content** The distortion of a voltage or current wave, expressed in the value of the various harmonic components relative to the power-frequency component [936].
- **Signature** Those characteristics of a waveform that help identify an event or conditions [C37.100].
- **Total demand distortion (TDD)**

  – The total root-sum-square harmonic current distortion, in percent, of the maximum demand load current (15 or 30 minute demand) [519].

  – The total rms current distortion in percent of maximum demand current [1250].
- **Total harmonic distortion** = distortion factor [1250].
- **Total harmonic distortion disturbance level** The level of a given electromagnetic disturbance caused by the superposition of the emission of all pieces of equipment in a given system [1159].
- **Voltage deviation**

  – The instantaneous difference between the actual instantaneous voltage and the corresponding value of the previously undisturbed waveform. Note: Voltage deviation amplitude is expressed in percent or per unit referred to the peak value of the previously undisturbed voltage [936]. Note that this definition corresponds to the definition of missing voltage as used in this book and proposed in [1159.2].

  – The ratio of the rms voltage to the average rms voltage of a signal [473].
- **Voltage distortion** Any deviation from the nominal sine wave of the ac line voltage [1159], [1250].
- **Voltage or current waveform** The voltage or current as a function of time.
- **Waveform distortion** A steady-state deviation from an ideal sine wave of power frequency [1159].

C.6 EQUIPMENT BEHAVIOR

- **Critical load**

  – That part of the load that requires continuous quality electric power for its successful operation [241].

– Devices and equipment whose failure to operate satisfactorily jeopardizes the health or safety of personnel, and/or results in loss of function, financial loss, or damage to property deemed critical by the user [1100].

– = sensitive load.

- **Dropout** A loss of equipment operation due to noise, sag, or interruption [1100], [1159].
- **Dropout voltage** The voltage at which a device ceases operation [446], [1100], [1159].
- **Electromagnetic susceptibility** = susceptibility.
- **Immunity** The ability of a device, equipment, or system to perform without degradation in the presence of an electromagnetic disturbance [IEC 61000-1-1].
- **Immunity level** The maximum level of a given electromagnetic disturbance, incident in a specified way on a particular device, equipment, or system, at which no degradation of operation occurs [IEC 61000-1-1].
- **Immunity limit** The minimum required immunity level [IEC 61000-1-1].
- **Immunity to interference** = immunity [IEEE Electromagnetic Compatibility Society].
- **Ridethrough capability** The ability of equipment, to withstand momentary interruptions of sags [1250]. The term voltage tolerance is preferred.
- **Sensitive (equipment or load)** Relating to equipment or load fed from the power supply, which experiences failure or maloperation due to voltage variations or events.
- **Spurious response** Any response, other than the desired response, of an electrical transducer or device [599].
- **Susceptibility** The inability of a device, equipment, or system to perform without degradation in the presence of an electromagnetic disturbance [IEC 61000-1-1].
- **Tripping of equipment** Unintended operation or failure of equipment, normally resulting in the equipment seizing operation.
- **Vulnerability** The characteristic of a device for being damaged by an external influence, such as a transient overvoltage [C62.45].

C.7 OTHER POWER QUALITY DISTURBANCES

- **AC power-line fields** Power frequency electric and magnetic fields produced by ac power lines [539].
- **Ambient noise** The all-encompassing noise associated with a given environment, usually a composite of contributions from many sources near and far [539].
- **Angle of retard unbalance** The load voltage/current unbalance due to unequal angles of retard either between positive and negative half cycles of a single ac wave or between two or more phases in a three-phase system [428].
- **Background noise** The total system noise independent of the presence or absence of radio noise from the power line or substation [430].
- **Chopped impulse wave** An impulse wave that has been caused to collapse suddenly by a flashover [Power Engineering Society], [IEC].

- **Common-mode noise** The noise voltage that appears equally and in phase from each signal conductor to ground [422], [525], [1050], [1100], [1143].
- **Common-mode overvoltage** An event in which the differential mode voltage does not exceed its normal operating range, but the common-mode voltage does [1057].
- **Common-mode voltage** The noise voltage that appears equally and in phase from current-carrying conductor to ground [1159], [IEC 61000-2-1].
- **Conducted radio noise** Radio noise propagated by conduction from a source through electrical connections [539].
- **Critical stroke magnitude** The amplitude of the current of the lightning stroke that, upon terminating on the phase conductor, would raise the voltage of the conductor to a level at which flashover is likely [998].
- **Current unbalance** A current variation for a three-phase load, in which the three current magnitudes or the phase-angle differences between them are not equal (this book).
- **DC offset** The presence of a dc voltage or current component in an ac power system [1159].
- **Differential mode voltage** The voltage difference between two phases of a balanced circuit [802.3], [802.12].
- **Flicker** = light flicker [1159], [1250], [IEC].
- **Frequency deviation** = voltage frequency variation [1100], [1159].
- **Full impulse voltage** An aperiodic transient voltage that rises rapidly to a maximum value and falls, usually less rapidly, to zero [4].
- **Full lightning impulse** A lightning impulse not interrupted by any type of discharge [4].
- **Geomagnetically induced currents** Currents induced in power systems by variations in the geomagnetic field. These variations, and thus the induced currents, have periods of several minutes [367].
- **Glitch** A perturbation of the pulse waveform of relatively short duration and of uncertain origin [4]. The use of this term should be avoided.
- **High-frequency transient** An oscillatory transient with an oscillation frequency above 500 kHz [1159].
- **Imbalance** = voltage unbalance [1159].
- **Impulse** A surge of unidirectional polarity, for example a 1.2/50 μs voltage surge [4], [28], [829], [1100], [1250], [C62.11], [C62.22].
- **Impulse noise** Noise characterized by transient disturbances separated in time by quiescent intervals [145], [539], [599].
- **Impulsive transient** A type of voltage of current transient, during which the deviation from the normal voltage is unidirectional; i.e., either always positive or always negative [1159].
- **Light flicker** A variation in intensity of lighting as perceived by a human observer. Light flicker can be due to voltage fluctuations.
- **Lightning overvoltage** A type of transient overvoltage in which a fast front voltage is produced by lightning or fault [1313.1].
- **Load voltage unbalance** = voltage unbalance [428].

- **Low-frequency transient** An oscillatory transient with an oscillation frequency less than 5 kHz [1159].
- **Mains marking signals** Mains signaling voltage consisting of superimposed short-time alterations at selected points of the voltage waveform [EN 50160].
- **Mains signaling voltage** A signal superimposed on the supply voltage for the purpose of transmission of information in the public distribution system and to customer's premises [EN 50160].
- **Medium-frequency transient** An oscillatory transient with an oscillation frequency between 5 and 500 kHz [1159].
- **Microsecond transient** An impulsive transient with a duration between 50 ns and one millisecond [1159].
- **Millisecond transient** An impulsive transient with a duration longer than one millisecond [1159].
- **Nanosecond transient** An impulsive transient with a duration less than 50 ns [1159].
- **Noise** Unwanted electrical signals with broadband spectral content lower than 200 kHz superimposed upon the power system voltage or current in phase or neutral conductors [1159], [1250]. The term noise in this definition is synonymous with the term "electromagnetic interference." Often the term "noise" is used to refer to those terms of the interference that are not covered by any of the other terms. The use of the term "noise" should be avoided.
- **Notch** A voltage disturbance lasting less than one half-cycle, which is initially of opposite polarity than the waveform [1100], [1250], [1159], [C62.48].
- **Oscillatory transient** A type of voltage or current transient, during which the deviation from the normal voltage oscillates around zero: the deviation reaches positive as well as negative values [1159].
- **Percent unbalance of phase voltages** The ratio of the maximum deviation of a phase voltage from the average of the total phases to the average of the phase voltages, expressed in percent [IEEE Aerospace and Electronic Systems Society]. The use of the term "unbalance" as in this definition should be avoided; the voltage (current) unbalance is normally quantified as the ratio of negative- and positive-sequence voltage (current).
- **Periodic frequency modulation** The periodic variation of the output frequency from its rated value [936].
- **Periodic output voltage modulation** The periodic variation of output voltage amplitude at frequencies less than the fundamental output frequency [936]. Note that this term corresponds to the term "voltage fluctuation."
- **Power frequency variation** = voltage frequency variation [1159].
- **Power-line carrier signals** Mains signaling voltages in the frequency range between 3 and 148.5 kHz [EN 50160].
- **Radiated radio noise** Radio noise that is propagated by radiation from a source into space in the form of electromagnetic waves [539].
- **Radio frequency disturbance** An electromagnetic disturbance having components in the radio frequency range [539].
- **Radio frequency interference** = radio frequency disturbance [539].
- **Radio noise** Radiated electromagnetic disturbances in the radio frequency range [430].

- **Rapid voltage change** A single rapid variation of the rms value of a voltage between two consecutive levels which are sustained for definite but unspecified durations [EN 50160]. This term corresponds to the term "voltage magnitude step."
- **Ripple control signals** Mains signaling voltages in the frequency range between 110 and 3000 Hz [EN 50160].
- **Slew rate**
  - The rate of change of ac voltage, expressed in volts per second [1159].
  - Rate of change of (ac voltage) frequency [1100].
- **Solar-induced currents** = geomagnetically induced currents.
- **Spike** = transient overvoltage [241].
- **Surge**
  - A transient wave of current, voltage, or power in an electric circuit [C62.1], [C62.11], [C62.22], [C62.41].
  - A transient voltage or current, which usually rises rapidly to a peak value and then falls more slowly to zero, occurring in electrical equipment or networks in service [4].
  - A transient wave of voltage or current [1250], [C62.34], [C62.48].
  - The term "surge" is also used in the meaning of "short overvoltage"; its use should be completely avoided.
- **Switching overvoltage** A transient overvoltage in which a slow front, short duration, unidirectional or oscillatory, highly damped voltage is generated by switching or by a fault [1313.1].
- **Switching surge** = switching overvoltage [524], [524a], [1048], [C62.22].
- **Temporary overvoltage** An undamped or only slightly damped overvoltage of relatively long duration [1313.1].
- **Three-phase unbalance** = voltage unbalance.
- **Transient**
  - = transient (voltage or current) disturbance.
  - A change in the steady-state condition of voltage or current, or both [382].
  - Any voltage or current event with a duration of less than a few cycles [1250].
  - A subcycle disturbance in the ac waveform that is evidenced by a sharp brief discontinuity of the waveform. May be of either polarity and may be additive to or subtractive from the nominal waveform [1100].
  - A disturbance lasting less than one half-cycle [UIE].
- **Transient (voltage or current) disturbance** A subcycle disturbance in the ac waveform that is evidenced by a sharp brief discontinuity of the waveform [1100].
- **Transient overvoltage**
  - Short-duration oscillatory or non-oscillatory overvoltage usually highly damped and with a duration of a few milliseconds or less [EN 50160].
  - Momentary excursion of voltage outside of the normal 60 Hz voltage wave [241].

– A voltage event in which the time-domain voltage is outside of the normal operating range for a very short duration, typically less than a few milliseconds [1313.1].

- **Transverse-mode voltage** The voltage between two conductors at a given location [C37.90], [C63.31], [C63.32].
- **Unbalance** = voltage unbalance.
- **Unbalance factor** The ratio of the negative sequence component to the positive sequence component of the voltage in a three-phase system [936].
- **Unbalance ratio** The difference between the highest and the lowest fundamental rms values in a three-phase system, referred to the average of the three fundamental rms values of current or voltages [936]. The use of the term "unbalance" in this meaning should be avoided.
- **Very fast front, short duration overvoltage** A transient overvoltage in which a short duration, usually unidirectional, voltage is generated (often by GIS disconnect switch operation or when switching motors). High-frequency oscillations are often superimposed on the unidirectional wave [1313.1]
- **Voltage change** A variation of the rms voltage between two consecutive levels sustained for definite but unspecified durations [1159], [IEC 61000-2-1]. This term corresponds to the term "voltage magnitude step."
- **Voltage flicker** Abbreviation of "voltage fluctuation leading to light flicker."
- **Voltage fluctuation**

  – A special type of voltage variation in which the voltage shows changes in magnitude and/or phase angle on a timescale of seconds or less. Severe voltage fluctuations lead to light flicker.

  – A series of voltage changes or a cyclical variation of the voltage envelope [1159], [61000-2-1].

  – Voltage variations of amplitude less than 10% of the nominal voltage [UIE].
- **Voltage frequency variation** A voltage variation in which the voltage frequency deviates from its ideal or nominal value (this book).
- **Voltage imbalance** = voltage unbalance [1159].
- **Voltage magnitude step** A voltage event in which the rms value of the voltage shows a fast rise or drop from one constant value to another constant value, both inside the normal operating range (this book).
- **Voltage unbalance** A voltage variation in a three-phase system in which the three voltage magnitudes or the phase-angle differences between them are not equal. The voltage unbalance is quantified as the ratio of the negative- and positive-sequence voltage.
- **Wave-shape fault** A voltage quality event with a duration less than one cycle. This term is used by some monitoring equipment to classify voltage disturbances.

Appendix D: Figures

Appendix E: Tables

Index

O

P

S

About the Author

Math H. J. Bollen received an M.Sc. in electrical engineering and a Ph.D. in technical science from Eindhoven University of Technology, The Netherlands, in 1985 and 1989, respectively.

From September 1989 to August 1992, Dr. Bollen was a research associate in the Group of Electrical Energy Systems, Eindhoven University of Technology, funded in cooperation with Tilburg University. Research included such areas as reliability, protection, and design of industrial power systems, with the main emphasis on reliability assessment. From September to December 1992, he was a visiting lecturer at the University of The Netherlands Antilles, Curaçao, Netherlands Antilles, where he researched the reliability aspects of insular power systems. Dr. Bollen lectured on PASCAL programming, telecommunications, and power system reliability. From January to June 1993, he continued research with the Group of Electrical Energy Systems funded by the University Board as a "highly-promising young researcher."

From July 1993 to August 1996, Dr. Bollen was a lecturer in the Department of Electrical Engineering and Electronics, Electrical Energy and Power Systems Group, University of Manchester Institute of Science and Technology (UMIST), Manchester, U.K. He lectured in such areas as electrical machines for mechanical engineering, power quality, insular power systems, and reliability of power systems. His research consisted of voltage sags in transmission and distribution systems, reliability, and power quality. Since September 1996, he has been an associate professor in the Department of Electric Power Engineering, Chalmers University of Technology, Gothenburg, Sweden.

A senior member of the Institute of Electrical and Electronics Engineers (IEEE), Dr. Bollen is a member of both the IEEE Power Engineering Society and the IEEE Industrial Applications Society. He is chairman of Reliability Analysis Techniques Working Group, co-chairman of the Gold Book Working Group and Power System Reliability Subcommittee, and vice-chairman of the Voltage Sag Working Group in IEEE's Industrial Applications Society. In October 1998, he was the recipient of the ABB Energy Prize (Gunnar Engström Stipendiet) for work on power quality and toward understanding of voltage sags.

Printed and bound by CPI Group (UK) Ltd, Croydon, CR0 4YY
16/09/2026
14963448-0008